TETRAHEDRON ORGANIC CHEMISTRY SERIES

*Series Editors:* J E Baldwin, FRS & P D Magnus, FRS

VOLUME 8

# Cycloaddition Reactions in Organic Synthesis

## Related Pergamon Titles of Interest

### BOOKS

*Tetrahedron Organic Chemistry Series:*

BLASZCZAK: Organotin Reagents in Organic Synthesis*
DAVIES: Organotransition Metal Chemistry: Applications to Organic Synthesis
DEROME: Modern NMR Techniques for Chemistry Research
DESLONGCHAMPS: Stereoelectronic Effects in Organic Chemistry
GALLAGHER: Electrophile-Mediated Cyclisations in Synthesis*
GIESE: Radicals in Organic Synthesis: Formation of Carbon-Carbon Bonds
GIESE: Radicals in Organic Synthesis 2: Formation of Carbon-Hydrogen and Carbon-Heteroatom Bonds*
HANESSIAN: Total Synthesis of Natural Products
HASSNER: Name Reactions and Unnamed Reactions*
MARKO: Challenges & Triumphs in Indole Alkaloid Total Synthesis*
PAULMIER: Selenium Reagents & Intermediates in Organic Synthesis
PERLMUTTER: Conjugate Addition Reactions in Organic Synthesis*
SIMPKINS: Sulfones in Organic Chemistry*
WILLIAMS: Synthesis of Optically Active Alpha-Amino Acids
WONG: Enzymes in Synthetic Organic Chemistry*

### JOURNALS

EUROPEAN POLYMER JOURNAL
JOURNAL OF PHARMACEUTICAL AND BIOMEDICAL ANALYSIS
TETRAHEDRON
TETRAHEDRON: ASYMMETRY
TETRAHEDRON COMPUTER METHODOLOGY
TETRAHEDRON LETTERS

*Full details of all Pergamon publications/free specimen copy of any Pergamon journal available on request from your nearest Pergamon office*

*In preparation

# Cycloaddition Reactions in Organic Synthesis

W. CARRUTHERS †
*Department of Chemistry*
*University of Exeter*

PERGAMON PRESS
Member of Maxwell Macmillan Pergamon Publishing Corporation
OXFORD · NEW YORK · BEIJING · FRANKFURT
SÃO PAULO · SYDNEY · TOKYO · TORONTO

| | |
|---|---|
| U.K. | Pergamon Press plc, Headington Hill Hall, Oxford OX3 0BW, England |
| U.S.A. | Pergamon Press, Inc., Maxwell House, Fairview Park, Elmsford, New York 10523, U.S.A. |
| PEOPLE'S REPUBLIC OF CHINA | Pergamon Press, Room 4037, Qianmen Hotel, Beijing, People's Republic of China |
| FEDERAL REPUBLIC OF GERMANY | Pergamon Press GmbH, Hammerweg 6, D-6242 Kronberg, Federal Republic of Germany |
| BRAZIL | Pergamon Editora Ltda, Rua Eça de Queiros, 346, CEP 04011, Paraiso, São Paulo, Brazil |
| AUSTRALIA | Pergamon Press Australia Pty Ltd., P.O. Box 544, Potts Point, N.S.W. 2011, Australia |
| JAPAN | Pergamon Press, 5th Floor, Matsuoka Central Building, 1-7-1 Nishishinjuku, Shinjuku-ku, Tokyo 160, Japan |
| CANADA | Pergamon Press Canada Ltd., Suite No. 271, 253 College Street, Toronto, Ontario, Canada M5T 1R5 |



First edition 1990

**Library of Congress Cataloging in Publication Data**

Carruthers, W.
Cycloaddition reactions in organic synthesis.
(Tetrahedron organic chemistry series; v. 8)
1. Ring formation (Chemistry) 2. Organic
compounds—Synthesis. I. Title. II. Series.
QD281.R5C37 1990 547.2 90–7838

**British Library Cataloguing in Publication Data**

Carruthers, W. (William)
Cycloaddition reactions in organic synthesis.
1. Organic compounds. Synthesis.
I. Title II. Series
547.2

ISBN 0–08–034713–4 Hardcover
ISBN 0–08–034712–6 Flexicover

*Printed in Great Britain by BPCC Wheatons Ltd, Exeter*

## PREFACE

Cycloadditions are among the most widely-used reactions in organic synthesis and their widespread application and value in synthesis form the justification for the preparation of this book. Pre-eminent among them is the Diels-Alder reaction by virtue of its versatility and the control of stereochemistry which is often possible in both inter- and intra-molecular reactions. It forms the topic of three chapters in the book. Other chapters are concerned with miscellaneous [4+2] cycloadditions, the ene reaction, 1,3-dipolar cycloaddition reactions and [2+2] cycloaddition reactions, all in constant and increasing use in synthesis at the present time. Discussion of the various topics is not exhaustive. My aim has been to bring out the synthetic usefulness of each reaction rather than to provide a comprehensive account. In general, reaction mechanisms are not discussed except in so far as is necessary for an understanding of the course and stereochemistry of a reaction.

I am indebted to Sandi and Paul Ellison for the preparation and printing of the text and to Robin Batten of this department for the drawings.

W. CARRUTHERS

March 1990

Bill Carruthers died unexpectedly of a heart attack on 25 April 1990 shortly after the completion of the typescript for this book; his daughter Mary was able to over-see the final stages that were involved in taking the script into its final form.

This book is the last of a series of widely acclaimed texts that Bill produced. The three editions of *Some Modern Methods of Organic Synthesis* have sold more than 25,000 copies throughout the world.

The clarity of thought and the attention to detail that are to be admired in this book typified Bill's approach to chemistry: his enthusiasm for the subject was always apparent and this comes over in his writing. Moreover his love of the topic was imparted to his many research students from whom he deservedly gained very great respect and admiration.

Bill Carruthers will be sorely missed by those of us who have worked with him as friends and colleagues at Exeter University. This text represents a lasting tribute to his endeavours.

The University of Exeter,
May 1990.

# CONTENTS

# 1 THE DIELS-ALDER REACTION - GENERAL ASPECTS

The Diels-Alder reaction is one of a general class of cyclo-addition reactions which includes also 1,3-dipolar cycloaddition reactions and [2+2]$\pi$ cyclo-additions. In the Diels-Alder reaction a conjugated diene reacts with a dienophile which may be a double or triple bond to form an adduct with a six-membered hydroaromatic or heteroaromatic ring with the formation of two new $\sigma$-bonds at the expense of two $\pi$-bonds in the starting materials.

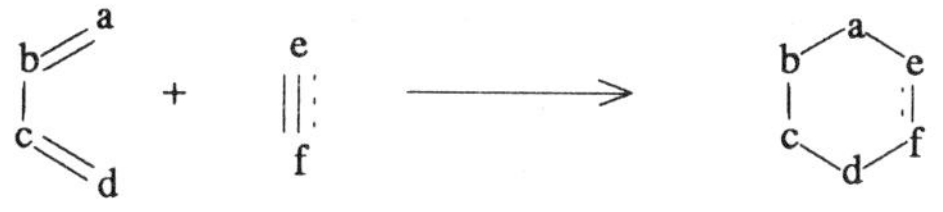

*Scheme 1*

Both inter- and intra-molecular reactions are widely used in synthesis. In many cases reactions take place easily at ambient or slightly elevated temperatures; reactions which are sluggish or which involve thermally unstable reactants or lead to unstable products can often be accelerated by catalysts or by conducting the reaction at high pressures. Most Diels-Alder reactions involve a diene carrying electron-donating substituents (e.g. alkyl, alkoxy) and a dienophile bearing electron-attracting substituents (for example, CO, CN). But there is another smaller group which involves reaction between an electron-rich dienophile and an electron-deficient diene.[1] These reactions with inverse electron demand, as they are called, also have their uses in synthesis. The essential feature is that the two components should have complementary electronic character. The vast majority of reactions involve an electron-rich diene and an electron-deficient dienophile.

The usefulness of the Diels-Alder reaction in synthesis arises from its versatility and its high regio- and stereo-selectivity. A large variety of dienes and dienophiles, bearing a variety of functional groups, can be used, and many different types of ring structures built up. Not all the atoms involved in the ring closure need be carbon atoms, so that both carbocyclic and heterocyclic rings can be made using the reaction. It is very frequently found, moreover, that although reaction could conceivably give rise to a number of structurally- or stereo-isomeric products, one isomer is formed exclusively or at least in preponderant amount.

## The Dienophile

Many different kinds of dienophile can take part in the Diels-Alder reaction. They may be derivatives of ethylene or acetylene or reagents in which one or both of the atoms is a heteroatom. All dienophiles do not react equally easily; the reactivity depends on the structure. In general, for the 'normal' reaction, the greater the number of electron-attracting substituents on the double or triple bond the more reactive is the dienophile.

The rate of a Diels-Alder reaction is determined largely by the degree of interaction between the highest occupied molecular orbital (HOMO) of one component and the lowest unoccupied molecular orbital (LUMO) of the other, and the smaller the energy separation between these orbitals the more readily the reaction proceeds. In a normal Diels-Alder reaction, that is one between an electron-deficient dienophile and an electron-rich diene, the main interaction is that between the HOMO of the diene and the LUMO of the dienophile. Electron withdrawing substituents on the double bond of the dienophile in a normal Diels-Alder reaction facilitate the reaction by lowering the energy of the LUMO and thus decreasing energy separation between the LUMO of the dienophile and the HOMO of a given diene. The more electron-withdrawing groups there are on the double bond of the dienophile the lower the energy of the LUMO and the more readily the reaction proceeds. Thus tetracyanoethylene is a very good dienophile. In the same way electron-donating substituents on the diene accelerate the reaction by raising the energy level of the highest occupied molecular orbital (HOMO). In Diels-Alder reactions with inverse electron demand it is the HOMO(dienophile) - LUMO(diene) interaction which controls the reaction.[2]

The most commonly encountered activating substituents for the 'normal' Diels-Alder reaction are CO, $CO_2R$, CN and $NO_2$, and dienophiles which contain one or more of these groups in conjugation with the double or triple bond react readily with dienes. Acrylic esters and nitriles, acetylenic esters, αβ-unsaturated ketones and quinones, for example, are widely employed as dienophiles. Alkyl substituents, on the other hand, may reduce the reactivity of dienophiles through a steric effect. Olefinic compounds such as allyl alcohol and its esters, and allyl halides are relatively unreactive although they can sometimes be induced to react with dienes under forcing conditions.

Vinyl ethers and enamines, in which the dienophile carries an electron-donating substituent take part in Diels-Alder reactions with inverse electron demand. They react with dienes bearing electron-withdrawing substituents and with other electron-deficient dienes such as αβ-unsaturated carbonyl compounds. The latter reaction leads to dihydropyrans which are easily hydrolised to glutaraldehydes. Thus, in a key step in a synthesis of secologanin the αβ-unsaturated aldehyde (1) and the enol ether (2) gave the dihydropyran derivative (3)[3] and an important step in the synthesis of the alkaloid (±)-minovine involved cycloaddition of the cyclic enamine (5) to the indolylacrylic ester (4) as shown in Scheme 2.[4]

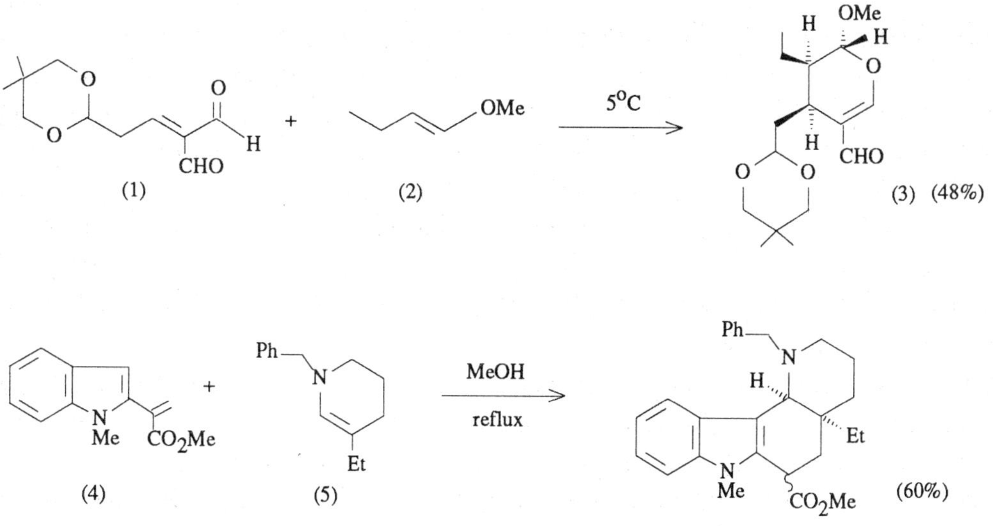

O O H CHO (1) + OMe (2) $5^oC$ H OMe H O H CHO O O (3) (48%)

N Me $CO_2Me$ (4) + Ph N Et (5) MeOH reflux Ph N H Et N Me $CO_2Me$ (60%)

*Scheme 2*

It is not certain that cycloadditions of enamines are concerted; they may be stepwise ionic reactions.[5]

Isolated carbon-carbon double or triple bonds do not usually take part in intermolecular Diels-Alder reactions under normal conditions, but a number of cyclic alkenes and alkynes with pronounced angular strain do. Some cyclic alkynes are powerful dienophiles, and arynes, such as dehydro-benzene undergo a variety of Diels-Alder additions.[6] Cyclopropene derivatives are also useful dienophiles and add to both electron-rich and electron-deficient dienes. Thus, the cyclopropenone ketal (6) and 1-methoxybutadiene gave the adduct (7) in 72 per cent yield in a normal HOMO (diene)-controlled Diels-Alder addition, but the electron-deficient diene (8) reacted equally well to give the adduct (9). In each case the adduct was a single stereoisomer, thought to be the *trans* compound formed by way of an *exo* transition state.[7]

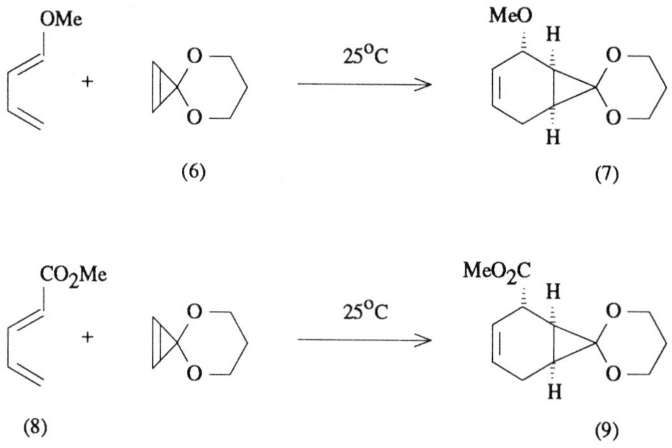


*Scheme 3*

Dienophiles in which the multiple bond is not activated by conjugation with an electron-withdrawing or an electron-donating substituent do not readily take part in Diels-Alder cycloaddition reactions under non-forcing conditions. However, this limitation can sometimes be circumvented by the use of 'synthetic equivalents', that is derivatives in which the double or triple bond is activated by the temporary attachment of an electron-withdrawing group which is removed after the cycloaddition has been effected. Thus, the readily available phenyl vinyl sulphone serves well as a synthetic equivalent of ethylene or terminal alkenes in Diels-Alder reactions. It reacts smoothly with a variety of dienes to give adducts, commonly crystalline, from which the phenylsulphonyl group is easily removed by reduction with sodium amalgam. Reactions with unsymmetrical dienes are highly regioselective, because of the strong directing effect of the sulphonyl group, and alkylation of the initial adduct before reductive cleavage of the sulphonyl group leads to products formed by formal addition of a terminal alkene to a diene.[8] Thus, reaction with 2,3-dimethylbutadiene leads initially to the adduct (10) which gives 1,2-dimethylcyclohexene on reductive cleavage, formed in effect by addition of ethylene to 2,3-dimethylbutadiene. Alkylation of (10) with benzyl bromide and subsequent desulphonylation gives 4-benzyl-1,2-dimethylcyclohexene corresponding to formal addition of allylbenzene to the diene (Scheme 4).

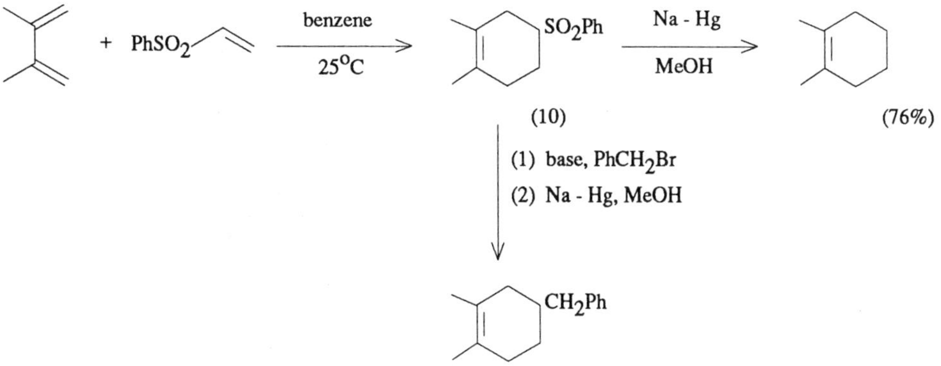

+ $PhSO_2$ benzene 25°C $SO_2Ph$ Na - Hg MeOH (76%)

(10)

(1) base, $PhCH_2Br$
(2) Na - Hg, MeOH

$CH_2Ph$

*Scheme 4*

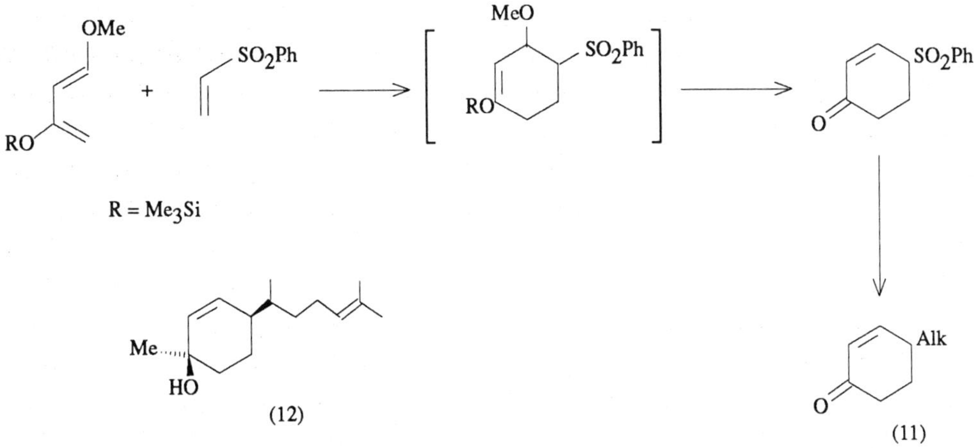

OMe RO + $SO_2Ph$ MeO $SO_2Ph$ RO $SO_2Ph$ O

R = $Me_3Si$

Me HO (12)

Alk O (11)

*Scheme 5*

In another application 1-methoxy-3-trimethylsilyloxybutadiene and phenyl vinyl sulphone gave specifically 4-phenylsulphonylcyclohexenone. Further manipulation of this provided 4-alkylcylclohexenones, which are notoriously difficult to obtain by direct alkylation of cyclohexenone itself. The sesquiterpene alcohol zinziberenol (12) was conveniently made by this route.

An alternative method for engaging unactivated alkenes as dienophiles employs the corresponding alkenyl phenyl sulphones obtained from the alkene with phenylselenenyl benzenesulphonate followed by elimination of selenium by oxidation with hydrogen peroxide.[9] Thus, 1-hexene provides specifically the vinyl sulphone (13). Because of the powerful directing effect of the sulphone group the cycloadditions are highly regioselective; 1-alkoxy-3-trimethylsilyloxydienes, for example, give specifically the valuable 5-alkylcyclohex-2-en-1-ones (Scheme 6).

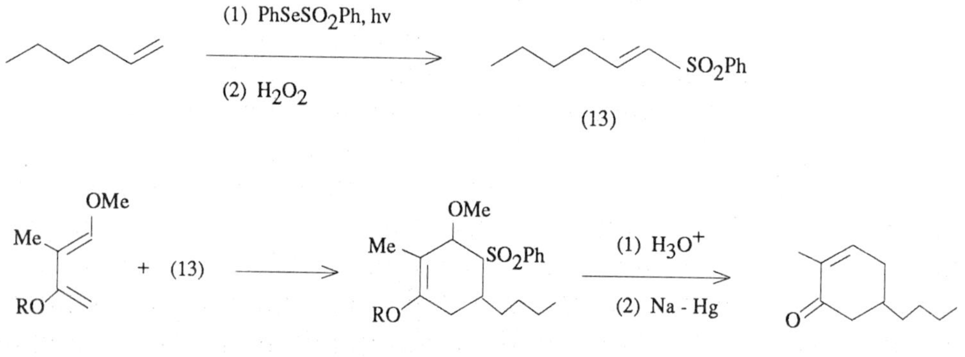

(1) $PhSeSO_2Ph$, hv
(2) $H_2O_2$
$SO_2Ph$
(13)

OMe
Me
RO
+ (13)
OMe
Me
$SO_2Ph$
RO
(1) $H_3O^+$
(2) Na - Hg
O

*Scheme 6*

The low reactivity of acetylene in Diels-Alder reactions, coupled with the synthetic usefulness of the adducts, has led to the development of a number of dienophilic acetylene equivalents.[10] Phenyl vinyl sulphoxide is one such reagent, although it reacts only with very reactive dienes; the initial adducts eliminate phenylsulphenic acid during reaction to generate a double bond. Thus, reaction of the sulphoxide with *trans,trans*-1,4-diphenylbutadiene forms *para*-terphenyl directly by elimination and dehydrogenation.[11] *Para*-tolyl ethynyl sulphone has also been used as an acetylene equivalent. With 1,4-diphenylbutadiene it forms the adduct (14) and thence, by reductive cleavage of the sulphone with sodium amalgam the cyclohexadiene (15).[12]

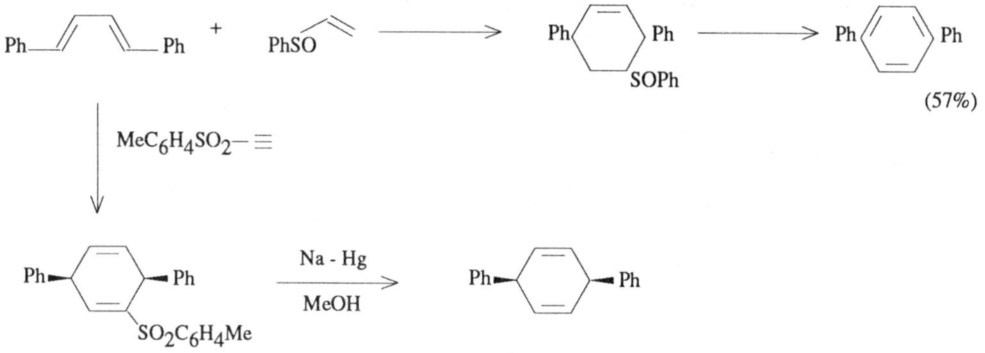

Ph Ph + PhSO → Ph Ph SOPh → Ph Ph (57%)

$MeC_6H_4SO_2$—≡

Ph Ph $SO_2C_6H_4Me$ Na - Hg / MeOH → Ph Ph

(14) (15)

*Scheme 7*

Also employed as acetylene equivalents have been norbornadiene,[13] *trans*-1-benzenesulphonyl-2-(trimethylsilyl)ethylene[14] and bis(phenylsulphonyl)ethylene.[15] 2-Benzyloxy-1-nitroethylene and 2-phenylsulphonyl-1-nitroethylene have served as nitro-acetylene equivalents.[16]

Nitro-olefins have been used as synthetic equivalents for both ethylenes and acetylenes in Diels-Alder reactions. Nitro-olefins are excellent dienophiles; they react under mild conditions and the nitro group controls the regiochemistry of the reactions. Reductive elimination of the nitro group from the adducts gives a product formed, effectively, by addition of an olefin to the diene, while elimination of the elements of nitrous acid gives an 'acetylene' adduct; the product (16), for example, is formally produced by addition of cyclohexynone to 2,3-dimethylbutadiene.[17]

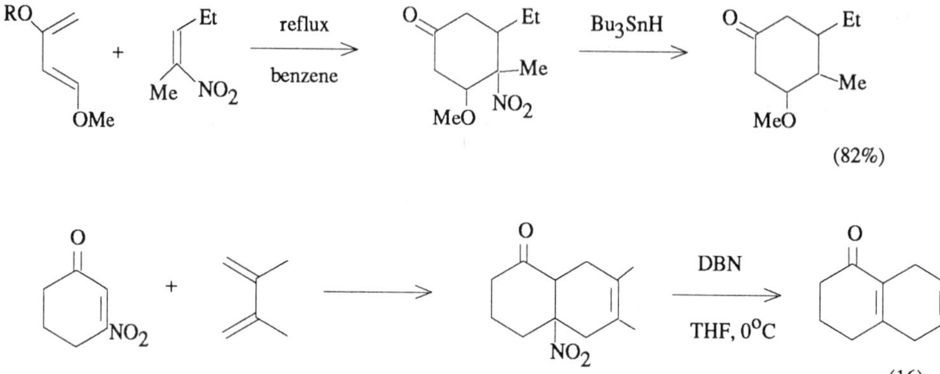

*Scheme 8*

Another potentially useful dienophile is ketene but, as is well known, it reacts preferentially with dienes in a [2π+2π] addition to form cyclobutanones rather than the desired 3-cyclohexenones. However, indirect methods have been developed to achieve the conversion corresponding to Diels-Alder addition of ketene to dienes. These involve addition of a suitably chosen vinyl compound to the diene followed by conversion of the initial adduct into the required ketone. A number of reagents have been used; among the best appear to be 2-chloroacryloyl chloride[18] and 2-chloroacrylonitrile.[19] Conversion of the initial adducts into the desired ketones is effected by hydrolysis.[20]

Also employed as ketene equivalents have been di-n-butyl vinylboronate,[21] nitroethylene,[22] and p-tolyl vinyl sulphoxide.[23] However, the vinylboronate is not a very reactive dienophile and the thermal instability of nitroethylene limits its application to reactions with reactive dienes. An attractive feature of vinyl sulphoxides as ketene equivalents is that they can be obtained in optically active form, making possible enantioselective Diels-Alder additions.[23] These ketene equivalents have been used with only a very limited range of dienes, and it remains to be seen whether they will be generally useful.

Methylthiomaleic anhydride,[24] methyl(phenylthio)propiolate,[25] methyl 3-bromopropiolate[26] and 1,3-dimethoxycarbonylallene[27] have all been used as synthetic equivalents of 1-methoxycarbonylketene (17), giving adducts which can be converted into cyclic β-keto-esters (Scheme 10).

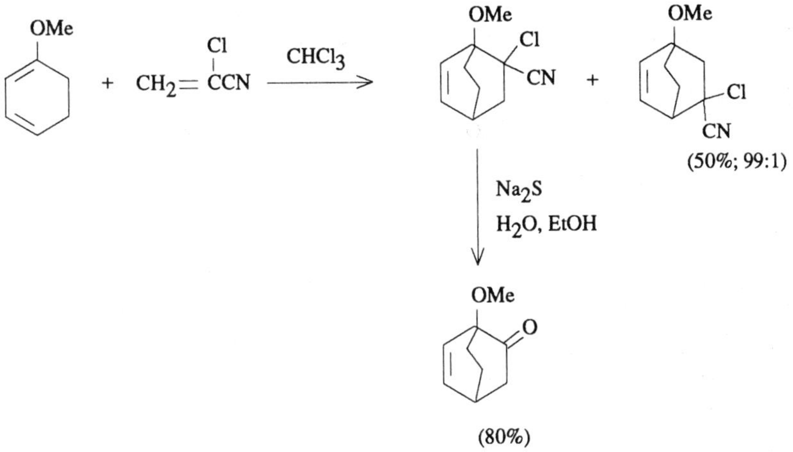

OMe

+ $CH_2{=}C(Cl)CN$ $\xrightarrow{CHCl_3}$ (OMe, Cl, CN adduct) + (OMe, Cl, CN adduct)

(50%; 99:1)

$Na_2S$, $H_2O$, EtOH

OMe, O

(80%)

*Scheme 9*

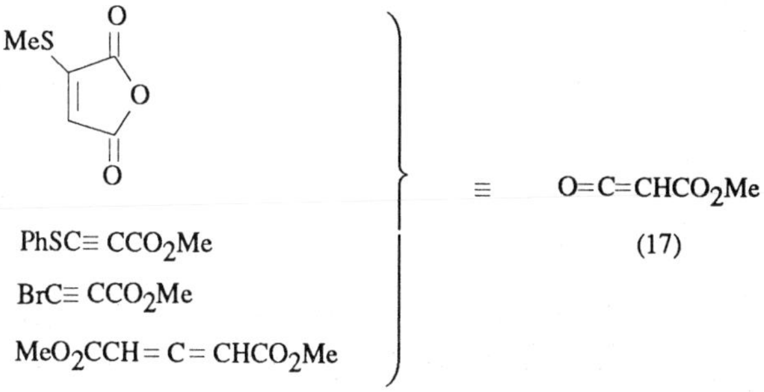

MeS (O, O, O)

$PhSC{\equiv}CCO_2Me$

$BrC{\equiv}CCO_2Me$

$MeO_2CCH{=}C{=}CHCO_2Me$

$\equiv$ $O{=}C{=}CHCO_2Me$

(17)

*Scheme 10*

## Heterodienophiles

One or both of the atoms of the multiple bond of a dienophile may be a heteroatom and such heterodienophiles are now being used increasingly in synthesis.[28] Carbonyl compounds, thiocarbonyl compounds, imines and nitroso compounds are particularly useful in this respect.

The thermal addition of certain aldehydes bearing electron-attracting substituents adjacent to the carbonyl group to conjugated dienes to give 4,5-dihydropyrans (Scheme 11) has been known for a long time and has been employed in the synthesis of hexoses and disaccharides.[29] But the scope of this reaction has been limited by the elevated temperatures required and the restricted range of aldehydes which take part.

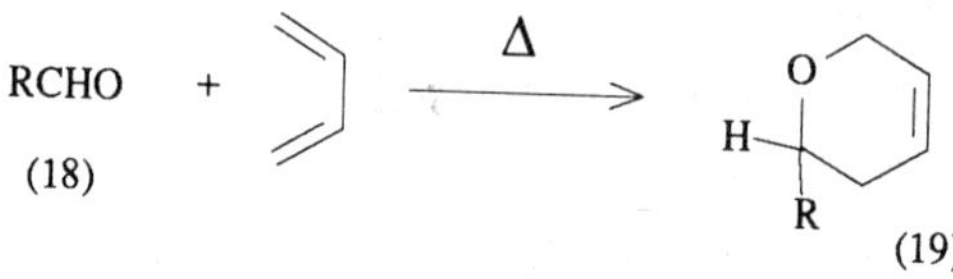

RCHO + (18) → Δ → (19)

*Scheme 11*

More recently, cycloaddition of aldehydes to reactive alkoxybutadienes has been effected at moderate temperatures under ultra-high pressures[30] (see p.49). The scope of reactions at ordinary pressures has been greatly widened by the finding that certain reactive 1,3-dienes, such as (20), bearing alkoxy or silyloxy substituents, react readily with a wide range of aldehydes in the presence of Lewis acid catalysts to give substituted 2,3-dihydro-4-pyrones (21). The catalysts used include magnesium bromide, zinc chloride, boron trifluoride etherate, the europium complexes $Eu(fod)_3$ and $Eu(hfc)_3$ and titanium tetrachloride, and each of these imparts a characteristic stamp on the course of the reaction. The reactions take place under mild conditions with a high degree of control over the stereochemistry of the 2,3-dihydropyrones formed, and they have been employed in the stereocontrolled syntheses of hexoses and, by cleavage of the dihydropyrone ring, of 2-alkyl-3-hydroxycarboxylic acids.[31]

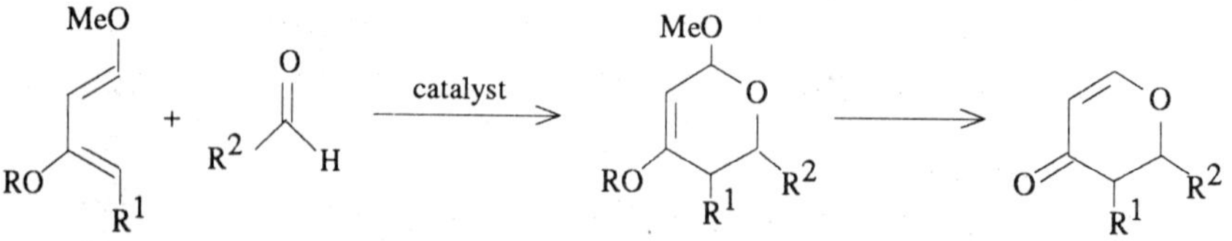

(20) + $R^2CHO$ → catalyst → → (21)

$R = Me_3Si$

*Scheme 12*

The course of the reactions and the stereochemistry of the dihydropyrones formed depends on the catalyst and also, to some extent, on the reaction conditions and the geometry and substitution pattern of the diene. Reactions catalysed by zinc chloride are true pericyclic reactions and proceed through 1:1 adducts (as 23) which can be isolated if desired.[32] With *trans,trans*-1,4-disubstituted dienes such as (22) the reactions are highly stereoselective and lead largely to the *cis*-2,3-disubstituted-4-pyrone (24) by way of an *endo* transition state, with the important exception of reactions with aldehydes bearing an α- or β- alkoxy substituent[33] (see below).

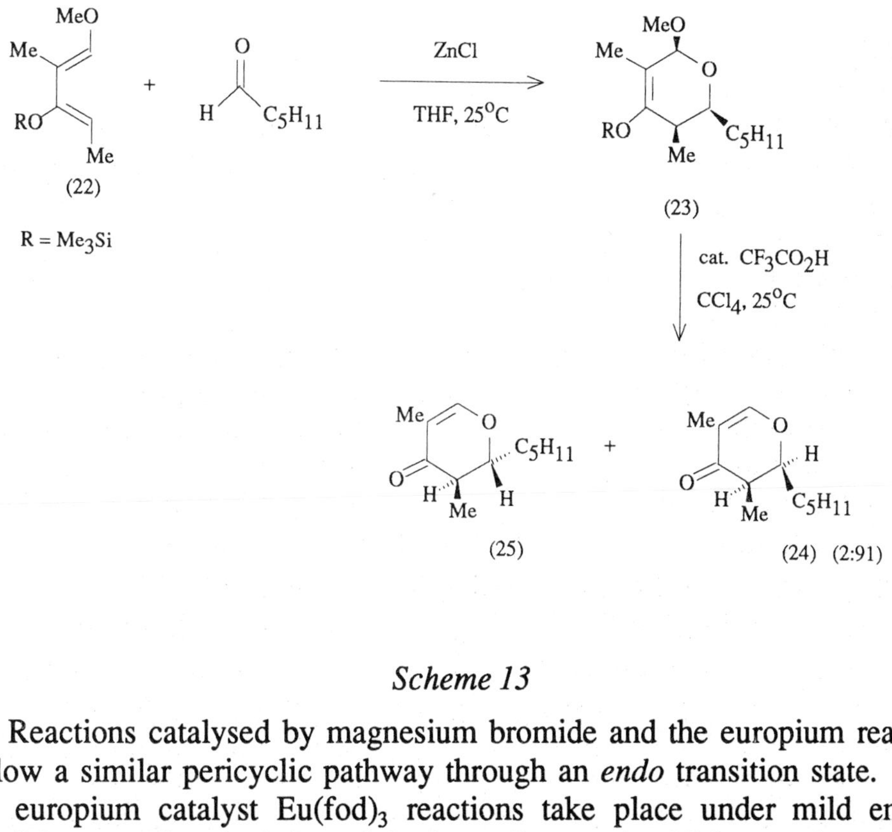

MeO, Me, RO, Me (22) + H, O, $C_5H_{11}$ — ZnCl, THF, 25°C → MeO, Me, O, RO, Me, $C_5H_{11}$ (23)

R = $Me_3Si$

cat. $CF_3CO_2H$, $CCl_4$, 25°C

Me, O, $C_5H_{11}$, O, H, Me, H (25) + Me, O, H, O, H, Me, $C_5H_{11}$ (24) (2:91)

*Scheme 13*

Reactions catalysed by magnesium bromide and the europium reagents follow a similar pericyclic pathway through an *endo* transition state. With the europium catalyst $Eu(fod)_3$ reactions take place under mild enough conditions to allow isolation of the immediate cycloaddition products. Thus reaction of acetaldehyde with diene (22) catalysed by $Eu(fod)_3$ takes place with virtually complete *endo* specificity to give the dihydropyran (26), with stereocontrolled formation of three chiral centres. A fourth centre is generated at C-5 by axial protonation of the silyl enol ether with methanol-triethylamine, to give the tetrahydropyrone (27). Alternatively, treatment with trifluoracetic acid leads to the *cis*-2,3-disubstituted dihydropyrone (28).[34]

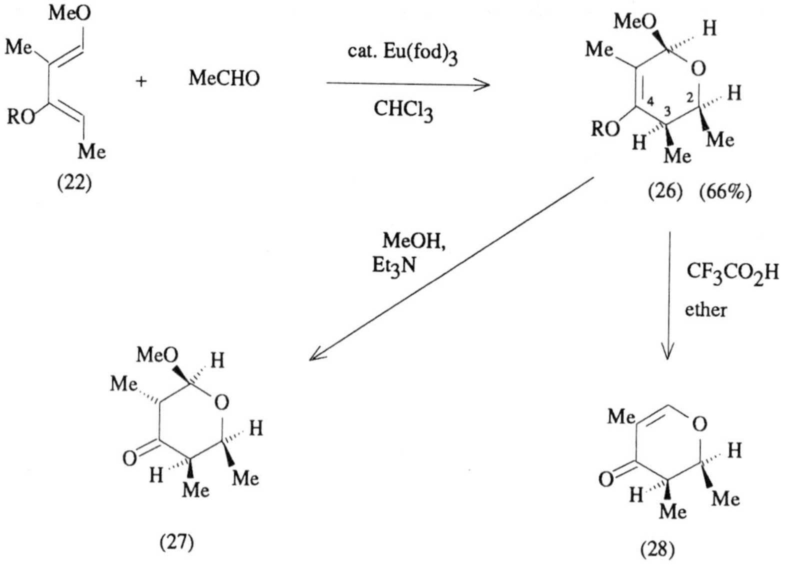

*Scheme 14*

With the chiral catalyst (+)-Eu(hfc)$_3$ and the correct choice of chiral auxiliary on the alkoxydienes, adducts of high enantiomeric purity have been obtained and have been employed in the synthesis of optically pure saccharides without recourse to formal resolution[35] (see p.72). The preference for the *endo* transition state in these reactions, leading to *cis*-2,3-disubstituted dihydropyrones is ascribed to the formation of a Lewis acid-aldehyde complex in which the catalyst binds *anti* to the R group of the aldehyde. In general the catalyst-solvent array is larger than R and preferentially adopts the *exo* position[36] (Scheme 15).

For reactions catalysed by $BF_3.OEt_2$ the situation is more complex. It appears that more than one reaction pathway may be followed, with different stereochemical consequences, and that the delicate balance between them may be influenced by a number of factors including the experimental conditions and the substitution pattern in the diene.[37] Reaction of the diene (22) with several aldehydes catalysed by $BF_3.OEt_2$ in methylene chloride solution led mainly to the corresponding *trans*-2,3-dihydro-4-pyrone, contrasting with the $ZnCl_2$-catalysed reactions which gave largely the *cis* isomers. In toluene solution however the main products were the *cis* isomers and these were formed in even larger amounts by substitution of the t-butyldimethylsilyl ether for the trimethylsilyl ether in (22)[38] (Scheme 16). It is suggested that the reactions catalysed by boron trifluoride take place, not by a pericyclic pathway, but through a siloxonium species like (29).

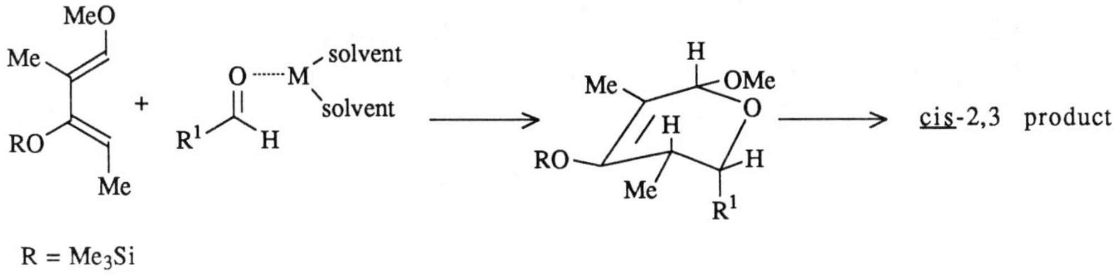

$\longrightarrow$ cis-2,3 product

R = $Me_3Si$

*Scheme 15*

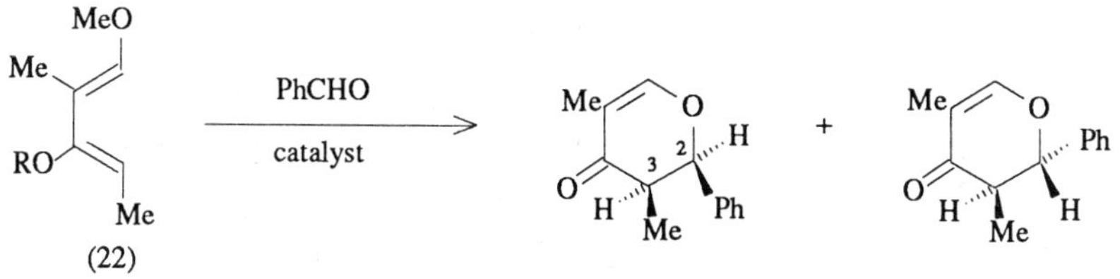

(22)

PhCHO, catalyst

| Catalyst | 2,3-cis product | 2,3-trans product |
|---|---|---|
| $BF_3 \cdot Et_2O$, $CH_2Cl_2$, -78°C | 1 | 4.6 |
| $BF_3 \cdot Et_2O$, toluene | 2.2 | 1 |
| $ZnCl_2$, THF | 40 | 1 |

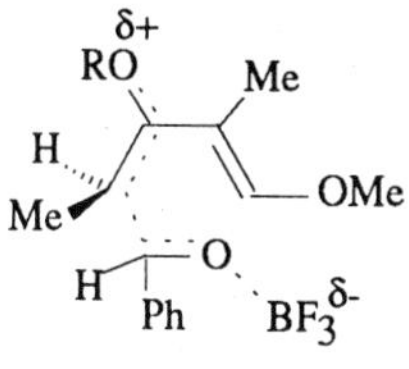

(29)

R = $Me_3Si$

*Scheme 16*

In the reactions of α-alkoxy aldehydes with alkoxy dienes the stereochemical outcome is different; reactions in the pericyclic mode now lead preferentially to the *trans*-2,3-disubstituted pyrone. A chelated complex (as 30) is formed in presence of the catalyst and the *exo* transition state is now preferred (Scheme 17).[39] Magnesium bromide is the best catalyst for these chelation-controlled reactions. Thus, reaction of diene (22) with the α-benzyloxy aldehyde (31) in the presence of magnesium bromide gave the 2,3-*trans*-4-pyrone (32) exclusively in 50 per cent yield.

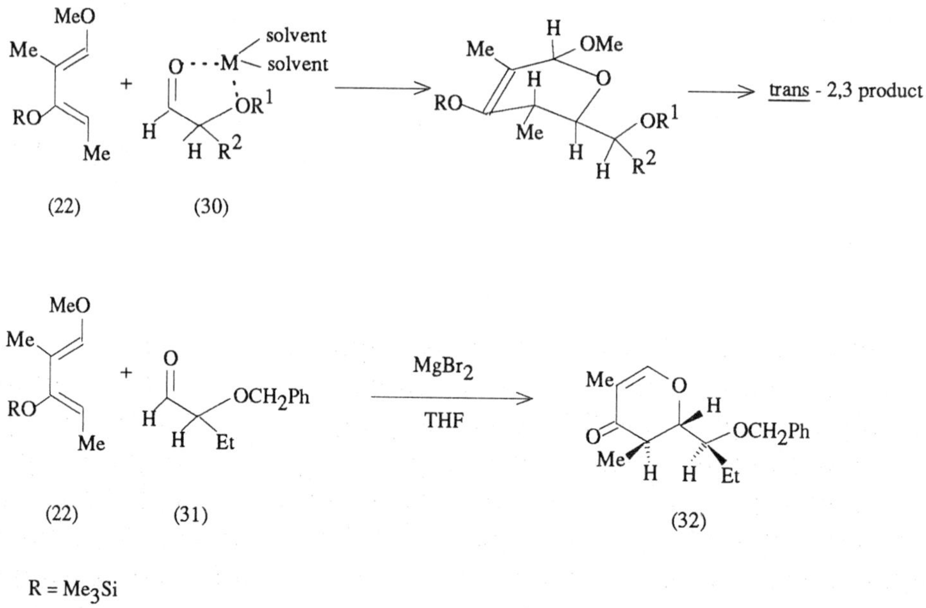

MeO, Me, RO, Me (22) + O···M (solvent, solvent), $OR^1$, H, H $R^2$ (30) → Me, H, OMe, H, O, RO, Me, $OR^1$, H, H $R^2$ → trans - 2,3 product

MeO, Me, RO, Me (22) + O, H, $OCH_2Ph$, H, Et (31) —$MgBr_2$, THF→ Me, O, H, $OCH_2Ph$, O, Me, H, H, Et (32)

$R = Me_3Si$

*Scheme 17*

With β-alkoxy aldehydes catalysis by magnesium bromide gave only poor selectivity. Excellent results were obtained with titanium tetrachloride as catalyst, leading to the *cis*-2,3-disubstituted 4-pyrones. The reactions now follow a Mukaiyama aldol-like course and not a pericyclic pathway. Thus, when diene (22) was reacted with the β-alkoxy aldehyde (33) and titanium tetrachloride and the resultant aldol treated with trifluoroacetic acid, the *cis*-2,3-disubstituted 4-pyrone (34) was obtained in 56 per cent yield. Similarly the α-alkoxy aldehyde (35) gave the *cis* disubstituted pyrone (36) in 93 per cent yield by way of the intermediate (37). None of the corresponding 2,3-*trans* isomer was detected in either case.[39]

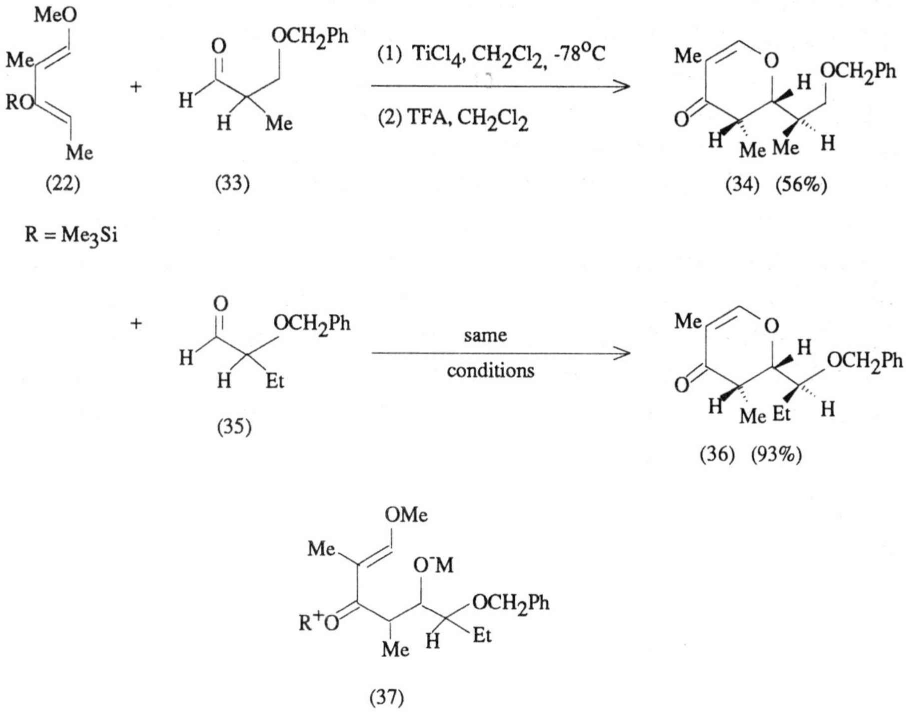

MeO Me RO Me (22) R = $Me_3Si$

+ O H H Me $OCH_2Ph$ (33)

(1) $TiCl_4$, $CH_2Cl_2$, -78°C

(2) TFA, $CH_2Cl_2$

Me O O H $OCH_2Ph$ H Me Me H (34) (56%)

+ O H H Et $OCH_2Ph$ (35)

same conditions

Me O O H $OCH_2Ph$ H Me Et H (36) (93%)

OMe Me $O^-M$ $R^+O$ Me H $OCH_2Ph$ Et (37)

*Scheme 18*

Thus, to summarise,[40] cycloadditions with α-alkoxy aldehydes, catalysed by magnesium bromide, lead preferentially to *trans*-2,3-disubstituted 4-pyrones *via* a chelation-controlled pericyclic pathway. With titanium tetrachloride as catalyst the corresponding *cis* isomers are obtained through a Mukaiyama aldol process. With β-alkoxy aldehydes catalysis by magnesium bromide results in only poor selectivity but with titanium tetrachloride high yields of 2,3-*cis*-4-pyrones are again obtained *via* a Mukaiyama aldol pathway.

As is implied in the reaction schemes above, catalysed addition of dienes to aldehydes follows the Cram selectivity rule, allowing the controlled development of three contiguous chiral centres with a high degree of selectivity as in (32) and (36). In another example the chiral aldehyde (38) and 1-methoxy-3-trimethylsilyloxybutadiene gave the adduct (39), which was converted in a number of steps into 2,4-dideoxy-D-glucose (40).[41]

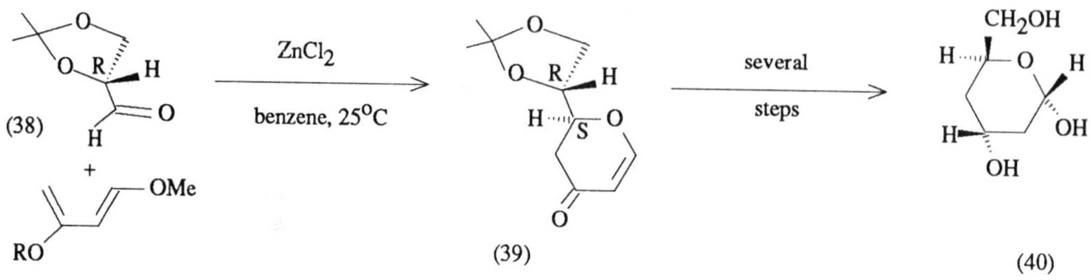

(38) + OMe, RO — $ZnCl_2$, benzene, 25°C → (39) — several steps → (40)

*Scheme 19*

The ease of the catalysed reactions between aldehydes and suitable 1,3-dienes, their flexibility and the high degree of stereocontrol possible, have led to their application in the stereocontrolled syntheses of a variety of hexose systems related to sugars, and of open-chain compounds derived by cleaving the dihydropyrone ring (see p.103).

**Thiocarbonyl compounds**

Thiocarbonyl compounds of all kinds are highly reactive dienophiles. They are more reactive than the corresponding carbonyl compounds and their cycloaddition to conjugated dienes provides a good route to six-membered sulphur heterocycles.[42] However, they have not so far been used much in complex syntheses.

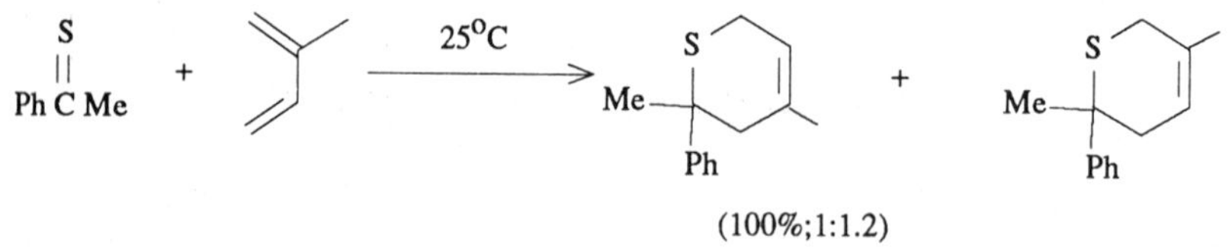

PhC(S)Me + diene — 25°C → (100%;1:1.2)

*Scheme 20*

Although all kinds of thiocarbonyl compounds can act as dienophiles they do not all do so with equal ease. Thio-esters are less reactive than thio-aldehydes and thio-ketones, but their reactivity is enhanced by an adjacent electron-withdrawing group. For example, cyanodithioformates,[43] dithio-oxalates[44] and dithiopyruvates[44] are reactive dienophiles. Alkyl- and aryl-dithiocarboxylates, on the other hand, react with dienes only at elevated temperatures.[45] The following order of reactivity of thiocarbonyl compounds is suggested by observation (Scheme 21).

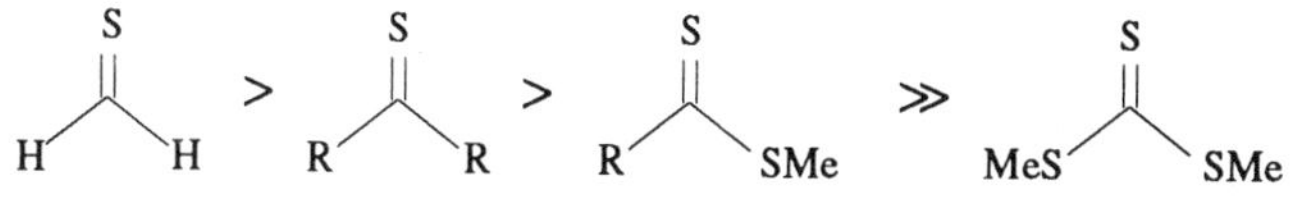

S=C(H)H > S=C(R)R > S=C(R)SMe ≫ S=C(MeS)SMe

*Scheme 21*

Early attempts to use thio-aldehydes as dienophiles were thwarted by their tendency to polymerise, but a number of methods have now been developed to produce them *in situ*. Thiobenzaldehyde and thioacetaldehyde, for example, may be generated by thermolysis of the appropriate thiosulph-inate, itself readily prepared by oxidation of the corresponding disulphide, and trapped *in situ* by reaction with a diene.[46] Thus, thiobenzaldehyde generated by thermolysis of thiosulphinate (41) was trapped efficiently by anthracene and 2,3-dimethylbutadiene to give the adducts (42) and (43).

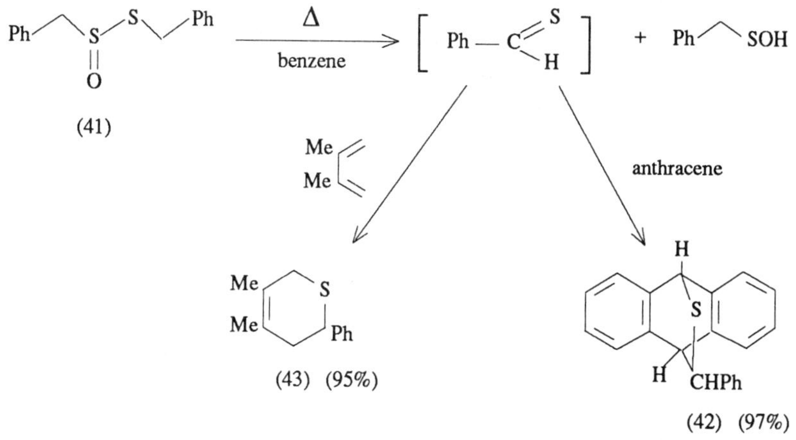

Ph S(=O) S Ph (41) —Δ, benzene→ [ Ph—C(=S)H ] + Ph SOH

Me, Me (diene) → (43) (95%) Me, Me, S, Ph

anthracene → (42) (97%) H, S, H, CHPh

*Scheme 22*

In an intramolecular example which shows the potential scope of the procedure, the thiosulphinate (44) heated in toluene solution gave the two thiabicyclononenes (46) and (47) in 40 per cent yield, by way of the thioaldehyde (45).[47]

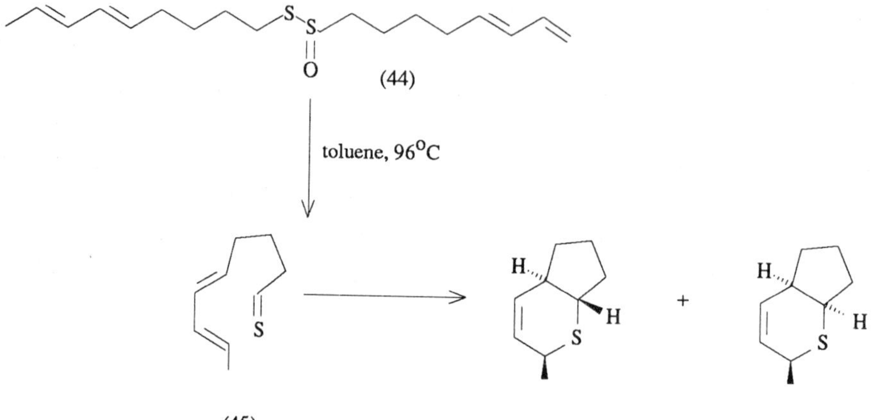

(44)

toluene, 96°C

(45) (46) (47)

*Scheme 23*

The anthracene adduct (42) is a good storable source of thiobenzaldehyde, and so are the readily available thioaldehyde-cyclopentadiene adducts.[48] Photofragmentation of phenacyl sulphides provides another good general route to thio-aldehydes. If the irradiation is conducted in the presence of a diene the adduct dihydrothiopyran is obtained directly.[49]

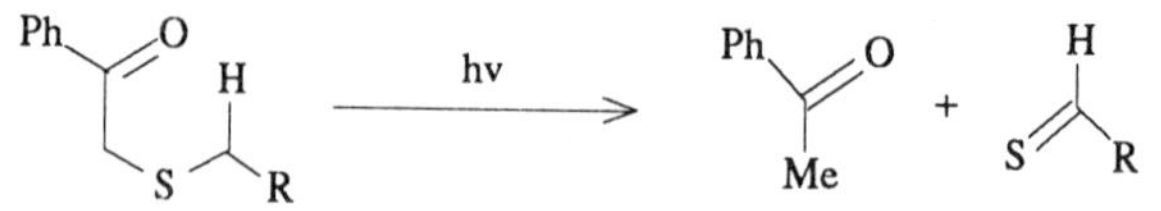

hv

*Scheme 24*

Transient thio-aldehydes have also been obtained by the action of triethylamine on ethoxycarbonylmethanesulphonyl chloride,[50] sodium thiosulphate S-esters[51] or α-sulphonyldisulphides[52] and by fluoride-induced elimination from α-silyldisulphides.[53]

The regiochemistry of addition of thio-aldehydes to unsymmetrical dienes depends on the nature of the group R of the thio-aldehyde. If R is an electron-withdrawing group, reaction with 1-substituted butadienes gives "ortho" adducts and 2-substituted butadienes give "para" adducts.[54] With the less reactive alkane-thials, the "meta" adducts are the main products (Scheme 25).

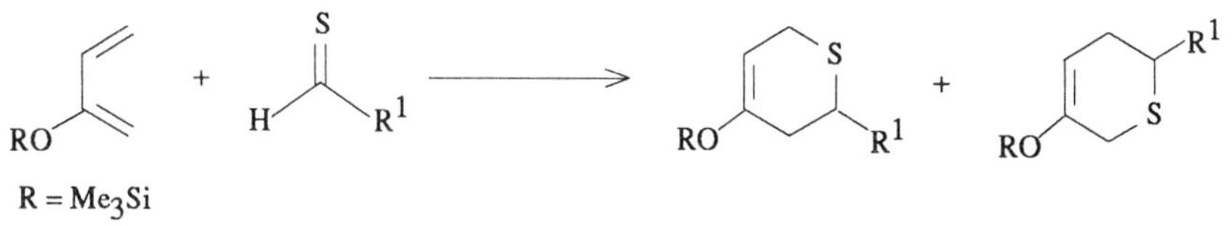

R = $Me_3Si$

| | | |
|---|---|---|
| $R^1 = CO_2Pr^i$ | 9% | 53% |
| $R^1 = CH_2OAc$ | 74% | <5% |

*Scheme 25*

Thiopyrans obtained by Diels-Alder addition of thio-aldehydes to dienes have been employed in an interesting way to prepare medium-ring compounds. Thus, the adduct (48) from benzoylthioformaldehyde and 2-ethoxybutadiene, on conversion into (49) followed by methylation and reaction with potassium t-butoxide gave the ring-enlarged thiacyclononenone (51) by way of the kinetically favoured ylide (50).[55] A similar sequence has been employed to prepare other functionalised medium ring sulphur heterocycles used as intermediates in prospective macrolide syntheses.[56]

Ring-contraction reactions have also been effected, to give functionalised cyclopentenes. For example the thiocyclohexene (52) on treatment with lithium di-isopropylamide followed by quenching with methyl iodide gave the cyclopentene (53) in 87 per cent yield as a mixture of stereoisomers.[57]

**N-Sulphinylsulphonamides**

Another useful group of sulphur-containing dienophiles are the N-sulphinylsulphonamides (54) which are readily obtained from arylsulphonamides and thionyl chloride. They add readily to conjugated dienes to give 3,6-dihydro-1,2-thiazine derivatives.[58]

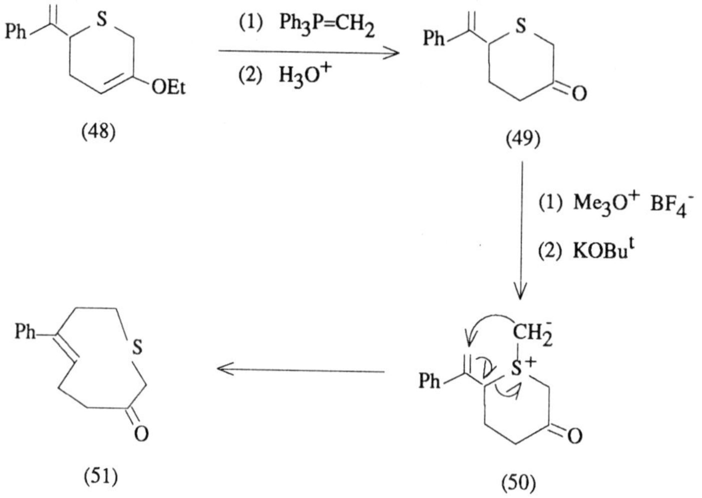

Ph, S, OEt, (48); (1) $Ph_3P{=}CH_2$, (2) $H_3O^+$; Ph, S, O, (49); (1) $Me_3O^+ BF_4^-$, (2) $KOBu^t$; $CH_2^-$, $S^+$, Ph, O, (50); Ph, S, O, (51)

*Scheme 26*

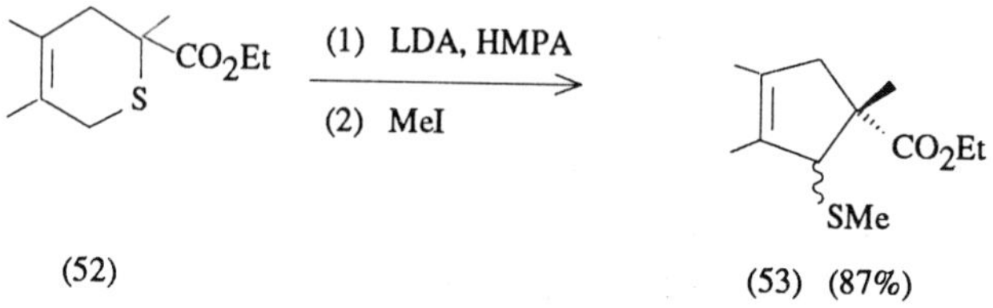

$CO_2Et$, S, (52); (1) LDA, HMPA, (2) MeI; $CO_2Et$, SMe, (53) (87%)

*Scheme 27*

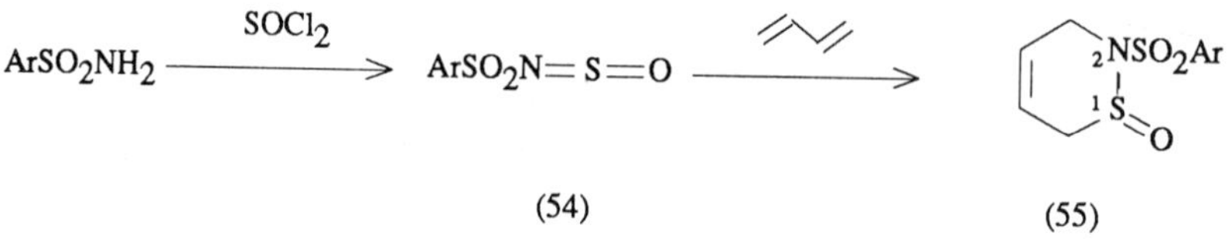

$ArSO_2NH_2$ —$SOCl_2$→ $ArSO_2N{=}S{=}O$ (54) → $_2NSO_2Ar$, $^1S{=}O$ (55)

*Scheme 28*

Addition of sulphinylsulphonamides to unsymmetrical dienes is highly regioselective. 2-Substituted dienes give thermally stable 5-substituted thiazines, while 1-substituted dienes, at low temperatures give 3-substituted thiazines; the latter, however, readily isomerise at higher temperatures to the more stable 6-substituted thiazines.[59]

The importance of these cycloadducts in synthesis lies in the fact that on alkaline hydrolysis the thiazine ring is cleaved to give a sulphinate which, on acidification, loses sulphur dioxide by a cyclic retro-ene pathway to form a derivative of a homoallylic amine[60] (Scheme 29).

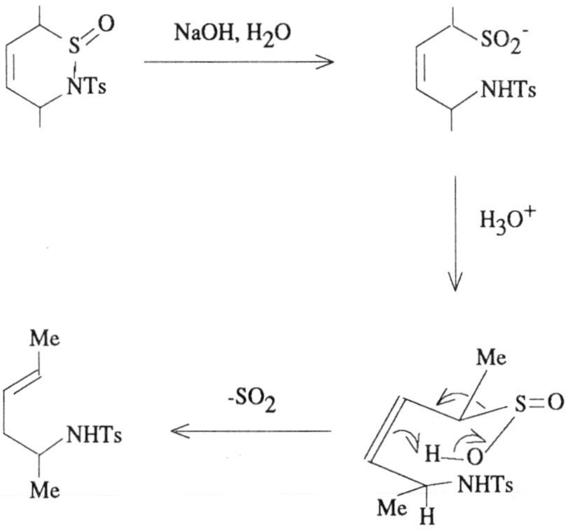

S=O NTs — $NaOH, H_2O$ → $SO_2^-$ NHTs

$H_3O^+$

Me NHTs Me ← $-SO_2$ — Me S=O H O NHTs Me H

*Scheme 29*

A strong preference for a chair-like transition state with an equatorial substituent on the carbon next to the sulphur explains the predominant formation of the (*E*)-alkene, as well as the diastereomeric induction at C-4 when the reaction is conducted in deuterated solvent. This sequence has been employed in the stereocontrolled synthesis of acyclic homoallylic amines and amino-alcohols (Chapter 2).

## Imines

Another useful group of heterodienophiles are the imines, containing the group C=N-Z where Z=$SO_2R$,COR etc. These react with 1,3-dienes to form 1,2,3,6-tetrahydropyridines[61] (Scheme 30).

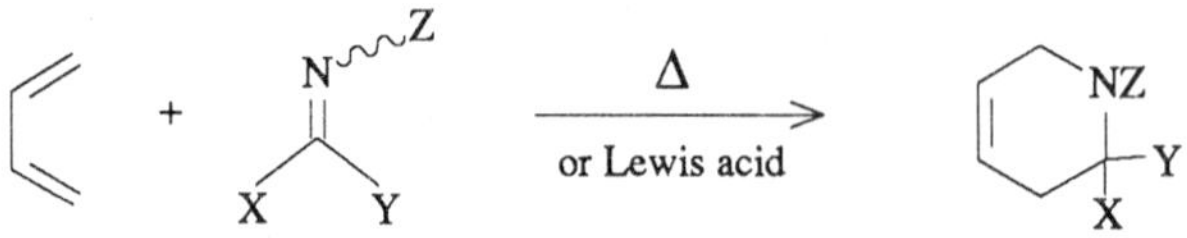

Δ or Lewis acid

$Z = SO_2Ar, COR$

$X,Y = H, CO_2R, Ar, CCl_3$

*Scheme 30*

Thus, the 'activated' imine (56) and 1-methoxy-3-trimethylsilyloxybutadiene gave the dihydropyridone (57) in 84 per cent yield.[62]

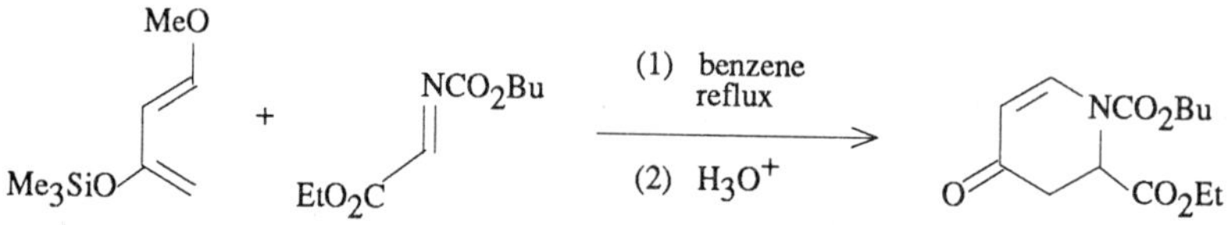

(1) benzene reflux (2) $H_3O^+$

(56) (57)

*Scheme 31*

The reactions are usually highly regio- and stereo-selective. The predominant regioisomer can often be predicted by consideration of the most stable hypothetical dipolar intermediate or transition state although, in fact, little is known about the detailed reaction pathway of imino Diels-Alder reactions. It appears that acyclic imines probably react by way of the (*E*)-isomer in most cases. However, many of the most useful electron-deficient imino dienophiles cannot easily be isolated and the easy (*Z-E*) isomerisation in these systems and nitrogen inversion in the adducts results in uncertainty about the geometry of the reacting imine. With imines bearing electron-attracting groups on both the carbon and nitrogen of the imine the stereochemistry of the product depends, of course, on which one adopts the *endo* configuration during reaction. It appears that π-substituents on nitrogen are better *endo* directors than equivalent substituents on the carbon of the dienophile. With cyclic imines, the geometry of the imine is unambiguous and the expected *endo* adduct is formed (Scheme 32).

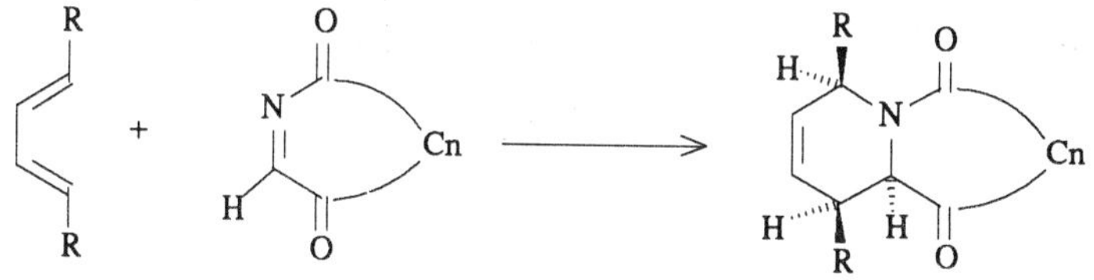

*Scheme 32*

Hitherto it has been believed that only 'activated' imines in which either the nitrogen or carbon atom of the imine, or both, carried an electron-withdrawing group, would undergo intermolecular imino Diels-Alder reactions readily. It has now been found that ordinary unactivated Schiff bases react smoothly with the reactive diene 1-methoxy-3-trimethylsilyloxy-butadiene in the presence of zinc chloride to form 2,3-dihydro-4-pyridones in good yield (Scheme 33).[63] Under the same conditions $\alpha\beta$-unsaturated imines reacted at the imino group[64] and cyclic imines and the appropriate dienes gave polycyclic compounds related to yohimbine.[65] Boron trifluoride etherate has also been used as catalyst in these reactions.[66]

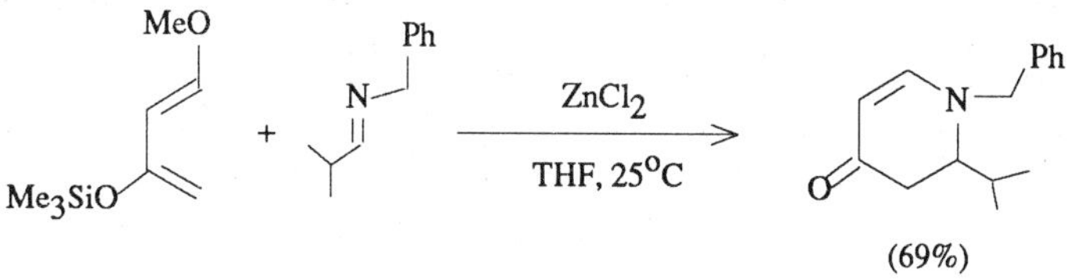

*Scheme 33*

It has also been found that simple unactivated iminium salts, generated *in situ* under Mannich-like conditions react smoothly with dienes in aqueous solution to form adducts in high yield.[67]

Intramolecular versions of the imino Diels-Alder reaction are being increasingly employed in the synthesis of alkaloids and other polycyclic nitrogenous compounds.[68] The intramolecular reactions in many cases are highly stereoselective and frequently take place in the absence of activating groups on the imino double bond. Thus, the acetate (58) on pyrolysis, affords the bicyclic lactam (59), subsequently converted into the alkaloid σ-coniceine.

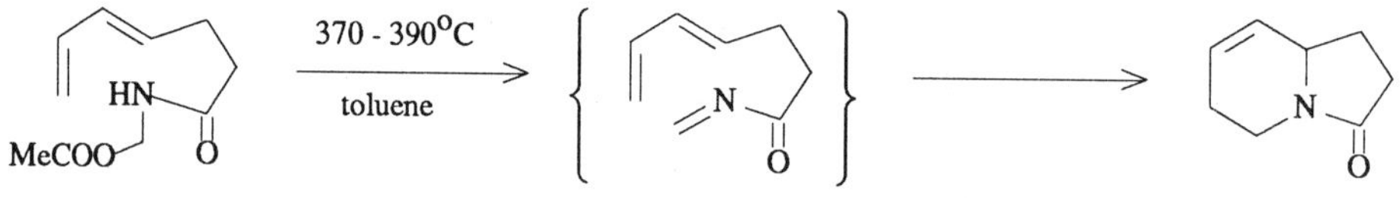

HN MeCOO O — 370 - 390°C, toluene → N O → N O

(58) (59)

*Scheme 34*

**Nitroso compounds**

Another useful group of heterodienophiles are the nitroso compounds. The most reactive are the aromatic nitroso compounds and those in which the nitroso group is directly linked to an electron-withdrawing group as in nitrosyl cyanide and the C-nitrosocarbonyl compounds, RCON=O, and C-nitrosoformate esters, ROCON=O. These all react readily with 1,3-dienes to form derivatives of 3,6-dihydro-1,2-oxazine[69] (Scheme 35).

MeO + MeO N=O — $CHCl_3$, 0°C → MeO N O OMe

(76%)

*Scheme 35*

C-Nitrosocarbonyl compounds are prepared *in situ* as transient intermediates by oxidation of hydroxamic acids with periodate,[70] and C-nitrosoformate esters in the same way from N-hydroxycarbamic esters[71] although neither type has yet been detected physically. Generated in the presence of 1,3-dienes they afford dihydro-oxazines in good yield. The C-nitroso-carbonyl compounds are conveniently purified and stored in the form of their adducts with 9,10-dimethylanthracene or cyclopentadiene, from which they are readily regenerated by warming in benzene solution. Nitrosyl cyanide is prepared from nitrosyl chloride and silver cyanide and is also best stored as its adduct with 9,10-dimethylanthracene.[72]

The very reactive C-nitrosocarbonyl compounds and C-nitrosoformate esters appear to react equally well with dienes carrying electron-withdrawing or electron-donating substituents.[73] Thus, in a key step in a synthesis of the glutamine synthetase inhibitor tabtoxine-β-lactam, reaction of benzyl nitrosoformate, generated *in situ* from the corresponding carbamic ester, with 1-cyanocyclohexadiene gave the single adduct (60) in 73 per cent yield.[74] Not all addition reactions of acylnitroso compounds are as selective as this.

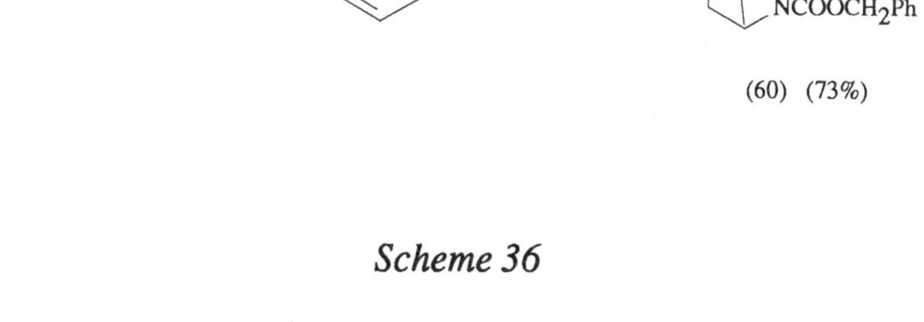

$PhCH_2OCONHOH$

$Et_4\overset{+}{N}\ IO_4^-$, $CH_2Cl_2$

$PhCH_2OCON{=}O$ + (1-cyanocyclohexa-1,3-diene, CN) ⟶ (adduct: CN, O, $NCOOCH_2Ph$)

(60) (73%)

*Scheme 36*

The reactivites of a series of acylnitroso compounds in reaction with the dimethylacetal of hexa-2,4-dienal fell in the order

$$ROCONO > RCONO > ArCONO > R_2NCONO.$$[75]

Another useful reagent is 1-chloronitrosocyclohexane. Unlike most tertiary nitrosoalkanes it forms adducts with conjugated dienes. Although sluggish, the reactions have preparative value because when they are carried out in alcoholic solution the initial adducts are solvolysed to give the corresponding 3,6-dihydro-1,2-oxazines formed, in effect, by addition of HN=O to the diene.[76] A similar transformation can be effected using benzyl nitrosoformate.[74]

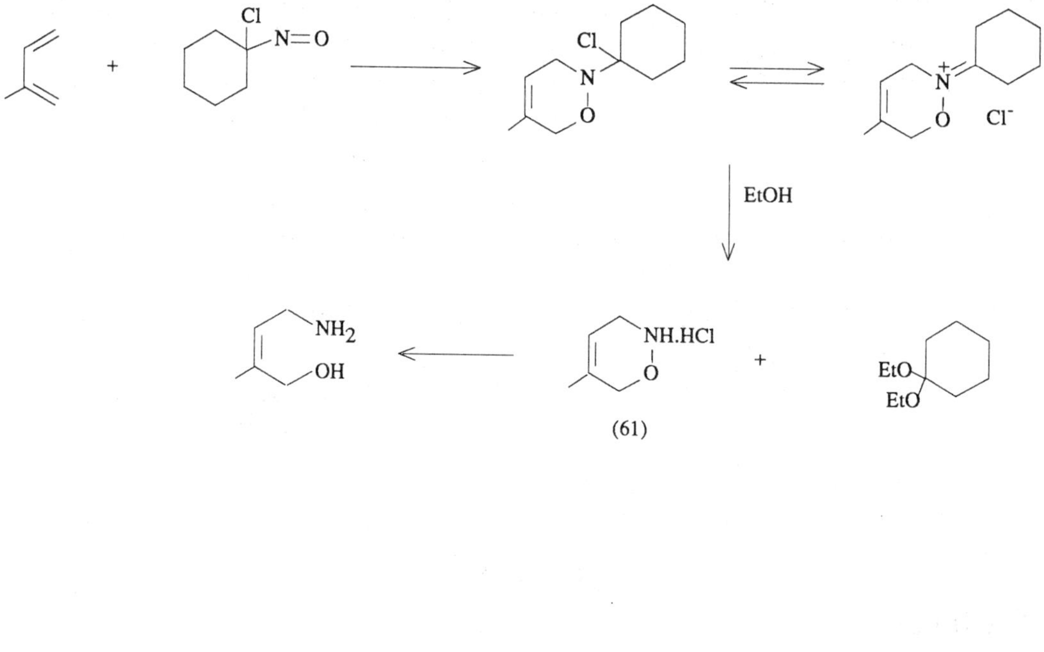

*Scheme 37*

The 1,2-oxazines formed in these reactions are valuable in synthesis because on cleavage of the N–O bond they give rise to 1,4-amino-alcohols. *cis*-1-Amino-4-hydroxy-3-methyl-2-butene, for example, was smoothly obtained by reductive cleavage of the dihydro-oxazine (61) from isoprene and chloronitrosocyclohexame[77] and the sequence has been used to prepare a number of substituted *cis*-1-amino-4-hydroxycyclohexanes with excellent control of stereochemistry (see p.115). Intra-molecular cycloadditions of nitroso compounds have also been employed in synthesis (see p.188).

The optically active enantiomerically pure α-chloronitroso compounds (62) and (63) have been prepared. The former was obtained from the oxime of epiandrosterone by oxidation with t-butyl hypochlorite. On reaction with cyclohexadiene in methanol it gave the adduct (64) in 69 per cent yield and >95 per cent enantiomeric excess, and thence the valuable cyclohexene derivative (65).[78] The compound (63) was obtained from mannose: it also reacted with cyclohexadiene, giving an adduct of opposite absolute configuration from (64).[79]

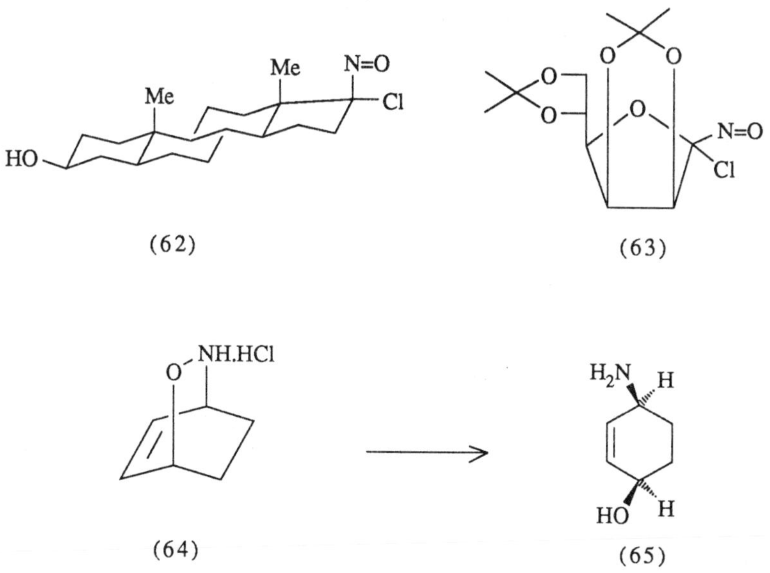

(62) (63)

(64) (65)

*Scheme 38*

## The Diene

A wide range of dienes takes part in the Diels-Alder reaction, including open-chain and cyclic dienes, and transiently formed dienes such as *ortho*-quinodimethanes. Heterodienes in which one or more of the atoms of the diene is a hetero-atom are also well-known and are much used in synthesis. An essential condition for reaction is that the diene has, or can adopt, a *cisoid* conformation, and most dienes which satisfy this condition undergo the reaction more or less easily depending on their structure. Substituents on the diene influence the rate of reaction electronically and through their steric effect on the conformational equilibrium. In a 'normal' Diels-Alder reaction the rate of addition is often increased by electron-donating substituents (for example Me, NHCOR, OMe) on the diene as well as by electron-withdrawing substituents on the dienophile. Reactions with inverse electron demand are favoured by electron-withdrawing substituents on the diene. Among open-chain dienes, substituents which discourage the diene from adopting a *cisoid* conformation hinder the reaction. Thus, (E)-substituted dienes usually react with dienophiles more readily than the corresponding (Z)-isomers.

Very many Diels-Alder reactions with alkyl- and aryl-substituted butadienes have been effected, and more recently dienes bearing hetero-atom substituents (NHCOR, OR, SR) have been used with conspicuous success.[80] Hetero-substituents on the diene have a controlling effect on the regioselectivity of the cycloadditions and allow further useful transformations of the initial cyclo-adducts. Alkoxy and trimethylsilyloxy substituents are particularly useful. Thus, 2-alkoxy- and 2-trimethylsilyloxy-1,3-dienes react easily with dienophiles to form adducts which are readily hydrolised to cyclohexanone derivatives[81] (Scheme 39).

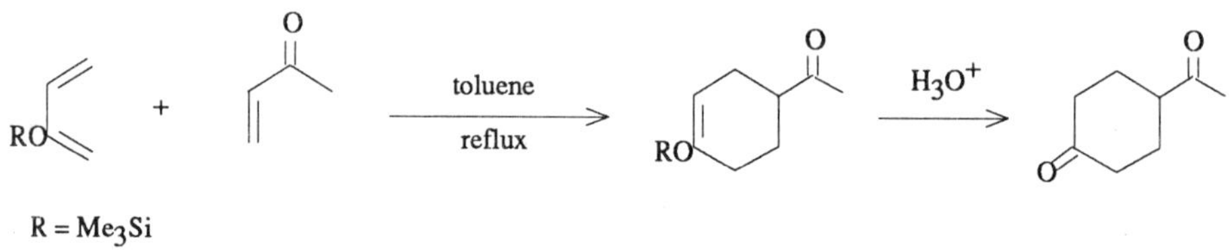

RO (diene) + (methyl vinyl ketone) —toluene, reflux→ RO (cyclohexene adduct) —$H_3O^+$→ (cyclohexanone product)

R = $Me_3Si$

*Scheme 39*

One of the most widely used dienes of this class is 1-methoxy-3-tri-methylsilyloxybutadiene (66) which is readily obtained from commercially available 1-methoxybuten-3-one[82] (Scheme 40). It owes its usefulness to its high reactivity, to the high regioselectivity shown in reactions with unsymmetrical dienophiles and to the diversity of products which can be obtained from the initial adducts. Cyclohexenones, cyclohexadienones and benzene derivatives can be prepared depending on the precise mode of operation, and the reagent has been employed in elegant syntheses of several natural products.[83]

The β-methoxycyclohexanone formed initially can sometimes be isolated, as in the reaction in Scheme 40, but frequently it is converted directly into the corresponding αβ-unsaturated ketone. Thus, reaction of the diene with 2-methylpropenal and acid hydrolysis of the initial adduct gave 4-formyl-4-methylcyclohexenone.[84]

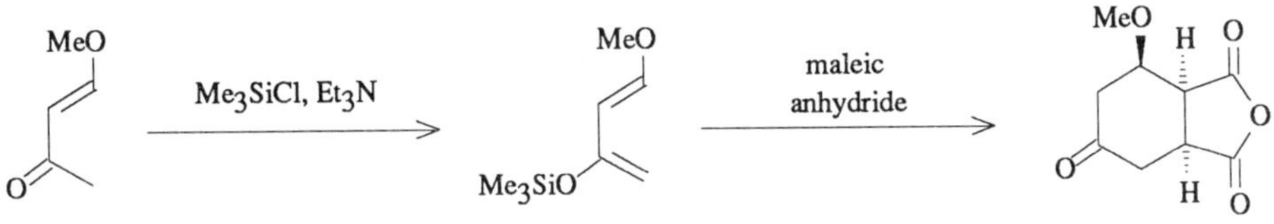

*Scheme 40*

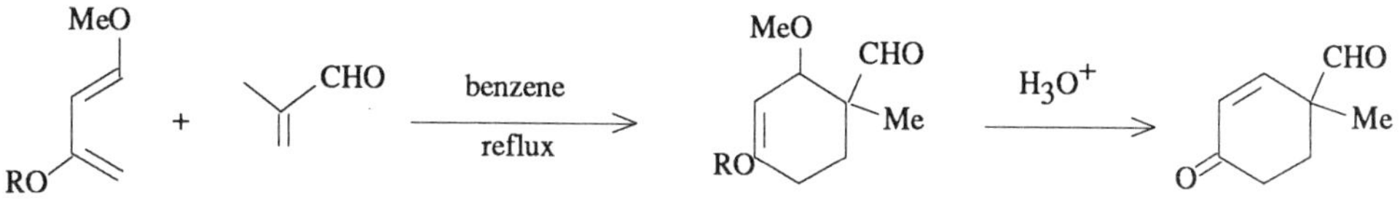

*Scheme 41*

Aromatic compounds have been obtained in two ways: by reaction of (66) with acetylenic dienophiles followed by acid hydrolysis, or by reaction with β-phenylsulphinyl-αβ-unsaturated carbonyl compounds followed by elimination of phenylsulphinic acid.[85]

The 2- and 4-methyl derivatives of 1-methoxy-3-trimethylsilyloxy-1,3-butadiene, and the 2,4-dimethyl derivative also undergo Diels-Alder addition to a range of dienophiles, providing rapid access to diversely functionalised aromatic compounds, cyclohexenones and cyclohexadienones.[86]

Another useful group of oxygenated butadienes are the vinylketene acetals such as 1,1-dimethoxy-3-trimethylsilyloxybutadiene [86,87] and 1-methoxy-1,3-bis(trimethylsilyloxy)butadiene (71).[88] These reactive dienes add readily to acetylenic dienophiles to form derivatives of resorcinol. With (71) cycloaddition to dimethyl acetylenedicarboxylate, for example, is followed by elemination of methanol to give the resorcinol (72). The dimethyl acetal affords derivatives of resorcinol monomethyl ether; methyl propiolate, for example, gave the monomethyl ether (68). An isomeric methyl ether (70) was obtained in a variation using methyl *trans*-β-nitroacrylate as dienophile. Here the orientation of the cyclo-addition is controlled by the nitro group and elimination of nitrous acid from the adduct (69) leads exclusively to (70).[89]

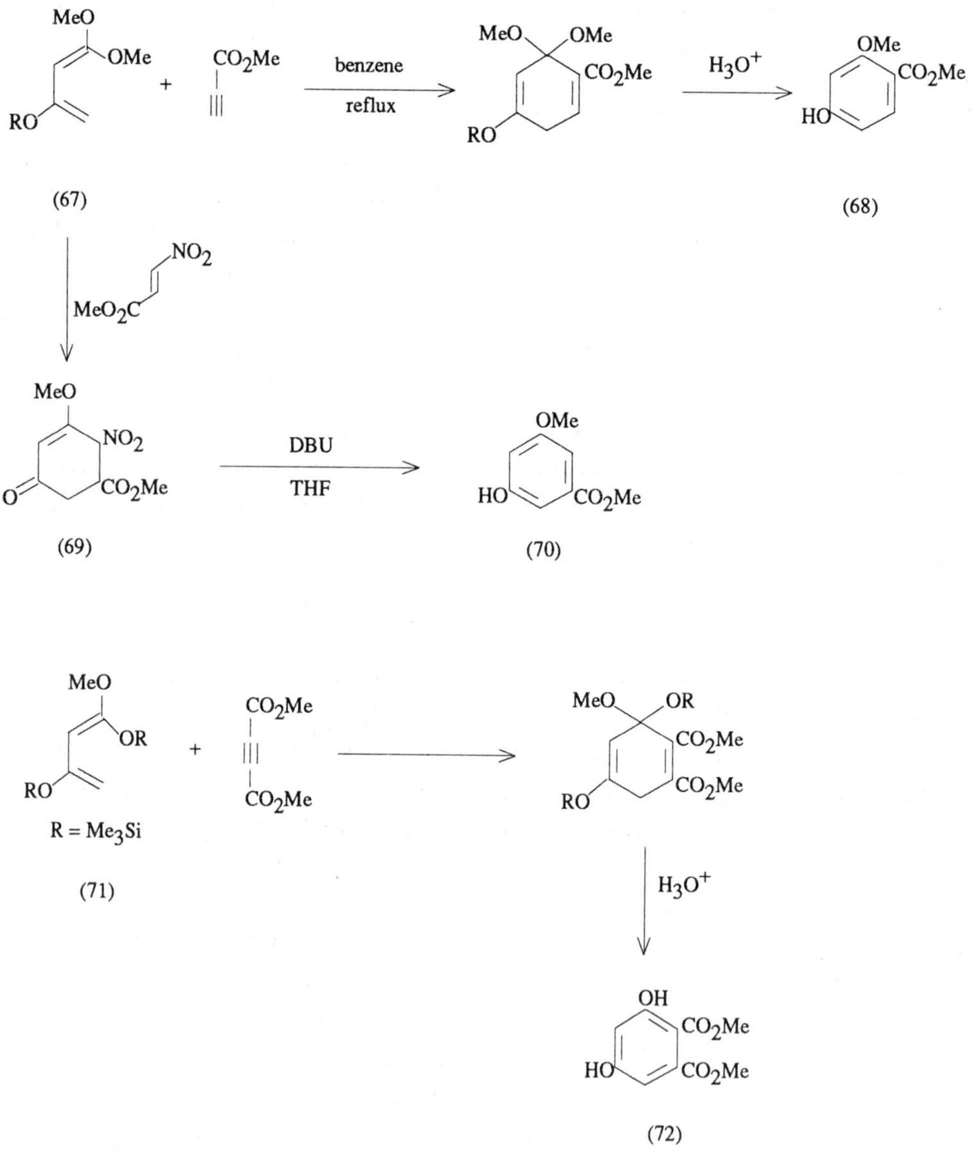

MeO OMe RO (67) + $CO_2Me$ —benzene, reflex→ MeO OMe $CO_2Me$ RO —$H_3O^+$→ OMe $CO_2Me$ HO (68)

(67) —$MeO_2C$ / $NO_2$→ MeO $NO_2$ $CO_2Me$ O (69) —DBU, THF→ OMe HO $CO_2Me$ (70)

MeO OR RO R = $Me_3Si$ (71) + $CO_2Me$ / $CO_2Me$ → MeO OR $CO_2Me$ $CO_2Me$ RO —$H_3O^+$→ OH $CO_2Me$ $CO_2Me$ HO (72)

*Scheme 42*

Another group of alkyl vinylketen acetals are obtained from the enolate ions of αβ-unsaturated esters and trimethylsilyl chloride. They react with α-chloroquinones to give adducts which have been converted into naphthaquinones and anthraquinones[90] (see p.124).

Butadienes bearing sulphide substituents are also reactive dienes; 1- and 2-phenylthiobutadiene, for example, form adducts with a variety of dienophiles. These adducts are versatile synthetic intermediates since the sulphide group can be reductively cleaved, or eliminated to form a new double bond after oxidation to the sulphoxide.[91] A number of 2-alkoxyphenylthiobutadienes form adducts with enones in which the enol ether function serves as a readily unmasked ketone. In reactions with unsymmetrical dienophiles the regiochemistry of the addition is controlled by the sulphide substituent, and advantage can be taken of this in preparing adducts of unusual orientation. Thus, 2-methoxy-1-phenylthiobutadiene[92] and methyl vinyl ketone in the presence of magnesium bromide catalyst, gave the adduct (74) with complete regio- and (*endo*) stereoselectivity[93] and 2-methoxy-3-phenylthiobutadiene gave predominantly the adduct (75). Hydrolysis of the latter and reductive removal of the phenylthio substituent then gave the acetylcyclohexanone (76) of different orientation from that obtained from 2-methoxybutadiene itself.[94]

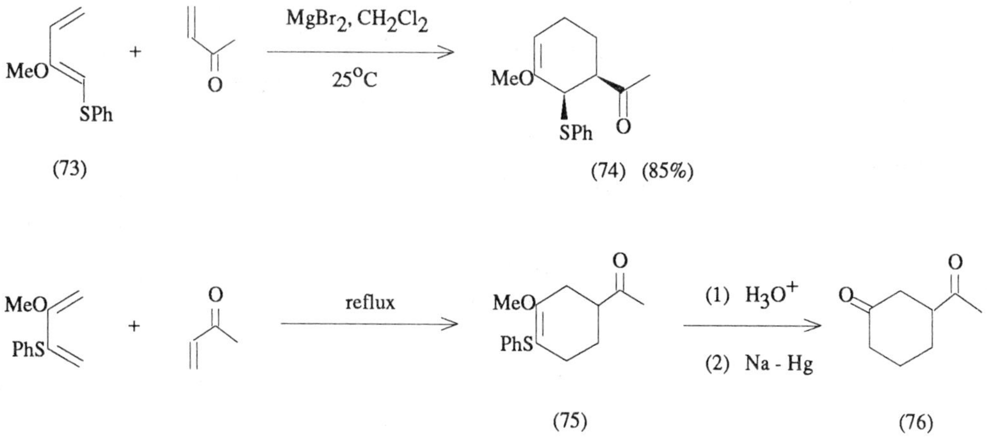

MeO, SPh (73) + O; $MgBr_2$, $CH_2Cl_2$, 25°C → MeO, SPh, O (74) (85%)

MeO, PhS + O; reflux → MeO, PhS, O (75); (1) $H_3O^+$ (2) Na - Hg → O, O (76)

*Scheme 43*

The sulphoxide of 1-phenylthiobutadiene, in contrast to the phenylthio compound itself, reacts with electron-rich dienophiles. Thus, in an approach to the hasubanan alkaloids, reaction of the sulphoxide with the enamine (77) gave the adduct (78) which was converted into the allylic alcohol (79) by sulphoxide-sulphenate rearrangement.[95]

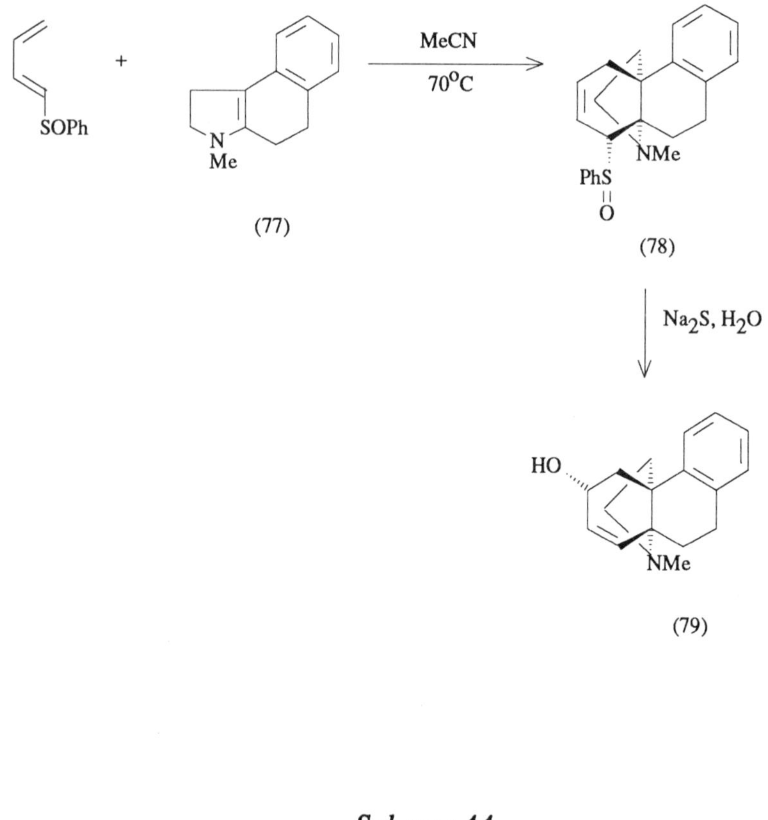

SOPh + (77) $\xrightarrow[70^oC]{MeCN}$ (78) $\xrightarrow{Na_2S, H_2O}$ (79)

*Scheme 44*

Nitrogen-substituted dienes include the 1-dialkylamino-1,3-dienes and 1-acylamino-1,3-dienes.[96] 1-Dialkylamino-1,3-dienes react readily with a variety of dienophiles giving adducts which afford derivatives of 1,2-dihydrobenzene on elimination of nitrogen[97] (Scheme 45). With acetylenic dienophiles benzene derivatives are obtained.

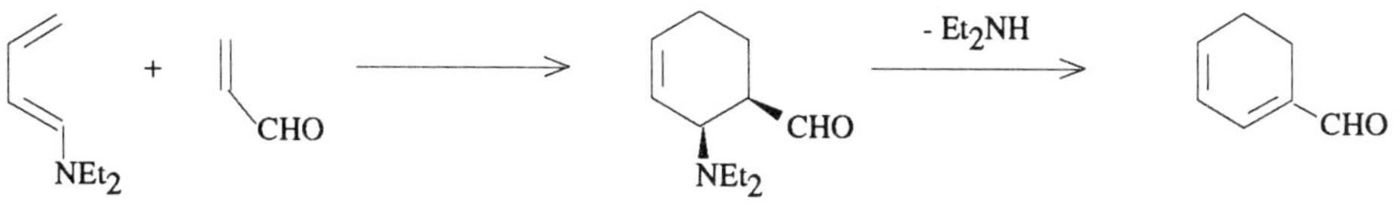

*Scheme 45*

1-Acylaminodienes have been used in the synthesis of a number of nitrogenous natural products (see p.98). They are readily obtained by acylation of vinylimines[98] or by Curtius rearrangement of dienoic acid azides[99] and they react with a broad range of dienophiles, providing access to a variety of amino functionalised cyclic systems in excellent yield.[100] Even poor dienophiles react readily. The reactions are highly regio- and *endo*-selective. The acylamino group is a powerful regiochemical directing group and only traces of regioisomeric adducts were detected in many cycloadditions with unsymmetrical dienes. Thus, reaction of the carbamate (80) with methyl acrylate was completely regioselective giving a 4:1 mixture of the *endo* and *exo* cycloadducts (81) and (82) in 90 per cent yield.

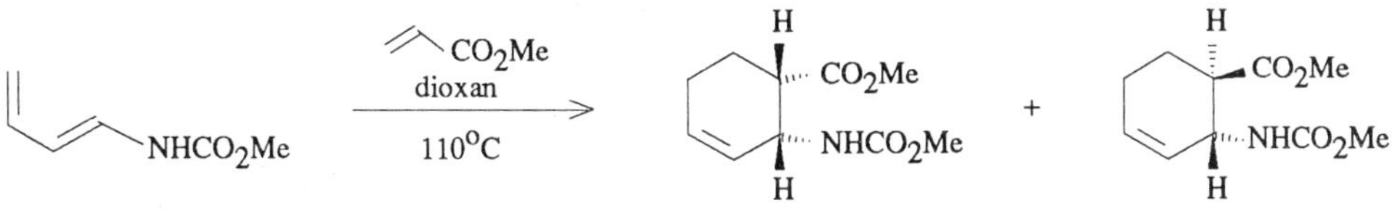

*Scheme 46*

2-Acylamino-1,3-dienes have also been used in synthesis, although so far less extensively than the 1-substituted isomers. They lead to 1-acylaminocyclohexenes which on hydrolysis give cyclohexanones.[101]

A useful group of dienes in this series are the N-alkoxycarbonyl-1,2-dihydropyridines. They react with dienophiles to form derivatives of 5,6-dehydroisoquinuclidine.[102] Derivatives of *cis*-1,2,5,8,9,10-hexahydroisoquinoline have also been obtained by reaction of N-methoxycarbonyl-1,2-dihydropyridine with acrolein followed by Cope rearrangement of a derived 1,5-diene[103] (Scheme 47).

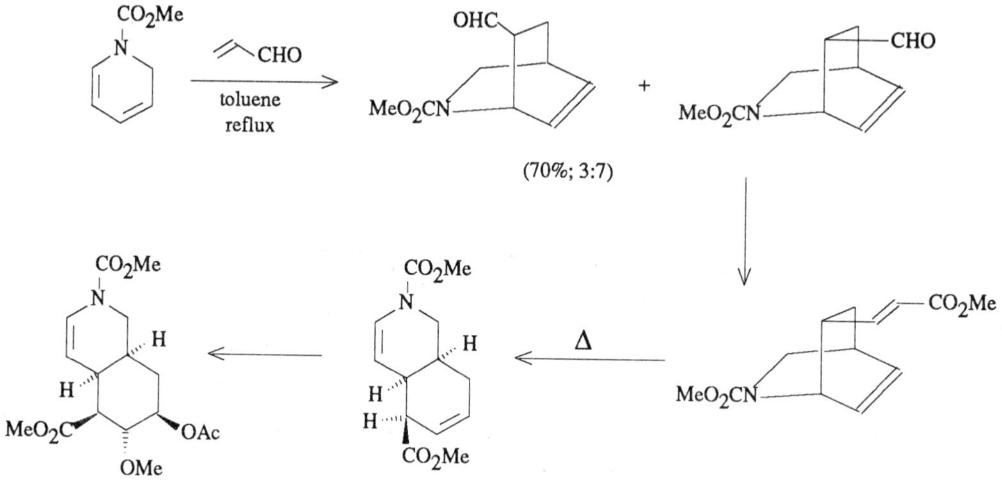

*Scheme 47*

A similar sequence was used to construct the D and E rings in a synthesis of reserpine,[104] and in a synthesis of (±)-catharanthine.[105] Intramolecular reactions have also been effected.[106]

Diels-Alder reactions of 1-trimethylsilylbutadienes have been employed to a limited extent. The trimethylsilyl group reduces somewhat the rate of Diels-Alder reactions and has if anything only a weak "*ortho*" directing effect. Other substituents on the diene thus tend to have the major influence on the regioselectivity of the additions.[107] The products of the reactions are allylsilanes which undergo clean deprotonation with acid and other useful transformations. Thus, 1-acetoxy-4-trimethylsilylbutadiene reacted smoothly with methyl acrylate in boiling xylene to give mainly the adduct (83). Perhydroxylation of the double bond followed by acid-catalysed elimination of trimethylsilanol led to (84) which by further manipulation was converted into (±)-shikimic acid.[108]

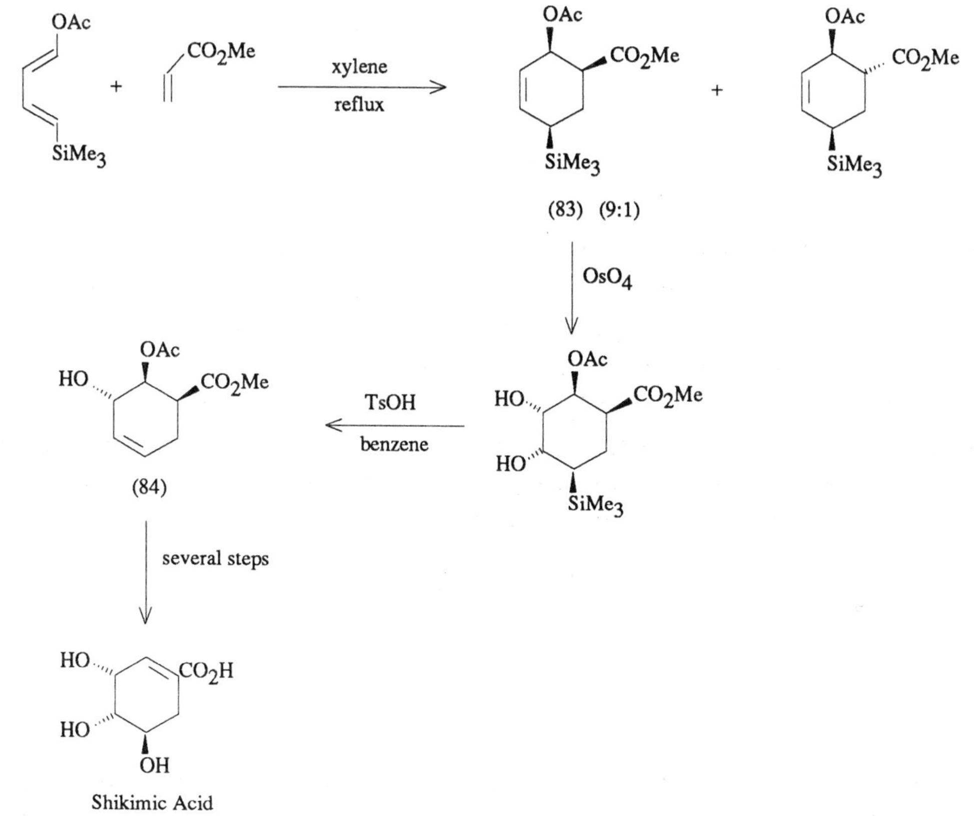

xylene, reflux; (83) (9:1); $OsO_4$; TsOH, benzene; (84); several steps; Shikimic Acid

*Scheme 48*

The trimethylsilyl substituent has also been used in Diels-Alder reactions in ways which exploit its steric effect.[109]

Reaction of 1,3-dienylboronates with dienophiles forms allylboronates which on further elaboration give cyclohexene derivatives.[110] Thus, the boronate (85) and maleic anhydride gave the adduct (86) which on oxidation formed the alcohol (87). With acetaldehyde the product (88) was obtained selectively and on hydrolysis furnished the bicyclic lactone (89).

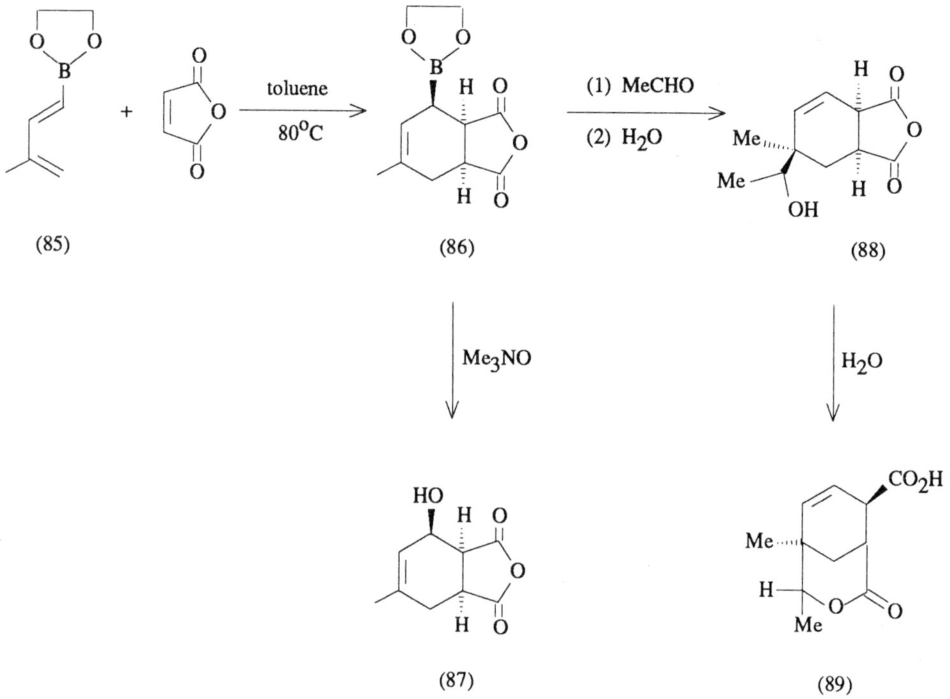

*Scheme 49*

The α-pyrone derivatives (90)-(92) are useful in synthesis as equivalents of butadienes. The pyrone (90) reacts with dienophiles, with loss of carbon dioxide, to form cyclic dienes, and with some electron-rich alkynes benzene derivatives are formed directly.[111] However, although N,N-diethylaminopropyne reacted readily to give (93) a number of other olefinic and acetylenic dienophiles gave complex mixtures of products.

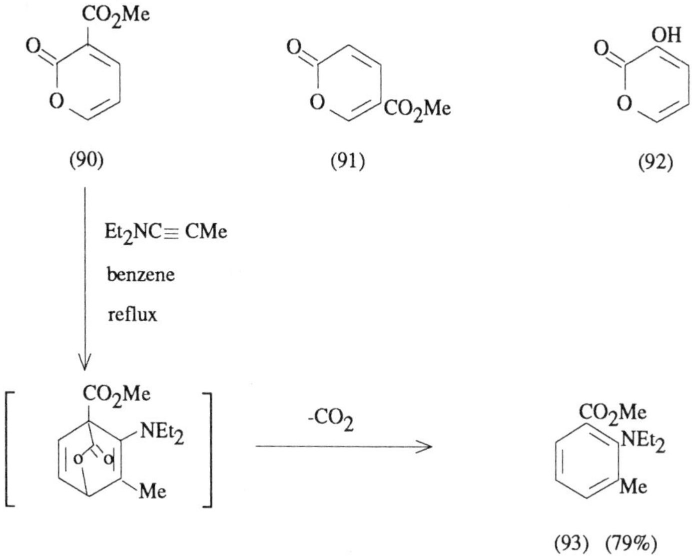

$CO_2Me$ (90) $CO_2Me$ (91) OH (92)

$Et_2NC{\equiv}CMe$, benzene, reflux

$CO_2Me$, $NEt_2$, Me; $-CO_2$

$CO_2Me$, $NEt_2$, Me (93) (79%)

*Scheme 50*

Coumalates (as 91) can serve either as dienes or dienophiles depending on their reaction partner.[112] Alkyl substituted α-pyrones have also been used; with acetylenedicarboxylic ester they give alkyl substituted phthalic esters directly.[113] The crystalline hydroxypyrone (92) is a useful synthetic equivalent of vinylketene. It reacts with a variety of dienophiles to give adducts which are converted easily into dihydrophenols or cyclohexenones by loss of carbon dioxide. Acetylenic dienophiles give aromatic compounds.[114]

## Ortho-quinodimethanes (Ortho-xylylenes)

Another valuable group of 1,3-dienes are the *ortho*-quinodimethanes or *ortho*-xylylenes (as 94). These are very reactive dienes and readily form Diels-Alder adducts with olefinic and acetylenic dienophiles.[115]

*Scheme 51*

*ortho*-Xylylenes are formed *in situ* by a number of routes, for example by thermolysis of benzocyclobutenes or photolysis of *ortho*-alkyl aromatic aldehydes or ketones;[116] generated in the presence of a dienophile they give the adducts directly. Thus, 1-hydroxybenzocyclobutene, heated in benzene in the presence of naphthaquinone, gave the adduct (96) by way of the *ortho*-xylylene intermediate (95) formed by thermally allowed conrotatory opening of the cyclobutene ring. The same intermediate (95) is formed by irradiation of *ortho*-methylbenzaldehyde.

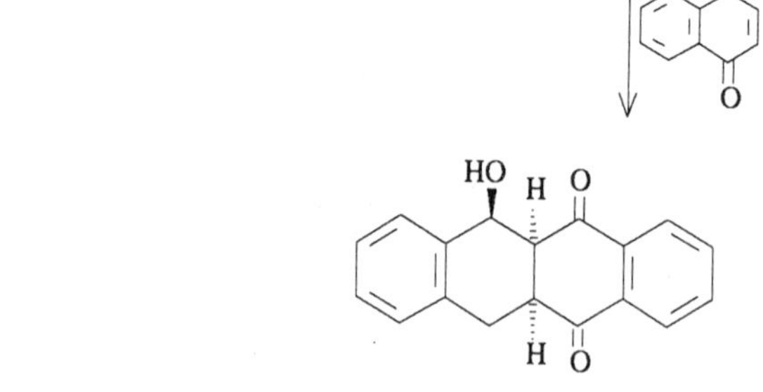

*Scheme 52*

A disadvantage of this approach to *ortho*-xylylenes is the comparative inaccessibility of appropriately substituted benzocyclobutenes. A novel route to these compounds is provided by the cobalt-catalysed co-cylisation of hexa-1,5-diynes with bis(trimethylsilyl)acetylene[117] (see p.194).

*Ortho*-xylylenes can also be obtained by pyrolytic elimination of sulphur dioxide from 1,3-dihydrobenzo[*c*]thiophen sulphones,[118] and by pyrolysis of *ortho*-alkylbenzyl chlorides[119] and homophthalic anhydride[120] but these methods require high temperatures. A milder approach is by 1,4-elimination from appropriate $\alpha,\alpha'$-disubstituted 1,2-dialkylbenzenes, for example by base induced elimination from methyl *ortho*-methylbenzyl ethers,[121] or iodide-induced debromination of $\alpha,\alpha'$-dibromo-*o*-xylenes.[122] Thus, the difficultly accessible anthracene-1,4-quinone was obtained in 43 per cent yield from reaction of sodium iodide with $\alpha,\alpha,\alpha',\alpha'$-tetrabromo-*o*-xylene in the presence of benzoquinone, by way of the *ortho*-xylylene (97). Other anthraquinones have been made in the same way.[123]

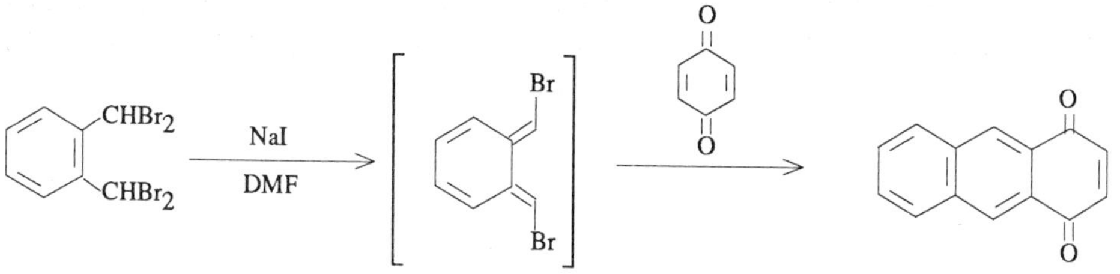

*Scheme 53*

Another mild route proceeds by elimination from *ortho*-($\alpha$-trimethylsilylalkyl)benzyltrimethylammonium salts (as 98) triggered by fluoride ion.[124] Thus, generated in the presence of methyl acrylate, the *ortho*-xylylene (99) gave the tetralin derivative (100) in 80 per cent yield. Intramolecular reactions are readily effected.[125]

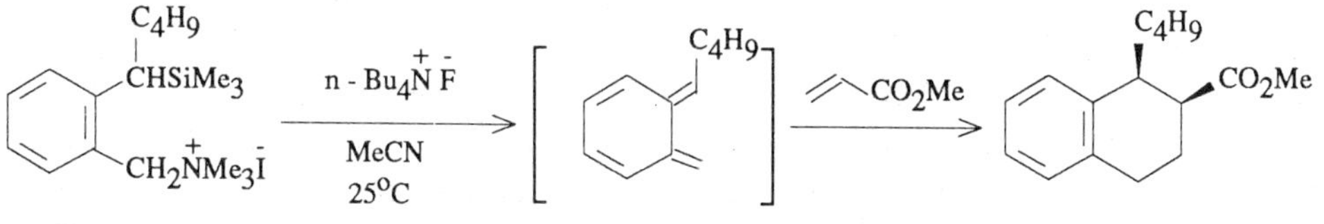

*Scheme 54*

N-Substituted *ortho*-quinonemethide imines (as 103) have been made in the same way. Thus, treatment of the quaternary salt (102) with fluoride ion gave the spirocyclic compound (104) by way of (103), and intramolecular reactions have been employed to make benzo[*c*]quinolizidines.[126] Polycyclic nitrogen compounds, including indole alkaloids, have also been made using indole-2,3-quinodimethanes.[127]

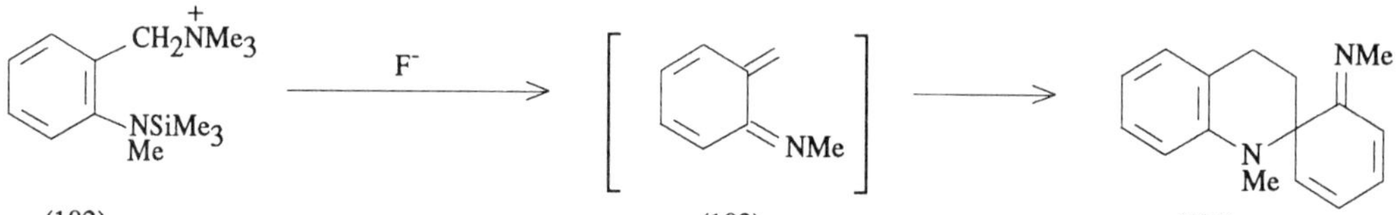

$CH_2\overset{+}{N}Me_3$ $NSiMe_3$ Me (102) $\xrightarrow{F^-}$ [ NMe (103) ] $\longrightarrow$ NMe N Me (104)

*Scheme 55*

## Heterodienes

A variety of heterodienes in which one or more of the atoms in the diene system is a hetero-atom, take part in Diels-Alder reactions, but so far they have not been extensively employed in synthesis. Probably the most widely used are αβ-unsaturated carbonyl compounds. They react readily as dienes with electron-rich dienophiles such as enol ethers and enamines in Diels-Alder reactions with inverse electron demand.[128] The reactions with enamines are apparently not direct cycloadditions and may involve the intermediate formation of zwitterions. With less reactive dienophiles dimerisation of the αβ-unsaturated carbonyl compound is a competing reaction. When elevated temperatures are used the theoretically favoured *endo* adducts are not the exclusive products and attempts to improve the stereospecificity by lowering the reaction temperature have led to the use of high pressure.[129] Electron-attracting groups in the α- or β-position to the carbonyl function also permit the use of lower reaction temperatures[130] and Lewis acids such as zinc chloride and soluble lanthanide complexes are effective catalysts, allowing the survival of fragile but valuable functionality in the addends and products.[131]

The reaction of αβ-unsaturated carbonyl compounds with alkyl vinyl ethers makes available various derivatives of 2-alkoxy-3,4-dihydro-2*H*-pyran, useful in the synthesis of carbohydrates,[132] thromboxanes[133] and spiroacetals. Thus, the cyclic enol ether (106), generated *in situ*, reacted with methacrolein to give the spiroacetals (107) and (108) in 72 per cent yield. The reactivity of the enol ether in (106) as a dienophile exceeds that of the αβ-unsaturated carbonyl system as a diene.[134]

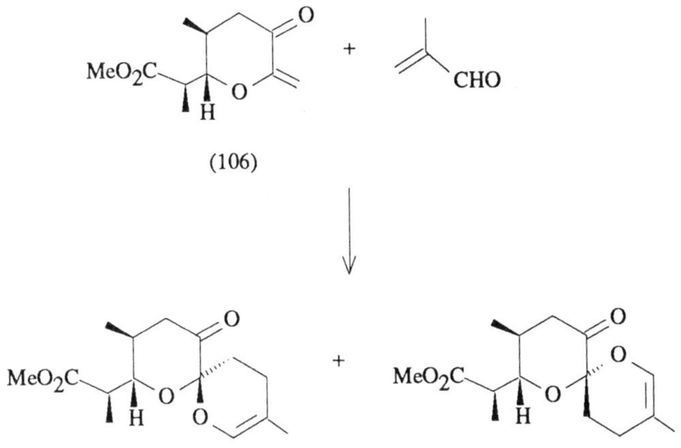

*Scheme 56*

Aza-dienes have not been extensively employed in synthesis because of their inaccessibility and their comparative unreactivity.[135] αβ-Unsaturated imines, for example, take part in Diels-Alder reactions preferentially through their enamine tautomers and not as 1-aza-1,3-dienes; where isomerisation is prevented, [2+2]cycloaddition generally intervenes.[136] N-Acyl-1-aza-1,3-dienes, however, formed *in situ* by gas-phase pyrolysis of N-acyl-O-acetyl-N-allylhydroxylamines, undergo intramolecular reactions readily to give derivatives of indolizidine and quinolizidine. Thus, the O-acetyl derivative (109) gave (110) in 70 per cent yield[137] and the reaction has been applied to make the optically active amide (111), a key intermediate in the synthesis of (–)-deoxynupharidine.[138]

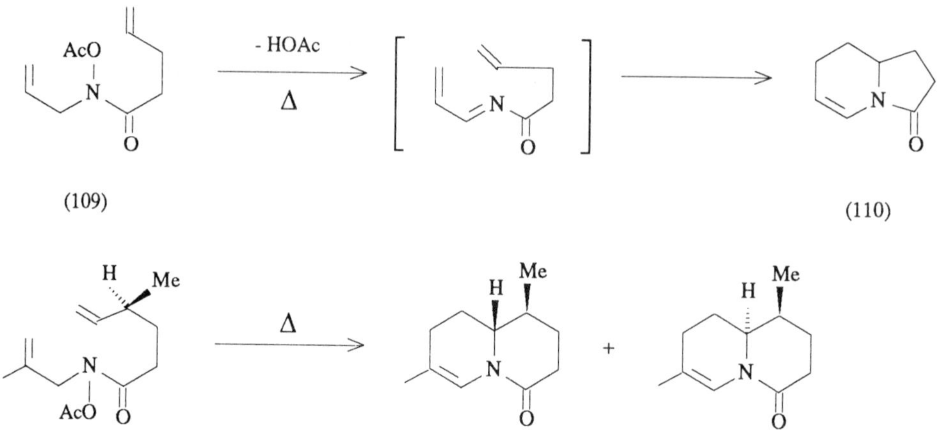

(109) (110) (111) (3:1)

*Scheme 57*

αβ-Unsaturated hydrazones, prepared from N,N-dimethylhydrazone and αβ-unsaturated aldehydes, react readily with electron-deficient dienophiles to give adducts which can be converted into substituted piperidines or pyrimidines.[139] Here the reactivity of the 1-azadiene system is increased by the electron-donating dimethylamino substituent.

2-Azadienes have been used in the synthesis of a number of heterocyclic systems, but not in a general way in big syntheses.[136] Most of the 2-aza-dienes employed carry strong electron-donating groups capable of enhancing their reactivity towards typical electron-deficient dienophiles; dialkylamino substituents are especially useful in this way.[140]

A number of other types of heterodiene have been used in Diels-Alder reactions including N-acylimines,[141] nitrosoalkenes,[142] nitroalkenes[143] and some heterocyclic compounds.[136] Particularly useful among the last class are oxazoles and pyrimidines, which can serve as synthetic equivalents of 2-aza-1,3-dienes. Diels-Alder addition of olefinic and acetylenic dienophiles to oxazoles has been used for the preparation of pyridine and furan derivatives, including derivatives of pyridoxine[144] (Scheme 58). The furanosesquiterpene evodone (114) was obtained by intramolecular addition to the oxazole (112) and loss of hyrdogen cyanide from the initial adduct (113) in a retro Diels-Alder step.[145]

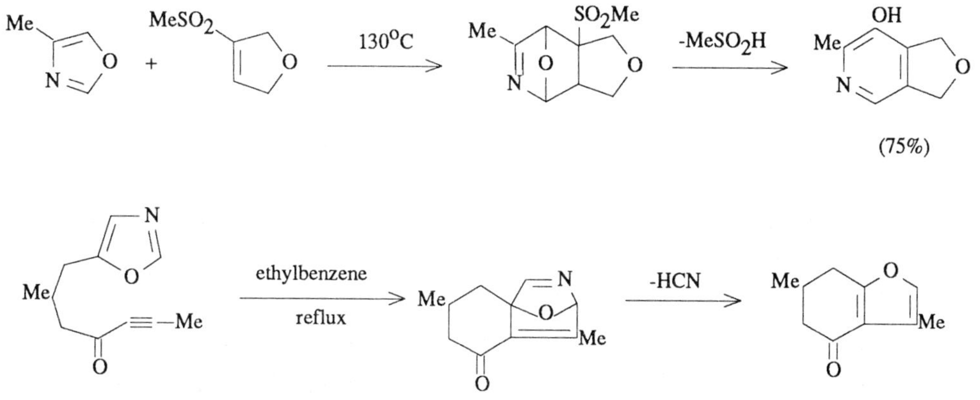

Me
N
O
+
$MeSO_2$
O
130°C
Me
N
O
$SO_2Me$
O
$-MeSO_2H$
OH
Me
N
O
(75%)

N
O
Me
—Me
O
(112)
ethylbenzene
reflux
Me
N
O
Me
O
(113)
-HCN
Me
O
Me
O
(114)

*Scheme 58*

However the ability of an oxazole to act as a heterodiene appears to depend on its substitution pattern.[146]

Electron-deficient pyrimidines, such as the 5-carboxylic esters, react with ynamines in Diels-Alder reactions with inverse electron demand to form pyridine derivatives. In each case the electron-rich dienophile adds selectively across C-2/C-5 of the pyrimidine nucleus and the orientation of the addition is guided by the pattern of electron-withdrawing substituents on the pyrimidine nucleus.[147] Electron-rich pyrimidines react with dimethyl acetylenedicarboxylate to give substituted pyridines, again by preferential addition of the dienophile across C-2/C-5 of the pyrimidine nucleus.[148] Intramolecular reactions of this kind have been effected. Thus the monoterpene alkaloid actinidine was synthesised by a sequence which incorporated the intramolecular cycloaddition of the dihydroxypyrimidine (115); the presumed intermediate (116) was too unstable for detection.[149]

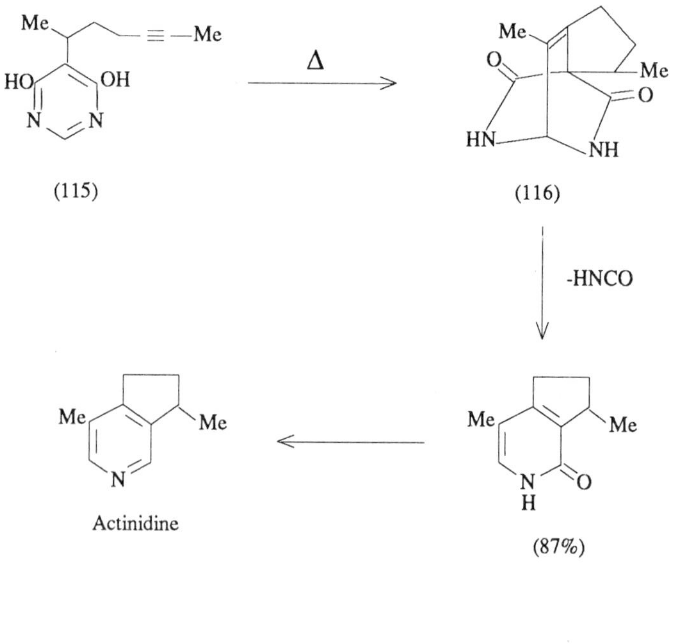

*Scheme 59*

Nitrosoalkenes and nitroalkenes also act as diene components in cycloaddition reactions with suitable alkenes but the reactions have not hitherto been much employed in synthesis. It has been found recently however that the intramolecular reactions take place readily and with high stereoselectivity to give dihydro-oxazine derivatives in good yield.[142,143] Thus the nitro-alkene (117, R=H) gave a mixture of the *cis* and *trans* bicyclic adducts (118) and (119) in a reaction catalysed by stannic chloride at –70°C; with tri-substituted nitro-alkenes (eg. 117, R=$CH_3$) much greater selectivity was observed. The initial adducts are usefully converted into fused lactones by the action of base followed by oxime exchange, presumably by way of the nitrile oxide.

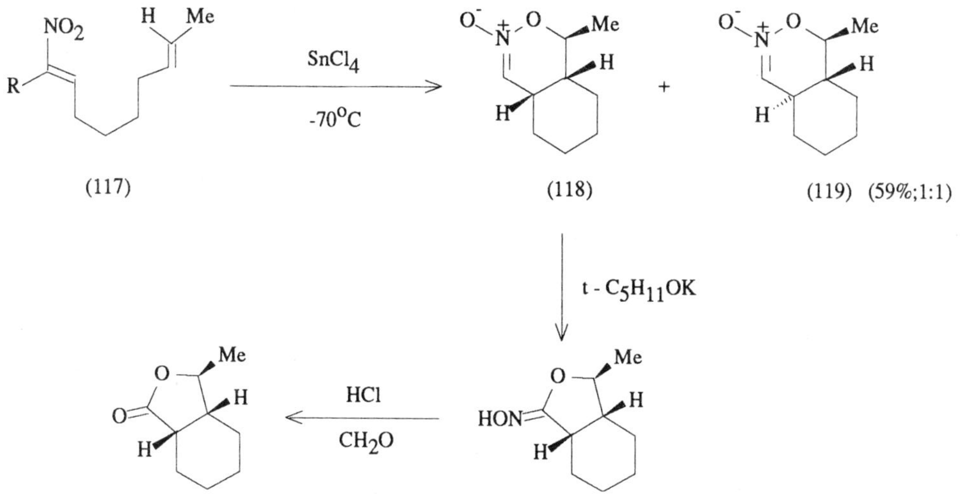

*Scheme 60*

## Reactions in Water

Diels-Alder reactions can often be greatly accelerated, and the regio- and stereo-selectivities enhanced, by proper choice of experimental conditions, for example by conducting the reactions in water, by high pressure and, notably, by Lewis acid catalysis.

Some remarkable increases in rates have been observed for Diels-Alder reactions conducted in water. Thus, reaction of cyclopentadiene with methyl vinyl ketone at 20°C in water was 700 times faster than in 2,2,4-trimethyl-pentane and the *endo*: *exo* ratio rose from about 4:1 normally found in organic solvents to more than 20:1. These effects persisted even with relatively insoluble substrates when two separate phases were present. It is suggested that the rate accelerations in these reactions are due to hydro-phobic interactions which bring the components close together, at the same time favouring the *endo* transition state.[150] High concentrations of the diene component in water have been realised by using the sodium salts of dienoic acids.[151] Thus, in the key step in a synthesis of the hydrindane (123), an intermediate in the synthesis of vitamin D3, reaction of the chiral diene acid (120), as the sodium salt, and methacrolein in water at 55°C gave the adducts (121) and (122) (85%) in the ratio of about 5:1 after sixteen hours. In contrast the corresponding methyl ester in excess of neat methacrolein at 55°C gave only a 10 per cent yield of 1:1 mixture of isomers after 63 hours.[152]

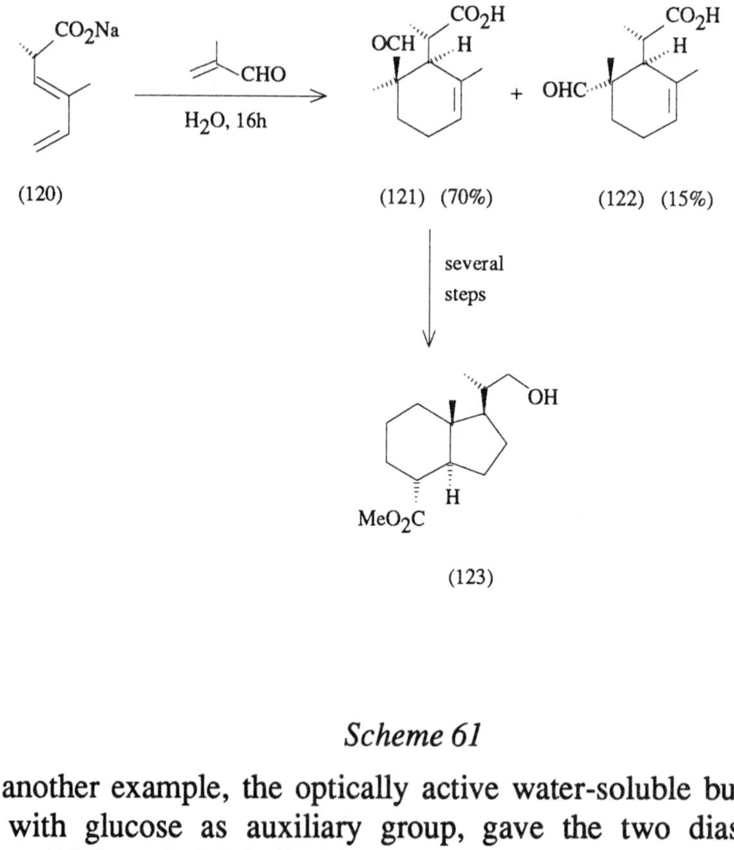

$CO_2Na$ CHO $H_2O$, 16h $CO_2H$ OCH H + OHC $CO_2H$ H

(120) (121) (70%) (122) (15%)

several steps

OH H $MeO_2C$

(123)

*Scheme 61*

In another example, the optically active water-soluble butadienyl ether (124), with glucose as auxiliary group, gave the two diastereoisomeric products (125) and (126) in 92 per cent yield in water after 3.5 hours. Further manipulation gave the enantiomerically pure cyclohexanols (127) and (128). In contrast, reaction of the tetra-acetate with methacrolein in toluene at 80°C required a week for completion.[153]

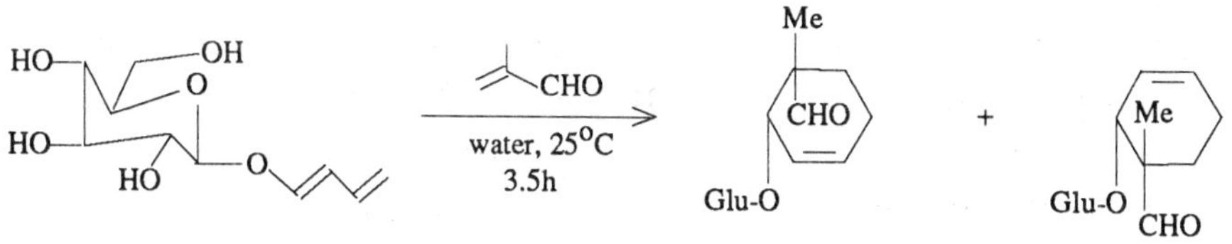

HO HO HO OH O O (124) CHO water, 25°C 3.5h Me CHO Glu-O (125) + Me Glu-O CHO (126) (92%;2:3)

OH OH Me (127) OH HO Me (128)

*Scheme 62*

Similarly, unactivated imminium salts generated *in situ* under Mannich-like conditions reacted smoothly with dienes in aqueous solution; simple imines are generally unreactive towards electron-rich dienes. Both inter- and intra-molecular reactions take place easily under mild conditions, affording a convenient route to a range of piperidine derivatives.[154] The reaction has been employed in the synthesis of a number of quinolizidine alkaloids[155] and the two optically active octahydroquinoline derivates (130) and (131) were obtained from the chiral diene (129).[156]

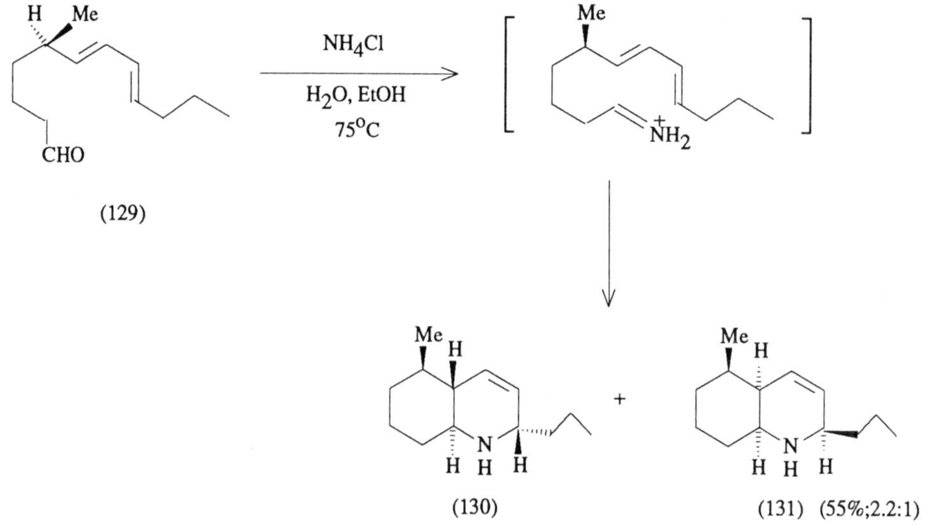

H Me

CHO

(129)

$NH_4Cl$

$H_2O$, EtOH

75°C

Me

$\overset{+}{N}H_2$

Me H

N

H H H

(130)

+

Me H

N

H H H

(131) (55%;2.2:1)

*Scheme 63*

Intermolecular reactions with good diastereoselectivities have been achieved using (*S*)-1-phenylethylamine hydrochloride[154] or hydrochlorides of amino acid esters[157] as the ammonium salts.

Rate accelerations of some intermolecular Diels-Alder reactions have also been observed in ethylene glycol and in formamide solution, although they are not as great as those in water.[158] These accelerations are ascribed to solvophobic binding of the reactants to each other in the polar solvents, but curiously, there is apparently no striking increase in *endo- exo* selectivity as there is in water.

## Reactions under High Pressure

Many Diels-Alder reactions are accelerated by very high pressures of 10-20kbar (1kbar = 986.9 atm.) because of the large decrease of 25-40 $cm^3mol^{-1}$ in the volume of activation in forming the transition state. Thus, typically, a cycloaddition which occurs at about 100°C at atmospheric pressure can be achieved at room temperature with a pressure of 9-10kbar.[159] The use of high pressures is particularly useful when steric hindrance to reaction or the thermal instability of a reactant or product precludes the use of conventional means to accelerate a reaction. Whereas an increase in temperature increases the rate of both forward and reverse reactions, pressure accelerates the forward reaction only. For example, in an important step in the synthesis of the diterpene jatropholone the furan (132) and the enone (133) gave the cycloadduct (134) in 80 per cent yield at 5kbar pressure at room temperature; reactions under thermal conditions or in the presence of Lewis acid catalysts were unsuccessful due to the instability of the diene.[160]

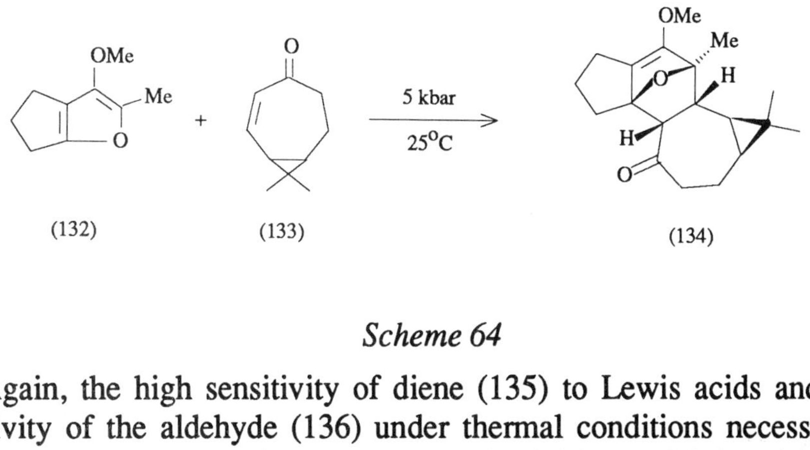

*Scheme 64*

Again, the high sensitivity of diene (135) to Lewis acids and the low reactivity of the aldehyde (136) under thermal conditions necessitated the use of high pressure to bring about the cycloaddition to (137) and (138).[161]

High pressure has also been employed to facilitate the addition of dienes to unreactive cyclohexenones[162] and crotonic esters[163] and to the ketonic group of glyoxylic esters.[164] The last reactions can be effected at lower pressures in the presence of the lanthanide catalyst $Eu(fod)_3$ as in Scheme 65.[165] Some highly diastereoselective additions to chiral non-racemic dienes[166] and sugar aldehydes[167] have also been facilitated under high pressure conditions. The increased *endo* selectivity of reactions conducted under high pressure can also be exploited in synthesis. Thus, in the cycloaddition of the enamino-ketone (139) and ethyl vinyl ether, which is of interest since it provides access to 3-amino sugars, the proportion of *endo* adduct (140) was greatly increased by conducting the reaction at low temperature under pressure.[168]

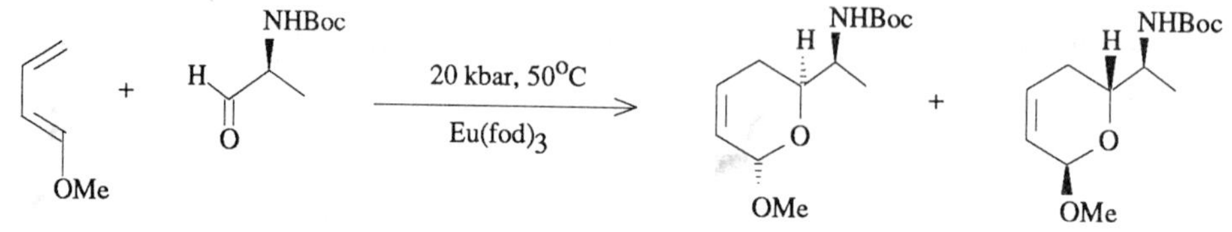

OMe + H, O, NHBoc → 20 kbar, 50°C, Eu(fod)$_3$ → NHBoc, H, O, OMe + NHBoc, H, O, OMe

(135) (136) (137) (138)

*Scheme 65*

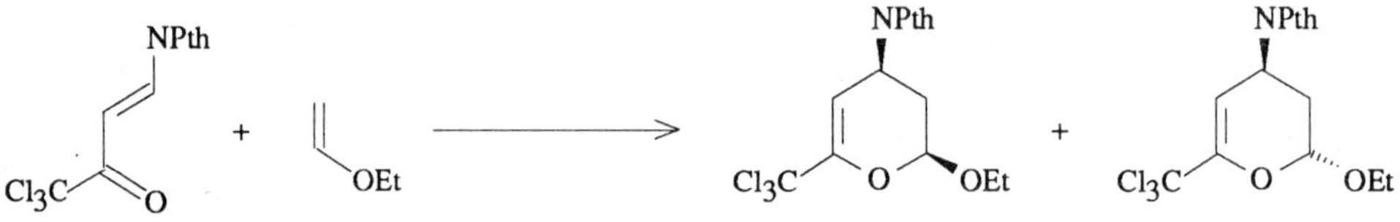

NPth, $Cl_3C$, O + OEt → NPth, $Cl_3C$, O, OEt + NPth, $Cl_3C$, O, OEt

(139) (140) (141)

| | (140) | (141) |
|---|---|---|
| 90°C, 1 bar, | ratio 1.67 | 1 |
| 0.5°C, 6 kbar, | ratio 13.6 | 1 |

*Scheme 66*

## Catalysis by Lewis Acids

Many Diels-Alder reactions are accelerated by Lewis acid catalysts such as boron trifluoride, tinIV chloride and zinc chloride[169] and the catalysed reactions, as well as proceeding faster, show increased regio- and stereo-selectivities over the uncatalysed reactions. Thus, in the reaction of methyl vinyl ketone with isoprene the proportion of *'para'* adduct increases from 71:29 in the uncatalysed reaction to 93:7 in reaction catalysed by tinIV chloride. Many other examples of this effect have been recorded.[170] Similarly, in the addition of acrylic acid to cyclopentadiene the proportion of *endo* adduct increased noticeably in the presence of aluminium chloride.[171]

These effects are ascribed to complex formation between the Lewis acid and the polar groups of the dienophile which brings about changes in the energies and orbital coefficients of the frontier orbitals of the dienophile.

It is well to bear in mind, however, that catalysed reactions may take an unexpected course. Thus, reaction of 2,3-dimethylbutadiene with methacrolein catalysed by tinIV chloride gave, not the expected (142) but the bridged bicyclic compounds (143) formed by rearrangement of the initial adduct.[172]

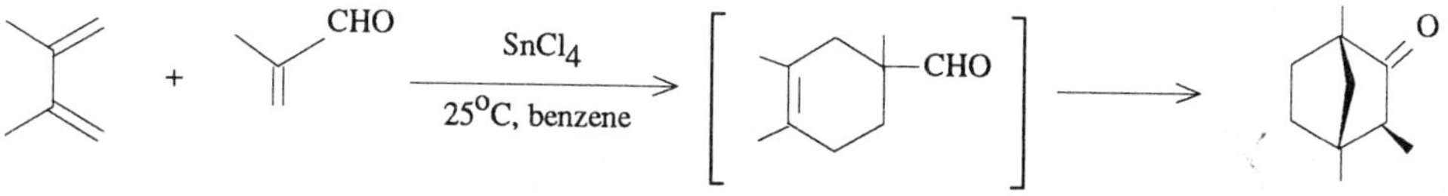

CHO + $SnCl_4$, 25°C, benzene → [CHO] → O

(142) (143)

*Scheme 67*

More recently the soluble lanthanide complexes $Eu(fod)_3$, $Eu(hfc)_3$ and $Yb(fod)_3$ have been used with great effect as mild catalysts in hetero Diels-Alder addition reactions of oxygenated butadienes; some examples have already been given (p.11).

Reaction of achiral dienes with aldehydes in the presence of the chiral lanthanide catalyst $Eu(hfc)_3$ gave adducts with only modest asymmetric induction[173] but with chiral l-menthyloxybutadienes (as 144) and $Eu(hfc)_3$ reaction with a variety of aldehydes led to optically active products with good diastereomeric excess,[174] favouring the "L-pyranoside" adduct (as 145) in each case. Such a reaction formed the key step in a synthesis of optically pure (146), a β-4-deoxy-L-glucoside.

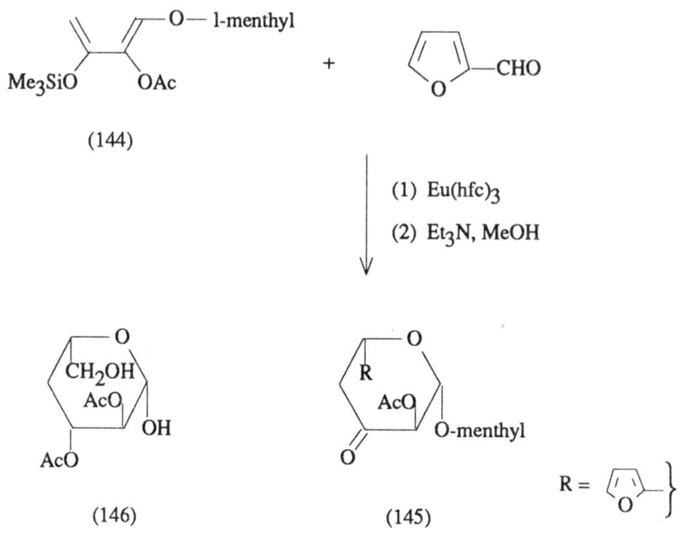

*Scheme 68*

The lanthanide $Yb(fod)_3$ catalyses the Diels-Alder reaction of acrolein with sensitive dienes other than oxygenated butadienes, with high stereoselectivity.[175]

It has been generally assumed that all Lewis acids promote the formation of the same regioisomer in a catalysed Diels-Alder reaction, but it appears that this is not always so. Thus, the catalysed reaction of isoprene with 5-hydroxy-1,4-naphthoquinone (juglone) gave two different regioisomers depending on whether the catalyst was boron trifluoride etherate or boron triacetate.[176] In another case involving juglone and 1-acetoxy-4-phenylthiobutadiene the regioselectivity was reversed on catalysis by boron trifluoride etherate.[177] No adequate general explanation for these effects is yet available.

Lewis acids in general catalyse Diels-Alder reactions of dienophiles bearing electron-withdrawing substituents. However conjugated dienes, styrenes and electron-rich alkenes also can function as powerful dienophiles when converted into cation-radicals by triarylaminium salts or by photosensitised electron transfer. Thus, the Diels-Alder dimerisation of cyclohexadiene was effected in 70 per cent yield at 0°C in the presence of tris(*p*-bromophenyl)aminium hexachlorostibnate, while the uncatalysed reaction gave only 30 per cent of dimer after 20 hours at 210°C[178] and a number of selective additions of cyclohexadiene to methyl substituted butadienes has been effected. These reactions were highly *endo* selective and showed the unusual feature of taking place more readily at the more heavily substituted double bond of the diene. The reactions are believed to involve the alkene cation-radicals which can be viewed as extremely electron-deficient and hence highly reactive dienophiles.[179] In some cases competitive or even selective formation of cyclobutane derivatives intervenes.[180] Intramolecular reactions can be employed with advantage, giving high yields of *endo* products.[181] In one example the tetra-ene (147) gave the cycloaddition products (148) and (149) in 42 per cent yield; in contrast, refluxing (147) in xylene for 48 hours gave no cycloaddition product.[182]

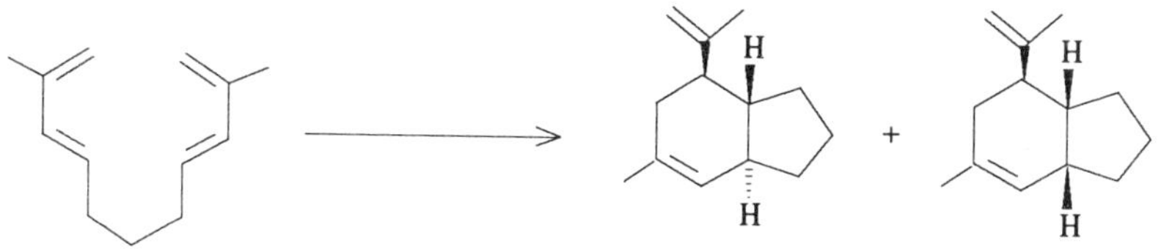

(147) (148) (149)

with $(p\text{-}BrC_6H_4)_3\overset{\cdot+}{N}\,Sb\overset{-}{C}l_6$, -72°C, ratio 2:1

with $CF_3SO_3H$ (148) only (88%)

*Scheme 69*

The same reaction is efficiently catalysed by trifluoromethanesulphonic acid and it has been suggested that in some reactions the aminium cation radical serves mainly as an indirect source of protons which bring about an acid-catalysed cyclisation,[183] but more recent work appear to confirm the general validity of the cation-radical chain mechanism.[184] The acid-catalysed reactions, such as that depicted in Scheme 69, are believed to take place by initial addition of an allylic cation to the diene. In agreement, cyclisation of unsymmetrical tetra-enes can be effected regio-selectively through the appropriate allyl alcohol which serves a the source of a specific allyl cation.[185] Acrolein acetals[186] and ethynyl ortho esters[187] also have been employed as precursors of allyl cations and propargyl cations respectively in ionic Diels-Alder reactions. The term Diels-Alder reaction refers only to the formal outcome of these reactions and does not imply that they necessarily follow a concerted Diels-Alder pathway.

## Regiochemistry

The value of the Diels-Alder reaction in synthesis is due in large measure to its high regio- and stereo-selectivity. Reaction of an unsymmetrical diene with an unsymmetrical dienophile could, in principle, give rise to two regioisomeric adducts but in practice one product generally predominates.[188] Thus, in the addition of acrylic acid derivatives to 1-substituted butadienes the "*ortho*" (1,2-) adduct is favoured, while with 2-substituted butadienes the "*para*" (1,4-) isomer predominates. These orientation effects have been shown to be governed largely by the atomic orbital coefficients at the termini of the conjugated systems concerned. The atoms with the larger terminal coefficients on each addend bond preferentially in the transition state. It turns out that in most cases this leads mainly to the 1,2- ('*ortho*') adduct with 1-substituted butadienes and to the 1,4- ('para') adduct with 2-substituted butadienes.[2,189] Secondary orbital interactions may also play a part, particularly when the primary interactions show little preference for one regio-isomer or the other.[190] These orienting forces are relatively weak, however, and in particular cases they may be overcome by steric factors. But in the presence of Lewis acids the proportion of the 'expected' isomer formed is frequently increased and under these conditions very high yields of a single isomer can often be obtained. Thus, in the addition of acrolein to isoprene the proportion of the '*para*' adduct was increased in the presence of tinIV chloride so that it became almost the exclusive product of the reaction. Similar effects have been observed in numerous other reactions.

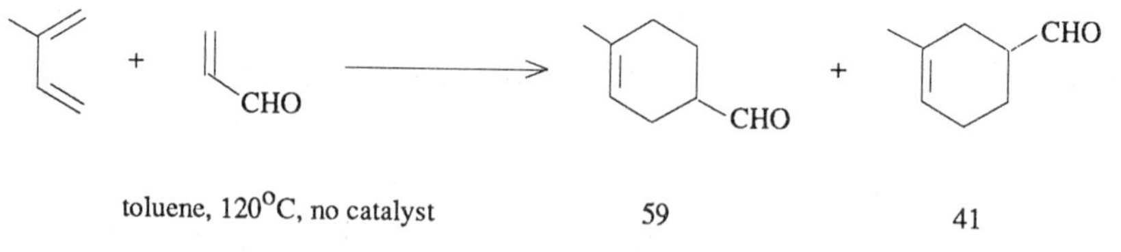

CHO + CHO ⟶ CHO + CHO

| | | |
|---|---|---|
| toluene, 120°C, no catalyst | 59 | 41 |
| benzene, 25°C, $SnCl_4 . 5H_2O$ | 96 | 4 |

*Scheme 70*

Substituents do not all have the same directing power. This can sometimes be exploited to obtain adducts of unusual orientation through the use of dienes or dienophiles bearing substituent groups which control the orientation of addition and are then removed after the adduct has been formed. The phenylthio and nitro groups are particularly useful in this way. Thus, 2-methoxy-1-phenylthiobutadiene and methyl vinyl ketone, in the presence of magnesium bromide catalyst, give the adduct (150) in which the methoxy group is '*meta*' to the carbonyl group of the dienophile. Here the regioselectivity is controlled by the phenylthio group; reaction of 2-methoxybutadiene itself with methyl vinyl ketone gives the '*para*' isomer.[191] Analogous regiocontrol by a sulphur substituent has been observed in the Diels-Alder reactions of several other alkoxy- and acyloxy-phenylthiobutadienes.[192] The sulphur substituent in the adducts can be removed by reductive cleavage or by conversion into the sulphoxide and elimination.

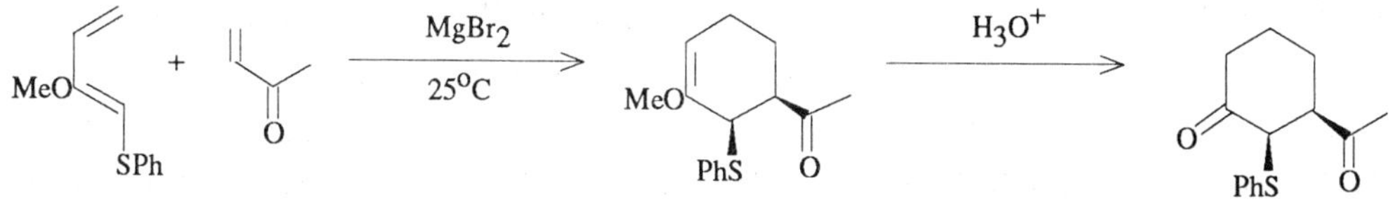

MeO, SPh, O + → ($MgBr_2$, 25°C) → MeO, PhS, O (150) → ($H_3O^+$) → O, PhS, O

*Scheme 71*

Similarly, in 3-nitrocyclohexenone the nitro group controls the orientation of addition so that reaction with 1,3-pentadiene affords the adduct (151), Reductive removal of the nitro group gave (152) of orientation different from that of the adduct (153) obtained from reaction of 1,3-pentadiene with cyclohexenone itself.[193]

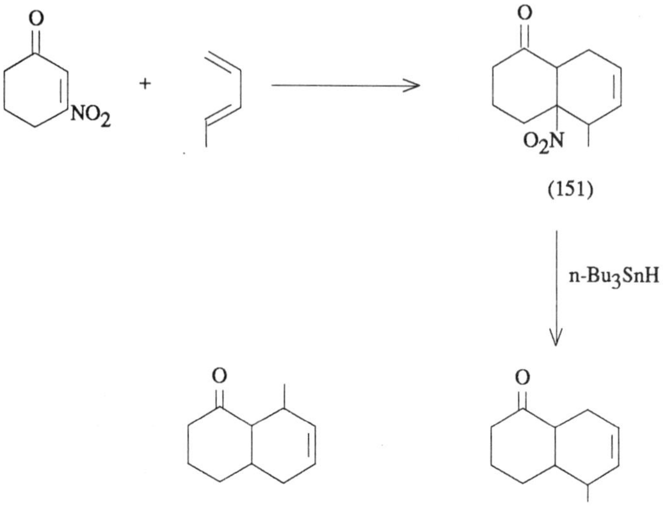

O NO2 + → O O2N (151) n-Bu3SnH O (153) O (152)

*Scheme 72*

**Stereochemistry**

Another valuable feature of the Diels-Alder reaction is its high stereoselectivity, and it is probably this factor more than any other which has led to its widespread application in the synthesis of complex natural products. Up to four new chiral centres may be set up in the reaction between a diene and a dienophile, but it is frequently found that one of the several possible racemates is formed in preponderant amount or even exclusively. The stereochemistry of the main product can generally be predicted on the basis of two well-known rules, the *cis*-rule, which states that the relative stereochemistry of the substituent groups in the diene and dienophile is maintained in the product of cyclo-addition, and the Alder *endo* rule according to which the diene and dienophile arrange themselves in parallel planes and the most stable transition state is that in which there is "maximum accumulation of double bonds" or, as we would now say, the maximum possibility of orbital overlap. The *cis*-rule is very widely followed. The *endo* rule appears to be strictly obeyed only in the addition of cyclic dienes to cyclic dienophiles, but it is a useful guide in other cases as well.[194] It should be noted, however, that the *endo* rule applies only to the kinetic products of reaction, and that its application to intramolecular cycloadditions is more uncertain. It is very frequently found that the proportion of *endo* isomer is increased in reactions catalysed by Lewis acids.

On the other hand, reactions may become less selective at elevated temperatures. The factors which determine the steric course of diene additions are not always completely clear. It appears that a number of different forces operate in the transition state and the precise steric composition of the product depends on the balance among these.

The simultaneous operation of the *cis* and *endo* rules is seen in the reaction of 1,4-diacetoxybutadiene with ethylene carbonate to give the adduct (154). In an intramolecular example the triene (155) gave almost exclusively the adduct (157) by way of the endo transition state (156).[195] Not all intramolecular Diels-Alder reactions are as *endo* selective as this.

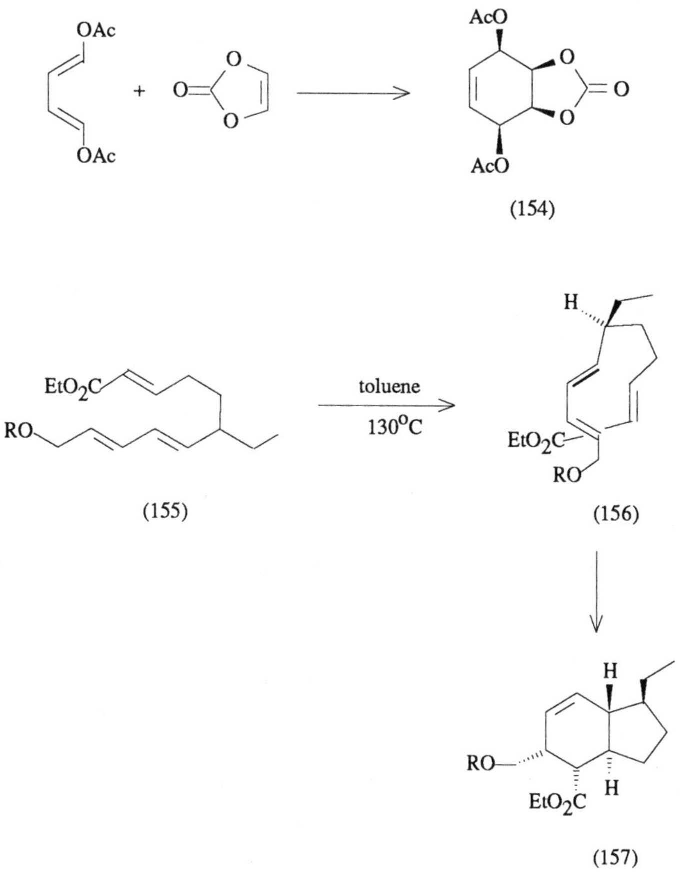

(154)

toluene
130°C

(155)

(156)

(157)

*Scheme 73*

In the reaction between a diene and a dienophile with a chiral centre next to the double bond, the stereochemical relationship between the newly-formed chiral centres in the ring of the adduct and the pre-existing centre in the dienophile is determined by the facial selectivity of the addition.

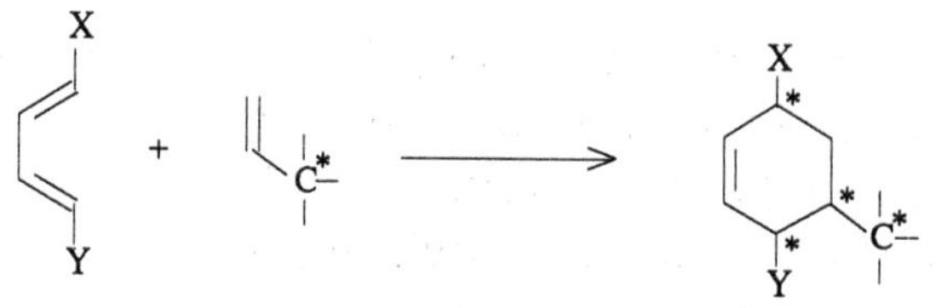

*Scheme 74*

Very high and predictable selectivity in this sense is observed in the catalysed addition of activated dienes to aldehydes, where addition to the carbonyl group very largely follows Cram's rule. Thus, addition of the diene (158) to 2-phenylpropanal in the presence of zinc chloride gave the single adduct (159),[196] and reaction of 1-methoxy-3-trimethylsilyloxy-butadiene with N-t-butoxycarbonylleucinal (163) gave mainly the adduct (164),[197] in each case by Cram selective addition of the diene to the carbonyl group. The *cis*-2,3-disubstitution pattern in (159) follows from the *endo* selectivity of the cycloaddition which operates under these conditions (see p.11). When the reaction between the diene (158) and 2-phenylpropanal was effected in chloroform with boron trifluoride etherate as catalyst the main product was the 2,3-*trans* pyrone (161) but the Cram relationship at C-2/C-7 still held.

Substituted dihydropyrones prepared stereoselectively in this way have been employed in the synthesis of a variety of hexoses and disaccharides (see p.106) and they have been exploited also in the stereocontrolled synthesis of open chain compounds by cleavage of the pyrone. The ring acts as a matrix whose stereochemistry is transferred to the acyclic product. Thus, oxidative cleavage of the pyrone ring in (164) gave the N-t-butoxy-carbonyl derivative of the novel β-hydroxy-γ-amino acid statine (166). Dihydropyrones like (159) and (161) can be considered as masked aldols. Thus, (159) on oxidative cleavage led smoothly and stereoselectively to the β-hydroxycarboxylic acid (160), while the C-2 epimer (161) gave the dia-stereomer (162) (Scheme 75).

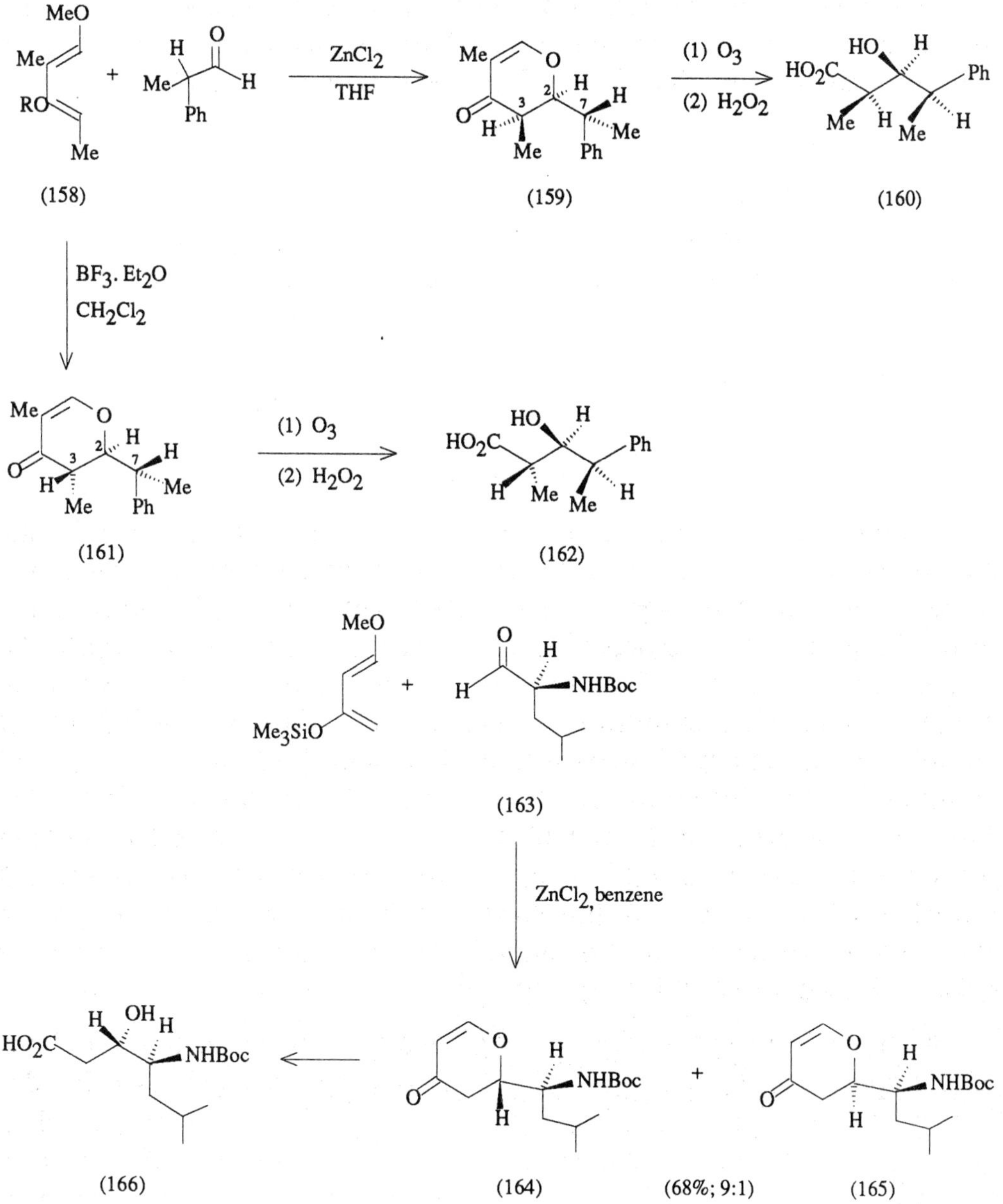

(158) + → $ZnCl_2$, THF → (159) → (1) $O_3$ (2) $H_2O_2$ → (160)

(158) → $BF_3.Et_2O$, $CH_2Cl_2$ → (161) → (1) $O_3$ (2) $H_2O_2$ → (162)

(163) → $ZnCl_2$, benzene → (164) + (165) (68%; 9:1) → (166)

*Scheme 75*

The α-phenylseleno group has been found to be a useful controlling element in Cram-type cycloadditions to carbonyl groups. Thus, in a synthesis of N-acetylneuraminic acid (171) reaction of the (*S*)-seleno-aldehyde (168), used as a synthetic equivalent of acrolein, with the diene (167) in the presence of boron trifluoride etherate gave a 5:1 mixture of *cis*- and *trans*-dihydropyrones (169) and (170) each of which was practically optically pure. The stereogenic centre bearing the phenylseleno group powerfully determined the stereochemical fate of the emerging centres at C-6. The overall sense of the cycloaddition in each case was in accordance with Cram's rule.[198]

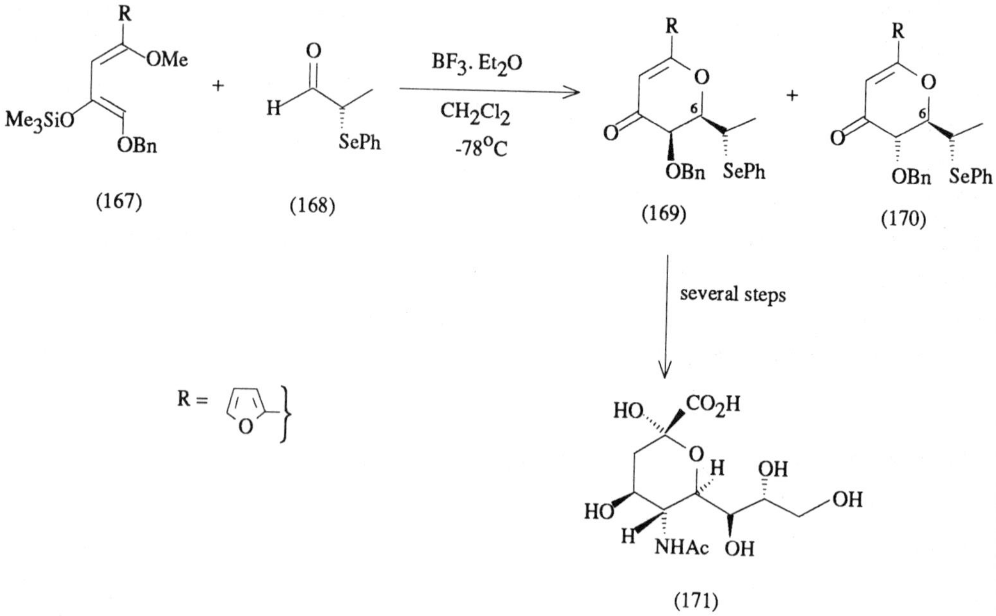

R, OMe, $Me_3SiO$, OBn (167) + O, H, SePh (168) → $BF_3 . Et_2O$, $CH_2Cl_2$, -78°C → R, O, 6, O, OBn SePh (169) + R, O, 6, O, OBn SePh (170)

(169) → several steps → HO, $CO_2H$, O, H, OH, OH, HO, H, NHAc, OH (171)

R = 2-furyl

*Scheme 76*

Cycloadditions to olefinic dienophiles and dienes with an allylic chiral centre are also frequently face selective. For example, addition of N-phenylmaleimide to the diene (172) affords the two diastereomeric adducts (173) and (174) in the ratio 5:1, by selective addition to one face of the diene in an *endo* transition state.

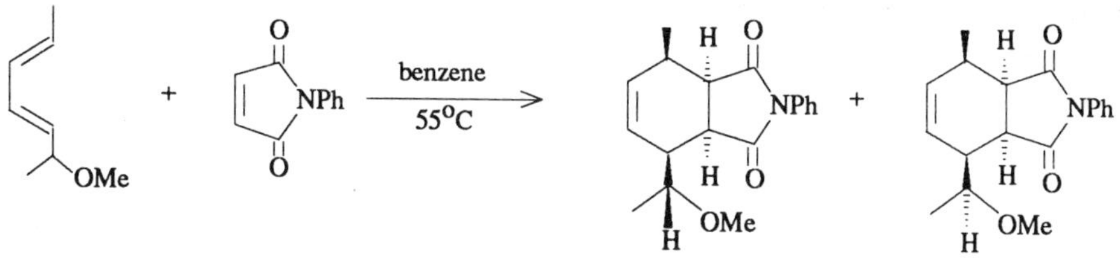

benzene, 55°C

(172) (173) (174) (5:1)

*Scheme 77*

However, it is not always clear how the selectivity in these reactions is controlled and it has not yet been possible to develop a predictive transition state picture which can accommodate all the known facts.[199] In some cases results can be accounted for by steric effects, in others steric effects by themselves are insufficient and stereoelectronic factors appear to come in to play, particularly where the allylic carbon carries a substituent with a lone pair of electrons, such as an alkoxy group. A major uncertainty is the conformation of the allylic group in the transition state[199,200] but this is evidently not the only factor, for the nature of the dienophile and even the solvent can influence the result.

## Asymmetric Diels-Alder Reactions

If either the diene or the dienophile is chiral and optically active, facially selective addition will give rise to a non-statistical mixture of optically active diastereomers. The more facially-slective the addition the more diastereoselective the reaction will be until, in the ideal case, a single diastereomer results. In one approach a stereogenic centre which is to form part of the product is incorporated in the diene or dienophile, usually at an allylic position. The products of cycloaddition are diastereomers and remain so because the original stereogenic centre is an element of the product. Thus, in the synthesis of statine (166) shown in Scheme 75, reaction of the optically active amino-aldehyde (163) with the 1,3-diene gave a mixture of diastereoisomers, the major component of which was converted into enantiomerically pure statine. In this synthesis a new asymmetric centre is generated in the cycloaddition, and the original asymmetric centre of the amino-aldehyde is an integral part of the product. However increasing attention is now being paid to the wider class of asymmetric Diels-Alder reactions using, in effect, non-optically active dienes and dienophiles by conducting the reactions in the presence of an optically active catalyst or, more successfully so far, by temporarily attaching to the diene or dienophile an optically active auxiliary group which is later removed from the diastereomerically pure adduct to give an enantiomerically pure product (Scheme 78).

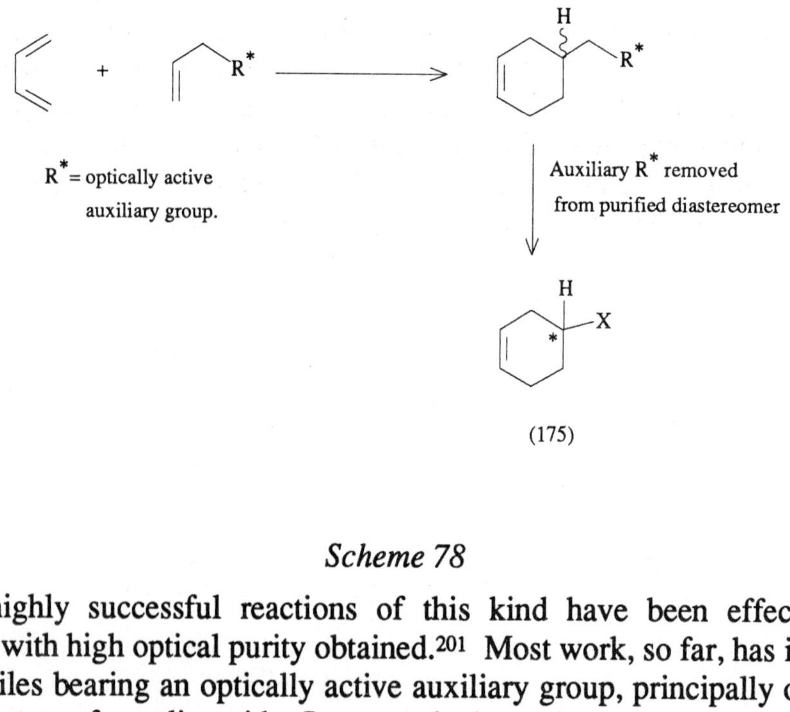

R* = optically active auxiliary group.

Auxiliary R* removed from purified diastereomer

(175)

*Scheme 78*

Some highly successful reactions of this kind have been effected and adducts with high optical purity obtained.[201] Most work, so far, has involved dienophiles bearing an optically active auxiliary group, principally optically active esters of acrylic acid. In general, thermal reactions give only low optical yields and the best results have been obtained in reactions catalysed by Lewis acids at low temperatures. In the catalysed reactions with acrylic esters the ester adopts the *transoid* conformation (176) in which the catalyst is co-ordinated with the carbonyl oxygen *anti* to the ether oxygen, rather than the more sterically hindered *cisoid* conformation (177).

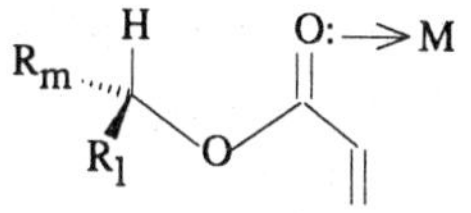

(176) 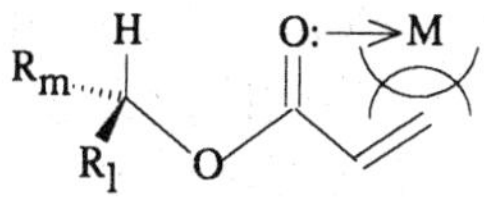 (177)

$R_m$ = medium substituent

$R_l$ = large substituent

*Scheme 79*

Approach of the diene is directed to the less hindered face of the double bond in this conformation (the back *re* face in 176) and the aim is to find conditions to make this reaction as selective as possible. A number of optically active alcohols have been employed as auxiliary groups in this sequence, including menthol and, better, (–)-8-phenylmenthol, but some of the best results so far have been obtained with the neopentyl alcohols (178) and (179), themselves prepared from (*R*)-(+)- and (*S*)-(–)-camphor.[202]

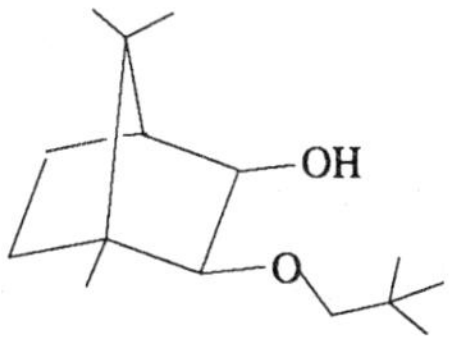

(178) 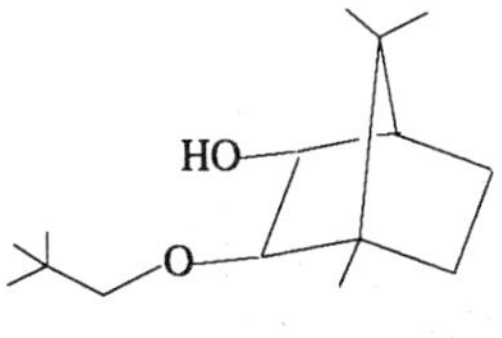 (179)

*Scheme 80*

In catalysed addition of the acrylate ester of (178) to cyclopentadiene the adduct (180) was obtained with almost complete diastereomeric selectivity. Reduction of the purified product with lithium aluminium hydride regenerated the auxiliary alcohol and gave the optically pure *endo* alcohol (181). Reaction is believed to take place by addition of the diene to the ester in the conformation (182) in which access to the front face of the double bond is hindered by the neopentyl group.

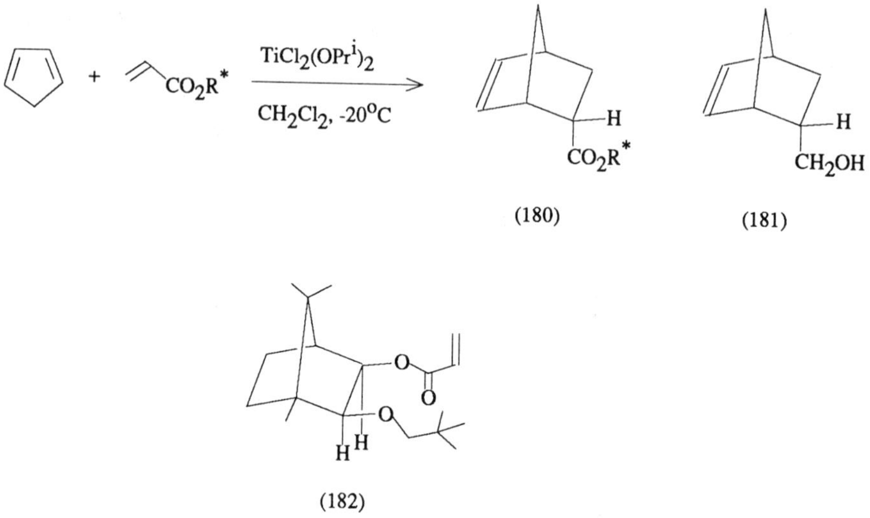

$TiCl_2(OPr^i)_2$

$CH_2Cl_2$, -20°C

$CO_2R^*$

H

$CO_2R^*$

H

$CH_2OH$

(180)

(181)

O

O

O

H H

(182)

*Scheme 81*

A disadvantage of this procedure is the relative inaccessibility of the two epimeric auxiliary alcohols. Another way to realise the rigid structure for the dienophile favourable for highly facially-selective addition is by metal co-ordination of appropriately placed functional groups. Chelate formation can amplify inherently weak diastereofacial discrimination to such an extent that good stereoselection may be achieved with simple chiral auxiliaries. Thus, high levels of diastereofacial selectivity have been obtained in Diels-Alder reactions of the acrylates of (*S*)-ethyl lactate and of (*R*)-pantolactone, catalysed by titanium tetrachloride.[203] The acrylate (183) for example, and isoprene gave the adduct (184) almost exclusively. Purification by crystallisation and hydrolysis provided the enantiomerically pure cyclohexene-4-carboxylic acid (186). The high efficiency of D-pantolactone as chiral auxiliary in this sequence is ascribed to strong chelation of the catalyst with the carbonyl groups of the lactone and the acrylate ester in the complex (187), forcing highly selective addition to the back face of the double bond.

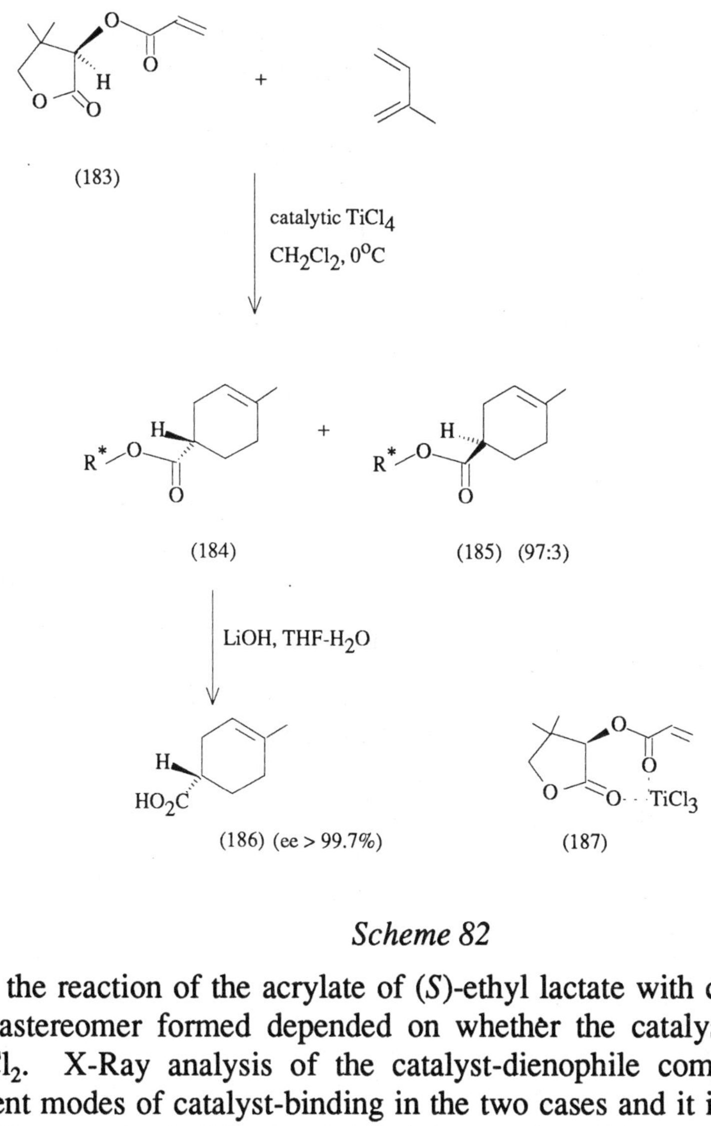

(183)

catalytic $TiCl_4$
$CH_2Cl_2$, $0^oC$

(184) (185) (97:3)

LiOH, THF-$H_2O$

(186) (ee > 99.7%) (187)

*Scheme 82*

In the reaction of the acrylate of (*S*)-ethyl lactate with cyclopentadiene, the diastereomer formed depended on whether the catalyst was $TiCl_4$ or $EtAlCl_2$. X-Ray analysis of the catalyst-dienophile complexes revealed different modes of catalyst-binding in the two cases and it is suggested that this influences the facial selectivity of the reactions.[204]

Very successful results have also been obtained with the αβ-unsaturated carboximides (188, $R^1$=H or $CH_3$) and (189); $R^1$=H or $CH_3$) prepared by N-acylation of the corresponding chiral 2-oxazolidones, themselves derived from (*S*)-valinol, (*S*)-phenylalanol or (1*S*,2*R*)-norephidrine.[205] In reactions catalysed by dimethyl- or diethyl-aluminium chloride the dienophiles reacted readily with several dienes to give adducts in high yield and with very high diastereoselectivity. In many cases the adducts were crystalline and could be obtained diastereomerically pure by crystallisation or chromatography. The chiral auxiliary groups are easily removed by transesterific-ation with lithium benzyloxide, affording the enantiomerically pure Diels-Alder adducts. Products of opposite absolute configuration are obtained from the isopropyl or benzyl carboximides (188) on the one hand and the norephedrine derived carboximides (189) on the other.

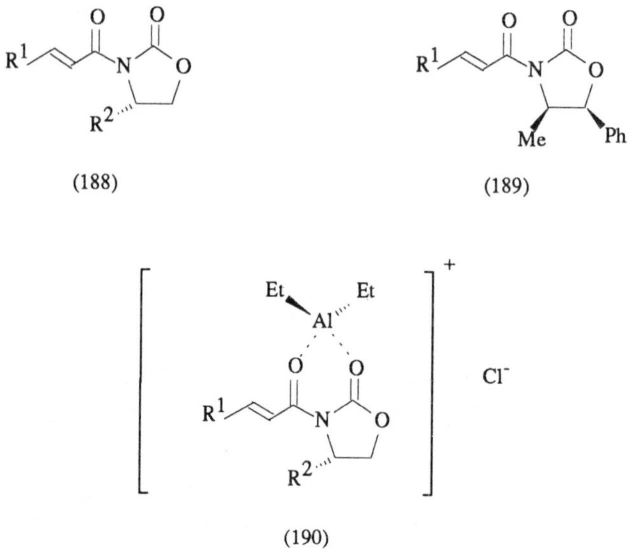

(188) (189)

(190)

*Scheme 83*

These reactions are believed to proceed by way of complexed ion-pairs (as 190) in which the substituents on the isoxazolidone ring shield one face of the double bond from attack by the diene. The exceptional reactivity of these dienophile-Lewis acid complexes allows reaction with less reactive acyclic dienes. For example, isoprene readily gave the adduct (191) which was converted into (*R*)-(+)-α-terpineol.[205]

Excellent results have also been obtained with N-acryloyl- and N-crotonyl-sultams derived from camphor. Again chelation plays an important part in directing the facial selectivity of the cycloadditions.[206] Other experiments have used optically active vinyl sulphoxides as dienophiles,[207] and, in reactions with inverse electron demand, optically active vinyl ethers[208] have been employed.

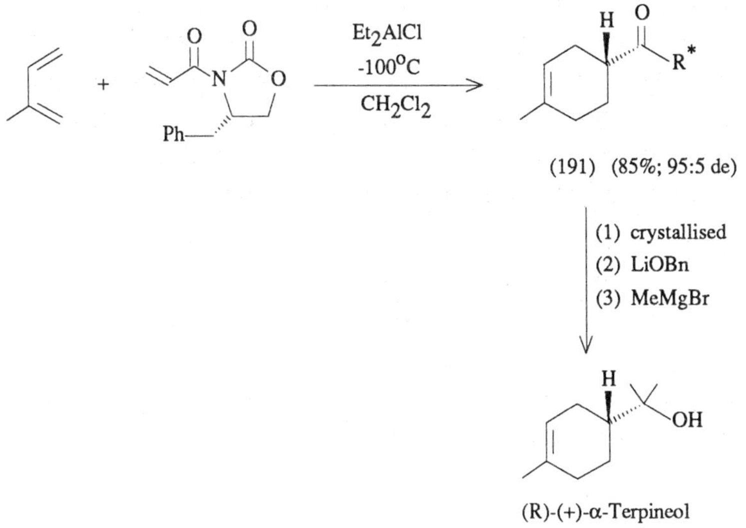

$Et_2AlCl$
-100°C
$CH_2Cl_2$

(191) (85%; 95:5 de)

(1) crystallised
(2) LiOBn
(3) MeMgBr

(R)-(+)-α-Terpineol

*Scheme 84*

Chiral ketols such as (192), in which the chiral auxiliary group is attached one atom closer to the three-carbon enone unit than in the acrylate esters described above have also given excellent results. With these reagents very high diastereofacial selectivities were realised in additions to a number of dienophiles, sometimes even in the absence of a Lewis acid catalyst. Reactions take place by way of the hydrogen-bonded or chelated form (193) with the enone fragment in the *cisoid* conformation.[209]

(192) (193)

*Scheme 85*

Thus, catalysed reaction of the t-butyl derivative (193) and 1,4-di-acetoxybutadiene gave the single *endo* adduct (194) with >98 per cent diastereomeric excess, subsequently converted into (–)-shikimic acid, and with 1-butadienyl phenylacetate the adduct (195) was obtained with a selectivity of more than 100:1.

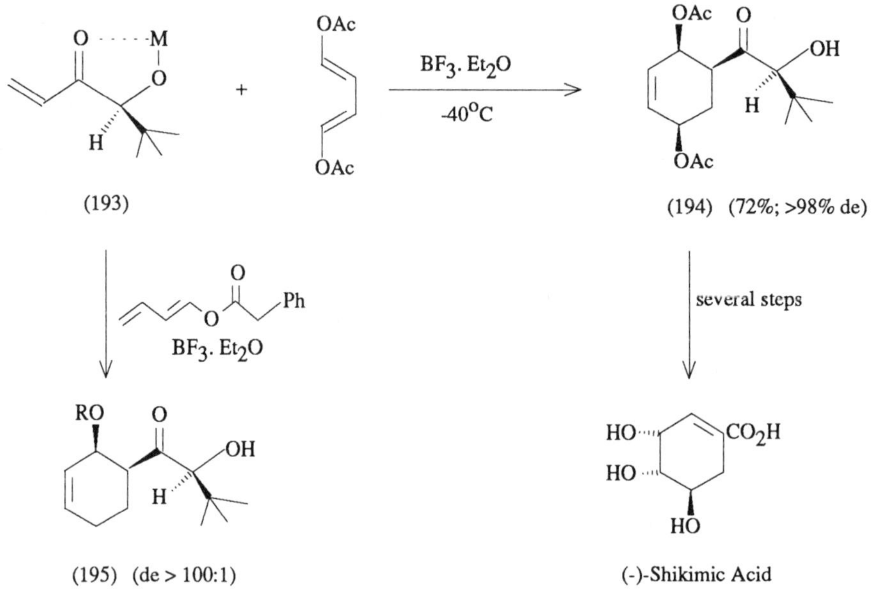

$BF_3.Et_2O$, -40°C

(193)

(194) (72%; >98% de)

several steps

$BF_3.Et_2O$

(195) (de > 100:1)

(-)-Shikimic Acid

*Scheme 86*

Interesting results were obtained on replacing the butadienyl phenylacetate in the last exmaple by the closely related optically active esters derived from (*R*)- and (*S*)-O-methylmandelic acid (196). These dienes themselves show only moderate diastereofacial selectivity in catalysed reactions with non-chiral dienophiles. Reaction of the (*S*)-t-butyl ketol (193) with the (*R*)-mandelate however gave the adducts (197a) and (197b) in the ratio 35:1 and with the (*S*)-mandelate the ratio rose to 130:1. In each case the absolute configuration at C-1 and C-2 of the main products, (197a) and (198a), was the same, and is directly correlated with that of the dienophile (193). Both compounds were converted into the same diol (199) after removal of the chiral auxiliary group. This outcome reflects the overwhelmingly large diastereofacial bias of the dienophile (193) compared with that of the chiral dienes (196), the different ratios (130:1 and 35:1) observed for the two cycloadditions corresponding to reaction of matched and mismatched pairs of the reactants.[210]

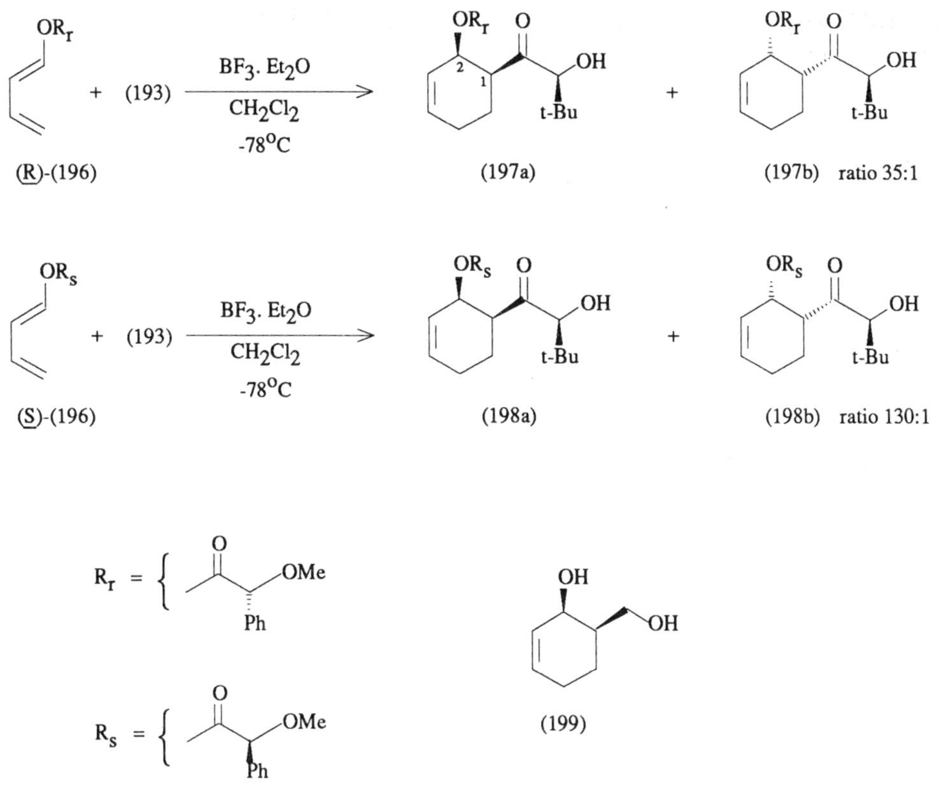

$OR_r$ + (193) $\xrightarrow[\text{CH}_2\text{Cl}_2,\ -78^\circ\text{C}]{\text{BF}_3\cdot\text{Et}_2\text{O}}$ (197a) + (197b)

(R)-(196) (197a) (197b) ratio 35:1

$OR_s$ + (193) $\xrightarrow[\text{CH}_2\text{Cl}_2,\ -78^\circ\text{C}]{\text{BF}_3\cdot\text{Et}_2\text{O}}$ (198a) + (198b)

(S)-(196) (198a) (198b) ratio 130:1

$R_r$ = { O, OMe, Ph

$R_s$ = { O, OMe, Ph

OH, OH

(199)

*Scheme 87*

The reactions of dienes bearing optically active auxiliary groups have not been so widely studied as those of dienophiles, and some of the examples recorded are not very efficient. Virtually complete asymmetric induction was observed in the catalysed reaction of juglone with the (*S*)-O-methylmandelyl ester of 1-hydroxybutadiene, but reaction with acrolein was much less selective.[211] The asymmetric diene (200) containing a tetra-acetyl-*D*-glucose residue as chiral auxiliary showed high diastereofacial selectivity in cycloaddition to benzoquinones[212] and it was used in the key step in a synthesis of (+)-4-demethoxydaunomycinone.[213] Greatly increased selectivities can be obtained on reaction of chiral dienes with "matched" chiral dienophiles, as discussed above.

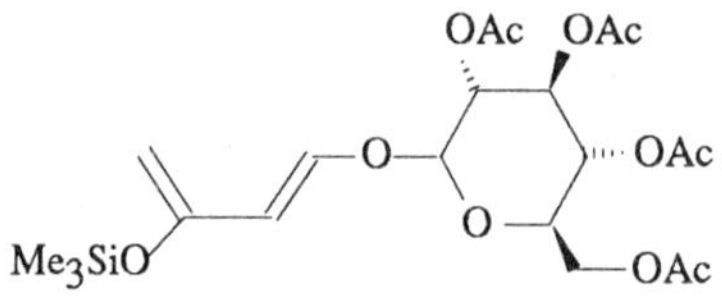

(200)

*Scheme 88*

Reactions with asymmetric quinodimethanes have been reported, but the diastereoselectivities obtained were modest.[214]

By far the most convenient way to bring about an enantioselective Diels-Alder reaction, if it could be achieved, would be by the use of an optically active catalyst. This would obviate the need for chiral auxiliary groups on the diene, or dienophile. Lewis acids and lanthanide complexes bearing chiral ligands have been used as catalysts, but so far results have been variable. Reaction of cyclopentadiene with methacrolein catalysed by (–)-menthyloxyaluminium dichloride gave a cycloadduct with 72 per cent enantiomeric excess, but other reactions were less selective.[215] Several mono- and di-isopinocampheylhalogenoboranes were examined as chiral catalysts for the same reation, but again only modest enantioselectivities were observed.[216] Somewhat better results were obtained in the reaction of cyclopentadiene with N-crotonoyl-4,4-dimethyl-1,3-oxazolidin-2-one (201) in the presence of Lewis acids complexed with optically active $\alpha$-glycols such as (202).[217]

Interesting results have been obtained in reactions of aldehydes with dienes carrying an auxiliary chiral substituent catalysed by the chiral lanthanides (+)- and (–)-Eu(hfc)$_3$ (Scheme 90).

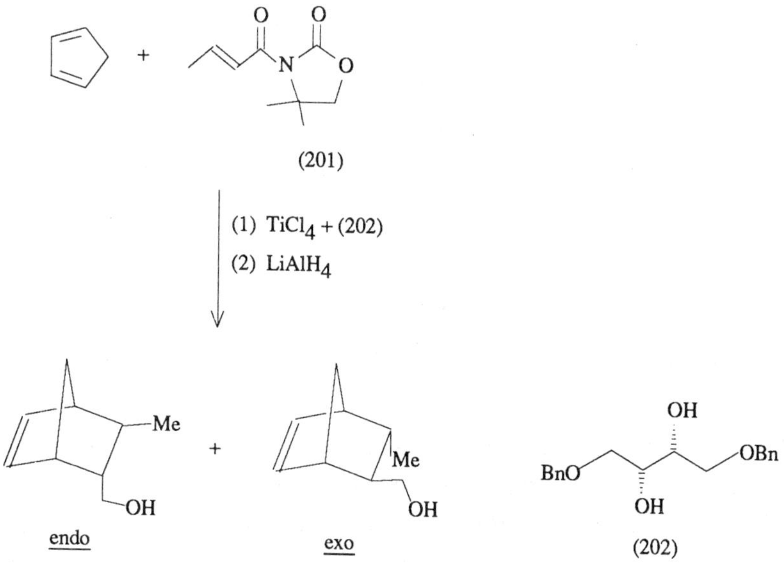

(endo : exo = 94:6. > 98% ee)

*Scheme 89*

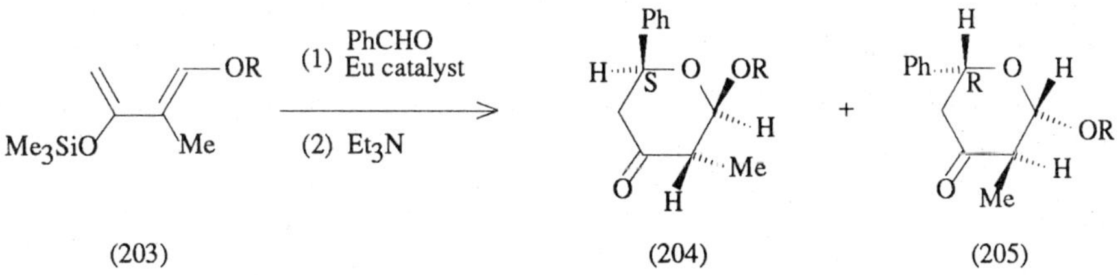

R = Me, (+)-Eu(hfc)$_3$ catalyst; ratio 25 : 75

R = l-menthyl, Eu(fod)$_3$ catalyst; 55 : 45

R = l-menthyl, (+)-Eu(hfc)$_3$ catalyst 8 : 92

R = d-menthyl, Eu(fod)$_3$ catalyst; 45 : 55

R = d-menthyl, (+)-Eu(hfc)$_3$ catalyst; 41 : 59

*Scheme 90*

Reaction of the non-chiral diene (203,R=Me) with benzaldehyde catalysed by (+)-Eu(hfc)$_3$ showed modest enantioselectivity in favour of the "L-pyranose" structure (205), showing that (+)-Eu(hfc)$_3$ is weakly "L-pyranose" selective. Reaction of the chiral diene (203,R=L-menthyl) with benzaldehyde catalysed by non-chiral Eu(fod)$_3$ was weakly selective in favour of the "D-pyranose" product (204). But, surprisingly, combination of the weakly D-favouring L-menthyloxydiene with the modestly L-favouring (+)-Eu(hfc)$_3$ catalyst led to a marked increase in selectivity in favour of the "L-pyranose" adduct (204); the D-menthyloxydiene with (+)-Eu(hfc)$_3$ did not show this effect, giving only weak selectivity in favour of the "L-pyranose" adduct.[218] The increase in facial selectivity here is not simply another instance of double diastereoselection in which two isolated steric biases provide mutual reinforcement. Here the inherent facial bias of the chiral diene is inverted upon interaction with the catalyst. The factors underlying this interactivity are not yet understood, but the phenomenon can be exploited for the synthesis of optically pure saccharides in the L-pyranose series without recourse to formal resolution.

Not many examples of highly enantioselective intramolecular Diels-Alder reactions have been reported. Promising results were obtained with the carboxylic acid derivatives (206) containing the optically active auxiliary 8-phenylmenthyloxy group[219] or the N-oxazolidones (209) and (210).[220] The best results were obtained with the imides which cyclised in the presence of dimethylaluminium chloride to give mainly the *endo* products (207) and (208) which were obtained in greater than 99 per cent diastereomeric excess after chromatography.

## The Retro Diels-Alder Reaction

Diels-Alder reactions are reversible, and many adducts dissociate into their components on heating. Chemical modification of the adducts before dissociation may lead to the generation of a diene and/or dienophile different from the original ones, and retro Diels-Alder reaction sequences of this kind have been widely used in synthesis.[221] Thus substituted fumaric esters have been obtained by stereoselective alkylation of the dianion (211), obtained from the Diels-Alder adducts of both maleic and fumaric esters with cyclopentadiene, as shown in Scheme 92.

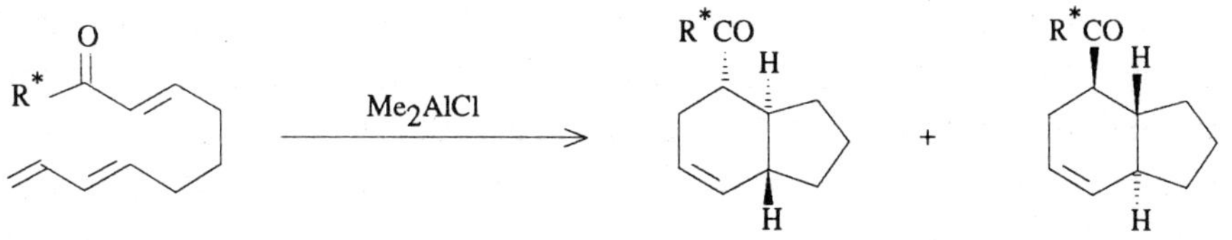

O
R*
Me$_2$AlCl
R*CO
H
H
+
R*CO
H
H

(206) (207) (208)

| | | | | |
|---|---|---|---|---|
| R = (209) | ratio: | 95 | 5 | (73% yield) |
| R = (210) | ratio: | 15 | 85 | (70% yield) |

O
O
N
Ph
O
O
N
Ph
Me

(209) (209)

*Scheme 91*

CO$_2$Me
CO$_2$Me
or
CO$_2$Me
CO$_2$Me
2LDA
(-)
(-)
CO$_2$Me
CO$_2$Me
(211)
RX
650$^o$C
vapour
phase
CO$_2$Me
R
CO$_2$Me
+
CO$_2$Me
MeO$_2$C
R
(95% for R = Me)

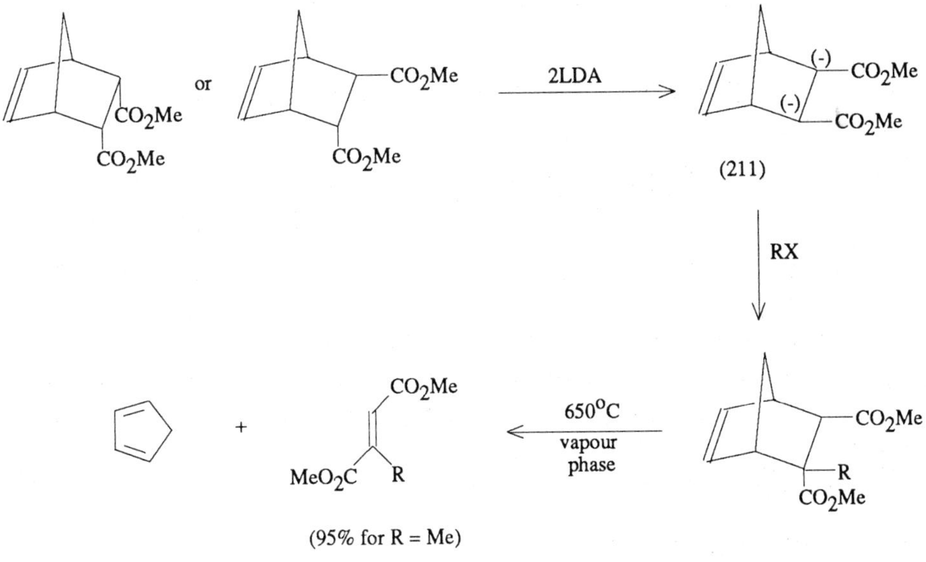

*Scheme 92*

In the above example the bonds cleaved in the retro reaction are the same as those formed in the initial Diels-Alder addition, but this is not necessary and many valuable retro reactions involve the cleavage of bonds other than those formed in the forward reaction.[222] Thus, otherwise relatively inaccessible 3-monosubstituted and 3,4-disubstituted furans have been prepared by way of a tandem Diels-Alder/retro Diels-Alder reaction of 4-phenyloxazole and alkylacetylenes, as in the synthesis of the derivative (212).[223]

The high temperatures required for most retro Diels-Alder reactions are not always convenient. A number of retro reactions involving anionic intermediates take place at more moderate temperatures.[224] Thus, the adduct (213) spontaneously fragmented on treatment with potassium hydride at room temperature, with generation of the cyclohexene (214) which was subsequently converted into conduritol A (215).[225] The unusual ease of this reaction is attributed to the generation of the resonance stabilised anion of 9-hydroxyanthracene.[226]

Although dramatic rate enhancements of forward Diels-Alder cycloadditions by Lewis acid catalysis is well known, similar use of Lewis acids to accelerate cycloreversions has not been common.[227] A study of substituent effects on the rate of fragmentation[228] has shown that retro Diels-Alder reaction of cyclopentadiene adducts is accelerated if there is a trimethylsilyl substituent on C-7 of the bicyclo[2,2,1]heptane system, that is on the cyclopentadiene methylene bridge. This can be useful where the retro reaction has to be effected under milder conditions than usual.[229]

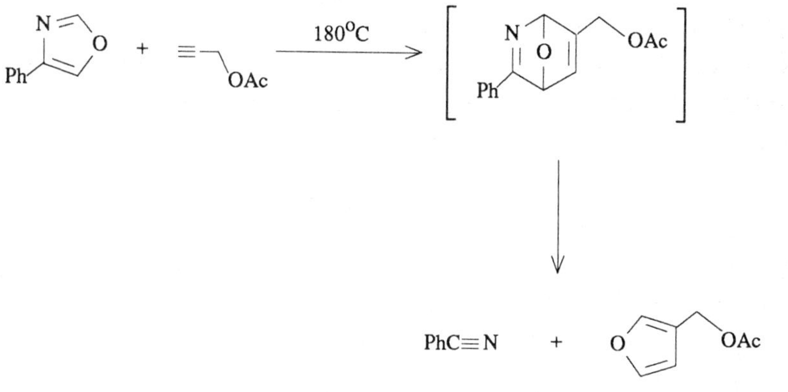

Ph N O + ≡ OAc —180°C→ [ N O Ph OAc ]

↓

PhC≡N + O OAc

(212)

*Scheme 93*

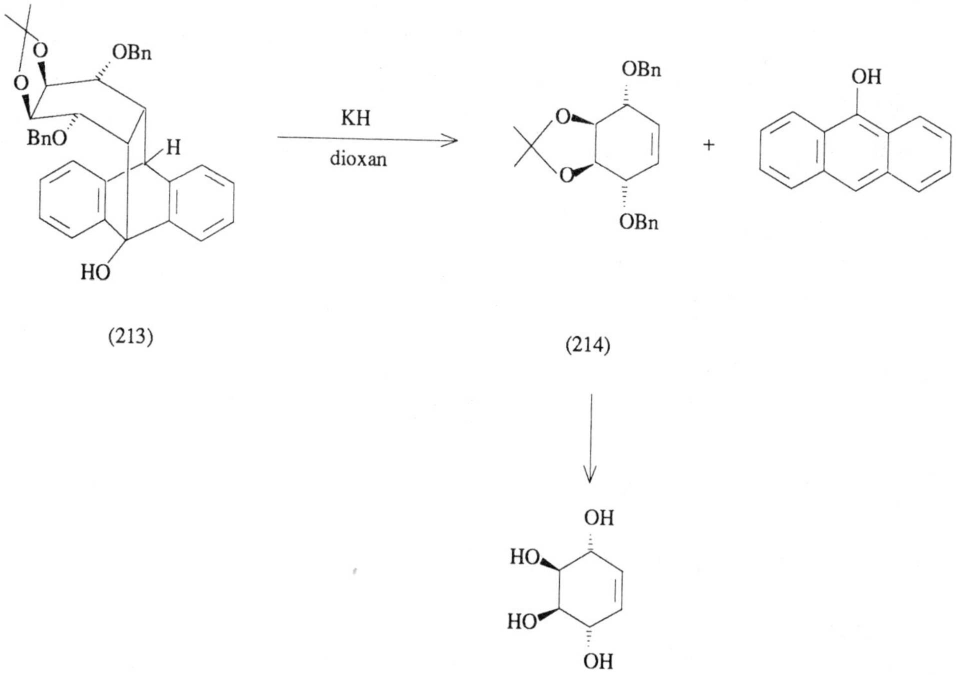

O O OBn BnO H HO

(213)

—KH, dioxan→

OBn O O OBn

(214)

+ OH

↓

OH HO HO OH

(215)

*Scheme 94*

## References


1 J. Sauer, *Angew. Chem. internat. edn.*, 1967, **6**, 16.

2 cf. I. Fleming, "Frontier Orbitals and Organic Chemical Reactions", Wiley, 1976.

3 L.F. Tietze, K-H. Glüsenkamp and W. Holla, *Angew. Chem. internat. edn.*, 1982, **21**, 793.

4 F.E. Ziegler and E.B. Spitzer, *J. Am. Chem. Soc.*, 1970 **92**, 3492.

5 I. Fleming and M.H. Kargar, *J. Chem. Soc. C.*, 1967 226.

6 G. Wittig, *Angew. Chem. internat. edn.*, 1962, **1**, 415; R.W. Hoffmann, *Dehydrobenzene and Cycloalkynes*, Academic Press, 1967.

7 D.L. Boger and C.E. Brotherton, *Tetrahedron*, 1986, **42**, 2777; see also M.L. Dean, *Synthesis*, 1972, 675.

8 R.V.C. Carr and L.A. Paquette, *J. Am. Chem. Soc.*, 1980, **102**, 853; L.A. Paquette and W.A. Kinney, *Tetrahedron Lett.*, 1982, **23**, 131.

9 L.A. Paquette and G.D. Crouse, *J. Org. Chem.*, 1983, **48**, 141; L.A. Paquette and W.A. Kinney, *Tetrahedron Lett.*, 1982, **23**, 5127.

10 O. De Lucchi and G. Modena, *Tetrahedron*, 1984 **40**, 2585.

11 L.A. Paquette, R.E. Moerck, B. Harirchian and P.D. Magnus, *J. Am. Chem. Soc.*, 1978, **100**, 1597.

12 A.P. Davis and G.H. Whitham, *J. Chem. Soc. Chem. Commun.*, 1980, 639.

13 R.N. Warrener, R.A. Russell, R. Solomon, I.G. Pitt and D.N. Butler, *Tetrahedron Lett.*, 1987, **28**, 6503.

14 L.A. Paquette and R.V. Williams, *Tetrahedron Lett.*, 1981, **22**, 4643.

15 O. de Lucchi, V. Lucchini, L. Pasquato and G. Modena, *J. Org. Chem.*, 1984, **49**, 596.

16 G.A. Kraus, J. Thurston, P.J. Thomas, R.A. Jacobs and Y. Su,

*Tetrahedron Lett.*, 1988, **29**, 1879; M.E. Jung and D. D. Grove, *J. Chem. Soc. Chem. Commun.*, 1987, 753.

17 N. Ono, H. Miyake and A. Kaji, *J. Chem. Soc. Chem. Commun.*, 1982, 33; E.J. Corey and H. Estreicher, *Tetrahedron Lett.*, 1981, 603.

18 E.J. Corey, T. Ravindranathan and S. Terashima, *J. Am. Chem. Soc.*, 1971, **93**, 4326.

19 D.A. Evans, W.L. Scott and L.K. Truesdale, *Tetrahedron Lett.*, 1972, 121.

20 N.C. Madge and A.B. Holmes, *J. Chem. Soc. Chem. Commun.*, 1980, 956.

21 D.A. Evans, W.L. Scott and L.K. Truesdale, *Tetrahedron Lett.*, 1972, 121.

22 P.A. Bartlett, F.R. Green and T.R. Webb, *Tetrahedron Lett.*, 1977, 33; D. Ranganathan, C.B. Rao, S. Ranganathan, A.K. Mehrotra and R. Iyengar, *J. Org. Chem.*, 1980, **45**, 1185.

23 C. Maignan and R.A. Raphael, *Tetrahedron*, 1983, **39**, 3245.

24 B.M. Frost and G. Lunn, *J. Am. Chem. Soc.*, 1977, **99**, 7079.

25 I. Gupta and P. Yates, *J. Chem. Soc. Chem. Commun.*, 1982, 1227.

26 P. Chamberlain and A.E. Rooney, *Tetrahedron Lett.*, 1979, 383.

27 A.P. Kozikowski, W.C. Floyd and M.P. Kuniak, *J. Chem. Soc. Chem Commun.*, 1977, 583; A.P. Kozikowski and A. Ames, *J. Am. Chem. Soc.*, 1981, **103**, 392.

28 S.M. Weinreb and R.R. Staib, *Tetrahedron*, 1982, **39**, 3087; S.M. Weinreb and J.J. Lewin, *Heterocycles*, 1979, **12**, 949.

29 A. Zamojski, A. Banaszek, G. Grynkiewecz, *Adv. Carbohydr. Chem. Biochem.*, 1982, **40**, 1; S. David, A. Libineau, J.M. Vatele, *Nouv. J. Chim.*, 1980, **4**, 547; M.E. Jung, K. Shishido, L. Light and L. Davis, *Tetrahedron Lett.*, 1981, 4607.

30 J. Jurczak, A. Golebiowski, M. Chmielewski and S. Filipek, *Synthesis*, 1979, 41; J. Jurczak, A. Golebiowski and T. Bauer, *Synthesis*, 1986, 928.

31 S.J. Danishefsky, E. Larson, D. Askin and N. Kato, *J. Am. Chem. Soc.*, 1985, **107**, 1246; S.J. Danishefsky, W.H. Pearson, D.F. Harvey, C.J. Maring and J.P. Springer, *J. Am. Chem. Soc.*, 1985, **107**, 1256; S.J. Danishefsky and C.J. Maring, *J. Am. Chem. Soc.*, 1985, **107**, 1269.

32 S.J. Danishefsky, E.R. Larson and D. Askin, *J. Am. Chem. Soc.*, 1982, **104**, 6457.

33 cf. S.J. Danishefsky and C.J. Maring, *J. Am. Chem. Soc.*, 1985, **107**, 1269.

34 M. Bednarski and S. Danishefsky, *J. Am. Chem. Soc.*, 1983, **105**, 3716.

35 M. Bednarski and S. Danishefsky, *J. Am. Chem. Soc.*, 1983, **105**, 6968.

36 S. Danishefsky, W.H. Pearson and D.H. Harvey, *J. Am. Chem. Soc.*, 1984, **106**, 2456.

37 S.J. Danishefsky, E. Larson, D. Askin and N. Kato, *J. Am. Chem. Soc.*, 1985, **107**, 1246.

38 S. Danishefsky, K-H. Chao and G. Schulte, *J. Org. Chem.*, 1985, **50**, 4650.

39 S.J. Danishefsky, W.H.Pearson and D.F. Harvey, *J. Am. Chem. Soc.*, 1984, **106**, 2456; S.J. Danishefsky, W.H. Pearson, D.F. Harvey, C.J. Maring and J.P. Springer, *J. Am. Chem. Soc.*, 1985, **107**, 1256.

40 S.J. Danishefsky, *Aldrichimica Acta*, 1986, **19**, 59.

41 S. Danishefsky, S. Kobayashi and J.F. Kerwin, *J. Org. Chem.*, 1982, **47**, 1981.

42 cf. S.M. Weinreb and R.R. Staib, *Tetrahedron*, 1982, **38**, 3087.

43 D.M. Vijas and G.W. Hay, *Canad. J. Chem.*, 1971, **49**, 3755.

44 E. Vedejs, M.J. Arnost, J.M. Dolphin and J. Eustache, *J. Org. Chem.*, 1980, **45**, 2601.

45 P. Beslin and P. Metzner, *Tetrahedron Lett.*, 1980, **21**, 4657.

46 J.E. Baldwin and R.C.G. Lopez, *Tetrahedron*, 1983, **39**, 1487.

47 J.E. Baldwin and R.C.G. Lopez, *Tetrahedron*, 1983, **39**, 1487.

48 E. Vedejs, T.H. Eberlein and R.G. Wilde, *J. Org. Chem.*, 1988, **53**, 2220.

49 E. Vedejs, T.D. Eberlein and D.L. Varie, *J. Am. Chem. Soc.*, 1982, **104**, 1445; E. Vedejs *et.al.*, *J. Org. Chem.*, 1986, **51**, 1556.

50 C.M. Bladon, I.E.G. Ferguson, G.W. Kirby, A.W. Lochead and D.C. McDougall, *J. Chem. Soc. Chem. Commun.*, 1983, 423.

51 G.W. Kirby, A.W. Lochead and G.N. Sheldrake, *J. Chem. Soc. Chem. Commun.*, 1984, 922.

52 G.W. Kirby, A.W. Lochead and G.N. Sheldrake, *J. Chem. Soc. Chem. Commun.*, 1984, 1469.

53 G.A. Kraft and P.T. Meinke, *Tetrahedron Lett.*, 1985, **26**, 1947.

54 E. Vedejs, D.A. Perry, K.N. Houk and N.G. Rondan, *J. Am. Chem. Soc.*, 1983, **105**, 6999; E. Vedejs, J.S. Stults and R.G. Wilde, *J. Am. Chem. Soc.*, 1988, **110**, 5452.

55 E. Vedejs, T.H. Eberlein and D.L. Varie, *J. Am. Chem. Soc.*, 1982, **104**, 1445; see also E. Vedejs, C.L. Fedde and C.E. Schwartz, *J. Org. Chem.*, 1987, **52**, 4269.

56 E. Vedejs, J.M. Dolphin and H. Mastalerz, *J. Am. Chem. Soc.*, 1983, **105**, 127; E. Vedejs, *Acc. Chem. Res.*, 1984, **17**, 358.

57 S.D. Larsen, *J. Am. Chem. Soc.*, 1988, **110**, 5932.

58 G. Kresze and W. Wuncherpfennig, *Angew. Chem. internat. edn.*, 1967, **6**, 149; S.M. Weinreb and R.R. Staib, *Tetrahedron*, 1982, **38**, 3087; S.M. Weinreb, *Acc. Chem. Res.*, 1988, **21**, 313.

59 G. Kresze and U. Wagner, *Annalen*, 1972, *762*, 93; cf. W.L. Mock and R.M. Nugent, *J. Am. Chem. Soc.*, 1975, **97**, 6521.

60 W.L. Mock and R.M. Nugent, *J. Org. Chem.*, 1978, **43**, 3443.

61 S.M. Weinreb and J.I. Levin, *Heterocycles*, 1979, **12**, 949; S.M. Weinreb and R.R. Staib, *Tetrahedron*, 1982, **38**, 3087; W. Oppolzer, *Angew. Chem. internat. edn.*, 1972, **11**, 1031.

62 M.E. Jung, K. Shishoda, L. Light and L.Davis, *Tetrahedron Lett.*, 1981, **22**, 4607.

63 J.F. Kerwin and S. Danishefsky, *Tetrahedron Lett.*, 1982, **23**, 3739.

64 S. Danishefsky and J.F. Kerwin, *J. Org. Chem.*, 1982, **47**, 3183.

65 S. Danishefsky, M.E. Langer and C. Vogel, *Tetrahedron Lett.*, 1985, **26**, 5983.

66 K.M. Ryan, R.A. Reamer, R.F. Volanti and I. Shinkai, *Tetrahedron Lett.*, 1987, **28**, 2103.

67 P.A. Grieco, S.D. Larsen and W.F. Fobare, *Tetrahedron Lett.*, 1986, **27**, 1975.

68 e.g. W. Oppolzer, E. Francotti and K. Bättig, *Helv. Chim. Acta*, 1981, **64**, 478; S.M. Weinreb, *Acc. Chem. Res.*, 1985, **18**, 16.

69 G.W. Kirby, *Chem. Soc. Rev.*, 1977, **6**, 1; S.M. Weinreb and R.R. Staib, *Tetrahedron*, 1982, **38**, 3087.

70 J.E.T. Corrie, G.W. Kirby and J.W.M. Mackinnon, *J. Chem. Soc. Perkin 1*, 1984, 883.

71 G.W. Kirby, H. McGuigan, J.W.M. Mackinnon, D.McLean and R.P. Sharma, *J. Chem. Soc. Perkin 1*, 1985, 1437.

72 P. Horsewood and G.W. Kirby, *J. Chem. Soc. Perkin 1*, 1980, 1587; G.W. Kirby, *Chem. Soc. Rev.*, 1977, **6**, 1.

73 cf. D.L. Boger, M. Patel and F. Takusagawa, *J. Org. Chem.*, 1985, **50**, 1911.

74 J.E. Baldwin, M. Otsuka and P.M. Wallace, *J. Chem. Soc. Chem. Commun.*, 1985, 1549; *Tetrahedron*, 1986, **42**, 3097.

75 A. Defoin, H. Fritz, G. Geffroy and J. Streith, *Tetrahedron Lett.*, 1986, **27**, 4727.

76 H. Labaziewicz and F.G. Riddell, *J. Chem. Soc. Perkin 1*, 1979, 2926.

77 N.J. Leonard, A.J. Playtis, F. Skoog and R.Y. Schmitz, *J. Am. Chem. Soc.*, 1971, **93**, 3056.

78 M. Saburi, G. Kresze and H. Braun, *Tetrahedron Lett.*, 1984, **25**, 5377.

79 H. Felber, G. Kresze, R. Prewo and A. Vasella, *Helv. Chim. Acta*, 1986, **69**, 1137.

80 cf. M. Petrzilka and J.I. Grayson, *Synthesis*, 1981, 753.

81 M.E. Jung, C.A. McCombs, Y. Takeda and Y-G. Pan, *J. Am. Chem. Soc.*, 1981, **103**, 6677.

82 S. Danishefsky, T. Kitahara and P.F. Schuda, *Organic Syntheses*, 1983, **61**, 147.

83 S. Danishefsky, *Acc. Chem. Res.*, 1981, **14**, 400.

84 S. Danishefsky, T. Kitahara, C.F. Yan and J. Morris, *J. Am. Chem. Soc.*, 1979, **101**, 6996.

85 S. Danishefsky, T. Harayama and R.K. Singh, *J. Am. Chem. Soc.*, 1979, **101**, 7008.

86 S. Danishefsky, C.F. Yan, R.K. Singh, R.B. Gammill, P.M. McCurry, N. Fritsch and J. Clardy, *J. Am. Chem. Soc.*, 1979, **101**, 7001.

87 J. Banville and P. Brassard, *J. Chem. Soc., Perkin 1*, 1976, 1852.

88 K. Yamamoto, S. Suzuki and J. Tsuji, *Chem. Lett.*, 1978, 649.

89 S. Danishefsky, M.P. Prisbylla and S. Hiner, *J. Am. Chem. Soc.*, 1978, **100**, 2918.

90 J. Savard and P. Brassard, *Tetrahedron Lett.*, 1979, 4911.

91 D.A. Evans, C.A. Bryan and C.L. Sims, *J. Am. Chem. Soc.*, 1972, **94**, 2891; M. Petrzilka and J.I. Grayson, *Synthesis*, 1981, 753.

92 T. Cohen, R.J. Ruffner, D.W. Shull, E.R. Fogel and J.R. Falck, *Organic Syntheses*, 1980, **59**, 202.

93 T. Cohen and Z. Kosarych, *J. Org. Chem.*, 1982, **47**, 4005.

94 B.M. Trost, W.C. Vladuchick and A.J. Bridges, *J. Am. Chem. Soc.*, 1980, **102**, 3554.

95 D.A. Evans, C.A. Bryan and C.L. Sims, *J. Am. Chem. Soc.*, 1972, **97**, 2891.

96 M. Petrzilka and J.I. Grayson, *Synthesis*, 1981, 753.

97 cf. S. Hünig and H. Kahanek, *Chem. Ber.*, 1957, **90**, 238; R.L. Snowden and M. Wüst, *Tetrahedron Lett.*, 1986, **27**, 699.

98 W. Oppolzer, L. Bieber and J. Francotte, *Tetrahedron Lett.*, 1979, 981.

99 cf. P.J. Jessup, C.B. Petty, J. Roos and L.E. Overman, *Organic Syntheses*, 1979, **59**, 1.

100 L.E. Overman, R.L. Freerks, C.B. Petty, L.A. Clizbe, R.K. Ono, G.F. Taylor and P.J. Jessup, *J. Am. Chem. Soc.*, 1981, **83**, 2816.

101 cf. L.E. Overman, *Acc, Chem. Res.*, 1980, **13**, 218.

102 see G.R. Krow, J.T. Carey, K.C. Cannon and K.J. Henz, *Tetrahedron Lett.*, 1982, **23**, 2527; R.J. Sundberg and J.D. Bloom, *J. Org. Chem.*, 1980, **45**, 3382.

103 P.A. Wender, J.M. Schaus and D.C. Torney, *Tetrahedron Lett.*, 1979, 2485.

104 P.A. Wender, J.M. Schaus and A.W. White, *J. Am. Chem. Soc.*, 1980, **102**, 6157.

105 S. Raucher, B.L. Bray and R.F. Lawrence, *J. Am. Chem. Soc.*, 1987, **109**, 442.

106 D.L. Commins, A.H. Abdulla and R.K. Smith, *Tetrahedron Lett.*, 1983, **24**, 2711; see also G.R. Krow, Y.B. Lee, S.W. Szczepanski and R. Raghavachari, *Tetrahedron Lett.*, 1985, **26**, 2617.

107 M.J. Carter, I. Fleming and A. Percival, *J. Chem. Soc., Perkin 1*, 1981, 2415; M.E. Jung and B. Gaede, *Tetrahedron*, 1979, **35**, 621.

108 M. Koreeda and M.A. Ciufolini, *J. Am. Chem. Soc.*, 1982, **104**, 2308.

109 See for example, R.L. Danheiser and H. Sard, *J. Org. Chem.*, 1980, **45**, 4810; R.K. Boeckman and T.E. Barta, *J. Org. Chem.*, 1985, **50**, 3423.

110 M. Vaultier, F. Truchet, B. Carboni, R.W. Hoffmann and I. Denne, *Tetrahedron Lett.*, 1987, **28**, 4169.

111 J.A. Gladysz, S.J. Lee, J.A.V. Tomasello and D.M. Donelson, *J. Org. Chem.*, 1977, **42**, 2930; R.E. Ireland *et al.*, *J. Am. Chem. Soc.*, 1983,

**105**, 1988.

112 M.E. Jung, L.J. Street and Y. Usui, *J. Am. Chem. Soc.*, 1986, **108**, 6810; M.E. Jung and J.A. Hagenah, *J. Org. Chem.*, 1987, **52**, 1889.

113 T. Ziegler, M. Layh and F. Effenberger, *Chem. Ber.*, 1987, **120**, 1347.

114 E.J. Corey and A.P. Kozikowski, *Tetrahedron Lett.*, 1975, 925, 2389.

115 J.L. Charlton and M.M. Alaudin, *Tetrahedron*, 1987, **43**, 2873.

116 P.G. Sammes, *Tetrahedron*, 1976, **32**, 405; J.J. McCullough, *Acc. Chem. Res.*, 1980, **13**, 270.

117 K.P.C. Vollhardt, *Angew. Chem. internat. edn.*, 1984, **23**, 539.

118 W. Oppolzer, D.A. Roberts and T.G.C. Bird, *Helv. Chim. Acta*, 1979, **62**, 2017; W. Oppolzer and D.A. Roberts, *Helv. Chim. Acta.*, 1980, **63**, 1703.

119 R. Gray, L.G. Harruff, J. Krymowski, J. Peterson and V. Boekelheide, *J. Am. Chem. Soc.*, 1978, **100**, 2892; P. Schiess, M. Heitzmann, S. Rutschmann and R. Staheli, *Tetrahedron Lett.*, 1978, 4569.

120 Y. Tamura, A. Wada, M. Sasho and Y. Kita, *Tetrahedron Lett.*, 1981, **22**, 4283.

121 T. Tuschka, K. Naito and B. Rickborn, *J. Org. Chem.*, 1983, **48**, 70.

122 M.P. Cava, A.A. Deana and K. Muth, *J. Am. Chem. Soc.*, 1959, **81**, 6458.

123 J.R. Wiseman, J.J. Pendery, C.A. Otto and K.G. Chiong, *J. Org. Chem.*, 1980, **45**, 516.

124 Y. Ito, M. Nakatsuka and T. Saegusa, *J. Am. Chem. Soc.*, 1980, **102**, 863; *ibid.*, 1982, **104**, 7609.

125 Y. Ito, M. Nakatsuka and T. Saegusa, *J. Am. Chem. Soc.*, 1981, **103**, 476.

126 Y. Ito, S. Miyata, M. Nakatsuka and T. Saegusa, *J. Am. Chem. Soc.*, 1981, **103**, 5250; see also Y. Mao and V. Boekelheide, *J. Org. Chem.*, 1980, **45**, 1547.

127 P. Magnus, T. Gallagher, P. Brown and P. Pappalardo, *Acc. Chem. Res.*, 1984, 35.

128 cf. G. Desimoni and G. Tacconi, *Chem. Rev.*, 1975, **75**, 651.

129 W.G. Dauben and H.O. Krabbenhoft, *J. Org. Chem.*, 1977, **42**, 282.

130 cf. R.A. John, V. Schmid and H. Wyler, *Helv. Chim. Acta*, 1987, **70**, 600.

131 S. Danishefsky and M. Bednarski, *Tetrahedron Lett.*, 1984, **25**, 721; S.S. Hall, G.F. Weber and A.J. Duggan, *J. Org. Chem.*, 1978, **43**, 667.

132 R.R. Schmidt and M. Maier, *Tetrahedron Lett.*, 1985, **26**, 2065; L.F. Tietze, E. Voss, K. Harms and G.M. Sheldrick, *Tetrahedron Lett.*, 1985, **26**, 5273; M. Maier and R.R. Schmidt, *Annalen*, 1985, 2261.

133 L.F. Tietze, K-H. Glüsenkamp, K. Harms and G. Remberg, *Tetrahedron Lett.*, 1982, **23**, 1147.

134 R.E. Ireland and J.P. Daub, *J. Org. Chem.*, 1983, **48**, 1303; R.E. Ireland, J.P. Daub, G.S. Mandel and N.S. Mandel, *J. Org. Chem.*, 1983, **48**, 1312.

135 D.L. Boger, *Chem. Rev.*, 1986, **86**, 781.

136 cf. D.L. Boger, *Tetrahedron*, 1983, **39**, 2869.

137 Y.S. Cheng, A.T. Lupo and F.W. Fowler, *J. Am. Chem. Soc.*, 1983, **105**, 7696.

138 Y.C. Hwang and F.W. Fowler, *J. Org. Chem.*, 1985, **50**, 2719.

139 B. Serckx-Poncin, A-M. Hesbain-Frisque and L. Ghosez, *Tetrahedron Lett.*, 1982, **23**, 3261.

140 F. Sainte, B. Serckx-Poncin, A-M. Hesbain-Frisque and L. Ghosez, *J. Am. Chem. Soc.*, 1982, **104**, 1428.

141 P.M. Scola and S.M. Weinreb, *J. Org. Chem.*, 1986, **51**, 3248.

142 S.E. Denmark, M.S. Dappen and J.A. Sternberg, *J. Org. Chem.*, 1984, **49**, 4741.

143 S.E. Denmark, M.S. Dappen and C.J. Cramer, *J. Am. Chem. Soc.*, 1986,

**108**, 1306.

144 R.A. Firestone, E.E. Harris and W. Reuter, *Tetrahedron*, 1967, **23**, 943; W. Böll and H. König, *Annalen*, 1979, 1657.

145 P.A. Jacobi, D.G. Walker and I.M.A. Odeh, *J. Org. Chem.*, 1981, **46**, 2065.

146 I.J. Turchi and M.J.S. Dewar, *Chem. Rev.*, 1975, **75**, 389; R. Lakham and B. Ternai, *Adv. Heterocyclic. Chem.*, 1974, **17**, 99.

147 H. Neunhoeffer and G. Werner, *Annalen*, 1974, 1190.

148 H. Neunhoffer and B. Lehmann, *Annalen*, 1975, 1113.

149 L.B. Davis, S.G. Greenberg and P.G. Sammes, *J. Chem. Soc, Perkin 1*, 1981, 1909.

150 D.C. Rideout and R. Breslow, *J. Am. Chem. Soc.*, 1980, **102**, 7816; R. Breslow, U. Maitra and D. Rideout, *Tetrahedron Lett.*, 1983, **24**, 1901; R. Breslow and U. Maitra, *Tetrahedron Lett.*, 1984, **25**, 1239.

151 eg. K. Yoshida and P.A. Grieco, *J. Org. Chem.*, 1984, **49**, 5257; S.E. Drewes, P.A. Grieco and J.C. Huffman, *J. Org. Chem.*, 1985, **50**, 1309; P.A. Grieco, P. Galatsis and R.F. Spohn, *Tetrahedron*, 1986, **42**, 2847.

152 E. Brandes, P.A. Grieco and P. Garner, *J. Chem. Soc. Chem. Commun.*, 1988, 500.

153 A. Lubineau and Y. Queneau, *Tetrahedron Lett.*, 1985, **26**, 2653; *J. Org. Chem.*, 1987, **52**, 1001.

154 S.D. Larsen and P.A. Grieco, *J. Am. Chem. Soc.*, 1985, **107**, 1768; P.A. Grieco, S.D. Larsen and W.F. Fobare, *Tetrahedron Lett.*, 1986, **27**, 1975..

155 P.A. Grieco and D.T. Parker, *J. Org. Chem.*, 1988, **53**, 3325.

156 P.A. Grieco and D.T. Parker, *J. Org. Chem.*, 1988, **53**, 3658.

157 H. Waldman, *Angew. Chem. internat. edn.*, 1988, **27**, 274.

158 T. Dunams, W.Hoekstra, M. Pentaleri and D. Liotta, *Tetrahedron Lett.*, 1988, **29**, 3745; R. Breslow and T. Guo, *J. Am. Chem. Soc.*, 1988, **110**, 5613.

159 W.G. Dauben and H.O. Krabenhaft, *J. Am. Chem. Soc.*, 1976, **98**, 1992; *J. Org. Chem.*, 1977, **42**, 282.

160 A.B. Smith, N.J. Liverton, N.J. Hrib, H. Swaramakrishnan and K. Winzenberg, *J. Am. Chem. Soc.*, 1986, **108**, 3040.

161 A. Golebiowski, U. Jacobsson and J. Jurczak, *Tetrahedron*, 1987, **43**, 3063.

162 A. Guingnant and J. d'Angelo, *Tetrahedron Lett.*, 1986, **27**, 3729.

163 C. Ferroud, G. Revial and J. d'Angelo, *Tetrahedron Lett.*, 1985, **26**, 3981.

164 J. Jurczak, A. Golebiowski and A. Rahm, *Tetrahedron Lett.*, 1986, **27**, 853.

165 J. Jurczak, A. Golebiowski and T. Bauer, *Synthesis*, 1986, 928.

166 cf. A.P. Kozikowski, T.R. Nieduzak, T. Konoike and J.P. Springer, *J. Am. Chem. Soc.*, 1987, **109**, 5167.

167 J. Jurczak, T. Bauer and S. Jarosz, *Tetrahedron*, 1986, **42**, 6477.

168 L.F. Tietze, T. Hübsch, E. Voss, M. Buback and W. Tost, *J. Am. Chem. Soc.*, 1988, **110**, 4065.

169 P. Yates and P. Eaton, *J. Am. Chem. Soc.*, 1960, **82**, 4436; J. Sauer, *Angew. Chem. internat. edn.*, 1967, **6**, 16.

170 W. Kreiser, W. Haumesser and A.F. Thomas, *Helv. Chim. Acta*, 1974, **57**, 164.

171 J. Sauer and J. Kredel, *Tetrahedron Lett.*, 1966, 731, 6359.

172 J.E. Baldwin and M.J. Lusch, *J. Org. Chem.*, 1979, **44**, 1923.

173 M. Bednarsky, C. Maring and S. Danishefsky, *Tetrahedron Lett.*, 1983, **24**, 3451.

174 M. Bednarsky and S. Danishefsky, *J. Am. Chem. Soc.*, 1983, **105**, 6968.

175 S. Danishefsky and M. Bednarski, *Tetrahedron Lett.*, 1985, **26**, 2507.

176 R.C. Gupta, D.A. Jackson and R.J. Stoodley, *J. Chem. Soc. Chem.*

*Commun.*, 1982, 929; see also T.R. Kelly and M. Montury, *Tetrahedron Lett.*, 1978, 4311; H-J. Liu and E.N.C. Browne, *Canad. J. Chem.*, 1981, **59**, 601; T. Poll, J.O. Mether and G. Helmchen, *Angew. Chem. internat. edn.*, 1985, **24**, 112.

177 B.M. Trost, J.Ippen and W.C. Vladuchick, *J. Am. Chem. Soc.*, 1977, **99**, 8117.

178 D.J. Belleville, D.D. Wirth and N.L. Bauld, *J. Am. Chem. Soc.*, 1981, **103**, 718; G.C. Cahoun and G.B. Schuster, *Tetrahedron Lett.*, 1986, **27**, 911; R.A. Pabon, D.J. Belleville and N.L. Bauld, *J. Am. Chem. Soc.*, 1983, **105**, 5158.

179 D.J. Belleville, N.L. Bauld, R. Pabon and S. Gardner, *J. Am. Chem. Soc.*, 1983, **105**, 3548; K.T. Lorenz and N.L. Bauld, *J. Am. Chem. Soc.*, 1987, **109**, 1157; see also D.J. Belleville and N.L. Bauld, *Tetrahedron*, 1986, **42**, 6167, 6189.

180 R.A. Pabon, D.J. Belleville and N.L. Bauld, *J. Am. Chem. Soc.*, 1984, **106**, 2730.

181 B. Harichian and N.L. Bauld, *Tetrahedron Lett.*, 1987, **28**, 927.

182 P.G. Gassman and D.A. Singleton, *J. Am. Chem. Soc.*, 1984, **106**, 6085.

183 cf. P.G. Gassman and D.A. Singleton, *J. Am. Chem. Soc.*, 1984, **106**, 7993.

184 D.W. Reynolds, K.T. Lorenz, H-S. Chion, D.J. Belleville, R.A. Pabon and N.L. Bauld, *J. Am. Chem. Soc.*, 1987, **109**, 4960.

185 P.G. Gassman and D.A. Singleton, *J. Org. Chem.*, 1986, **51**, 3075.

186 P.G. Gassman, D.A. Singleton, J.J. Wilwerding and S.P. Chavan, *J. Am. Chem. Soc.*, 1987, **109**, 2182.

187 P.G. Gassman and S.P. Chavan, *Tetrahedron Lett.*, 1988, **29**, 3407.

188 J. Sauer, *Angew. Chem. internat. edn.*, 1967, **6**, 16.

189 K.N. Houk, *Acc. Chem. Res.*, 1975, **8**, 361; R. Sustman, *Pure Appl. Chem.*, 1974, **40**, 569.

190 K.N. Houk, *et al.*, *J. Am. Chem. Soc.*, 1978, **100**, 6531; P.V. Alston, R.M. Ottenbrite and T. Cohen, *J. Org. Chem.*, 1978, **43**, 1864.

191 T. Cohen and Z. Kosarych, *J. Org. Chem.*, 1982, **47**, 4005.

192 B.M. Trost, J. Ippen and W.C. Vladuchik, *J. Am. Chem. Soc.*, 1977, **99**, 8117; B.M. Trost, W.C. Vladuchik and A.J. Bridges, *J. Am. Chem. Soc.*, 1980, **102**, 3554.

193 H. Ono, H. Miyake and A. Kaji, *J. Chem. Soc. Chem. Commun.*, 1982, 33.

194 cf. J.G. Martin and R.K. Hill, *Chem. Rev.*, 1961, **61**, 537; J. Sauer, *Angew. Chem. internat. edn.*, 1967, **6**, 16.

195 M.P. Edwards, S.V. Ley, S.G. Lister, B.D. Palmer and D.J. Williams, *J. Org. Chem.*, 1984, **49**, 3503; K.C. Nicolaou and R.L. Magolda, *J. Org. Chem.*, 1981, **46**, 1506.

196 S. Danishefsky, E.R. Larson and D. Askin, *J. Am. Chem. Soc.*, 1982, **104**, 6457.

197 S. Danishefsky, S. Kobayashi and J.F. Kerwin, *J. Org. Chem.*, 1982, **47**, 1981.

198 S.J. Danishefsky, M.P. De Ninno and S. Chen, *J. Am. Chem. Soc.*, 1988, **110**, 3929.

199 cf. R. Tripathy, R.W. Franck and K.D. Onan, *J. Am. Chem. Soc.*, 1988, **110**, 3257.

200 K.N. Houk, S.R. Moses, Yun-Dong Wu, N.G. Rondan, V. Jäger, R. Schohe and F.R. Fronczek, *J. Am. Chem. Soc.*, 1984, **106**, 3880; S.D. Kahn and W.J. Hehre, *J. Am. Chem. Soc.*, 1987, **109**, 663; R.W. Franck, T.V. John and K. Olejniczak, *J. Am. Chem. Soc.*, 1982, **104**, 1106; R.W. Franck, S. Argade, C.S. Subramanian and D.M. Frichet, *Tetrahedron Lett.*, 1985, **26**, 3187.

201 See W. Oppolzer, *Angew. Chem. internat. edn.*, 1984, **23**, 876; L.A. Paquette in *Asymmetric Synthesis*, Vol. 3, Chapter 4, Academic Press, 1984, ed. J.D. Morrison; G. Helmchen, R. Karge and J. Weetman in *Modern Synthetic Methods*, Vol. 4, p.261, ed. R. Scheffold, Springer-Verlag, Berlin, 1986.

202 W. Oppolzer, C. Chapius, G.M. Dao, D.R. Reichlin and T. Godel, *Tetrahedron Lett.*, 1982, **23**, 4781; G. Helmchen and R. Schmierer, *Angew. Chem. internat. edn.*, 1981, **20**, 205; see also W. Oppolzer and C. Chapius, *Tetrahedron Lett.*, 1983, **24**, 4665; W. Oppolzer, *Tetrahedron*, 1987, **43**, 1969.

203 T. Poll, G. Helmchen and B. Bauer, *Tetrahedron Lett.*, 1984, **25**, 2191; T. Poll, A. Sobczak, H. Hartmann and G. Helmchen, *Tetrahedron Lett.*, 1985, **26**, 3095.

204 T. Poll, J.O. Metter and G. Helmchen, *Angew. Chem. internat. edn.*, 1985, **24**, 112.

205 D.A. Evans, K.T. Chapman and J. Bisaha, *J. Am. Chem. Soc.*, 1984, **106**, 4261; *J. Am. Chem. Soc.*, 1988, **110**, 1238.

206 W. Oppolzer, C. Chapius, and G. Bernardinelli, *Helv. Chim. Acta*, 1984, **67**, 4261; W. Oppolzer, *Tetrahedron*, 1987, **43**, 1969.

207 S.D. Kahn and W.J. Hehre, *Tetrahedron Lett.*, 1986, **27**, 6041; C. Maignan and R.A. Raphael, *Tetrahedron Lett.*, 1983, **39**, 3245; T. Koizumi, Y. Arai and H. Takayama, *Tetrahedron Lett.*, 1987, **28**, 3689.

208 G.H. Posner and D.G. Wettlaufer, *J. Am. Chem. Soc.*, 1986, **108**, 7373; *Tetrahedron Lett.*, 1986, **27**, 667.

209 W. Choy, L.A. Reed and S. Masamune, *J. Org. Chem.*, 1983, **48**, 1137; S. Masamune, L.A. Reed, J.T. Davis and W. Choy, *J. Org. Chem.*, 1983, **48**, 4441.

210 cf. S. Masamune, W. Choy, J.S. Petersen and L.R. Sita. *Angew. Chem. internat. edn.*, 1985, **24**, 1.

211 B.M. Trost, D. O'Krongly and J.L. Belletire, *J. Am. Chem. Soc.*, 1980, **102**, 7595; R.C. Gupta, P.A. Harland and R.J. Stoodley, *J. Chem. Soc. Chem. Commun.*, 1983, 754.

212 R.C. Gupta, A.M.Z. Slawin, R.J. Stoodley and D.J. Williams, *J. Chem. Soc. Chem. Commun.*, 1986, 668.

213 R.C. Gupta, P.A. Harland and R.J. Stoodley, *Tetrahedron*, 1984, **40**, 4657.

214 J.L. Charlton, *Tetrahedron Lett.*, 1985, **26**, 3413.

215 S-I. Hashimoto, N. Komeshima and K. Koga, *J. Chem. Soc. Chem. Commun.*, 1979, 437; see also H. Takemura, N. Komeshima, I. Takahashi, S-I. Hashimoto, N. Ikota, K. Tomioka and K. Koga, *Tetrahedron Lett.*, 1987, **28**, 5687.

216 G. Bir and D. Kaufmann, *Tetrahedron Lett.*, 1987, **28**, 777.

217 C. Chapius and J. Jurczak, *Helv. Chim. Acta*, 1987, **70**, 436; see also D. Seebach, A.K. Beck, R. Imwinkelried, S. Roggo and A. Wonnacott, *Helv. Chim. Acta*, 1987, **70**, 954.

218 M. Bednarski and S. Danishefsky, *J. Am. Chem. Soc.*, 1986, **108**, 7060.

219 W.R. Roush, H.R. Gillis and A.I. Ko, *J. Am. Chem. Soc.*, 1982, **104**, 2269.

220 D.A. Evans, K.T. Chapman and J. Bisaha, *Tetrahedron Lett.*, 1984, **25**, 4071; D.A. Evans, K.T. Chapman and J. Bisaha, *J. Am. Chem. Soc.*, 1988, **110**, 1238.

221 J-L. Ripoll, A. Rouessac and F. Rouessac, *Tetrahedron.*, 1978, **34**, 19; M-C. Lasne and J-L. Ripoll, *Synthesis*, 1985, 121; M. Karpf, *Angew. Chem. internat. edn.*, 1986, **25**, 414.

222 See M-C. Lasne and J-L. Ripoll, *Synthesis*, 1985, 121.

223 M.F. Ansell, M.P.L. Caton and P.C. North, *Tetrahedron Lett.*, 1981, 1727; H. König, F. Graf and V. Weberndörfer, *Annalen*, 1981, 668; D. Liotta, M. Saindane and W. Ott, *Tetrahedron Lett.*, 1983, **24**, 2473.

224 T.V. Rajanbabu, D.F. Eaton and T. Fukunaga, *J. Org. Chem.*, 1983, **48**, 652; W.H. Bunnelli and W.R. Shangraw, *Tetrahedron*, 1987, **43**, 2005.

225 M. Karpf, *Angew. Chem. internat. edn.*, 1986, **25**, 414; A. Ichihara, *Synthesis*, 1987, 207.

226 S. Knapp, R.M. Ornaff and K.E. Rodrigues, *J. Am. Chem. Soc.*, 1983, **105**, 5494.

227 cf. W.H. Bunnelli and W.R. Shangraw, *Tetrahedron*, 1987, **43**, 2005.

228 P. Nanjappan and A.W. Czarnik, *J. Org. Chem.*, 1986, **51**, 2851.

229 P. Magnus, P.M. Cairns and J. Moursounides, *J. Am. Chem. Soc.*, 1987, **109**, 2469.

# 2 INTERMOLECULAR DIELS-ALDER REACTIONS

The Diels-Alder reaction owes its importance in synthesis principally to its versatility and its high regio- and stereo-selectivity. Both inter- and intramolecular reactions are possible, leading to the formation of a wide range of six-membered carbocyclic and heterocyclic structures. Aromatic compounds can be obtained by dehydrogenation of the initial adducts or by elimination of suitably placed functional groups earlier incorporated in the diene or dienophile, and open-chain compounds containing a number of chiral centres can be synthesised in a stereocontrolled way by cleavage of the Diels-Alder adduct, perhaps after further stereocontrolled transformations.

## Stereoselectivity

Because of this versatility, the Diels-Alder reaction forms an important step in the synthesis of numerous natural products. Both inter- and intramolecular reactions have been employed. Intermolecular reactions usually form an early step in the syntheses and the stereocentres set up in the cycloaddition are used to control the introduction of additional chiral centres at a later stage of the syntheses.

A well-known early example is found in the Woodward synthesis of (±)–reserpine (3)[1] which illustrates several of these points. The first step of the synthesis involved Diels-Alder reaction of vinyl acrylic acid with benzoquinone to give specifically the adduct (1) according to the Alder *endo* rule. The cyclohexene ring in this product was destined to become ring E of reserpine and already had three of the chiral centres of that ring in place with the correct relative orientation. Stereoselective reaction at the double bond of the cyclohexene, controlled by the stereocentres already present, was exploited to generate two other chiral centres, and cleavage of the ene-dione ring then led to the key intermediate (2) with five contiguous chiral centres, which was finally taken on to resperine.

Twenty years later the same principle was applied in a stereospecific total synthesis of (±)-gibberellic acid (9).[2] Two different Diels-Alder reactions were used in this synthesis, one intermolecular and the other intramolecular. The intermolecular reaction was early in the synthesis. Reaction of the quinone (4) with *trans*-2,4-pentadienol gave the single crystalline adduct (5) in 91 per cent yield by way of the *endo* transition state.

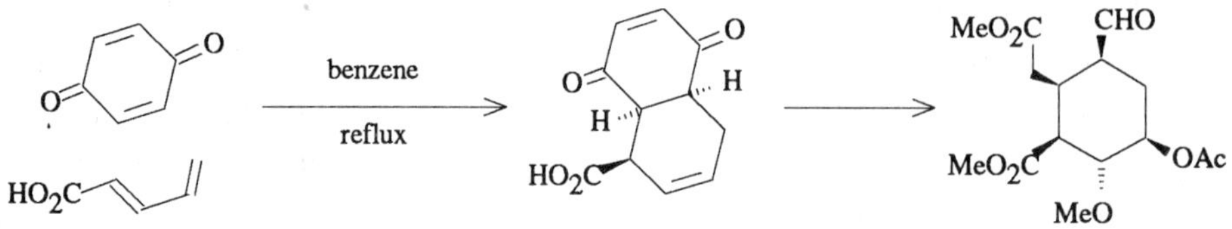

O
O
$HO_2C$
benzene
reflux
O
O
H
H
$HO_2C$
$MeO_2C$
CHO
$MeO_2C$
OAc
MeO
(1)
(2)

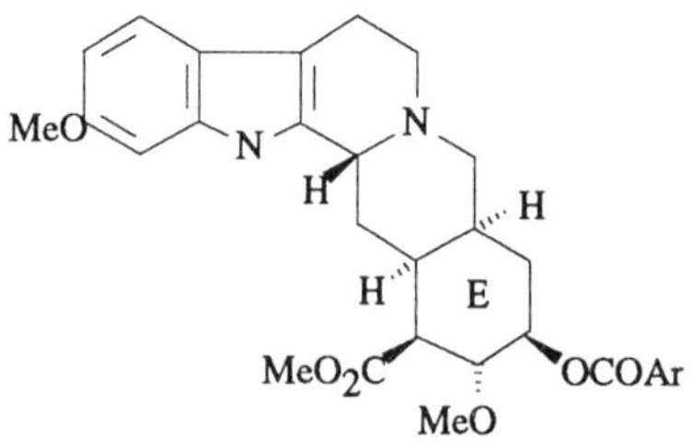

MeO
N
N
H
H
H
E
$MeO_2C$
OCOAr
MeO
(3)

*Scheme 1*

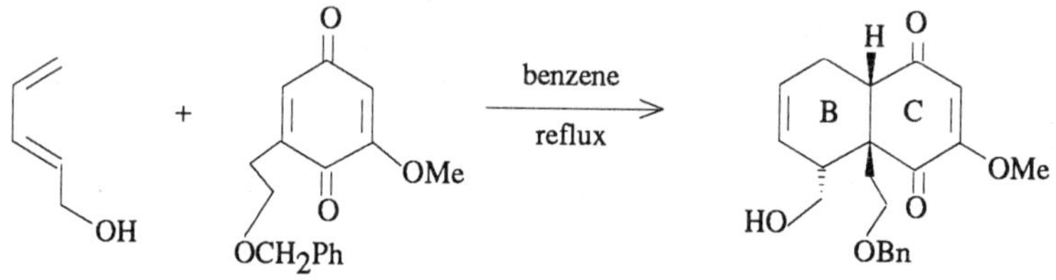

OH
+
O
OMe
O
$OCH_2Ph$
benzene
reflux
H
O
B
C
OMe
HO
O
OBn
(4)
(5)

*Scheme 2*

The *cis* ring fusion of rings B and C of gibberellic acid is set up in this reaction and in fact the stereochemistry of all the chiral centres in the final product follows from that in (5). The newly-formed six-membered ring in (5) eventually becomes the five-membered ring B of gibberellic acid by cleavage of the double bond and aldol recyclisation. An intramolecular Diels-Alder reaction was employed at a later stage of the synthesis. After extensive manipulation (5) was converted into the tricyclic intermediate (6) and thence into the ester (7). When (7) was heated in benzene at 160°C the pure crystalline lactone (8) was obtained in 55 per cent yield after crystallisation, formed by intramolecular addition to the α-face of the diene by way of an *endo* transition state; no other stereoisomer was detected (intramolecular reactions do not always favour the *endo* transition state so markedly). Two additional chiral centres of the gibberellic acid are formed in this step and the chlorine serves to generate another double bond in ring A; several more steps gave (±)-gibberellic acid (9).

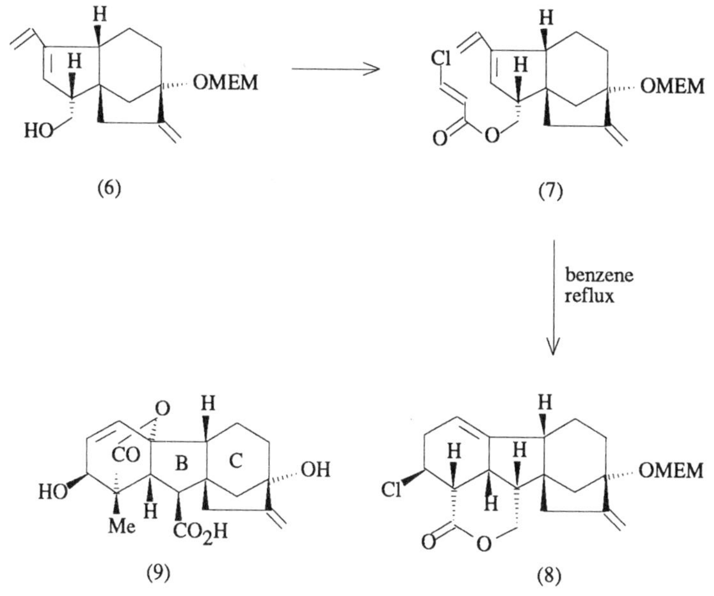

(6) (7) benzene reflux (9) (8)

*Scheme 3*

Optically active products can be obtained in Diels-Alder reactions by use of optically active components which eventually become an integral part of the target structure, or by incorporation of a chiral "auxiliary group" in the diene or dienophile.[3]

A good example of the first procedure is seen in the synthesis of cytochalasin B (14).[4] In this synthesis five of the chiral centres in the A-B bicyclic ring system are introduced with the correct relative and absolute configuration by an intermolecular Diels-Alder reaction between the optically active dienophile (10) and the diene (11); the chiral tertiary carbon and secondary hydroxyl group in (11) were derived respectively from (+)-citronellol and (+)-malic acid. Reaction took place only at the terminal diene system of (11) to give mainly (12) with highly selective *endo* addition to the less-hindered face of the double bond of the dienophile. Further manipulation led stereoselectively to (13) and thence after several steps to cytochalasin (14).

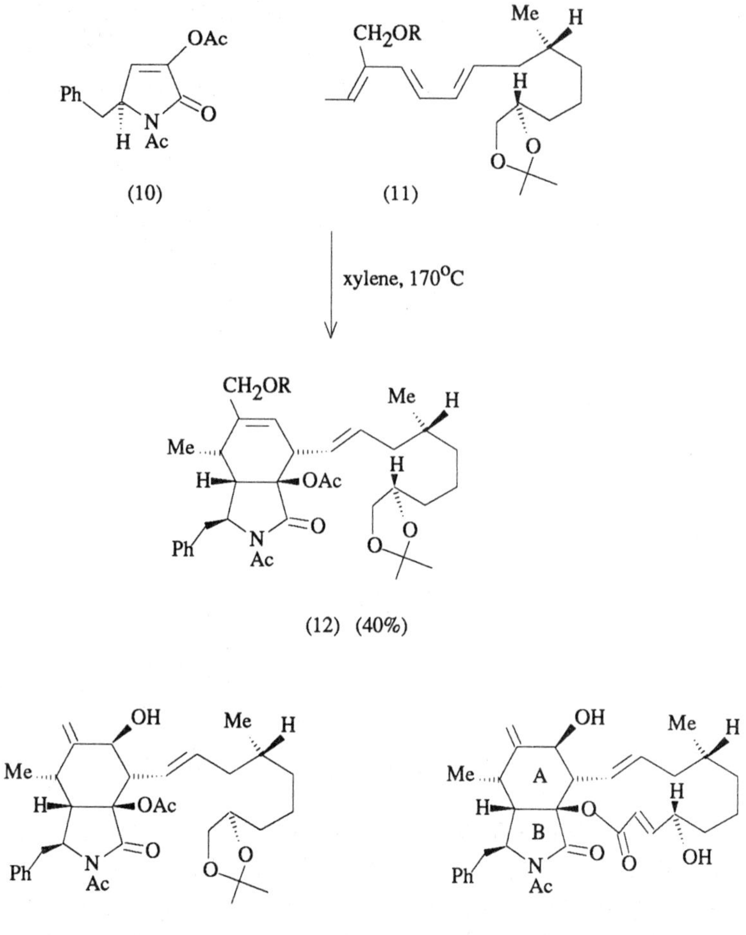

(10) (11)

xylene, 170°C

(12) (40%)

(13) (14)

*Scheme 4*

A diasteroselective Diels-Alder reaction using a chiral auxiliary group forms the key step in a synthesis of (–)-β-santalene (18).[5] Catalysed reaction of the optically active allenic ester (15) with cyclopentadiene gave the adduct (16), almost entirely as the *endo* isomer, by selective attack on the front less-hindered face of the allene in (15). Several further steps with the purified *endo* compound gave (17) and thence optically pure (–)-β-santalene.

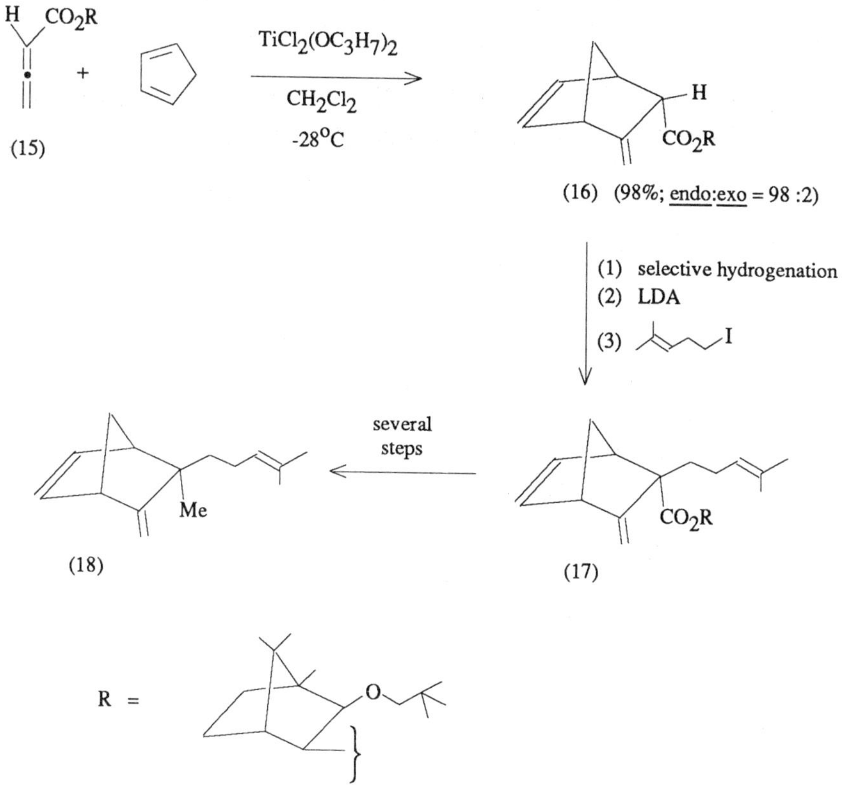

H $CO_2R$

(15)

$TiCl_2(OC_3H_7)_2$

$CH_2Cl_2$

-28°C

H

$CO_2R$

(16) (98%; endo:exo = 98 :2)

(1) selective hydrogenation

(2) LDA

(3) I

several steps

Me

(18)

$CO_2R$

(17)

R = O

*Scheme 5*

While the formation of a ring structure may be the principal objective of the cycloaddition, in many cases the initial cyclo-adduct is modified at a later stage in the synthesis by ring-cleavage or rearrangement. For example, the bicyclo[2,2,2]octanes obtained by Diels-Alder reaction of 1,3-cyclohexadienes with dienophiles may be transformed by subsequent molecular reorganisation involving a fragmentation to give cyclohexanes[6] or by a pinacolinic rearrangement to give bicyclo[3,2,1]octanes.[7] Since the structures of a number of types of natural products incorporate a bicyclo[3,2,1]octane ring system a general method for its construction by rearrangement of bicyclo-[2,2,2]octanes obtained in a Diels-Alder reaction would be useful. One such sequence is admirably illustrated by the conversion of the modified adduct (19) to the bicyclo[3,2,1]octane (20) which formed a key step in an elegant synthesis of stachenone.[8] In this synthesis α-chloroacrylonitrile serves as a ketene equivalent in the Diels-Alder reaction.

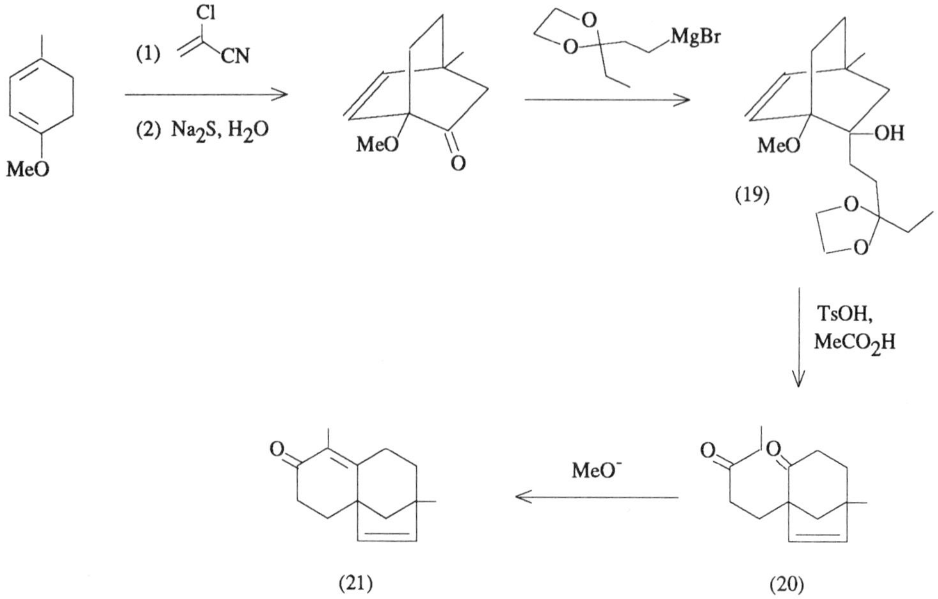

*Scheme 6*

## Hetero-substituted Dienes and Dienophiles

Hetero-atom substituents on the diene or dienophile lead, of course to hetero-substituted adducts, but they have additional value in that they can sometimes be used to introduce other functional groups in the adduct.[9] One way in which a 1-phenylthio substituent on the diene can be employed is illustrated in the synthesis of the hexahydronaphthalene (27), a key intermediate in an enantioselective synthesis of the hypocholesterolemic agent (+)-compactin.[10] The strategy here centred round the Diels-Alder reaction between the optically-active dienophile (22) and the optically-active diene (23). This led principally to the adduct (24) by *endo* addition to the α-face of (22). The phenylthio substituent was now used as a prop for the introduction of the methyl substituent at C-2 of (26). Oxidation with *meta*-chloroperbenzoic acid and subsequent treatment with trimethyl phosphite led stereospecifically to the allylic alcohol (25) via a sulphoxide-sulphenate rearrangement. Reaction of the inverted acetate (Mitsunobu) with lithium dimethylcuprate gave exclusively (26) which was subsequently converted by a fragmentation reaction into (27) with four contiguous chiral centres in the hydronaphthalene unit and thence into (+)-compactin. The adduct (24) contains potentially all four chiral centres of (27). In addition to the centres at C-1 and C-10, the bridge oxygen eventually becomes the hydroxyl substituent at C-9 and the phenylsulphide is the source of the chiral centre at C-2. In addition the carboxylate anion is well placed to assist in the fragmentation of the C6-O bond with simultaneous introduction of the C5-C6 double bond.

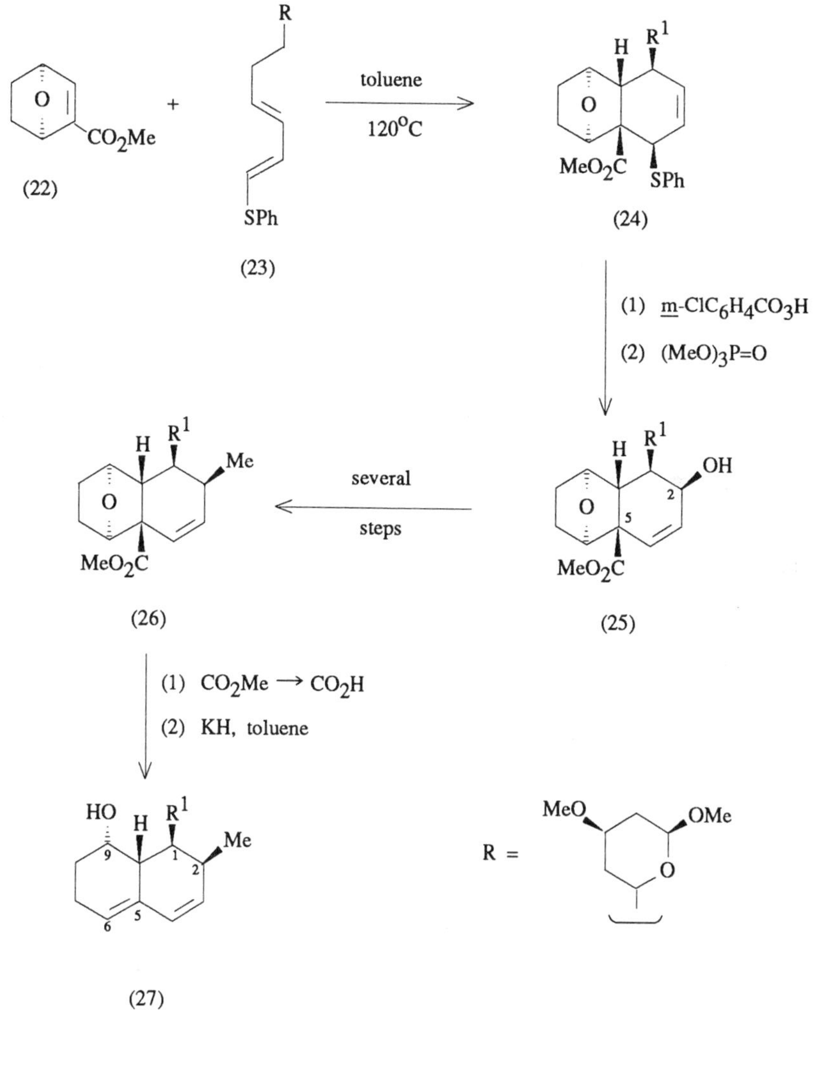

O
$CO_2Me$
(22)
+
R
SPh
(23)
toluene
120°C
H $R^1$
O
$MeO_2C$
SPh
(24)
(1) m-$ClC_6H_4CO_3H$
(2) $(MeO)_3P{=}O$
H $R^1$
OH
O
2
5
$MeO_2C$
(25)
several
steps
H $R^1$
Me
O
$MeO_2C$
(26)
(1) $CO_2Me \rightarrow CO_2H$
(2) KH, toluene
HO H $R^1$
Me
9 1 2
6 5
(27)
R =
MeO
OMe
O

*Scheme 7*

1-Acylamino-1,3-dienes also react with a broad range of dienophiles, providing access to a variety of amino functionalised systems in excellent yield. Their use in the stereoselective synthesis of alkaloids is illustrated in the synthesis[11] of (±)-pumiliotoxin C, one of the poison arrow alkaloids, shown by X-ray analysis of the crystalline hydrobromide to have the unusual *cis*-decahydroquinoline structure (32).

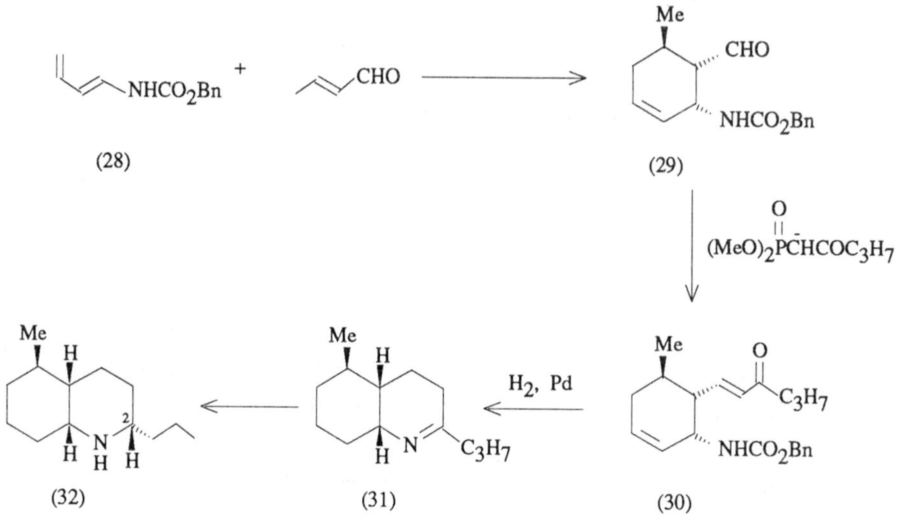

*Scheme 8*

The plan here was to take advantage of the stereoselectivity of the Diels-Alder reaction to establish the three chiral centres of the carbocyclic ring with the correct relative orientation and to use these centres to control the formation of the remaining chiral centre at C-2 of pumiliotoxin. The initial Diels-Alder reaction gave almost entirely the *endo* product (29). The final hydrogenation-hydrogenolysis brought about several transformations culminating in the selective addition of hydrogen to the convex (β) face of the intermediate *cis*-octahydroquinoline (31) to give pumiliotoxin. A similar sequence was employed in the synthesis of several other *cis*-decahydroquinolines.[12]

Another route to *cis*-hydroquinolines proceeds from N-carbomethoxy-1,2-dihydropyridine by a Diels-Alder/Cope rearrangement sequence, as described on p.33, and has been used in syntheses of (±)-resperine and (±)-catharanthine.

1-t-Butoxycarbonylamino-1,3-butadiene has been employed in an interesting synthesis of isogabaculine (34), an isomer of the γ-aminobutyrate enzyme inhibitor gabaculine (33).[13]

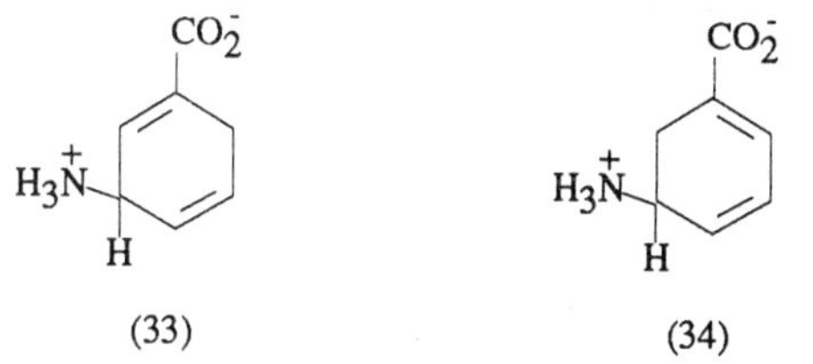

$CO_2^-$ $H_3\overset{+}{N}$ H (33)

$CO_2^-$ $H_3\overset{+}{N}$ H (34)

*Scheme 9*

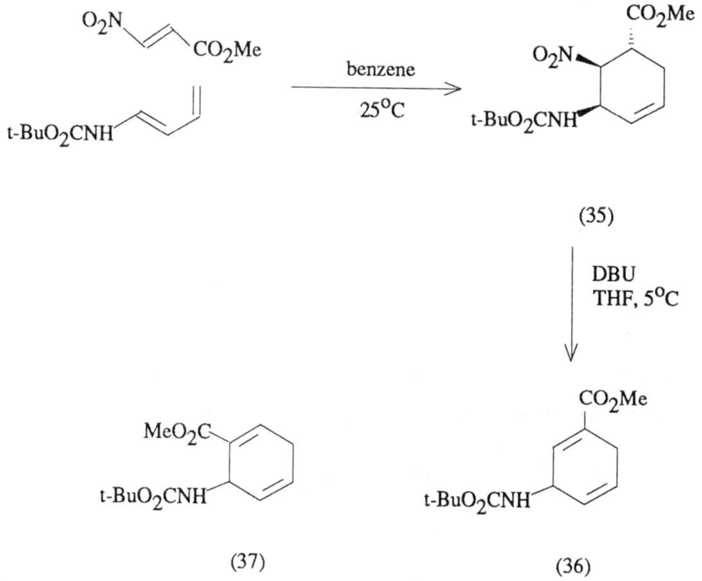

$O_2N$ $CO_2Me$ t-$BuO_2CNH$

benzene, 25°C

$CO_2Me$ $O_2N$ t-$BuO_2CNH$ (35)

DBU, THF, 5°C

$CO_2Me$ t-$BuO_2CNH$ (36)

$MeO_2C$ t-$BuO_2CNH$ (37)

*Scheme 10*

A feature of this synthesis is the use of methyl β-nitroacrylate as dienophile. It acts as an operational equivalent of methyl propiolate in the Diels-Alder reaction, leading to a dihydrobenzene derivative, but with a substitution pattern different from that obtained with methyl propiolate itself because the regiochemistry of addition is now controlled by the nitro group. Thus, reaction of the 1-acylaminobutadiene with the nitro-acrylate in benzene at room temperature led almost exclusively to the crystalline adduct (35). Elimination of nitrous acid by treatment with diazabicycloundecene then gave the 1,4-dihydrobenzene (36) which was converted into isogabaculine. Reaction of the diene with methyl propiolate would have given the "ortho" adduct (37).

### Syntheses with Alkyloxy- and Silyloxy-dienes

Another useful group of hetero-substituted dienes are the alkoxy- and silyloxy-butadienes. 2-Alkoxy- and 2-silyloxybutadienes on reaction with dienophiles form cyclic enol ethers which give cyclohexanones on hydrolysis. Particularly valuable are 1-alkoxy-3-trimethylsilyloxydienes such as (39). These are more easily prepared than 1,3-dialkoxybutadienes and have been widely used in synthesis.[14] They are reactive dienes, since the activating effects of the two substituents reinforce each other, and with ordinary carbon-carbon dienophiles bearing electron-attracting substituents they can be used to make cyclohexenones, 4-acylcyclohexenones, 4,4-disubstituted cyclohexadienones or benzene derivatives, depending on the precise mode of operation and the nature of the diene and dienophile (cf. p.28). They have also been used to make anthraquinones and related compounds by reaction with quinones (p.124), and with the carbonyl group of aldehydes they give derivatives of dihydro-γ-pyrone (p.10).

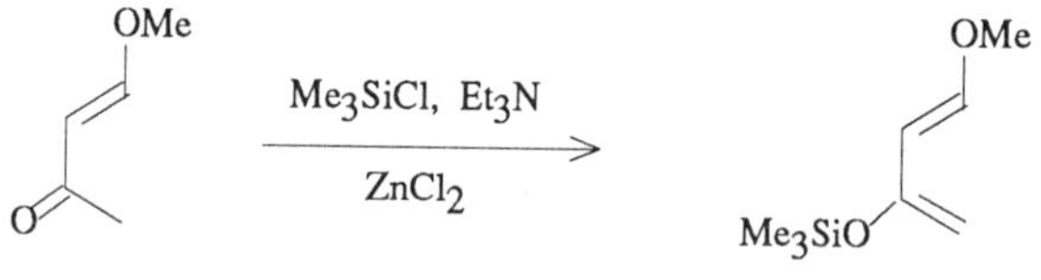

OMe

O

(38)

$Me_3SiCl$, $Et_3N$

$ZnCl_2$

OMe

$Me_3SiO$

(39)

*Scheme 11*

1-Methoxy-3-trimethylsilyloxybutadiene itself (39) is obtained by enol silylation of the methoxy-enone (38)[15] and has been employed in the synthesis of a variety of natural products.[16] For example, reaction with the phenylsulphinyl dienophile (40) formed the key step in an elegant synthesis of prephenic acid (45)[17] (Scheme 12).

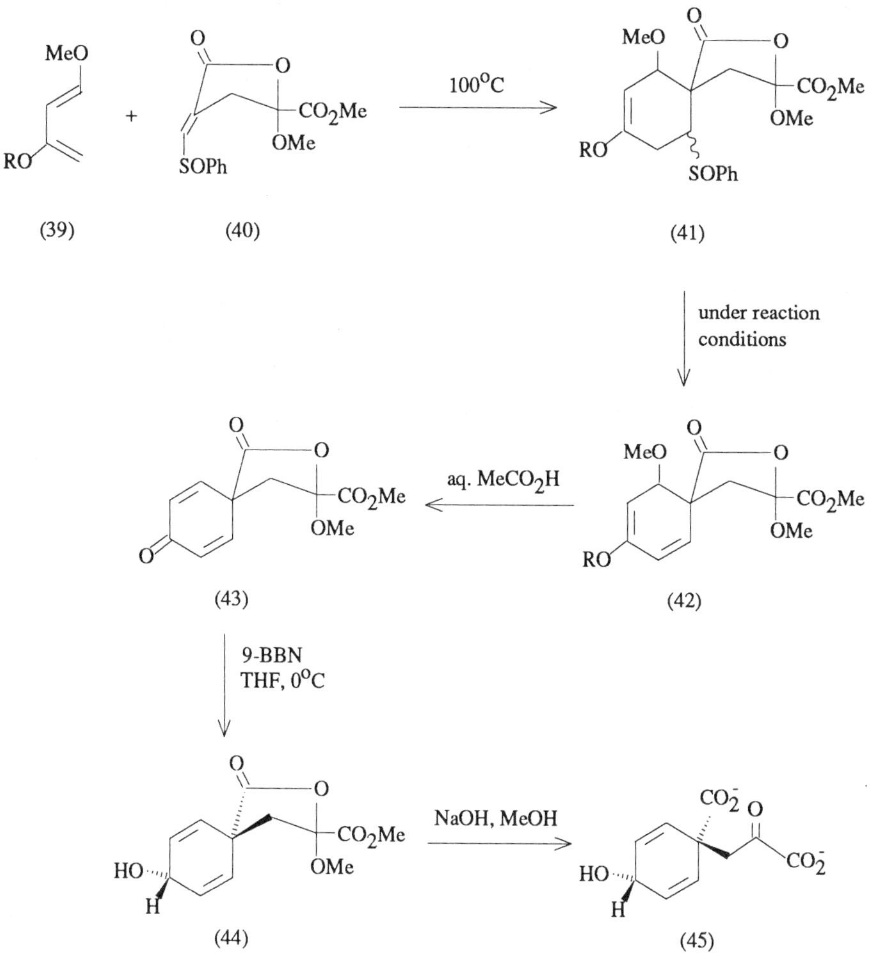

*Scheme 12*

The phenylsulphinyl group here helps to activate the olefinic bond of the dienophile but the regio-chemistry of the cycloaddition is controlled by the carbonyl group of the lactone. The main purpose of the phenylsulphinyl group is to provide a source of one of the double bonds in the dienone (43) by sulphoxide elimination. A closely similar sequence was used to synthesise optically pure pretyrosine (47) from the dienophile (46), itself derived from (–)-glutamic acid.[18] With dienophile (46) and the siloxydiene (48) the aromatic L-DOPA (49) was obtained.[19] Here a benzene derivative, and not a dihydrobenzene, is formed by elimination of carbon dioxide from the initial Diels-Alder adduct.

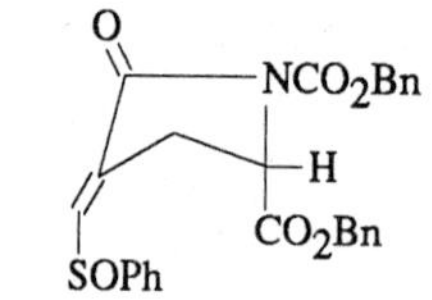

O
NCO2Bn
H
CO2Bn
SOPh

(46)

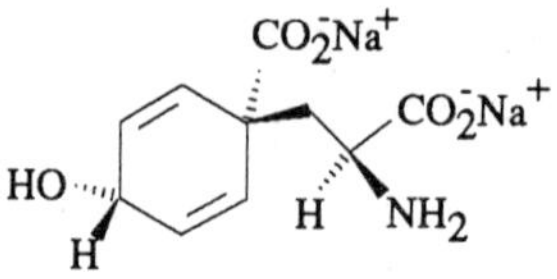

$CO_2^-Na^+$
$CO_2^-Na^+$
HO
H
H
$NH_2$

(47)

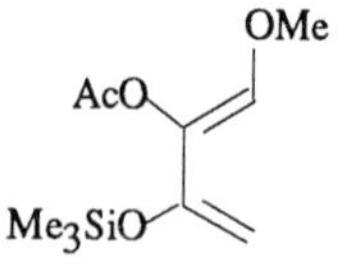

OMe
AcO
$Me_3SiO$

(48)

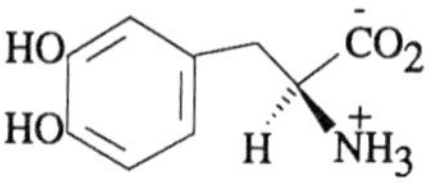

HO
HO
$CO_2^-$
H
$\overset{+}{N}H_3$

(49)

*Scheme 13*

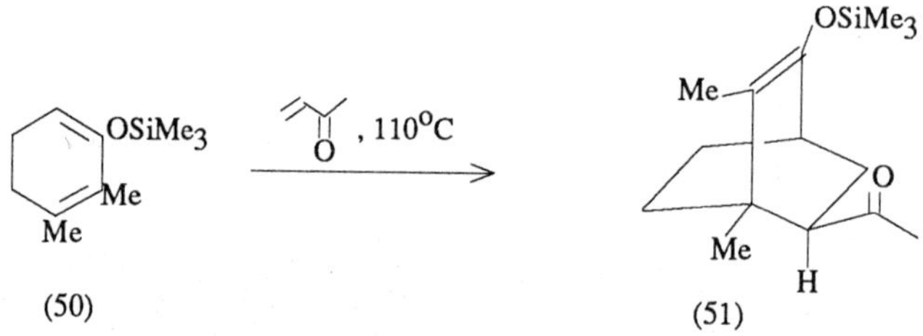

OSiMe3
Me
Me

(50)

, 110°C
O

$OSiMe_3$
Me
O
Me
H

(51)

Me
Me
Me
H

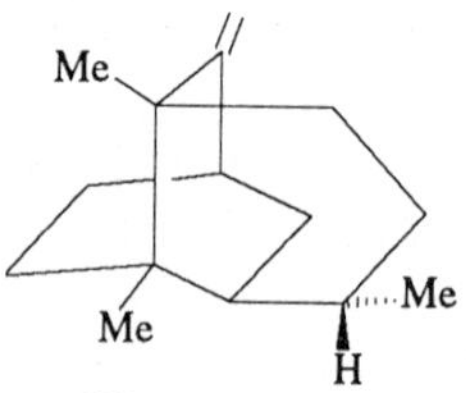

(52)

*Scheme 14*

The siloxy-dienes used in these reactions need not be open-chain compounds. Thus, the cyclohexadiene derivative (50), which is obtained by enol silylation of 2,3-dimethylcyclohexenone, reacts with methyl vinyl ketone to give the bicyclo[2,2,2]octane derivative (51), a key intermediate in a synthesis of (±)-seychellene (52).[20] Reaction again goes *via* the *endo* transition state and conveniently provides an adduct diketone monoprotected at one of the carbonyl groups, allowing selective reaction at the other with vinylmagnesium bromide.

Equally valuable synthetically are Lewis acid-catalysed reactions of siloxydienes with the carbonyl group of aldehydes, to form 2,3-dihydro-γ-pyrones.

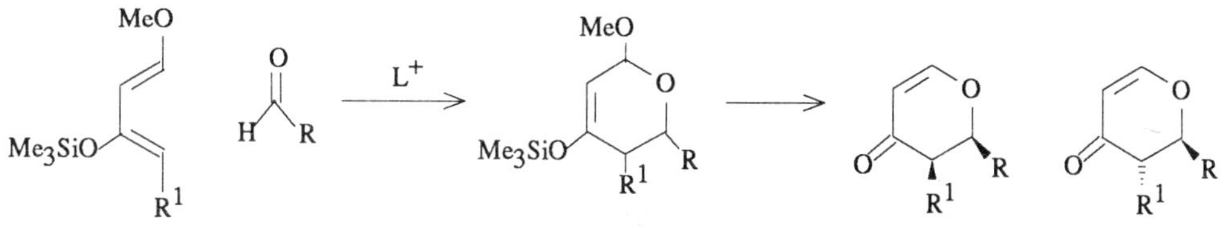

(53) (54) (55) (56)

*Scheme 15*

A valuable feature of these reactions is that with dienes of the type (53), carrying a substituent at C-4, conditions can be arranged to give either the *cis*- or *trans*-5,6-disubstituted pyrone (55) or (56) (see p.10). The double bond and the carbonyl group in the dihydropyrones can then be used to introduce three further chiral centres, providing a route for the stereo-controlled synthesis of a range of substituted pyrans, including hexoses of the mannose, glucose, talose and galactose series. Alternatively the cyclic compound can be cleaved after the stereochemical array has been set up, to give an open-chain compound, an apt example of the well-known procedure where a ring is used as a matrix to control stereochemistry which is then transferred to an acyclic compound. Means are also available for relating the stereochemistry at C-6 in (55) and (56) with that at a chiral α-carbon in R, and some highly enantioselective additions to aldehydes have been effected (see p.12). All of these features of the catalysed reaction of aldehydes with activated dienes have been incorporated in the stereoselective syntheses of a number of highly oxygenated natural products, including carbon-linked disaccharides, hexoses, (±)-fucose, (±)-duanosamine, alkenyl-arabinopyranosides and polypropionates.[21] It is true that the Diels-Alder reaction generally forms only one early step in these syntheses, but it is a key step in which the stereochemical bias of the sequence is introduced, to be exploited in a series of subsequent stereoselective reactions.

A good application of the topological control possible in reactions of α-alkoxyaldehydes with dienes in the presence of magnesium bromide or titanium tetrachloride catalysts is found in the synthesis of the mouse androgen dehydrobrevicomin (61).[22] Here, reaction of the diene (57) with α-benzyloxybutanal gave specifically the adduct (58) by Cram-type addition to the carbonyl group. Reduction of the ketone with di-isobutylaluminium hydride gave the glycal (59) which, with mercuric acetate, was converted into (60) in an intramolecular oxymercuration. Reverse oxymercuration in a different mode then gave (61).

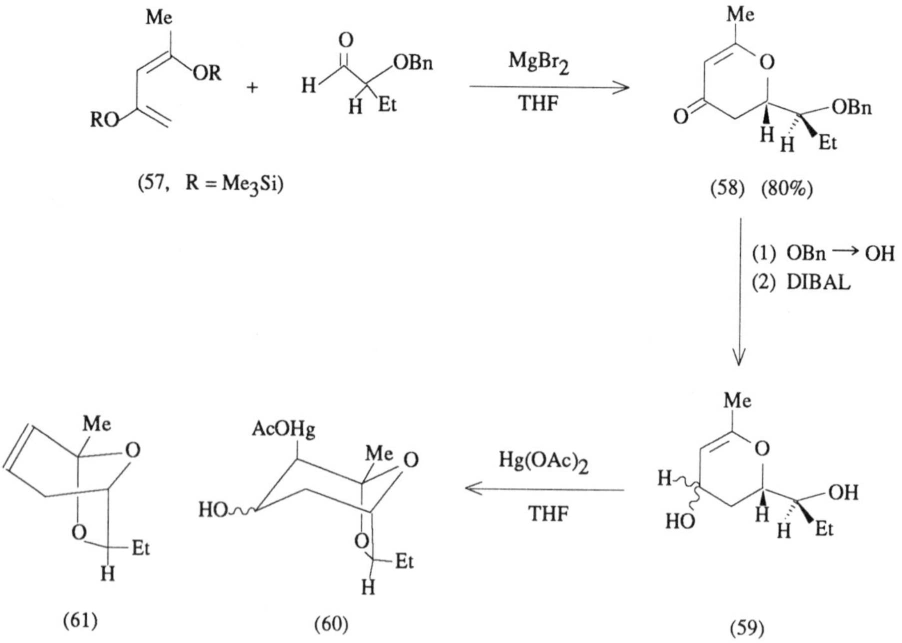

(57, R = $Me_3Si$) (58) (80%) (59) (60) (61)

*Scheme 16*

Systems such as (62), derivable by cyclo-condensation of aldehydes with alkoxydienes in a stereocontrolled and flexible way, can serve as convenient substrates for the synthesis of open-chain compounds (as 63) by cleavage of the pyran ring.

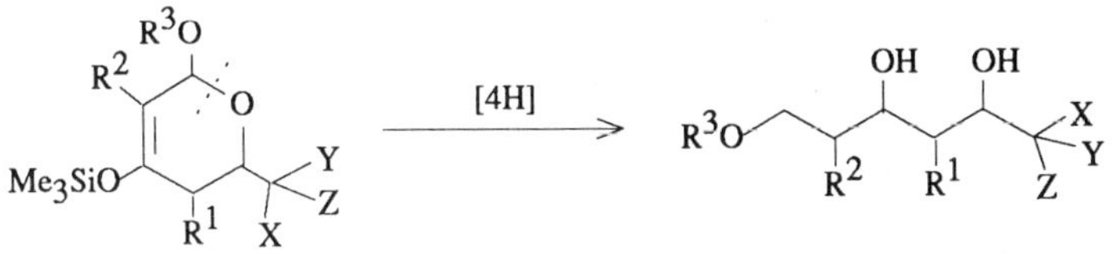

$R^3O$ $R^2$ O $Me_3SiO$ $R^1$ X Y Z — [4H] → OH OH $R^3O$ $R^2$ $R^1$ X Y Z

(62) (63)

*Scheme 17*

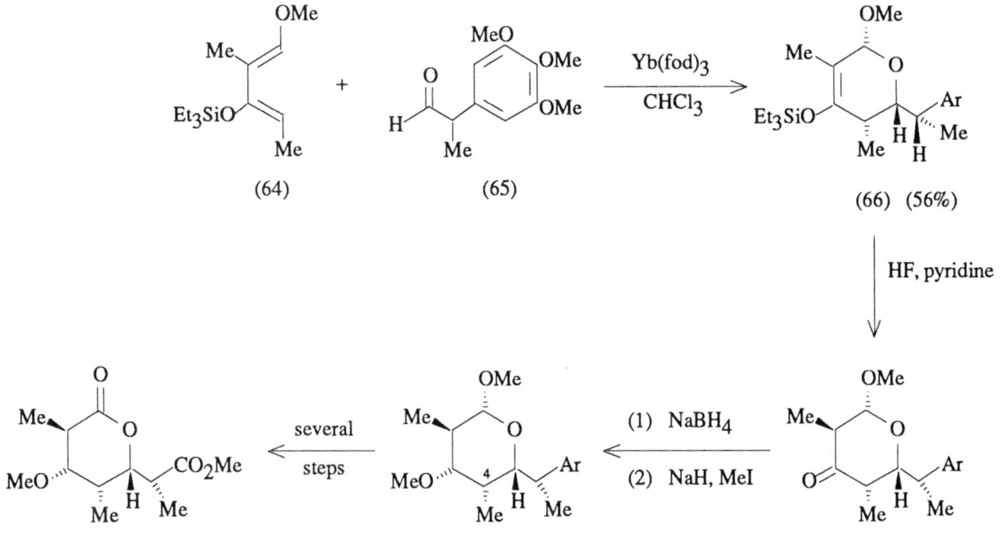

OMe Me $Et_3SiO$ Me (64) + MeO OMe OMe O H Me (65) — $Yb(fod)_3$, $CHCl_3$ → OMe Me O $Et_3SiO$ Ar Me H H Me (66) (56%)

HF, pyridine

OMe Me O O Ar Me H Me (67) — (1) $NaBH_4$ (2) NaH, MeI → OMe Me O MeO 4 Ar Me H Me (68) — several steps → O Me O MeO $CO_2Me$ Me H Me (69)

*Scheme 18*

A good illustration is provided by the synthesis of the "polypropionate" monensin lactone (69) a degradation product of monensin.[23] Cram-type addition of siloxydiene (64) to aldehyde (65), catalysed by $Yb(fod)_3$, gave (66) as the only observed product, formed by way of the *endo* transition state. This key product which already contains three contiguous chiral centres of the target molecule with the correct relative orientation, was converted into (69) in a series of stereoselective steps. The derived ketone (67), on reduction with sodium borohydride gave specifically the equatorial alcohol in which only the C-4 methyl group is axial. Cleavage of the methyl glycoside linkage, simultaneous oxidation of the lactol and the aromatic ring with catalytic ruthenium dioxide and sodium metaperiodate and final esterifiction of the acid with diazomethane led to the lactone (69).

δ-Lactones have also been obtained from 1,1,3-trioxybutadienes (as 71). These are very reactive dienes and in the presence of zinc chloride or the lanthanide $Eu(fod)_3$ they react readily with the carbonyl group of aldehydes and ketones. The initial cycloadducts, with or without isolation, give δ-lactones on hydrolysis.[24] Such a sequence, using the optically active aldehyde (70) was employed in a synthesis of (–)-pestolatin (72).[24] The optically pure compound was obtained after one recrystallisation.

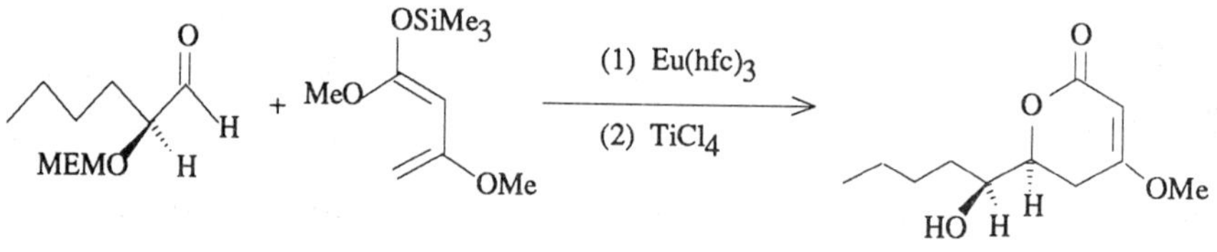

(1) $Eu(hfc)_3$
(2) $TiCl_4$

(70) (71) (72) (83% d.e.)

*Scheme 19*

Some highly diastereoselective Diels-Alder additions to carbonyl dienophiles have been achieved using the appropriate combination of an optically active diene and a chiral lanthanide catalyst. Thus, cycloaddition of the alkoxydiene (73) and benzaldehyde in the presence of the catalyst (+)-$Eu(hfc)_3$ gave very largely the adduct (74) of the L-pyranose series. This was purified by crystallisation and converted into the optically pure dihydropyran (76) which was employed in a synthesis of L-glucose.[25]

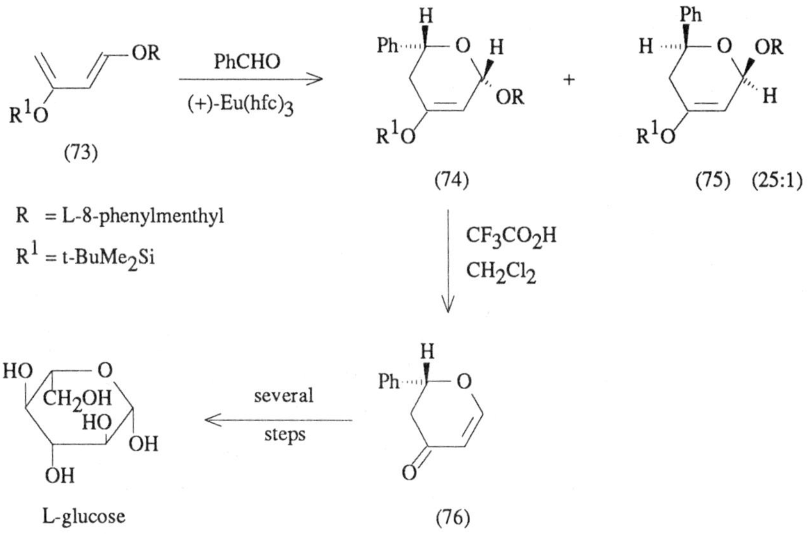

(73) PhCHO, (+)-Eu(hfc)$_3$ → (74) + (75) (25:1)

R = L-8-phenylmenthyl
$R^1$ = t-BuMe$_2$Si

(74) → $CF_3CO_2H$, $CH_2Cl_2$ → (76) → several steps → L-glucose

*Scheme 20*

Similarly, reaction of the diene (77) with furfural in the presence of Eu(hfc)$_3$ gave a product from which the main component (78) was easily obtained optically pure. In a series of further transformations this was converted into (79) and thence into optically pure (80), an intermediate in the synthesis of indanomycin.[26]

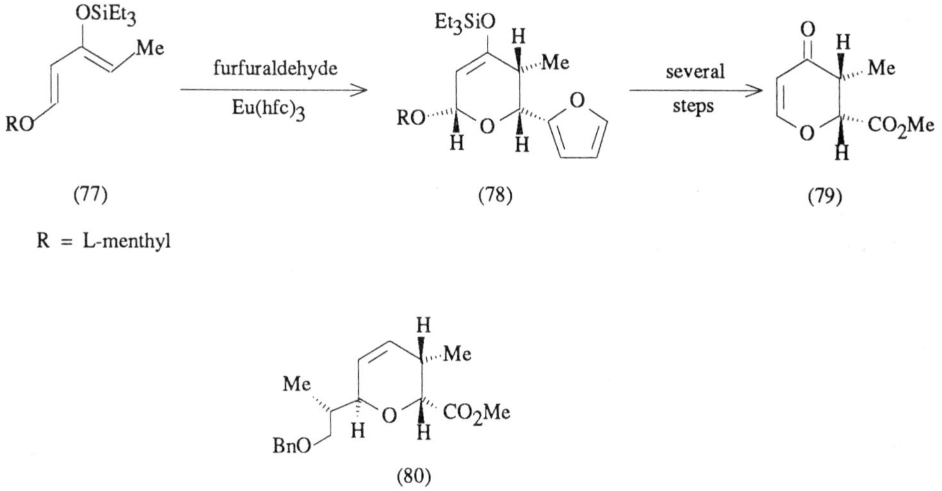

(77) furfuraldehyde, Eu(hfc)$_3$ → (78) → several steps → (79)

R = L-menthyl

(80)

*Scheme 21*

An alternative Diels-Alder approach to pyranose derivatives employing 1,4-diacetoxybutadiene as diene component is of more limited scope. Because of the lower reactivity of the diene it reacts only with activated carbonyl compounds such as glyoxylic esters.[27]

## αβ-Unsaturated Carbonyl Compounds as Dienes

3,4-Dihydro-2*H*-pyrans are obtained in Diels-Alder reactions with inverse electron demand between αβ-unsaturated carbonyl compounds as diene component and enol ethers or enamines. Reactions of this kind are well known[28] but they have not been widely employed hitherto in synthesis because of the high temperatures required.[29] With αβ-unsaturated carbonyl compounds carrying an electron-withdrawing substituent at the α-position, however, the reactions take place much more easily, sometimes at room temperature, opening the way to a potentially valuable route for the synthesis of derivatives of 3,4-dihydro-2*H*-pyrans.[30] Thus, the αβ-unsaturated aldehyde (81) and 1,2-dihydrofuran gave a mixture of the *endo* and *exo* adducts (82) in 85 per cent yield in a reaction at 25°C.[31]

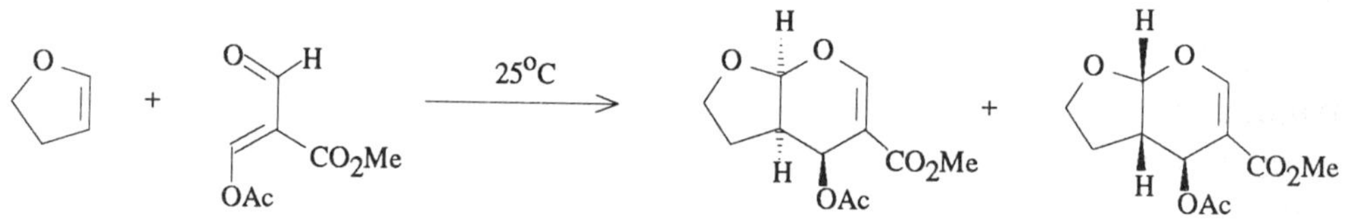

O + O H $CO_2Me$ OAc 25°C H O O H $CO_2Me$ OAc + H O O H $CO_2Me$ OAc

(81) (82) (85%; 1:2)

*Scheme 22*

Unfortunately the *endo* selectivity in these reactions is often low, and this limits their usefulness. With the β-acetoxy-α-phenylthio derivative (83), however, reaction with ethyl vinyl ether gave largely the *endo* adduct (84). Selective cleavage of the phenylthio group with Raney nickel, anti-Markownikov hydration of the double bond *via* hydroboration and removal of protecting groups then led to L-olivose (85).[32]

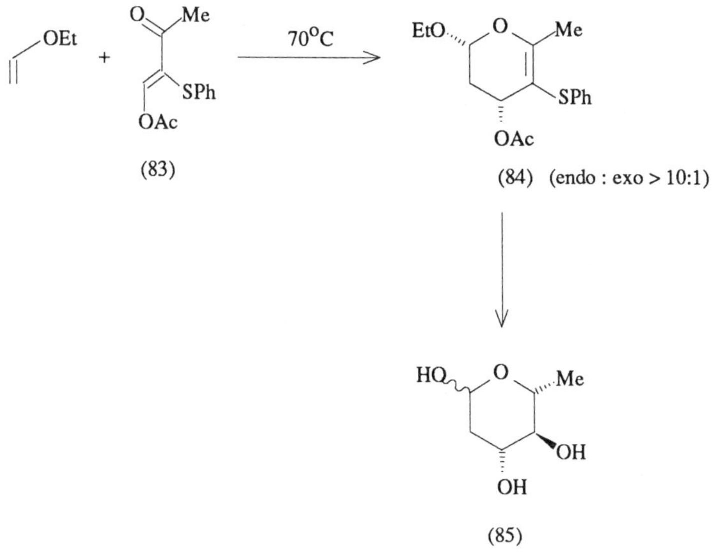

(83)

(84) (endo : exo > 10:1)

(85)

*Scheme 23*

Lewis-acid catalysis may provide a method for facilating these reactions and controlling the mode of addition. Thus, the αβ-unsaturated ketone (86), reacted readily with ethyl propenyl ether at room temperature in the presence of $Eu(fod)_3$ to give the single adduct (87) with complete stereospecificity. Several other vinyl ethers gave similar results. Reactions took place *via* the *endo* transition state by approach of the dienophile to the less hindered face of the heterodiene.[33]

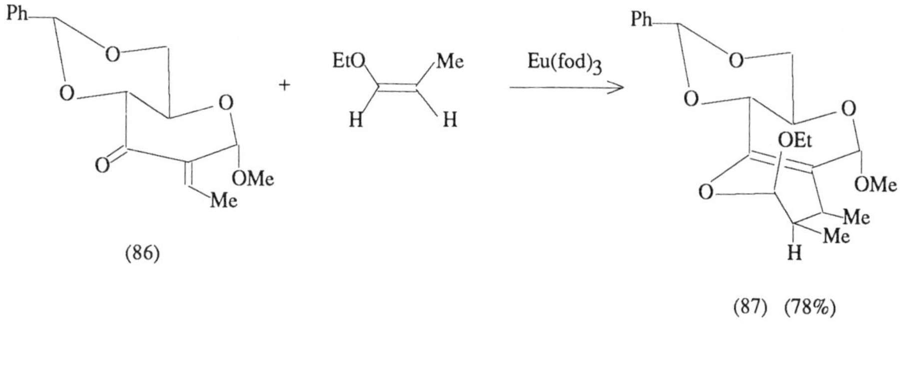

(86)

(87) (78%)

*Scheme 24*

N-Acyl derivatives of enamine-carbaldehydes form another group of excellent dienes for hetero Diels-Alder reactions with enol ethers and have been used in the synthesis of branched aminosugars[34] of the garosamine and daunosamine type.

In a different approach 2-aminoglycosides have been obtained by cycloaddition of dibenzyl azodicarboxylate to carbohydrate-derived glycals. Thus, the glycal (88) on irradiation in cyclohexane in the presence of dibenzyl azodicarboxylate gave the single cycloadduct (90) in 71 per cent yield. The diastereofacial selectivity of the cycloaddition in this and other examples appears to be controlled by the stereochemistry at C-3 of the dienophile. Treatment of the cycloadduct with a catalytic amount of toluenesulphonic acid in methanol occurred exclusively with inversion at C-1 to give (91) which on further manipulation was converted into the 2-aminoglycoside derivative (92).[35] This is an application of the well-known [4+2]cyclo-addition of azodicarboxylates to vinyl ethers.[36] The formation of dihydro-oxadiazines in these reactions is accelerated by illumination; this has been shown to be due to the photo-production of the *cis* azodicarboxylates (89) which react faster than the *trans* isomers.[37]

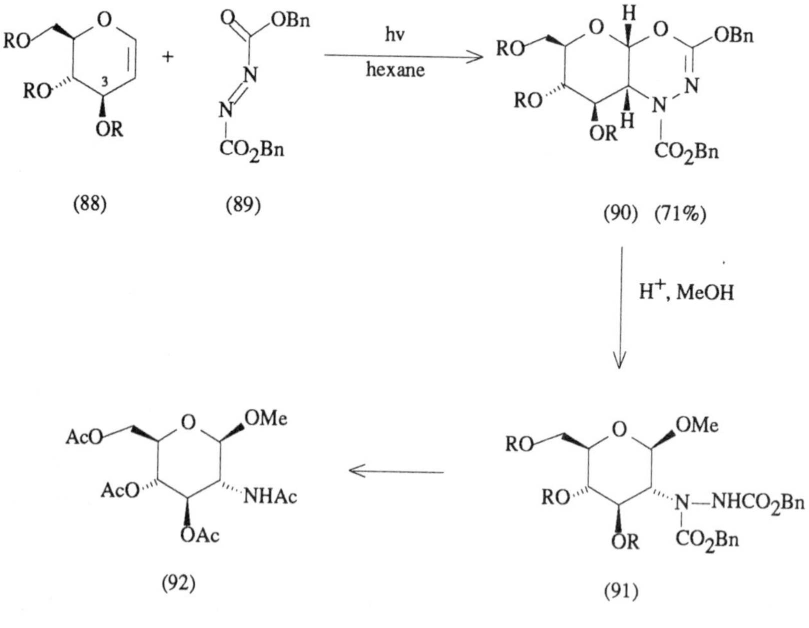

(88) (89) hv, hexane → (90) (71%)

(90) $H^+$, MeOH → (91) → (92)

*Scheme 25*

**Imines as Dienophiles**

Diels-Alder reactions of imines have been useful in approaches to the synthesis of alkaloids and other nitrogen-ring compounds. For example, the reaction between the imine (93) and 2-trimethylsilyloxycyclohexa-1,3-diene was used to make the azabicyclo-octanone (94) which, by further manipulation was converted into the triol (95) the starting point for the synthesis of several *Prosopis* alkaloids.[38]

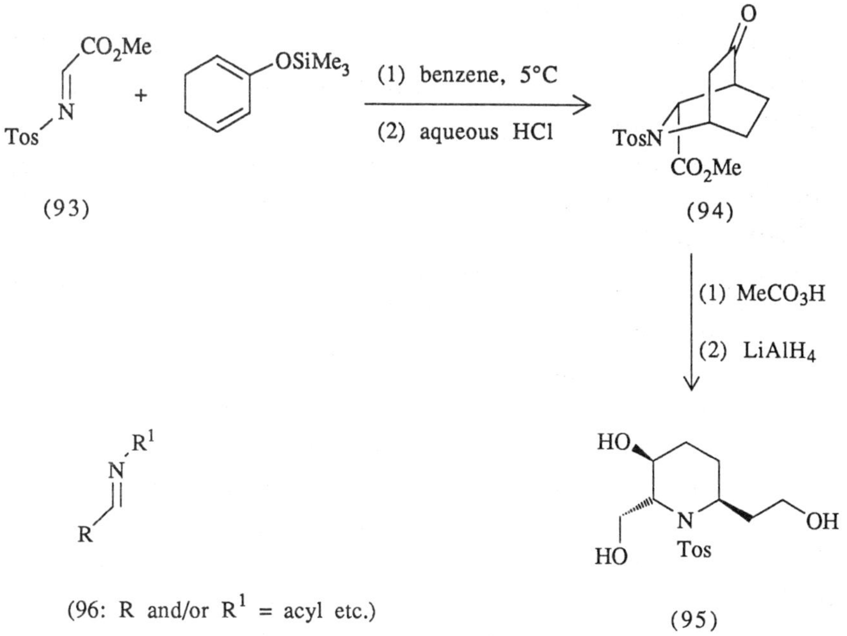

$CO_2Me$ Tos N + $OSiMe_3$ (1) benzene, 5°C (2) aqueous HCl O TosN $CO_2Me$

(93) (94)

(1) $MeCO_3H$

(2) $LiAlH_4$

$R^1$ N R HO HO N OH Tos

(96: R and/or $R^1$ = acyl etc.) (95)

*Scheme 26*

Much of the earlier work on the 'imino' Diels-Alder reaction involved activated glyoxal imines, acyl imines or doubly activated imines of the type (96)[39] and intramolecular reactions of such imines have been elegantly exploited in alkaloid synthesis (see chapter 3). More recently it has been found, that appropriately activated dienes (for example 1,3-dioxygenated dienes), in the presence of Lewis acids, react readily with non-activated imines to give 2,3-dihydro-4-pyridones.[40] Thus, 1-methoxy-3-trimethylsilyloxybutadiene and benzalaniline in the presence of zinc chloride catalyst gave the pyridone (97) in 62 per cent yield, and in a more complex example, reaction of the diene (99), available from Hagemann's ester, with the optically active imine (98), derived from L-tryptophan, took place without a catalyst to give specifically the adduct (100) by attack of the diene *anti* to the carbomethoxy functional group. In this way a straightforward entry to highly functionalised optically pure yohimbine progenitors is available.[41] A related sequence was used to make[42] (±)-ipalbidine.

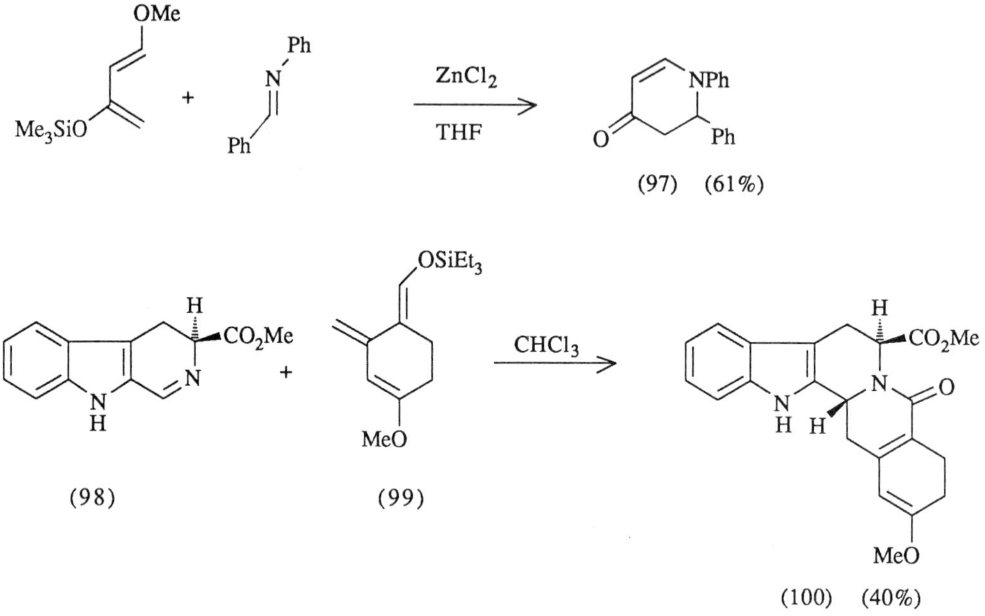

*Scheme 27*

**Nitroso Compounds as Dienophiles; Synthesis of 4-Amino-alcohols**

Another useful group of heterodienophiles are nitroso compounds. They react with 1,3-dienes to form derivatives of 1,2-oxazine which, on cleavage of the nitrogen-oxygen bond afford 4-amino-alcohols. Reactions of this kind have been used to prepare a series of polyhydroxy-piperidines and polyhydroxyamino-cyclohexanes. Both acylnitroso compounds and α-chloronitroso compounds have been used. Thus, in a synthesis of the tetra-acetylated aminoallose (105), reaction between the dimethylacetal of hexa-2,4-dienal and the reactive acylnitroso compound (101), generated *in situ* from the corresponding hydroxamic acid and tetrabutylammonium periodate, led specifically to the adduct (102). The exceptional regioselectivity here is ascribed to steric factors. *Cis*-hydroxylation of the double bond gave (103) stereospecifically, which on cleavage of the N–O bond and acetylation formed (104). Hydrolysis of the acetal group with aqueous formic acid followed by spontaneous cyclisation then gave (105).[43]

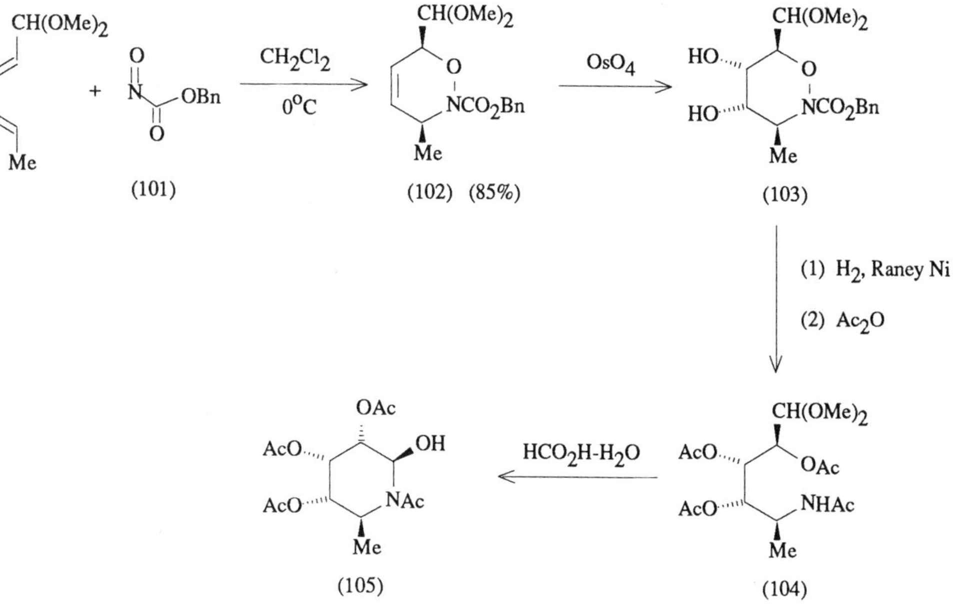

*Scheme 28*

Polyhydroxyaminopiperidines can also be obtained from the adducts formed from acylnitroso compounds and the dihydropyridine derivative (106). Thus, the adducts (107) and (108), obtained by reaction of (106) with the benzyloxycarbonyl nitroso compound (101) were used to make the peracetyl diamino-dideoxylyxopyranose derivatives (109) and (110).[44] (Scheme 29)

In an open-chain example, cleavage of both the double bond and the N-O bond of the Diels-Alder adduct (111) gave a product which was incorporated in a synthesis of tabtoxin (112), providing the hydroxyl and amino groups at C-2 and C-5 with the correct relative orientation.[45]

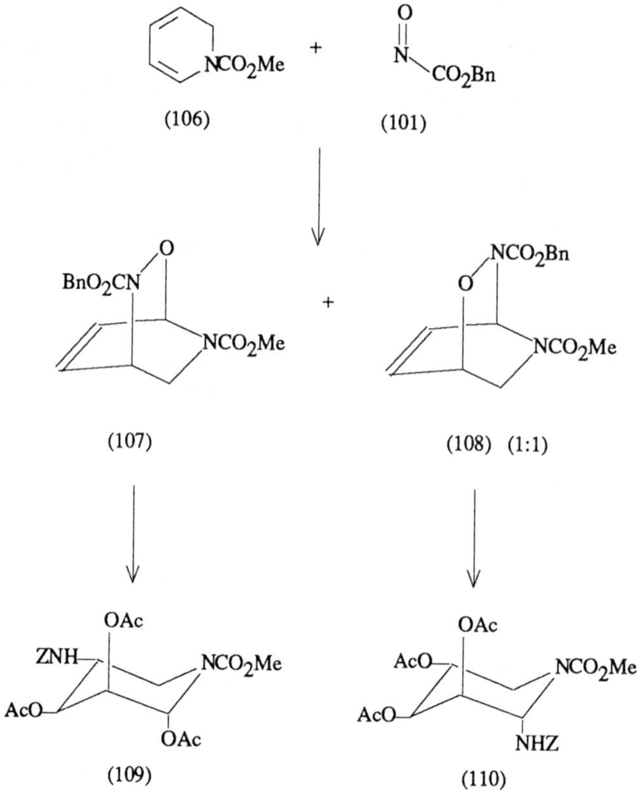

(106) + (101) → (107) + (108) (1:1)

(107) → (109)

(108) → (110)

*Scheme 29*

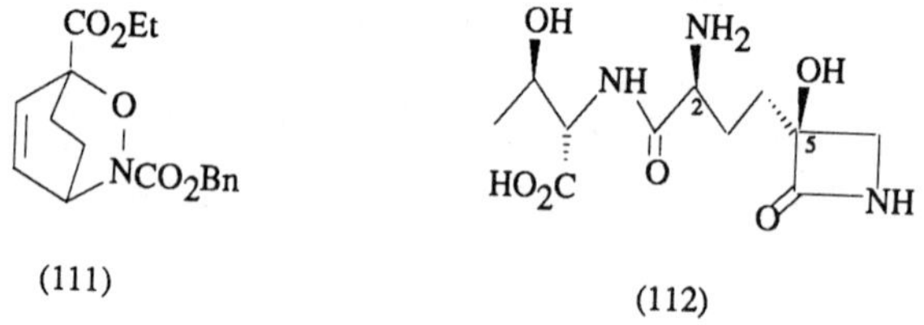

(111)

(112)

*Scheme 30*

α-Chloronitroso compounds have also been extensively employed in synthesis. They are thermally unstable so that their reactions have to be effected at low temperatures and require long reaction times, but they have the advantage that the cycloaddition to dienes leads to N-unsubstituted 3,6-dihydro-2*H*-1,2-oxazines which are easily converted into 1,4-amino-alcohols. The Diels-Alder addition of chloronitrosocyclohexane to cyclohexadienes forms the key step in the synthesis of a number of polyhydroxyaminocyclohexanes.[46] Thus, reaction with *trans*-5-acetylamino-6-acetoxy-1,3-cyclohexadiene gave specifically the cycloadduct (113) which was converted into the hexa-acetate (115).

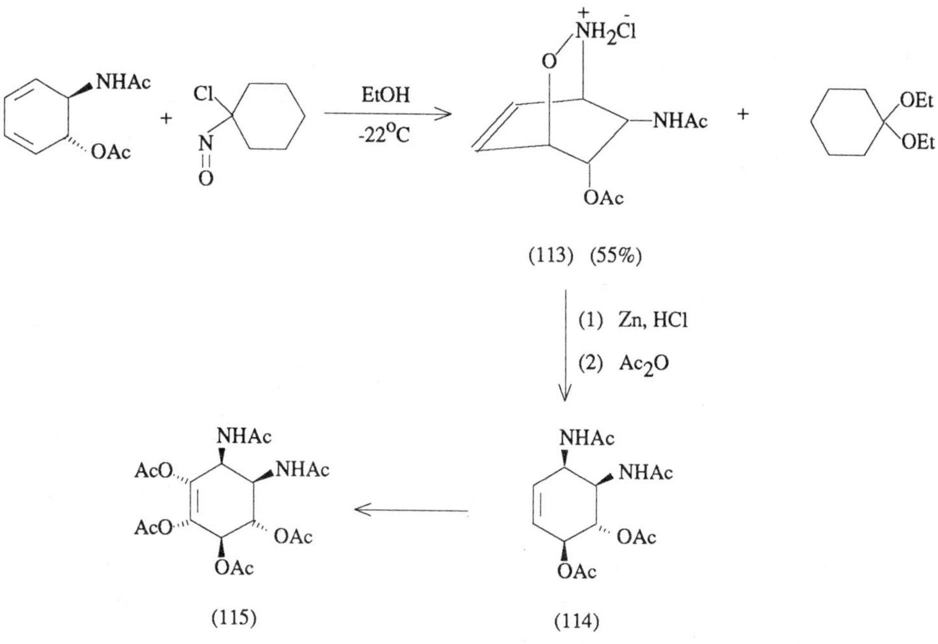

NHAc
OAc
Cl
N
O
EtOH
$-22^{\circ}C$
$\overset{+}{N}H_2\overset{-}{Cl}$
O
NHAc
OAc
OEt
OEt
(113) (55%)
(1) Zn, HCl
(2) $Ac_2O$
NHAc
AcO
NHAc
AcO
OAc
OAc
(115)
NHAc
NHAc
OAc
OAc
(114)

*Scheme 31*

Another elegant example is provided by the synthesis of (±)-5-amino-5,6-dideoxyallonic acid (116) from methyl *trans,trans*-sorbate.[47]

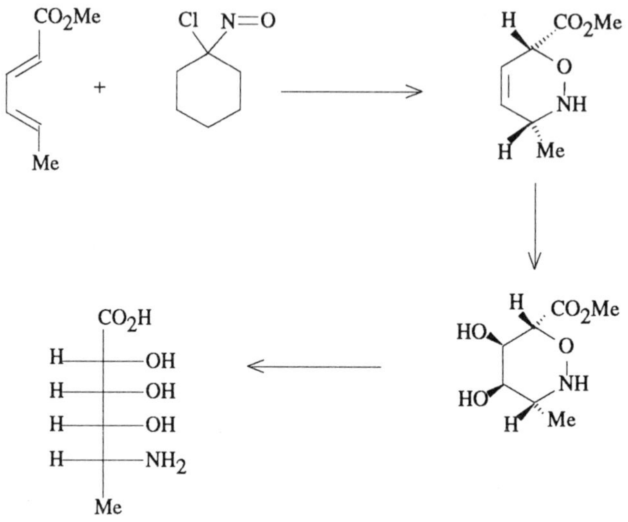

$CO_2Me$ Cl N=O + → H $CO_2Me$ O NH H Me

H $CO_2Me$ HO O NH HO H Me

$CO_2H$
H—OH
H—OH
H—OH
H—$NH_2$
Me

(116)

*Scheme 32*

**N-Sulphinylsulphonamides; Synthesis of Vicinal Amino-alcohols**

A further good method for the stereoselective formation of amines proceeds from the 3,6-dihydro-1,2-thiazine derivatives (117) formed by cycloaddition of N-sulphinysulphonamides to 1,3-dienes (see p.19).

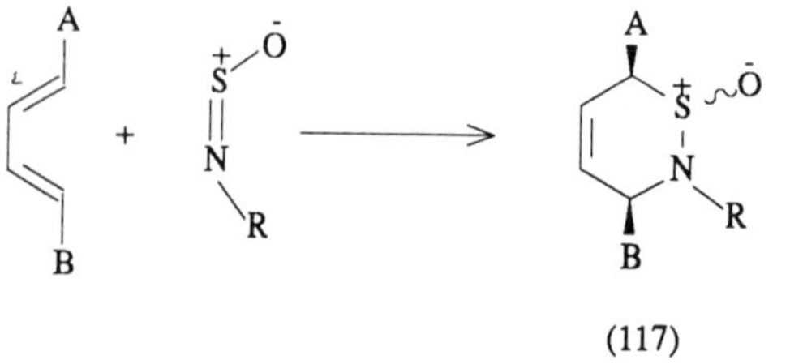

A B + $\overset{+}{S}$ $\bar{O}$ N R → A $\overset{+}{S}$ $\bar{O}$ N R B

(117)

R = $SO_2R$, $CO_2R$ etc.

*Scheme 33*

These adducts can be used in the stereocontrolled synthesis of homoallylic amine derivatives having predictable stereochemistry and double bond geometry. Thus, reaction of (*E*,*E*)-1,2,3,4-tetramethylbutadiene with N-sulphinyltoluenesulphonamide gave the Diels-Alder adduct (118) which was hydrolised to give (120) as a single stereoisomer. Similarly, the (*E*,*Z*)-diene gave the *trans*,*syn*-isomer (123) exclusively.[48]

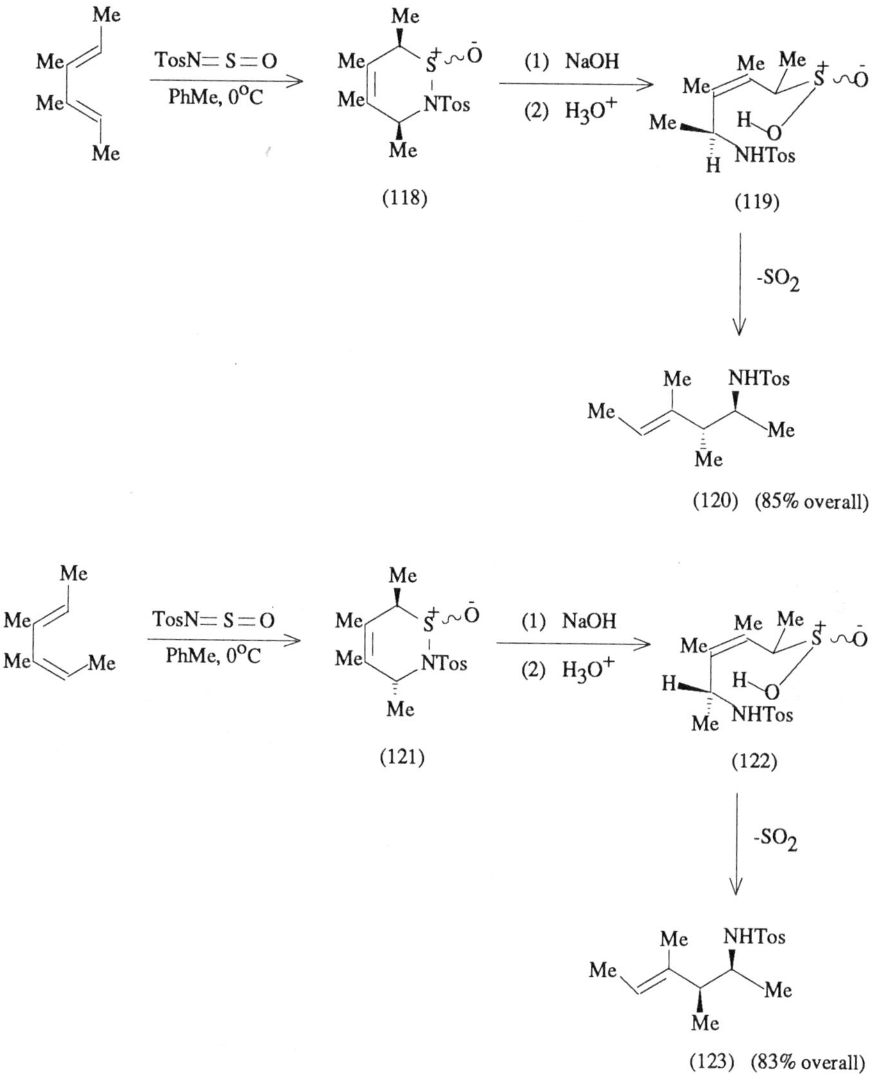

TosN=S=O, PhMe, 0°C → (118) → (1) NaOH (2) $H_3O^+$ → (119) → $-SO_2$ → (120) (85% overall)

TosN=S=O, PhMe, 0°C → (121) → (1) NaOH (2) $H_3O^+$ → (122) → $-SO_2$ → (123) (83% overall)

*Scheme 34*

The conversions (118) to (120) and (121) to (123) are completely stereospecific and are best explained by a concerted ene mechanism through envelope-like transition states of the form (119) and (122).[49] The important feature of this mechanism is that the substituent on the sulphur-bearing carbon acts as an equatorial "anchor" and thus controls the direction of proton transfer to one face of the olefinic bond.

By using the Evans' rearrangement of allylic sulphoxides to introduce a hydroxyl group stereoselectively, this sequence has been extended to the synthesis of unsaturated vicinal amino-alcohols with complete control of both the relative configuration of the amino and hydroxyl groups and the double bond geometry.[50] Thus, the adduct (125), from (*E,E*)-2,4-hexadiene and the N-sulphinyl carbamate (124), on treatment with phenylmagnesium bromide gave an intermediate allylic sulphoxide (126) which on heating with trimethyl phosphite afforded the *trans,syn*-unsaturated amino-alcohol (128) as a single stereoisomer. The (*E,Z*)-hexadiene similarly gave exclusively the *trans,anti*-isomer (129).

The high specificity of chirality transfer in the formation of the amino-alcohol derivatives is ascribed to specific transfer of oxygen to one face of the olefinic double bond in intermediates of the form (126) controlled by the 'anchor' effect of the pseudoequatorial methyl group on the sulphur-bearing carbon atom. An intramolecular version of this sequence has been exploited in stereospecific syntheses of the sphingolipid bases *threo*-sphingosine and *erythro*-sphingosine (see Chapter 3).

The same general procedure has been employed in the stereocontrolled synthesis of unsaturated vicinal diamines from Diels-Alder adducts of sulphur dioxide bis(imides) (130) and 1,3-dienes.[51] (Scheme 36)

**Synthesis of Benzene Derivatives**

Diels-Alder reactions are most frequently employed for the synthesis of alicyclic compounds, but they have on occasion been used for the *de novo* synthesis of benzene derivatives which are difficult to make by direct substitution. Aromatisation of the initial Diels-Alder adducts can be effected by straightforward dehydrogenation, by elimination of suitably placed substituents or by a retro Diels-Alder step with loss of a small molecule such as carbon dioxide or nitrogen. The groups to be eliminated during aromatisation are generally incorporated in the diene component of the Diels-Alder addition. One of the first dienes to be employed in this way was *trans,trans*-1,4-diacetoxy-1,3-butadiene, which reacts with a variety of olefinic and

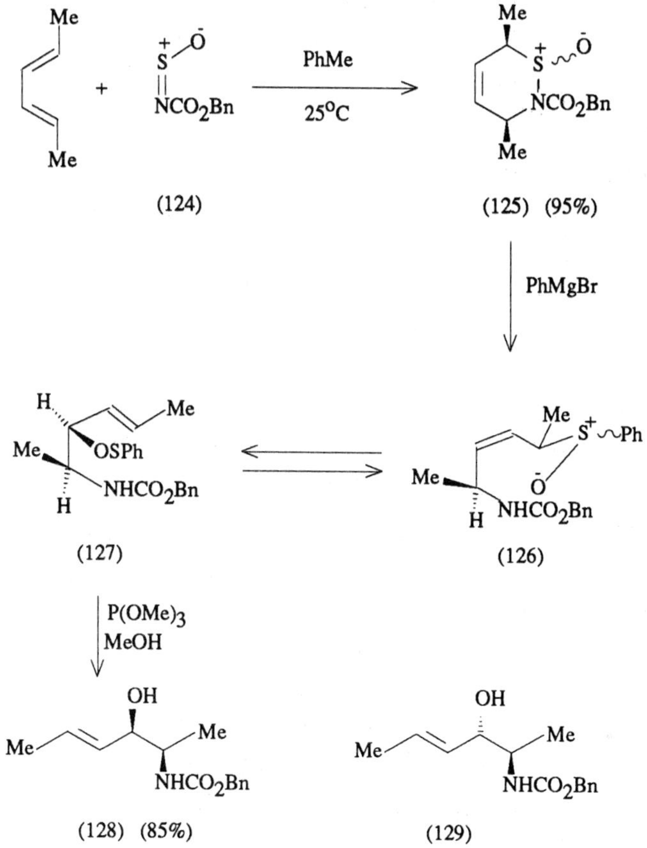

Me, Me, $\overset{+}{S}$, $\bar{O}$, $NCO_2Bn$, PhMe, 25°C, (124), (125) (95%), PhMgBr, H, Me, OSPh, $NHCO_2Bn$, (127), $S^+$, Ph, (126), $P(OMe)_3$, MeOH, OH, (128) (85%), (129)

*Scheme 35*

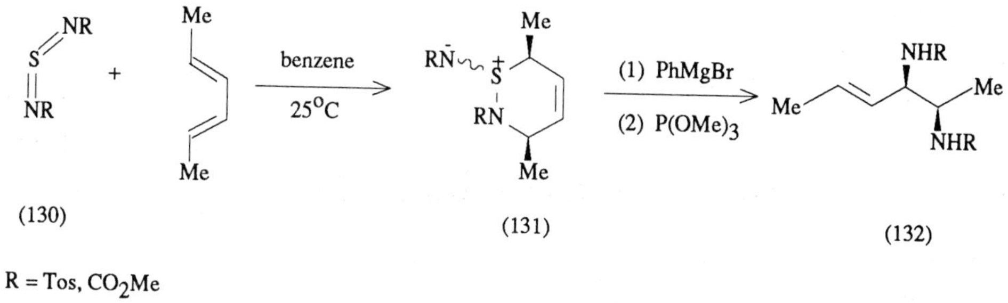

NR, S, NR, benzene, 25°C, $R\bar{N}$, RN, (1) PhMgBr, (2) $P(OMe)_3$, NHR, (130), (131), (132)

R = Tos, $CO_2Me$

*Scheme 36*

acetylenic dienophiles to give adducts which are readily aromatised by loss of acetic acid.[52] More recently, trimethylsilyloxy dienes have been used in the ring synthesis of substituted phenols.[53] Thus, synthesis of the benzenoid derivative (136), the aromatic unit of milbemycin β3, was achieved by reaction of the diene (133) with (134). Ring aromatisation with concomitant oxidation of the side chain was effected by treatment of the crude adduct with Jones' reagent.[54]

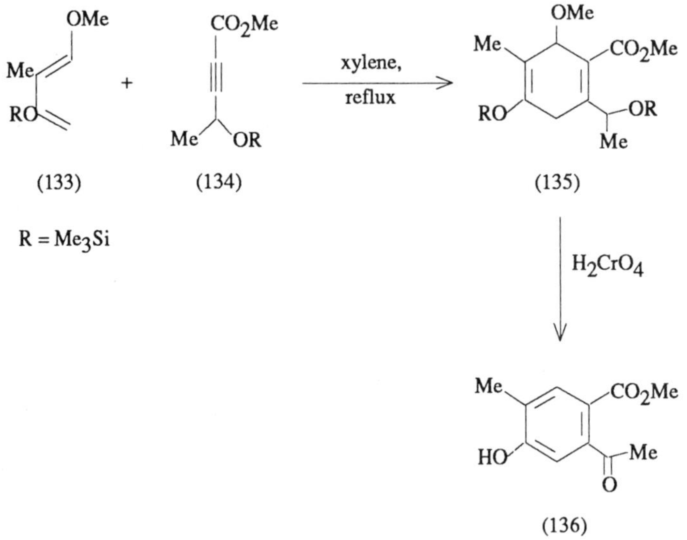

OMe, Me, RO (133); $CO_2Me$, Me, OR (134); xylene, reflux; OMe, Me, $CO_2Me$, RO, OR, Me (135)

R = $Me_3Si$

$H_2CrO_4$

Me, $CO_2Me$, HO, Me, O (136)

*Scheme 37*

More highly oxygenated derivatives of benzene are obtained from the reactive 1,1-dimethoxy-3-trimethylsilyloxybutadiene (137).[55] With acetylenic dienophiles derivatives of resorcinol methyl ether are obtained directly. Reaction with the acetylenic ester (138) was used to make the resorcinol derivative (139), a key intermediate in a synthesis of the plant growth inhibitor lasiodiplodin (140).[56]

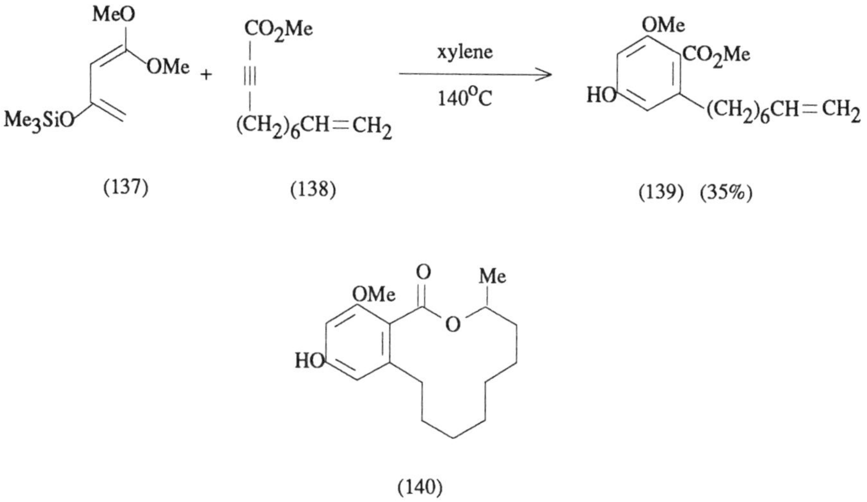

MeO OMe + $CO_2Me$ xylene 140°C OMe $CO_2Me$

$Me_3SiO$ $(CH_2)_6CH{=}CH_2$ HO $(CH_2)_6CH{=}CH_2$

(137) (138) (139) (35%)

OMe O Me O HO

(140)

*Scheme 38*

A good example of the ring synthesis of a benzene derivative difficult to obtain by direct substitution, through a sequence involving a Diels-Alder reaction followed by a retro reaction, is provided by the synthesis of the aldehyde (145), a building block in the synthesis of the ionophore lasalocid A, from the α-pyrone (143).[57] α-Pyrones act as butadiene equivalents in Diels-Alder reactions and with acetylenic dienophiles they give benzene derivatives directly by loss of carbon dioxide from the initial adduct. The α-pyrone (143) required for the synthesis of (145) was obtained by reaction of the parent pyrone (141) with the optically active Grignard reagent (142) and oxidation of the resulting dihydropyrone with manganese dioxide. Reaction of this pyrone with N,N-dibenzyl-1-amino-1-propyne in boiling benzene, gave directly the benzene derivative (144) which was converted in several more steps into the required phenol. None of the alternative regio-isomer was detected in the Diels-Alder reaction. Attempts to effect the cycloaddition with oxypropynes, such as 1-methoxypropyne, so as to obtain the phenol directly, were unsuccessful.

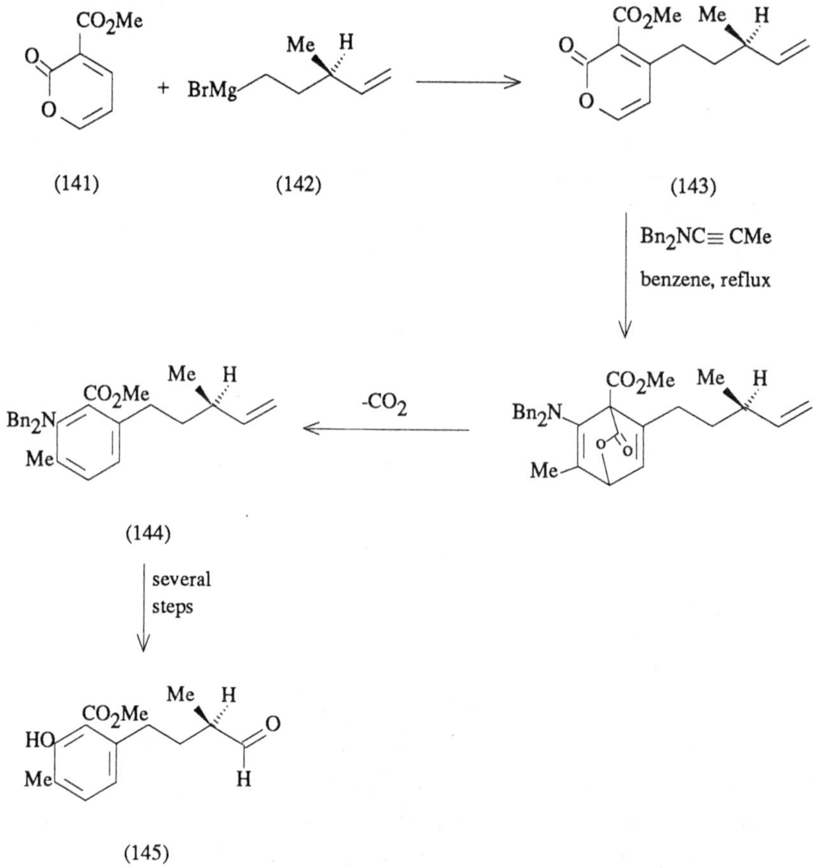

(141) (142) (143)

$Bn_2NC{\equiv}CMe$

benzene, reflux

$-CO_2$

(144)

several steps

(145)

*Scheme 39*

In another example leading to the construction of the B and C rings in the phosphodiesterase inhibitor PDE II (150), two hetero azadiene Diels-Alder reations were employed.[58] The reaction of dimethyl 1,2,4,5-tetrazine-3,6-dicarboxylate with 4,4-dimethoxybut-3-en-2-one gave the adduct (146) and thence, by loss of nitrogen, the diazine (147) which was converted into the acetylenic precursor (148). A second, intramolecular, cycloaddition with extrusion of nitrogen gave the indoline (149) which was converted in several more steps into (150).

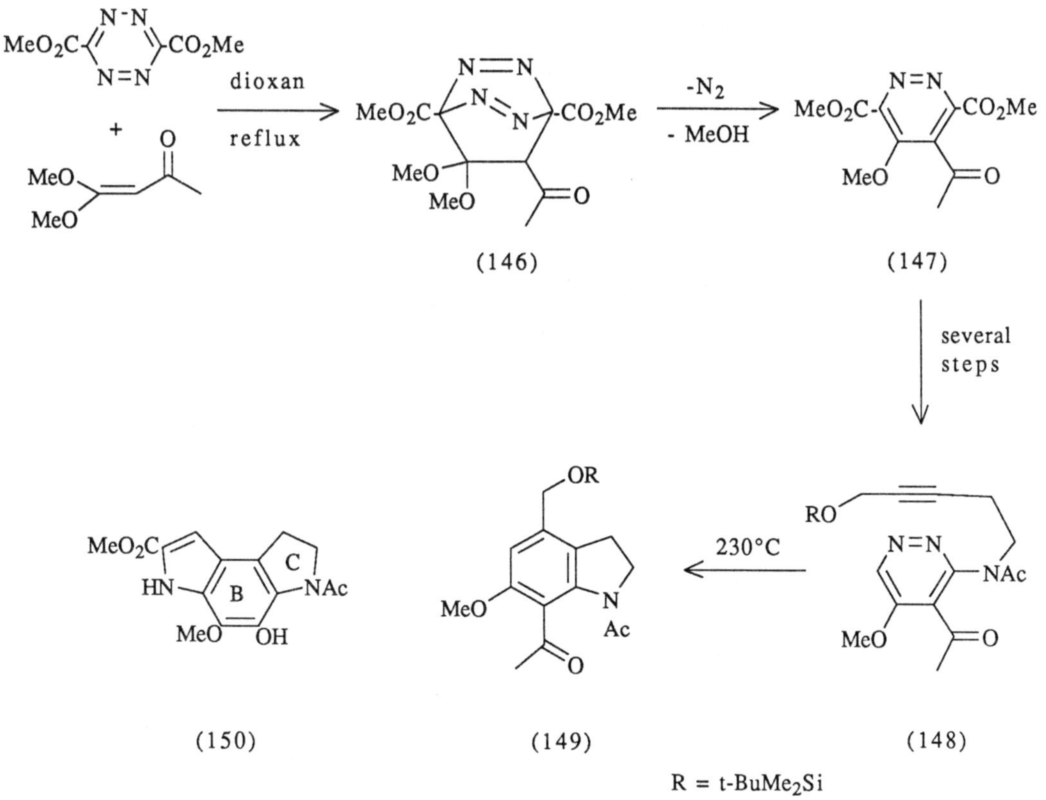

MeO$_2$C N-N N=N CO$_2$Me + MeO MeO O dioxan reflux MeO$_2$C N=N N N CO$_2$Me MeO MeO O (146) -N$_2$ - MeOH MeO$_2$C N=N CO$_2$Me MeO O (147) several steps

MeO$_2$C HN B C NAc MeO OH (150) OR MeO N Ac O (149) 230°C RO N=N NAc MeO O (148)

R = t-BuMe$_2$Si

*Scheme 40*

The Diels-Alder reaction can also be used to make derivatives of pyridine. The synthesis of vitamin B6 from 4-methyloxazole is well-known.[59] In a more recent example the 4-arylpyridine-fragment (155) of the antibiotic streptonigrin was obtained from the diene (151) and the imidazolidindione (152) in an imino Diels-Alder reaction. Here the final aromatisation was effected by dehydrogenation.[60]

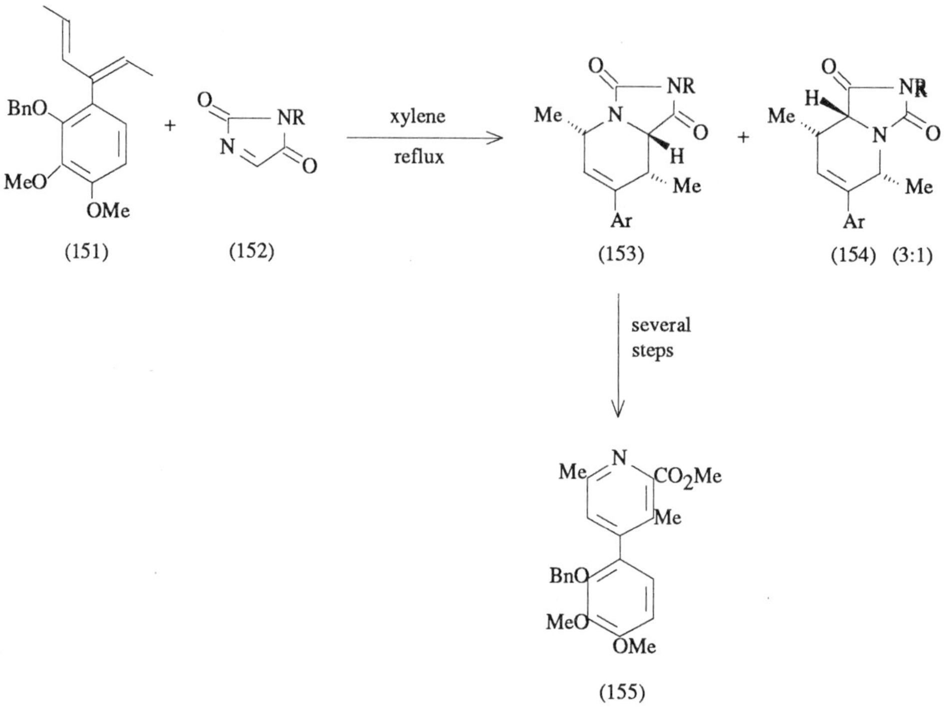

BnO MeO OMe (151) + O NR N O (152) xylene reflux Me N O NR O H Me Ar (153) + Me H O NR N O Me Ar (154) (3:1)

several steps

Me N $CO_2Me$ Me BnO MeO OMe (155)

*Scheme 41*

Pyridine derivatives have also been obtained from acetylenic dienophiles and 2-aza-1,3-dienes carrying an amino group at C-1.[61] 1,3-Bis(silyloxy)-2-aza-1,3-dienes and acetylenic dienophiles give 2-pyridones.[62]

The fragment eliminated in aromatisation of a Diels-Alder adduct may be incorporated in the dienophile instead of in the diene as in the examples given above, and several useful synthetic sequences exploit this variation. In one, which has been used to prepare *para*-acylphenols or *para*-hydroxybenzoic acids, an αβ-unsaturated ester or αβ-unsaturated ketone carries a β-phenylsulphide or β-phenylsulphoxide substituent as the eventual leaving group.[63]

## Synthesis of Quinones

A wide range of naphthaquinones and anthraquinones has been obtained by cyclo-addition of 1,3-dioxygenated butadienes to appropriate chloro-benzoquinones and chloronaphthaquinones.[64] The chloro substituents in the quinone dienophiles facilitate the reactions and also serve to direct the regiochemistry of the addition. The best results have been obtained with mixed vinyl ketene acetals (such as 156) prepared from αβ- or βγ-unsaturated esters by O-silylation of the derived anions with chlorotrimethylsilane.

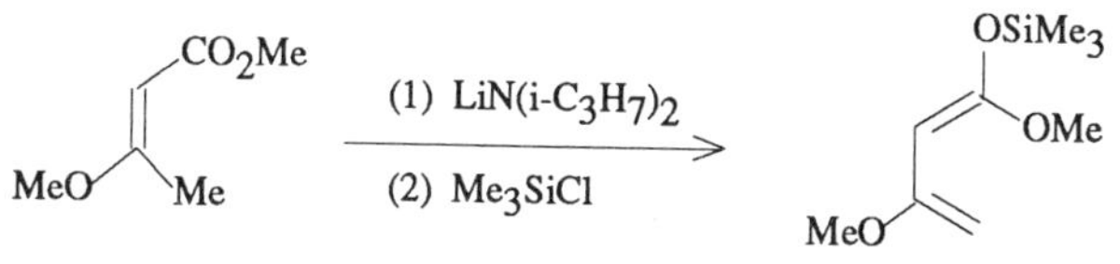

$CO_2Me$ MeO Me (1) $LiN(i\text{-}C_3H_7)_2$ (2) $Me_3SiCl$ → $OSiMe_3$ OMe MeO

(156)

*Scheme 42*

Cycloaddition of these mixed acetals to quinones takes place readily at room temperature. Aromatisation of the initial adduct is effected by pyrolysis, with evolution of hydrogen chloride, or, better, by percolation through silica gel. These reactions could apparently give different products depending on which of the acetal oxygen functions is eliminated during aromatisation, but in practice it is found that the methoxy substituent is eliminated preferentially, giving a phenol as the preponderant product. Thus chrysophanol (160) was obtained in one step from 3-chlorojuglone (157) and the acetal (158) but the isomeric 2-chlorojuglone gave ziganein (161), illustrating the well-established regiospecificity of these reactions.[65]

Benzoquinones also react readily, affording naphthaquinones in good yield. Many naturally-occurring naphthaquinones and anthraquinones have been synthesised by this convenient procedure.[66] Tetracyclic compounds related to the anthracyclines have also been obtained by reaction of the cyclic ketene acetal (162) with chloronaphthaquinones. Thus, with 3-chlorojuglone the adduct (163) was obtained exclusively in 55 per cent yield at room temperature.[67] Although these reactions are formally Diels-Alder cycloadditions this has not been rigorously established, and it may be that they proceed by Michael-type reaction of the keten acetals with the quinone, followed by rapid cyclisation.

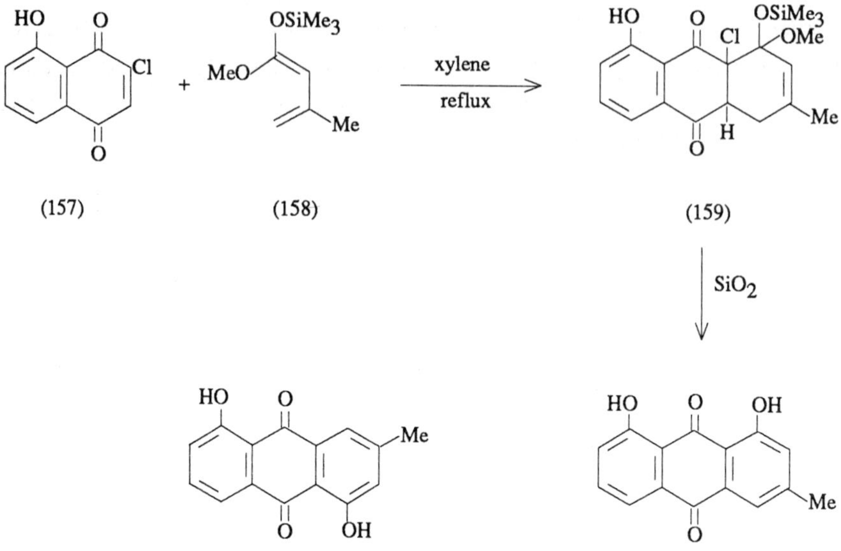

HO O Cl O (157) + MeO OSiMe$_3$ Me (158) xylene reflux → HO O Cl OSiMe$_3$ OMe H Me O (159)

$SiO_2$

HO O Me O OH (161) HO O OH Me O (160) (63%)

*Scheme 43*

HO O Cl O (157) + EtO OSiMe$_3$ Me (162) $CH_2Cl_2$ 0 - 20$^o$C → HO O OH Me O (163)

*Scheme 44*

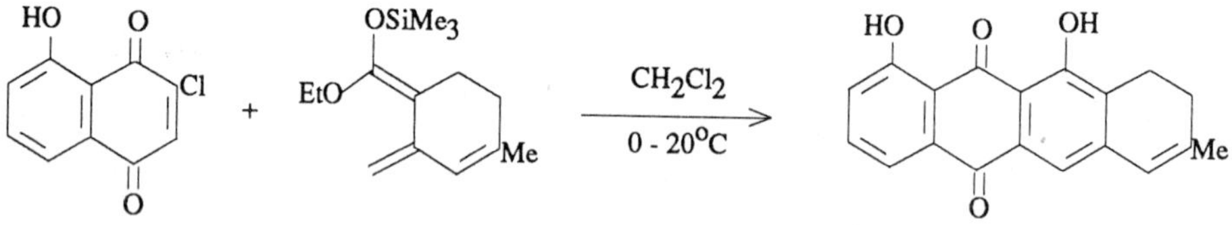

The influence of Lewis acid catalysts on the regiochemistry of Diels-Alder reactions is well illustrated in a synthesis of the natural product bostrycin (167) from the naphthopurpurin derivative (164). Straightforward reaction of (164) with 3-methyl-1-trimethylsilyloxybutadiene gave a mixture of regio-isomers, but in the presence of tetra-acetyl diborate the adduct (166) was obtained regiospecifically. Routine elaboration of (166) gave (±)-bostrycin with almost complete stereospecificity.[68]

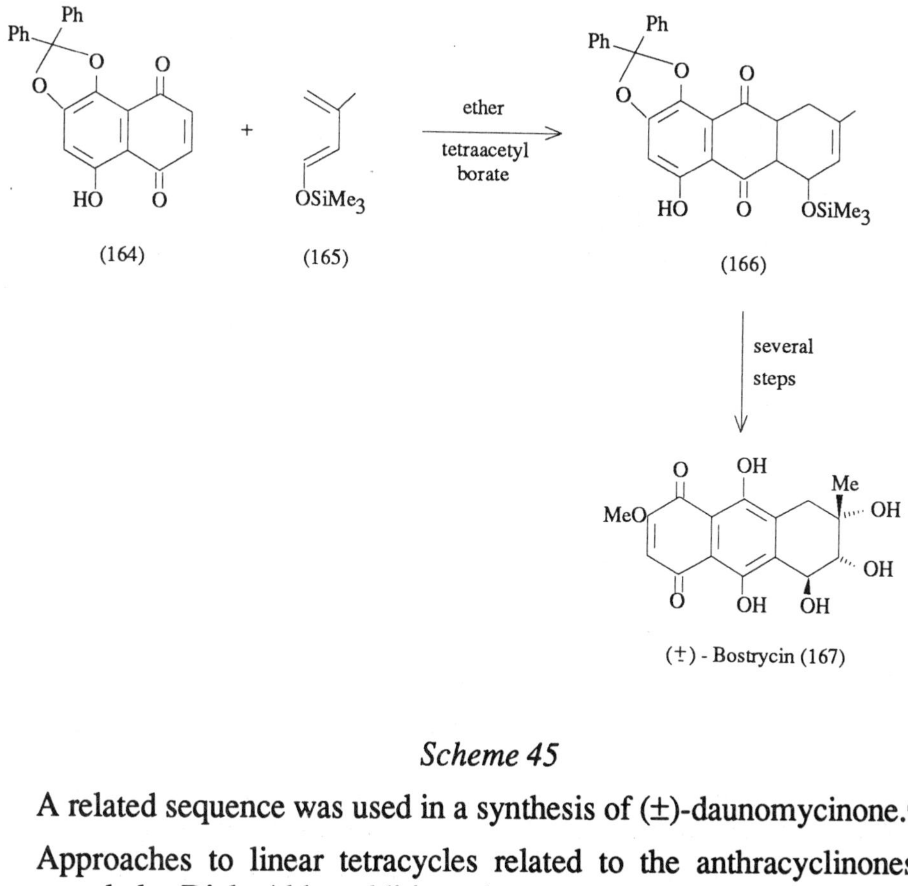

*Scheme 45*

A related sequence was used in a synthesis of (±)-daunomycinone.[69]

Approaches to linear tetracycles related to the anthracyclinones have been made by Diels-Alder addition of oxygenated butadienes to appropriate substituted 1,4-anthraquinones[70] resulting in one case in an enantiocontrolled synthesis of (+)-4-demethoxydaunomycinone.[71]

Anthraquinones have also been obtained by reaction of *ortho*-quinodimethanes with substituted benzoquinones. Again, halogen substituents in the quinone serve to control the regiochemistry of the cycloaddition. Thus, the unsymmetrical *ortho*-quinodimethane (169) and 2-bromo-6-methyl-benzoquinone gave the adducts (170) and (171) in the ratio 92:8; with the 3-bromo-6-methyl quinone the same products were obtained in the ratio 2:98. The main products were easily purified by crystallisation and converted in several steps into the isomeric anthraquinones islandicin (172) and digitopurpone (173).[72]

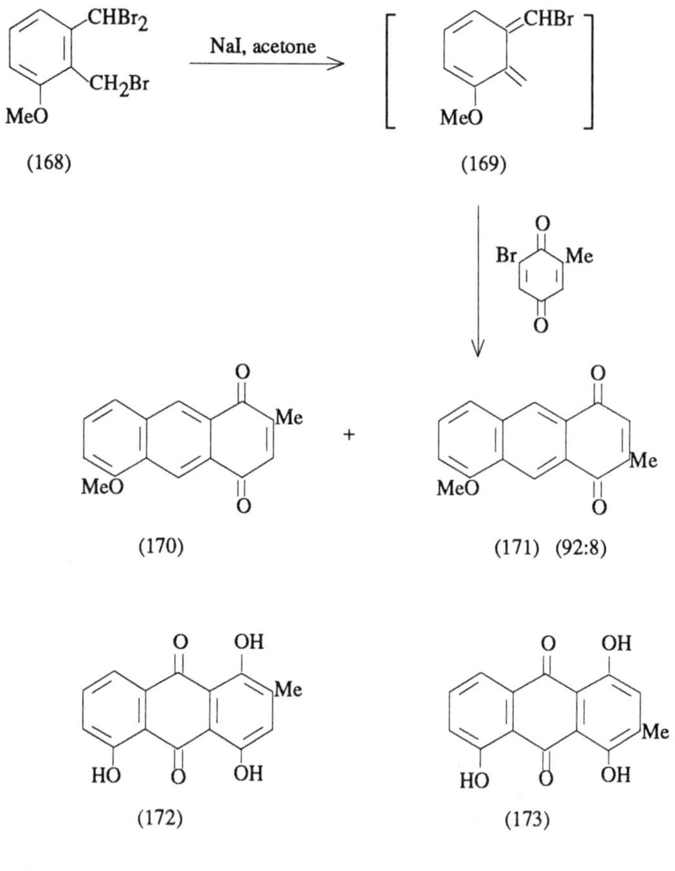

$CHBr_2$ $CH_2Br$ MeO (168) NaI, acetone → [ CHBr MeO ] (169)

Br Me O O

O Me MeO O (170) + O Me MeO O (171) (92:8)

O OH Me HO O OH (172) O OH Me HO O OH (173)

*Scheme 46*

In a related sequence homophthalic anhydride (174, R=H) and the methoxy derivative (174, R=OMe) reacted with the chloroquinone (175) regiospecifically to give adducts (176, R=H or OMe) which were converted into 4-demethoxydaunomycinone and daunomycinone respectively.[73] These reactions presumably take place by Diels-Alder addition to the dienol isomer (177) followed by spontaneous extrusion of carbon dioxide and hydrogen chloride from the adduct (178). Again the regiochemistry of the addition is controlled by the halogen substituent on the quinone; the nucleophilic end of the diene selectively reacts at the unsubstituted olefinic carbon of the chloroquinone.

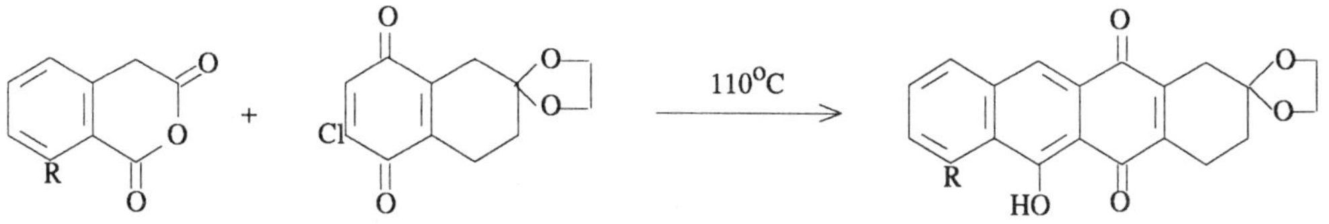

(174) (175) 110°C (176)

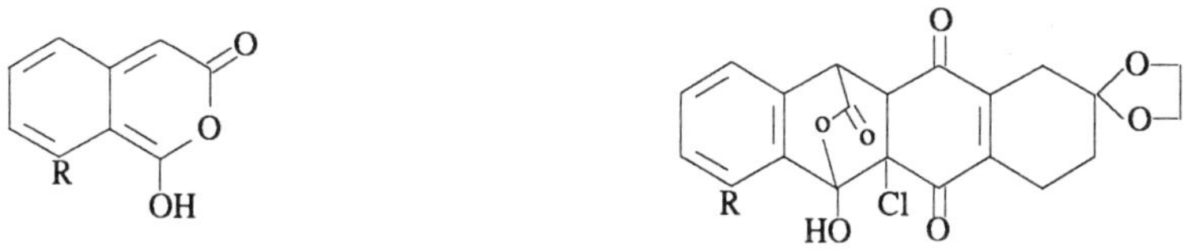

(177) (178)

*Scheme 47*

Interestingly the use of the lithium salts of (174 R=H,OMe) in the cycloadditions dramatically improved the yields of the adducts. Lithiation of (174,R=H) with lithium di-isopropylamide followed by reaction with (175) gave a nearly quantitative yield of (176 R=H); for (174 R= OMe) the yield was increased to 65 per cent.

Another group of aromatic natural products which have been synthesised by a Diels-Alder route from *ortho*-quinodimethanes are the lignans. The lignan nucleus is built up by addition of the appropriate dienophiles to aryl *ortho*-quinodimethanes, themselves produced by photolysis of *ortho*-methylbenzophenones[74] or by thermolysis of appropriate precursors. Thus, heating the sulphone (179) in di-n-butyl phthalate in the presence of maleic anhydride gave the adduct (181) in 70 per cent yield by way of transient *ortho*-quinodimethane (180)[75] and in an intramolecular approach to the synthesis of podophyllotoxin by the *ortho*-quinodimethane route, a mixture of (183) and (184) was obtained by heating the benzocyclobutene derivative (182) in toluene. The *trans*-lactone formed by way of the *endo* transition state, was the main product.[76]

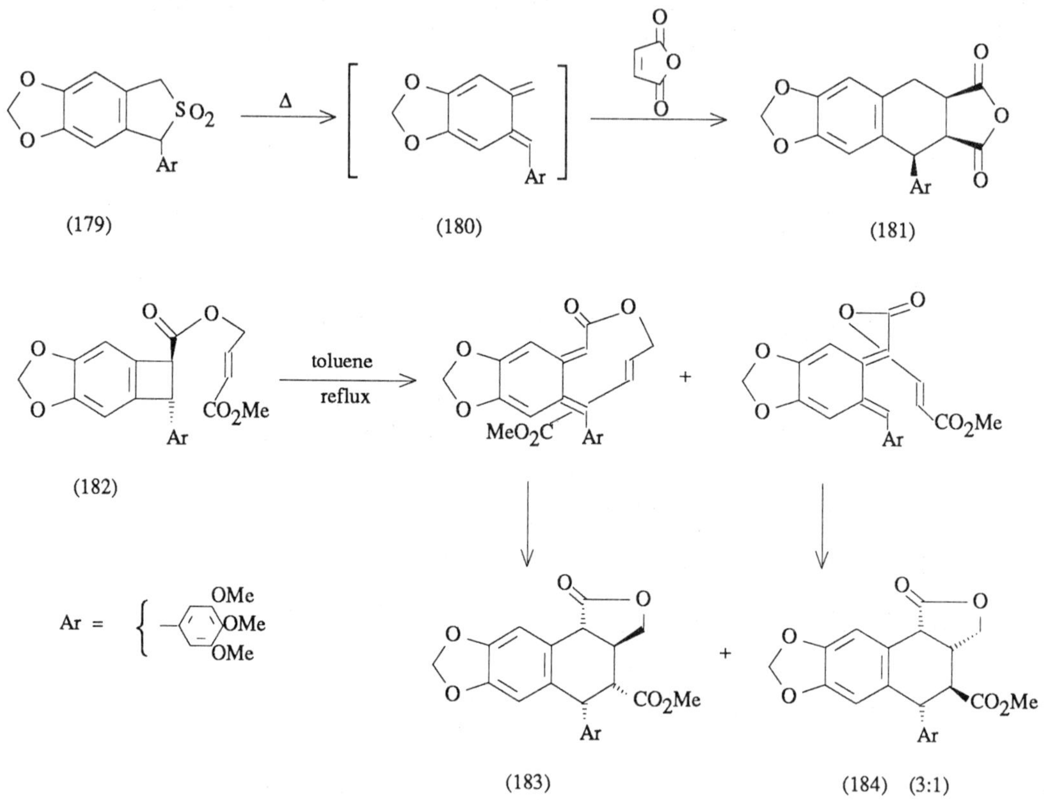

(179) (180) (181)

toluene reflux

(182) (183) (184) (3:1)

Ar = OMe OMe OMe

*Scheme 48*

## Retro Diels-Alder Reaction

The retro Diels-Alder reaction also has its uses in synthesis, and is frequently employed to uncover a reactive double bond by elimination of a diene such as cyclopentadiene from a modified adduct, or for the unmasking of a diene system by elimination of a dienophile, usually a small molecule such as carbon dioxide, nitrogen or ethylene.[77] Thus, in a synthesis of (+)- and (–)-sarcomycin methyl ester the starting material was the readily available enantiomerically pure adduct (185) which was elaborated in a number of steps to the diastereomers (186) and (187). Retro Diels-Alder reaction of (186) by flash vacuum pyrolysis gave optically pure (+)-(*S*)-sarcomycin methyl ester (188) in excellent yield, with elimination of cyclopentadiene. The isomer (187) similarly gave (–)-(*R*)-sarcomycin methyl ester.[78]

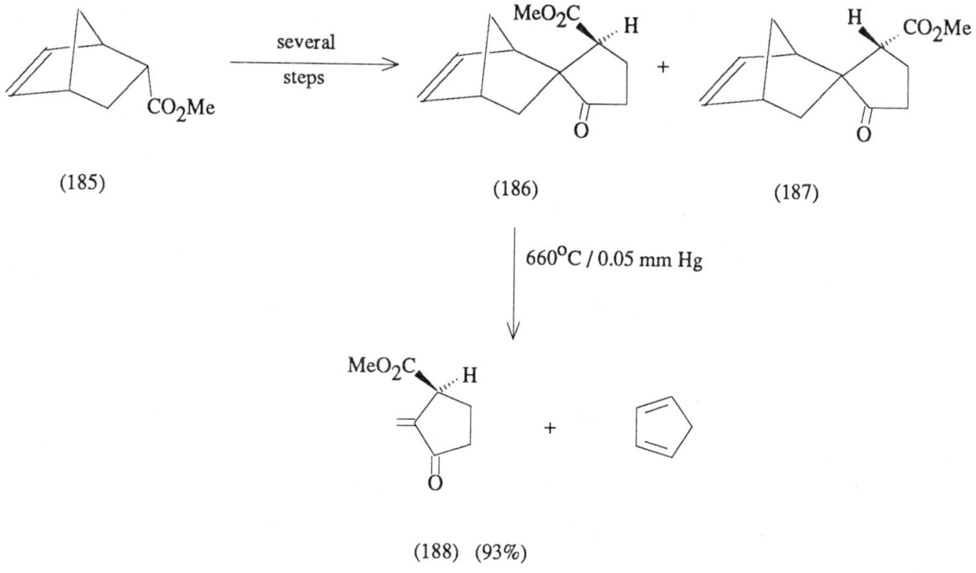

$CO_2Me$ several steps $MeO_2C$ H + H $CO_2Me$ O O

(185) (186) (187)

660°C / 0.05 mm Hg

$MeO_2C$ H + O

(188) (93%)

*Scheme 49*

In another example a retro Diels-Alder reaction formed the key step in syntheses of reactive derivatives of cyclopentadienone epoxide. Thus, the aldehyde (190), itself obtained from the adduct (189) of cyclopentadiene and benzoquinone,[79] on flash vacuum pyrolysis was converted into the cyclopentadiene mono-epoxide (191) which gave the mould metabolite terrein (192) in several more steps.[80] An analogous sequence was used in the synthesis of the antibiotic pentenomycin (194) from (193)[81]

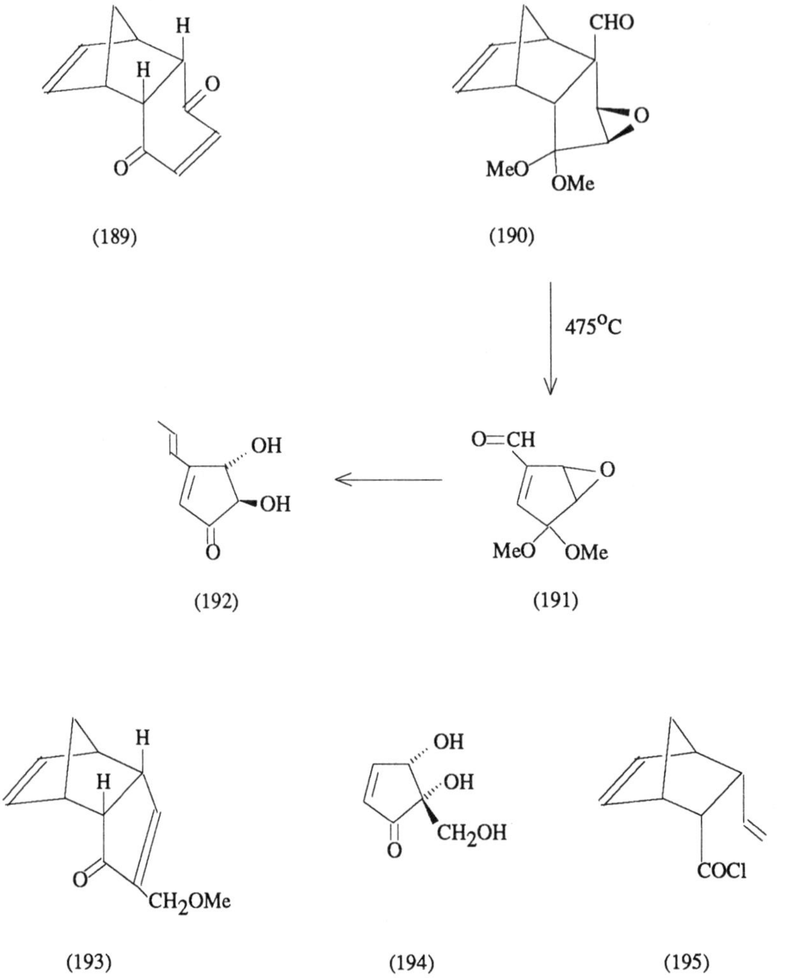

(189) (190)

475°C

(192) (191)

(193) (194) (195)

*Scheme 50*

In these syntheses the retro Diels-Alder step leads to the generation of highly reactive cyclopentadienone derivatives under non-chemical conditions where otherwise they would be likely to undergo chemical reaction. At the same time a carbon-carbon double bond is uncovered which has been masked in the Diels-Alder adduct, allowing selective chemical manipulation elsewhere in the molecule.

Another imaginative application of the retro Diels-Alder reaction is seen in a synthesis of aspidosperma-type alkaloids, where (195) serves as a synthetic equivalent of 2,4-pentadienoic acid chloride which is forced to undergo Diels-Alder reaction at the terminal double bond.[82]

## References


1 R.B. Woodward, F.E. Bader, H. Bickel, A.J. Frey and R.W. Kierstead, *Tetrahedron*, 1958, **2**, 1.

2 E.J. Corey, R.L. Danheiser, S. Chandrasekaran, P. Siret, G.E. Keckrand, J-L. Gras, *J. Am. Chem. Soc.*, 1978, **100**, 8031; E.J. Corey, R.L. Danheiser, S. Chandrasekaran, G.E. Keck, B. Gopalan, S.D. Larson, P. Siret and J-L. Gras, *loc.cit.*, p.8034.

3 See G. Helmchen, R. Karge and J. Weetman "Modern Synthetic Methods", Vol. 4. Ed. R. Scheffold. Springer-Verlag, Berlin, 1986.

4 G. Stork, Y. Nakahara, Y Nakahara and W.J. Greenless, *J. Am. Chem. Soc.*, 1978, **100**, 7775.

5 W. Oppolzer and C. Chapius, *Tetrahedron Lett.*, 1983, **24**, 4665.

6 e.g. K.P. Dastur, *J. Am. Chem. Soc.*, 1974, *96*, 2605; A.J. Birch and B.McKague, *Aust. J. Chem.*, 1970, **23**, 341.

7 e.g. S.A. Monti, S-C. Chen, Y-L. Yang, S-S. Yuan and O.P. Bourgeois, *J. Org. Chem.*, 1978, **43**, 4062.

8 S. Monti and Y-L. Yang, *J. Org. Chem.*, 1979, **44**, 897.

9 cf. M. Petrzilka and J.I. Grayson, *Synthesis*, 1981, 753.

10 P.A. Grieco, B. Lis, R.E. Zeller and J. Finn, *J. Am. Chem. Soc.*, 1986, **108**, 5908.

11 L.E. Overman and P.J. Jessup, *J. Am. Chem. Soc.*, 1978, **100**, 5179.

12 L.E. Overman and C. Fukaya, *J. Am. Chem. Soc.*, 1980, **102**, 1454.

13 S. Danishefsky and F.M. Hershenson, *J. Org. Chem.*, 1979, **44**, 1180.

14 cf. P. Brownbridge, *Synthesis*, 1983, **1**, 85; S. Danishefsky, *Acc. Chem. Res.*, 1981, **14**, 400; R. Schmidt, *Acc. Chem. Res.*, 1986, **19**, 250.

15 S. Danishefsky, T. Kitahara and P.F. Schuda, *Organic Syntheses*, 1983, **61**, 147.

16 S. Danishefsky, *Acc. Chem. Rev.*, 1981, **14**, 400.

17 S. Danishefsky, M. Hirama, N. Fritsch and J. Clardy, *J. Am. Chem. Soc.*, 1979, **101**, 7013.

18 S. Danishefsky, J. Morris and L-A. Clizbe, *J. Am. Chem. Soc.*, 1981, **103**, 1602.

19 S. Danishefsky, J. Morris and L-A. Clizbe, *Heterocycles*, 1986, **15**, 1205.

20 M.E. Jung, C.A. McCombs, Y. Taked and Y-G. Pann, *J. Am. Chem. Soc.*, 1981, **103**, 6677.

21 see S. Danishefsky and M.P. DeNinno, *Angew. Chem. internat. edn.*, 1987, **26**, 15; S. Danishefsky, *Aldrichimica Acta*, 1986, **19**, 59.

22 S.J. Danishefsky, W.H. Pearson, D.F. Harvey, C.J. Maring and J.P. Springer, *J. Am. Chem. Soc.*, 1985, **107**, 1256.

23 S. Danishefsky and D.F. Harvey, *J. Am. Chem. Soc.*, 1985, **107**, 6647.

24 M.M. Midland and R.S. Graham, *J. Am. Chem. Soc.*, 1984, **106**, 4294; S. Castellino and J.J. Sims, *Tetrahedron Lett.*, 1984, **25**, 2307.

25 M. Bednarski and S. Danishefsky, *J. Am. Chem. Soc.*, 1986, **108**, 7060.

26 S.J. Danishefsky, S. DeNinno and P. Lartey, *J. Am. Chem. Soc.*, 1987, **109**, 2082.

27 cf. R.R. Schmidt, *Acc. Chem. Res.*, 1986, **19**, 250.

28 G. Desimoni and G. Tacconi, *Chem. Rev.*, 1975, **75**, 651.

29 For suggestive exceptions see R.E. Ireland and J.P. Daub, *J. Org. Chem.*, 1983, **48**, 1303; R.E. Ireland, J.P. Daub, G.S. Mandel and N.S. Mandel, *J. Org. Chem.*, 1983, **48**, 1312.

30 B.B. Snider, *Tetrahedron*, 1980, **21**, 1133; B.B. Snider, D.M. Roush

and T.A. Killinger, *J. Am. Chem. Soc.*, 1979, **101**, 6023; R.R. Schmidt and M. Maier, *Tetrahedron Lett.*, 1982, **23**, 1789; R.R. Schmidt, *Acc. Chem. Res.*, 1986, **19**, 250; L-F. Tietze, K-H. Glusenkamp, K. Harms, G. Remberg and G.M. Sheldrick, *Tetrahedron Lett.*, 1982, **23**, 1147.

31 L-F. Tietze and K-H. Glusenkamp, *Angew. Chem. internat. edn.*, 1983, **22**, 887.

32 R. Schmidt and M. Maier, *Tetrahedron Lett.*, 1985, **26**, 2065; R. Schmidt, *Acc. Chem. Res.*, 1986, **19**, 250.

33 Y. Chapleur and M-N. Euvrard, *J. Chem. Soc. Chem. Commun.*, 1987, 884.

34 L-F. Tietze, E. Voss, K. Harms and G.M. Sheldrick, *Tetrahedron Lett.*, 1985, **26**, 5273; L.F. Tietze and E. Voss, *loc.cit.*, 6181.

35 B.J. Fitzsimmons, Y. Leblanc and J. Rokach, *J. Am. Chem. Soc.*, 1987, **109**, 285.

36 cf. J. Firl and S. Sommer, *Tetrahedron Lett.*, 1971, 4193.

37 E.K.V. Gustorf, D.V. White, B. Kim. D. Hess and J. Leitich, *J. Org. Chem.*, 1970, **35**, 1155.

38 A.B. Holmes, J. Thompson, A.J-G. Baxter and J. Dixon, *J. Chem. Soc. Chem. Commun.*, 1985, 37; T.N. Birkinshaw and A.B. Holmes, *Tetrahedron Lett.*, 1987, 813.

39 S.M. Weinreb and J.I. Levin, *Heterocycles*, 1979, **7**, 949; S.M. Weinreb, *Tetrahedron*, 1982, **38**, 3087.

40 J.F. Kerwin and S. Danishefsky, *Tetrahedron Lett.*, 1982, **23**, 3739.

41 S. Danishefsky, M.E. Langer and C. Vogel, *Tetrahedron Lett.*, 1985, **26**, 5983.

42 S.J. Danishefsky and C. Vogel, *J. Org. Chem.*, 1986, **51**, 3915.

43 A. Defoin, H. Fritz, G. Geffroy and J. Streith, *Tetrahedron Lett.*, 1986, **27**, 4727.

44 A. Defoin, C. Schmidlin and J. Streith, *Tetrahedron Lett.*, 1984, **25**, 4515; A. Defoin, H. Fritz, C. Schmidlin and J. Streith, *Helv. Chim. Acta*, 1987, **70**, 554; G. Augelmann, J. Streith and H. Fritz, *Helv. Chim. Acta*, 1985, **68**, 95.

45 J.E. Baldwin, P.D. Bailey, G. Gallagher, M. Otsuka, K.A. Singleton, P.M. Wallace, K. Prout and W.M. Wolf, *Tetrahedron*, 1984, **40**, 3695; J.E. Baldwin, M. Otsuka and P.M. Wallace, *Tetrahedron*, 1986, **42**, 3097.

46 G. Kresze and E. Kysela, *Annalen*, 1981, 202; G. Kresze, E. Kysela and W. Dittel, *Annalen*, 1981, 210; G. Kresze, W. Dittel and H. Melzer, *Annalen*, 1981, 224; G. Kresze, M.M. Weiss and W. Dittel, *Annalen*, 1984, 203.

47 B. Belleau and Yum-Kin Au-Yaung, *J. Am. Chem. Soc.*, 1963, **85**, 64.

48 R.S. Garigipati, J.A. Morton and S.M. Weinreb, *Tetrahedron Lett.*, 1983, **24**, 987.

49 W.L. Mock and R.M. Nugent, *J. Org. Chem.*, 1978, **43**, 3433; R.S. Garigipati, A.J. Freyer, R.R. Whittle and S.M. Weinreb, *J. Am. Chem. Soc.*, 1984, **108**, 7861.

50 R.S. Garigipati and S.M. Weinreb, *J. Am. Chem. Soc.*, 1983, **105**, 4499.

51 H. Natsugari, R.R. Whittle and S.M. Weinreb, *J. Am. Chem. Soc.*, 1984, **106**, 7867.

52 R.K. Hill and R.M. Carlson, *Tetrahedron Lett.*, 1964, 1157; cf. also H. Hiranuma and S.I. Miller, *J. Org. Chem.*, 1982, **47**, 5083.

53 cf. S. Danishefsky, *Acc. Chem. Res.*, 1981, **14**, 410.

54 R. Baker, V.B. Rao, P.D. Ravenscroft and C.J. Swain, *Synthesis*, 1983, 572.

55 S. Danishefsky, R.K. Singh and R.B. Gammill, *J. Org. Chem.*, 1978, **43**, 379; S. Danishefsky, Cheng-Feng Yan, R.K. Singh, R.B. Gammill, P.M. McCurry, N. Fritsch and J. Clardy, *J. Am. Chem. Soc.*, 1979, **107**, 7001.

56 S. Danishefsky and S.J. Etheridge, *J. Org. Chem.*, 1979, **44**, 4716.

57 R.E. Ireland, R.C. Anderson, R. Badoud, B.J. Fitzsimmons, G.J. McGarvey, S. Thaisrivongs and C.S. Wilcox, *J. Am. Chem. Soc.*, 1983, **105**, 1988.

58 D.L.Boger and R.S. Coleman, *J. Am. Chem. Soc.*, 1987, **109**, 2717.

59 W. Böll and H. König, *Annalen*, 1979, 1657.

60 S.M. Weinreb, F.Z. Basha, S. Hibino, N.A. Khatri, D. Kim, W.E. Pye and T.T. Wu, *J. Am. Chem. Soc.*, 1982, **104**, 536.

61 A. Demoulin, A-M. Hesbain-Frisque and L. Ghosez, *J. Am. Chem. Soc.*, 1982, **104**, 1428.

62 F. Sainte, B. Serckx-Poncin, A-M. Hesbain-Frisque and L. Ghosez, *J. Am. Chem. Soc.*, 1982, **104**, 1428.

63 S. Danishefsky, T. Harayama and R.K. Singh, *J. Am. Chem. Soc.*, 1979, **101**, 7008; S.A. Attwood, A.G.M. Barrett and J.C. Florent, *J. Chem. Soc. Chem. Commun.*, 1981, 556.

64 P. Brownbridge, *Synthesis*, 1983, 85.

65 J. Savard and P. Brassard, *Tetrahedron*, 1984, **40**, 3455.

66 see for example J-L. Grandmaison and P. Brassard, *J. Org. Chem.*, 1978, **43**, 1435; G. Roberge and P. Brassard, *J. Chem. Soc. Perkin 1*, 1979, 1041; C. Brisson and P. Brassard, *J. Org. Chem.*, 1981, **46**, 1810; D.W. Cameron, G.I. Feutrill, G.B. Gamble and J. Stavrakis, *Tetrahedron Lett.*, 1986, **27**, 4999.

67 J-P. Gesson, J-C. Jacquesy and B. Renoux, *Tetrahedron*, 1984, **40**, 4743; *Tetrahedron Lett.*, 1983, **24**, 2757, 2761; J-P. Gesson, J-C. Jacquesy and M. Mondon, *Tetrahedron Lett.*, 1981, **22**, 1337.

68 T.R. Kelly, J.K. Saha and R.R. Whittle, *J. Org. Chem.*, 1985, **50**, 3680.

69 T.R. Kelly, *et.al. Tetrahedron*, 1984, **40**, 4569; see also A. Echavarren, P. Prados and F. Farîna, *Tetrahedron*, 1984, **40**, 4561.

70 e.g. D.W. Cameron, C. Conn, M.J. Crossley, G.I. Feutrill, M.W. Fisher, P.G. Griffiths, B.K. Merrett and D. Pavlatos, *Tetrahedron Lett.*, 1986, **27**, 2417.

71 R.C. Gupta, P.A. Harland and R.J. Stoodley, *Tetrahedron*, 1984, **40**, 4657; D.W. Cameron, G.I. Feutrill, P.G. Griffiths and B.K. Merritt, *Tetrahedron Lett.*, 1986, **27**, 2421.

72 J.R. Wiseman, J.J. Pendery, C.A. Otto and K.G. Chiong, *J. Org. Chem.*, 1980, **45**, 516.

73 Y. Tamura, A. Wada, M. Sasho and Y. Kita, *Tetrahedron Lett.*, 1981, **22**, 4283; Y. Tamura, A. Wada, M. Sasha, K. Fukunaya, H. Maeda and Y. Kita, *J. Org. Chem.*, 1982, **47**, 4378.

74 cf. E. Block and R. Stevenson, *J. Chem. Soc. Perkin 1*, 1973, 308.

75 J. Mann and S.E. Piper, *J. Chem. Soc. Chem. Commun.*, 1982, 430.

76 M.E. Jung, P.Y-S. Lam, M.M. Mansuri and L.M. Speltz, *J. Org. Chem.*, 1985, **50**, 1087.

77 M. Karpf, *Angew. Chem. internat. edn.*, 1986, **25**, 414; A. Ischihara, *Synthesis*, 1987, 207.

78 G. Helmchen, K. Ihrig and H. Schindler, *Tetrahedron Lett.*, 1987, **28**, 183.

79 A.J.H. Klunder, W. Bos, J.M.M. Verlaak and B. Zwanenburg, *Tetrahedron Lett.*, 1981, **22**, 4553.

80 A.J.K. Klunder, W. Bos and B. Zwanenburg, *Tetrahedron Lett.*, 1981, **22**, 4557.

81 J.M.J. Verlaak, A.J.H. Klunder and B. Zwanenburg, *Tetrahedron Lett.*, 1982, **23**, 5463.

82 P. Magnus and P.M. Cairns, *J. Am. Chem. Soc.*, 1986, **108**, 217.

# 3 INTRAMOLECULAR DIELS-ALDER REACTIONS

In intramolecular Diels-Alder reactions the diene and dienophile form part of the same molecule and are connected by a chain which may be an all-carbon chain or may contain one or more heteroatoms. Cyclisation leads to a new polycyclic system, with generation of up to four new chiral centres. With precursors in which the diene unit is tethered at C-1, cyclisation could, in principle give rise to a fused ring system (1) or a bridged ring system (2), but in practice nearly all reactions of this kind lead to the formation of fused-ring compounds (1). It is only where the connecting chain is sufficiently long, or with some cyclic dienes, that formation of the alternative bridged-ring product (2) becomes geometrically feasible.

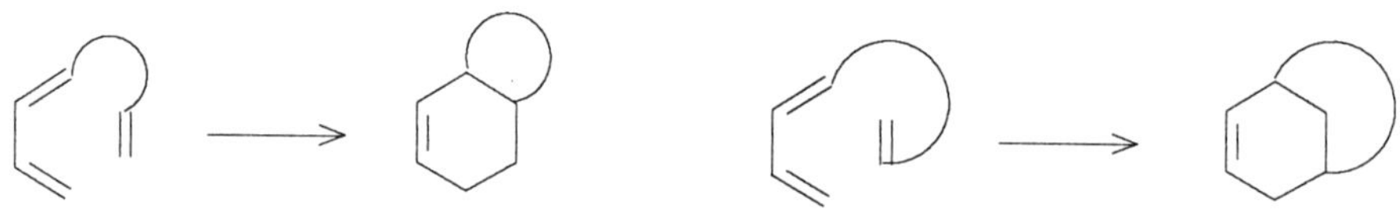

(1)

(2)

*Scheme 1*

With dienes tethered at C-2 intramolecular reactions lead regioselectively and stereoselectively to bridged-ring compounds with bridgehead double bonds. The strain energy in the bridgehead alkenes manifests itself in the more vigorous conditions which are frequently required in purely thermal intramolecular reactions with dienes of this kind.[1] In many cases, however, the reactions are catalysed by Lewis acids and they can then be effected under much milder conditions.[2] (Scheme 2)

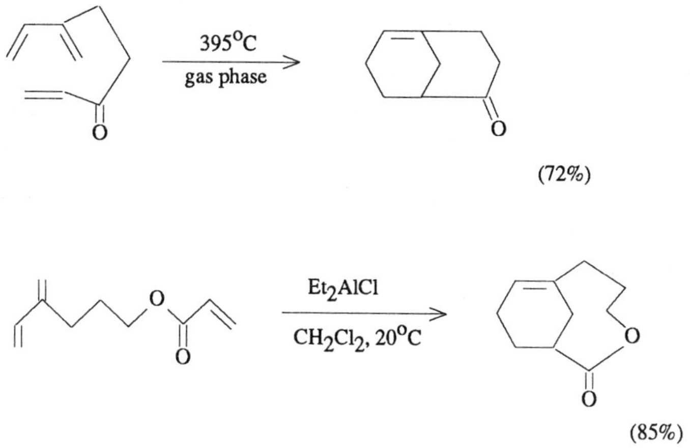
395°C
gas phase
O
O
(72%)
Et2AlCl
CH2Cl2, 20°C
O
O
O
O
(85%)

*Scheme 2*

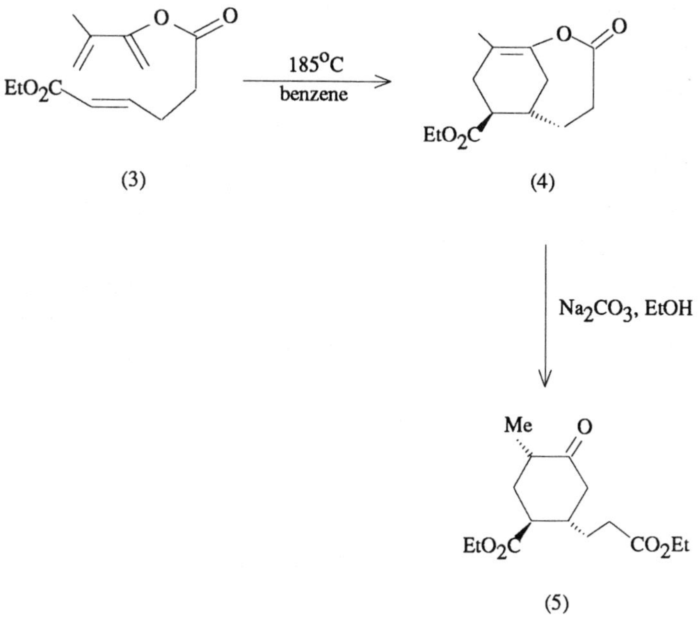
O
O
EtO2C
185°C
benzene
O
O
EtO2C
(3)
(4)
Na2CO3, EtOH
Me
O
EtO2C
CO2Et
(5)

*Scheme 3*

Reactions of this kind have been used to make a series of bridgehead enol lactones which are useful starting materials for the synthesis of stereochemically defined cyclohexanones and decalones with control of the relative stereochemistry of up to four contiguous asymmetric centres.[3] Thus, heating the enol ether (3) in benzene at 185°C gave the single cycloadduct (4) in 90 per cent yield, which was smoothly converted into the cyclohexanone derivative (5).

However, most applications of the intramolecular Diels-Alder reaction in synthesis have employed trienes in which the diene unit is tethered at C-1.

Because they provide ready access to polycyclic compounds with control of stereochemistry, intramolecular Diels-Alder reactions have been increasingly used in the synthesis of natural products.[4] In contrast to intermolecular reactions intramolecular Diels-Alder reactions generally form a late stage in the syntheses, and the preparation of the requisite trienes for cyclisation is frequently itself a considerable synthetic undertaking.

Nearly all the recorded examples of intramolecular Diels-Alder reactions give rise to hydroindenes or hydronaphthalenes, or to heterocyclic analogues of these ring systems, from precursors in which the diene and dienophile moieties are connected by a chain of three or four atoms at C-1 of the diene. Bicyclo[4,2,0]octenes cannot be made by the intramolecular Diels-Alder procedure[5] and very few reactions which lead to the formation of medium-sized rings have been recorded. If the connecting chain is long enough, the diene and dienophile react almost as if they were independent entities and bridged ring products may then be formed. Thus, the triene (6) on heating in benzonitrile gave a mixture of the two fused-ring products (7) and (8) and the bridged ring isomer (9).[6]

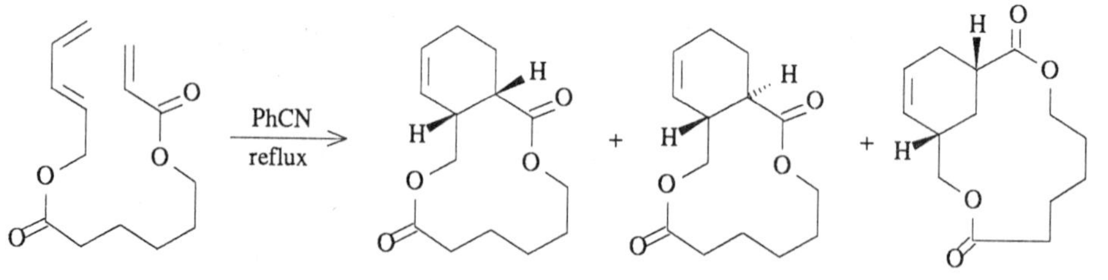

PhCN reflux

(6) (7) (8) (9) (77%; 6.2:6.8:1)

*Scheme 4*

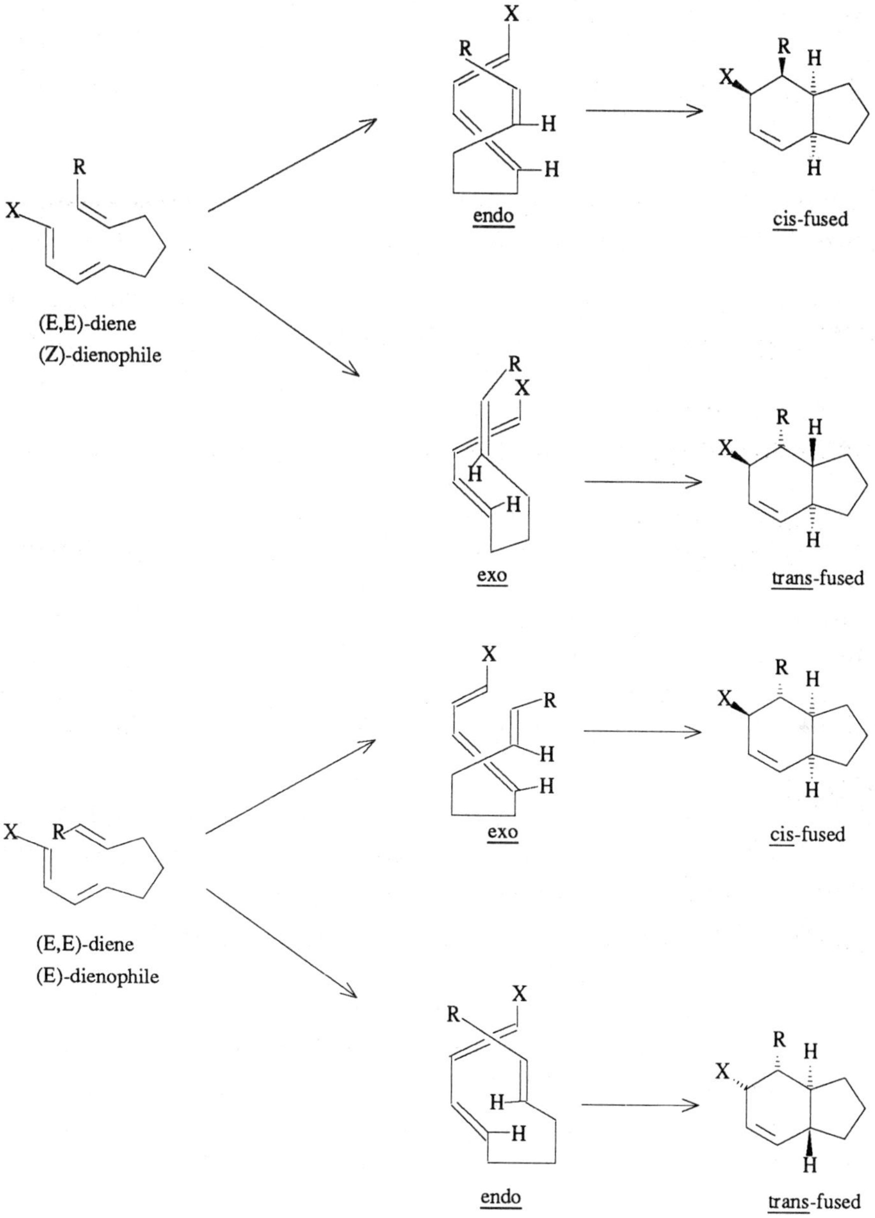

(E,E)-diene
(Z)-dienophile

endo

cis-fused

exo

trans-fused

(E,E)-diene
(E)-dienophile

exo

cis-fused

endo

trans-fused

*Scheme 5*

### Stereoselectivity

In reactions leading to hydroindenes and hydronaphthalenes from 1,3,8-nonatrienes and 1,3,9-decatrienes an important consideration for synthesis is foreknowledge of the stereochemistry of the new chiral centres, particularly of the ring fusion, but it is not always easy to predict what this will be. Frequently mixtures are produced, although one isomer may predominate. The stereochemistry of the products is the result of competing conformational, steric and electronic effects in the various possible transition states, and these are influenced by the structure of the substrate, particularly the length of the chain connecting the diene and dienophile units, and the nature and position of substituent groups on the diene and dienophile and on the connecting chain. In the intermolecular Diels-Alder reaction the Alder *endo* rule, although not universally followed, often provides a useful guide to the stereochemistry of the main product. In the intramolecular reaction this may not be so. Frequently the orienting influences due to secondary orbital interactions in an *endo* transition state are outweighed by the geometrical constraint of the connecting chain and by non-bonded interactions of substituent groups.[7]

Thus, in principle, 1,3,8-nonatrienes with an (*E*,*E*)-diene may cyclise *via* two alternative transition states for each of the two isomers with (*E*)- and (*Z*)-dienophile moieties, to form *cis*- or *trans*-hydroindenes (Scheme 5).

In practice, the ratio of *cis*- and *trans*-fused products depends on the reaction conditions and on the disposition of substituents on the diene and the connecting chain. For thermal reactions the *trans*-fused hydroindene is generally the main product, although mixtures are usually produced, and this is independent of the configuration of the dienophile. Thus, cyclisation of the methyl decatrienoate (10) in toluene at 150°C gave the *cis* and *trans* hydroindenes in practically the same ratio as the isomer (13) with the *cis*-dienophile moiety.[8]

Selectivity for the *trans*-fused product here is essentially independent of the stereochemistry of the dienophile and the *endo* rule clearly fails to predict the outcome of the cyclisation of (13). Secondary orbital interactions appear not to be significantly involved in the cyclisation of the *trans* isomer (10) either. In each case the distribution of products is determined by a balance of forces in the possible transition states. In reactions catalysed by Lewis acids, however, the *trans* dienophile (10) gave the *endo* product (11) exclusively, but selectivity with the *cis* dienophile (13) was not improved.

Apparently increased secondary orbital control resulting from interaction between the ester carbonyl and the Lewis acid, which presumably accounts for the improved selectivity with the *trans* ester (10), is insufficient to overcome the transition state preference for the *trans*-fused product in the reaction of (13).

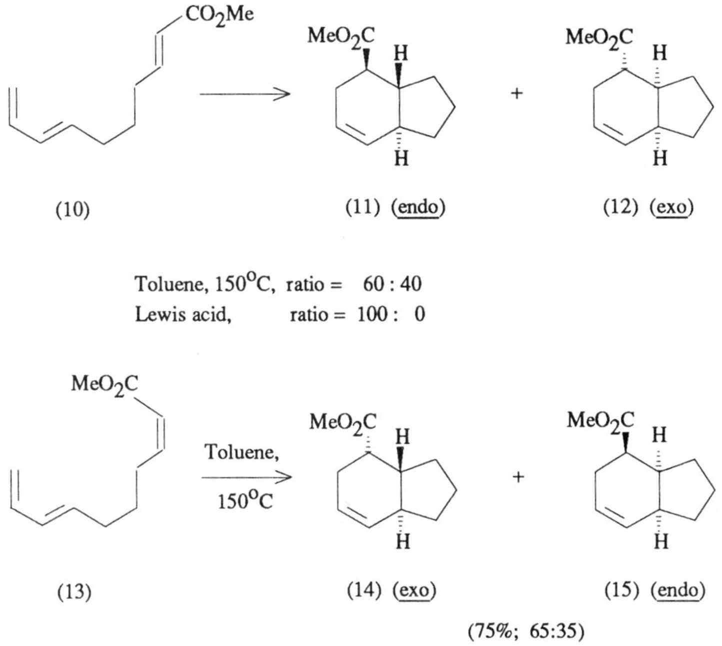

$CO_2Me$ $MeO_2C$ H H + $MeO_2C$ H H

(10) (11) (endo) (12) (exo)

Toluene, 150°C, ratio = 60 : 40
Lewis acid, ratio = 100 : 0

$MeO_2C$ Toluene, 150°C $MeO_2C$ H H + $MeO_2C$ H H

(13) (14) (exo) (15) (endo)

(75%; 65:35)

*Scheme 6*

The way in which the ratio of *cis/trans*-fused hydroindenes formed by intramolecular Diels-Alder cyclisation of a nonatriene can vary with the nature of the triene and the reaction conditions is illustrated in Schemes 7 and 8. In the thermal cyclisation of the trienoate (16), four stereoisomeric adducts were formed, but the main products arose by way of the *endo* transition state. The proportion of *endo* products was increased when the reactions were performed at as low a temperature as possible compatible with reaction and by conversion of the hydroxyl group into its trimethylsilyl or benzyl ether. In contrast, the corresponding *cis*-αβ-unsaturated esters gave products mainly formed by way of *exo* transition states.[9]

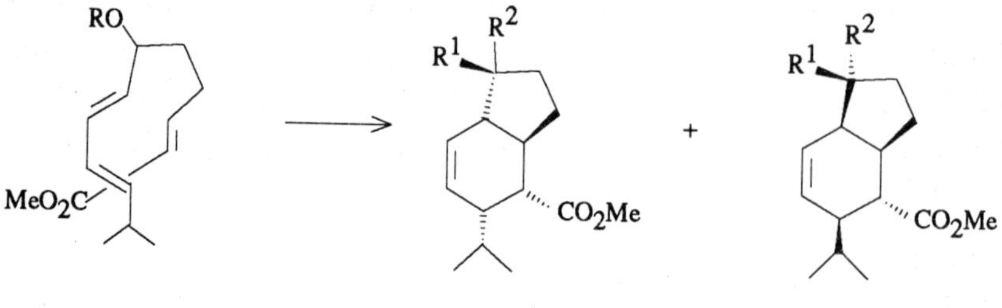

(16) (17) (endo) (18) (exo)

(a) $R^1$ = H, $R^2$ = OR
(b) $R^1$ = OR, $R^2$ = H

| | T°C | Yield% | 17a, | 17b, | 18a, | 18b | endo:exo |
|---|---|---|---|---|---|---|---|
| R = H | 150 | 71 | 37 | 37 | 4 | 26 | 70:30 |
| R = $PhCH_2$ | 150 | 78 | 31 | 51 | 4 | 14 | 82:18 |
| R = $Me_3Si$ | 150 | 83 | 52 | 31 | 4 | 17 | 79:21 |
| R = $Me_3Si$ | 115 | 82 | 53 | 32 | 2 | 15 | 83:17 |

*Scheme 7*

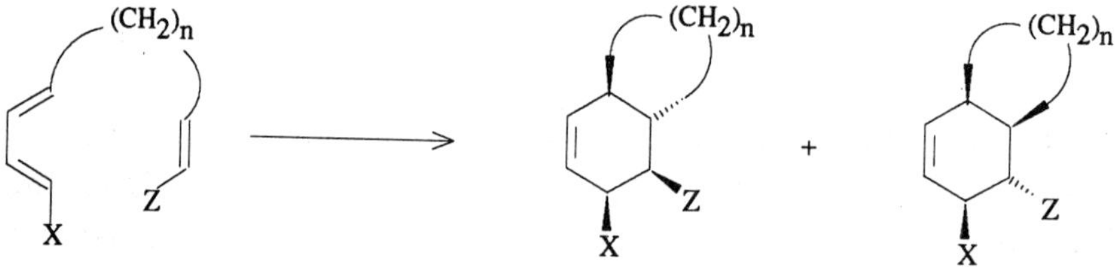

(19) trans-fused (endo) cis-fused (exo)

| Substituents | | Nonatrienes, n=3 | | Decatrienes, n=4 | |
|---|---|---|---|---|---|
| X | Z | T°C, | trans, cis | T°C, | trans, cis |
| H | H | 190 | 25 : 75 | 190 | 47 : 53 |
| H | $CO_2Et$ | 150 | 60 : 40 | 155 | 51 : 49 |
| i-Pr | $CO_2Et$ | 150 | 72 : 28 | 155 | 50 : 50 |
| $Et_2N$ | $CO_2Et$ | 60 | 85 : 15 | 40 | 55 : 45 |
| i-Pr | $CO_2Et$, $Et_2AlCl$ | 25 | >98 : 2 | 22 | 88 : 12 |

*Scheme 8*

In the related derivatives (19, X=$NEt_2$) where cyclisation took place at a considerably lower temperature, there was improved *trans* selectivity in the nonatriene series, but in the decatriene series the effect was negligible.[10] As expected the catalysed reactions gave much improved yields of the *trans*-fused (*endo*) product in each case.

Trans-fused hydroindenes are thus readily accessible by intramolecular Diels-Alder reactions of 1,3,8-nonatrienes. To ensure stereoselective formation of *cis*-fused hydroindenes the simplest procedure is to use nonatrienes with a *cis* internal double bond in the diene moiety. Because of geometrical constraints such dienes can only cyclise to give *cis*-fused hydr-indenes. The main snag appears to be the intervention of 1,5-hydrogen shifts which sometimes results in reduced yields of Diels-Alder products.[11]

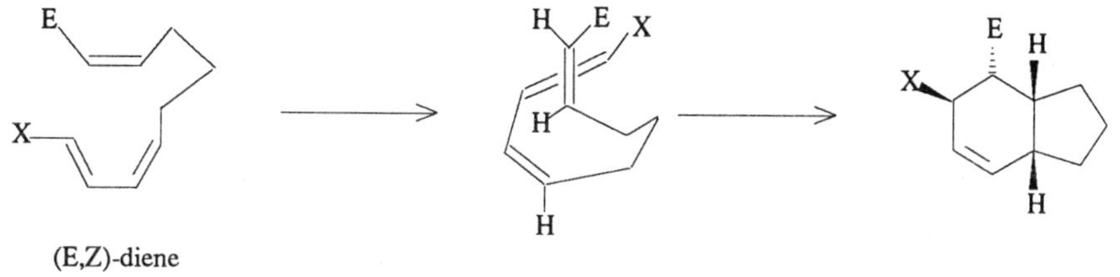

E
X
(E,Z)-diene
H E X
H
H
E H
X
H

*Scheme 9*

Thus in a key sequence in a synthesis of marasmic acid (22) the *cis*-hydrindene derivative (21) was obtained in 80 per cent yield from the triene (20) after treatment of the initial Diels-Alder adduct with potassium t-butoxide; no other isomeric cyclo-adduct was detected.[12] The corresponding (*E,E*)-diene gave mainly the *trans*-hydrindene; it cyclised at a lower temperature than (20).

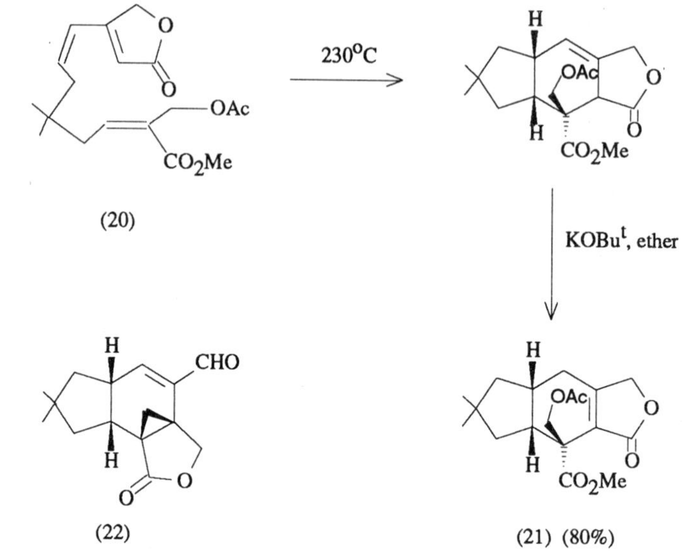

(20)

230°C

KOBu$^t$, ether

(22)

(21) (80%)

*Scheme 10*

Similarly, in a synthesis of (+)-actinobolin the chiral (Z)-diene (23) gave the *cis*-fused product (24) 'almost exclusively'.[13]

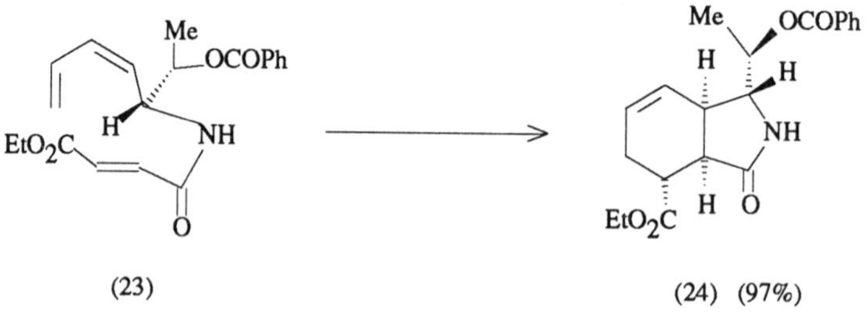

(23)

(24) (97%)

*Scheme 11*

In the cyclisation of 1,3,9-decatrienes to hydronaphthalenes (octalins) the extra methylene group in the linking chain makes the adoption of both *endo* and *exo* transition states easier than in the nonatrienes, so that mixtures of *cis*- and *trans*-fused adducts are frequently obtained. As with the nonatrienes the ratio of products is influenced by the nature and disposition of substituents in the triene, and by other factors such as the presence of heteroatoms or carbonyl groups in the chain linking the diene and dienophile moieties.[14]

The well-established preference for six-membered alicyclic rings to adopt a chair conformation and the tendency of substituents to take up an equatorial rather than an axial orientation frequently manifests itself in a highly ordered transition state in intramolecular Diels-Alder reactions of 1,3,9-decatrienes; the dominant conformation is usually the one with the least bond angle strain and the fewest non-bonded interactions. In many cases, however, two chair forms are possible, resulting in the formation of both the *trans*- and *cis*-fused octalin systems (Scheme 12).

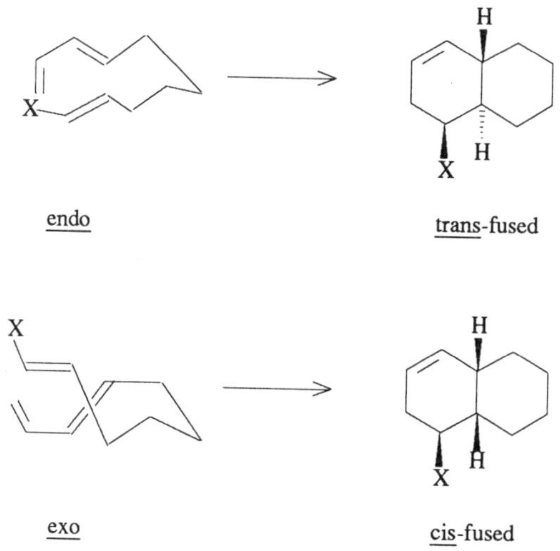

endo → trans-fused

exo → cis-fused

*Scheme 12*

As in the reactions with 1,3,8-nonatrienes the stereochemistry of the ring fusion is largely independent of the geometry of the dienophile moiety, suggesting that here again secondary orbital interactions favouring an *endo* transition state do not play a dominant rôle in controlling the stereochemistry of the transition state. Thus, the (*E,E*)- and (*Z,E*)-trienoate esters (25) and (26) gave the same ratio of *cis*- and *trans*-fused octalins on reaction at 150°C,[15] and very similar results were obtained with the corresponding 11-isopropyl compounds.[16]

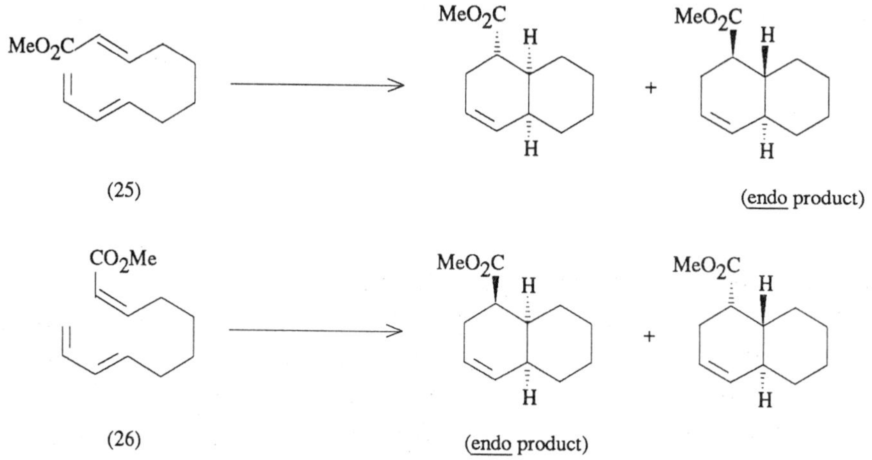

*Scheme 13*

As with the corresponding decatrieneoates, Lewis acid-catalysed reaction of the *trans*-dienophile (25) led to a greatly increased proportion of the *trans*-fused (*endo*) product, but unlike the decatrienoate the *cis*-dienophile (26) also gave a much higher proportion of the *endo* product, the *cis*-octalin in this case. Evidently, in the Lewis acid-catalysed reactions secondary orbital interactions become much more important and in the undecatrienoates are not outweighed by geometrical constraints in the *cis*-dienophile (26).

In purely thermal reactions wide differences in stereochemistry sometimes arise in reactions with closely related compounds. Thus, in contrast to 1,3,9-decatriene itself (27,R=H), the 3-methyl derivative (27,R=Me) cyclised at 160°C to give the *trans*-2-methyloctalin (28,R=Me) almost exclusively.[17] The highly preferred formation of the *trans* octalin here is ascribed to a destabilising non-bonded interaction of the methyl substituent in the transition state for the *cis*-fused octalin.

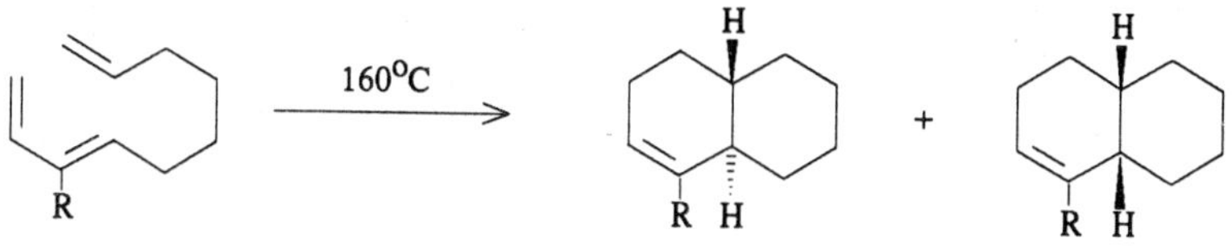

H H

160°C +

R R H R H

(27) (28)

| | |
|---|---|
| R = H | 55 : 45 |
| R = Me | 94 : 6 |

*Scheme 14*

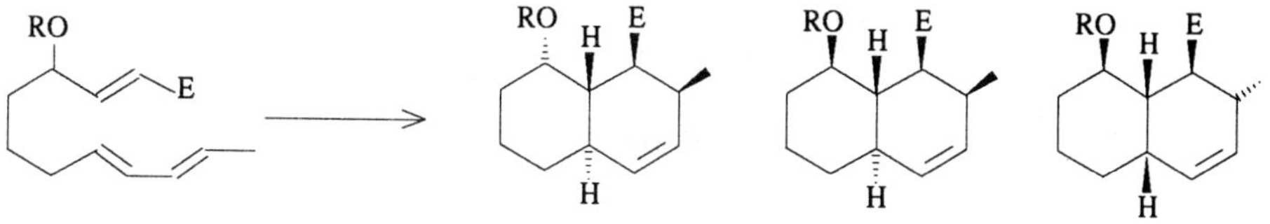

RO E RO H E RO H E RO H E

H H H

(29), R = H
(33), R = t-BuMe$_2$Si

(30) (31) (32)

| Triene | Conditions | Yield% | Ratio 30 | 31 | 32 |
|---|---|---|---|---|---|
| 29 | 155°C, 6h. | 53 | 12 | 41 | 47 |
| | 25°C, 1 mol. EtAlCl$_2$ | 55 | 55 | 45 | 1 |
| 33 | 140°C, 12h. | 65 | 65 | 13 | 22 |
| | 25°C, 1 mol. EtAlCl$_2$ | 65 | 98 | 2 | 0 |

*Scheme 15*

A good illustration of the way in which the stereochemistry of the cyclisation can sometimes be controlled by the appropriate choice of substituents and reaction conditions, is seen in the cyclisation of the alcohol (29) and the corresponding t-butyldimethylsilyl ether.[18] Heating (29) in bromobenzene gave a mixture of the three products (30,R=H), (31,R=H) and (32,R=H); the proportion of the desired *trans* (*endo*) compound (30,R=H) was increased in the Lewis acid-catalysed reaction. With the corresponding t-butyldimethylsilyl ether (33), however, (30,R=t-Bu$Me_2$Si) was almost the exclusive product.

The *trans*-fused compounds are formed by way of chair-like *endo* transition states of the forms (34) and (35) and these are favoured in the presence of the Lewis acid. With the bulky t-butyldimethylsilyl ether in the presence of Lewis acid the form (34) with the equatorial silyloxy substituent is favoured even more, leading to almost exclusive formation of (30,R=t-Bu$Me_2$Si).

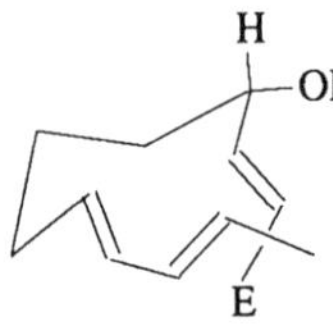

H OR E

(34)

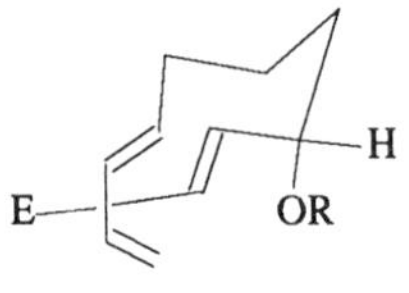

E H OR

(35) E = $CO_2Et$

*Scheme 16*

However, not all intramolecular Diels-Alder reactions of deca-1,3,9-trienes take place through transition states with a chair-like conformation of the linking chain. In an approach to the synthesis of nargenicin $A_1$, heating the optically active triene (37) in toluene at 110°C, which was expected to give the optically active octalin derivative (39) by way of the chair-like transition state (38), in fact led exclusively to the isomer (41).[19] For the *cis*-fused diastereomer (41) to be obtained the Diels-Alder reaction must have been constrained to proceed through the transition state (40) with a boat-like conformation of the connecting chain.

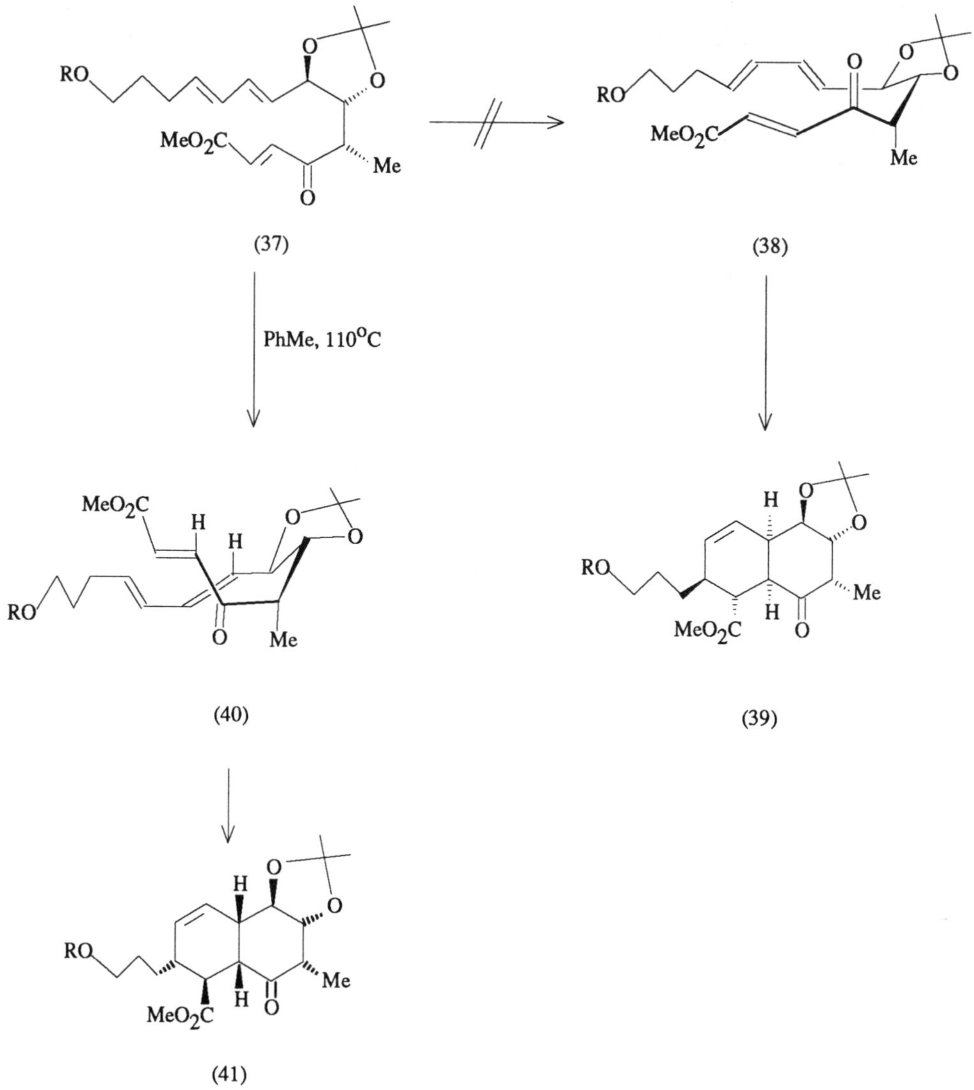

RO MeO2C O O Me O (37) RO MeO2C O O O Me (38)

PhMe, 110°C

MeO2C H H O O RO O Me (40) H O O RO MeO2C H O Me (39)

RO H O O O MeO2C H O Me (41)

*Scheme 17*

Judicious placement of a carbonyl group in the linking chain can sometimes cause a change in product distribution by preferred adoption of the transition state conformation in which orbital overlap is best preserved. *Trans*-fused products often predominate when the carbonyl group is in conjugation with the diene. Thus, the triene (42) gave mainly the *trans*-octalone (43) when heated in toluene at 160°C.[20]

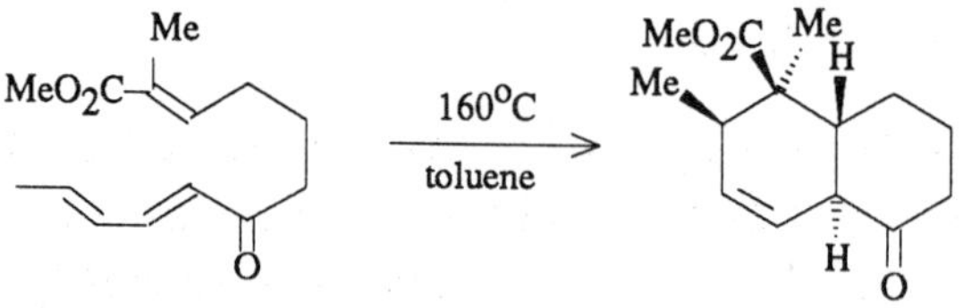

(42) (43)

*Scheme 18*

In the *endo* transition state leading to the observed *trans*-fused product (43), the carbonyl group remains coplanar with the diene, whereas in the *exo* transition state it is twisted out of the plane of the diene. Replacement of the carbonyl group by the cyclic acetal in analogues of (42) results in the formation of *cis*-fused products because of destabilising non-bonded interactions in the transition state leading to the *trans* isomers.[21]

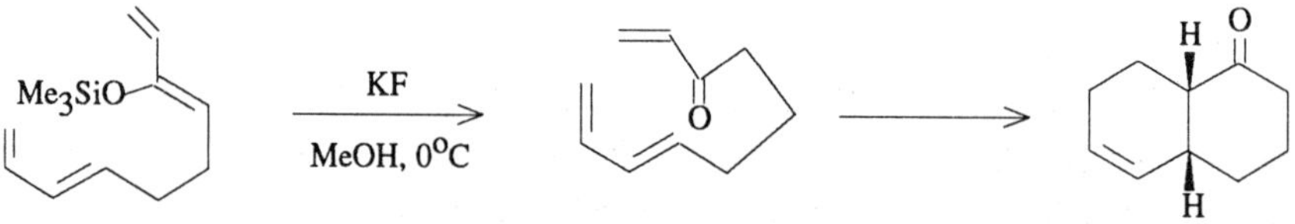

(44) (45) (46) (78%)

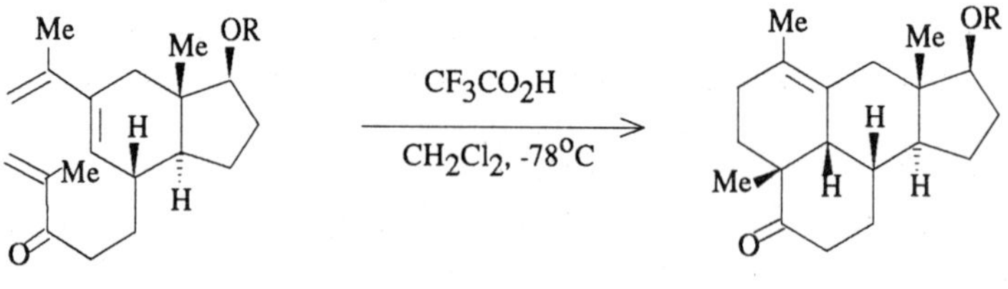

(47) (48) (68%)

R = t-BuMe$_2$Si

*Scheme 19*

When a carbonyl group in the connecting chain is conjugated with the dienophile, *cis*-fused products usually predominate, formed by way of the favoured *endo* transition state as with (37). These reactions generally take place under mild conditions. Thus, the decatriene (45) generated from its silyl enol ether, cyclised at 0°C to give the *cis*-fused product (46) exclusively[22] and in a key step in an approach to 11-oxygenated steroids the triene (47) gave very largely the *endo cis*-fused adduct (48) on catalysis with trifluoroacetic acid at –78°C.[23]

All the atoms in the linking chain need not be carbon atoms. Numerous examples are known in which the chain contains amino, ether or sulphide links, and the stereochemical course of these reactions largely follows that of the all-carbon analogues. Thus, the (*E,Z,E*)-triene (49) gave exclusively the *cis*-fused products (50) and (51) in which the main product arises by way of the preferred transition state (52) with an equatorial C-6 methyl substituent,[24] and the triene-ether (53) gave the products (54) and (55).[25]

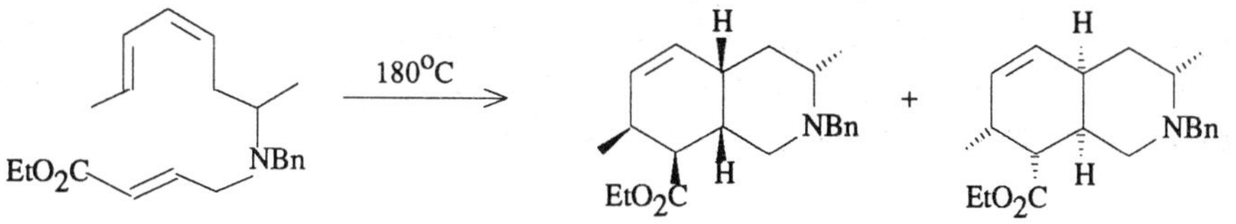

$EtO_2C$ NBn (49) —180°C→ H NBn $EtO_2C$ H (50) + H NBn $EtO_2C$ H (51) (3:1)

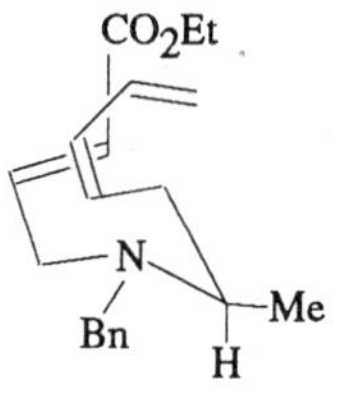

$CO_2Et$ N Bn Me H (52)

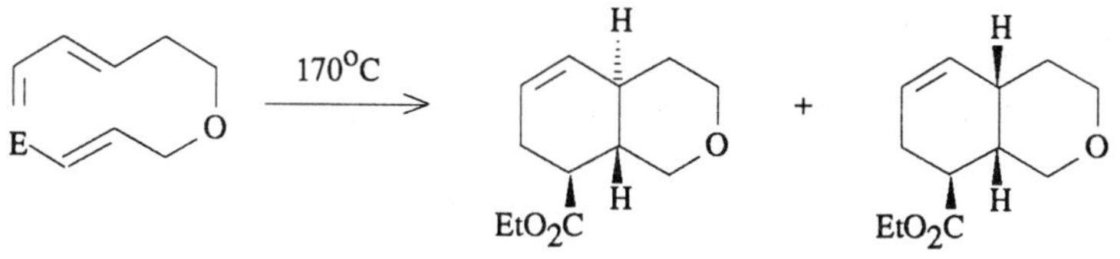

E O (53) $E = CO_2Et$ —170°C→ H O $EtO_2C$ H (54) + H O $EtO_2C$ H (55) (86%; 3:2)

*Scheme 20*

If the nitrogen is present in the form of an amide, the group has a variable effect on the stereoselectivity depending on its position relative to the diene and dienophile and the length of the connecting chain. Other things being equal that transition state is favoured which preserves the amide resonance.[26] Similarly with esters; the ester (56), for example, cyclised readily to give mainly the *trans*-fused product (57), but the non conjugated ester (58) in contrast to the ether (53), was resistant to cyclisation and isomerised to (56) before reacting.[25] It is suggested that the geometry of the transition state required for cyclisation of (58) prevents complete overlap of the lone pair electrons of the ethereal oxygen with the π-system of the carbonyl group and that reaction is thereby kinetically disfavoured in (58) compared with (53) where no such effect operates.

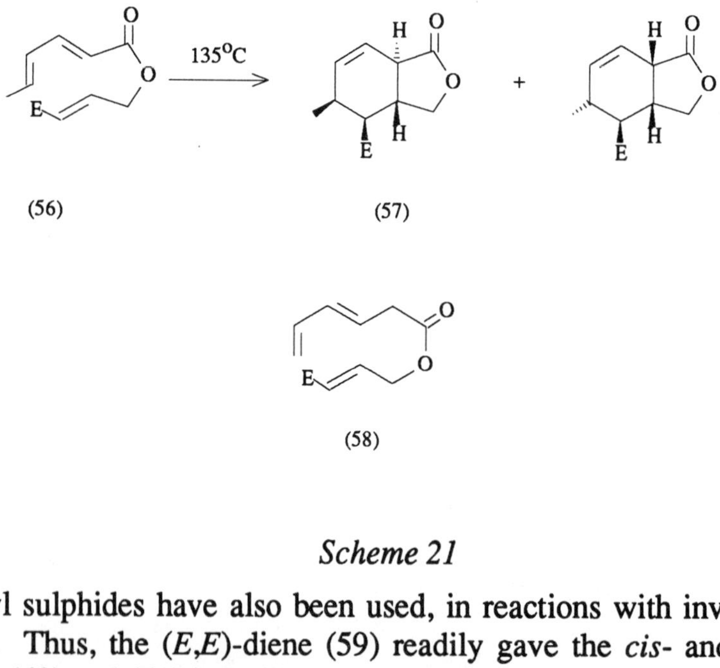

135°C

(56) (57) (58)

*Scheme 21*

Vinyl sulphides have also been used, in reactions with inverse electron demand. Thus, the (*E,E*)-diene (59) readily gave the *cis*- and *trans*-fused products (60) and (61) in boiling *ortho*-dichlorobenzene,[27] and in a different series, catalysed cyclo-addition to the *cis* enal system of (62) gave exclusively the *cis*-fused products (63) and (64). The reaction with the *trans* enal was less stereoselective.[28] In these reactions the vinyl sulphides are superior to enol ethers as electron-rich dienophiles. It is believed that the catalysed reaction proceeds by way of the zwitterion (65) and is not a true cycloaddition.

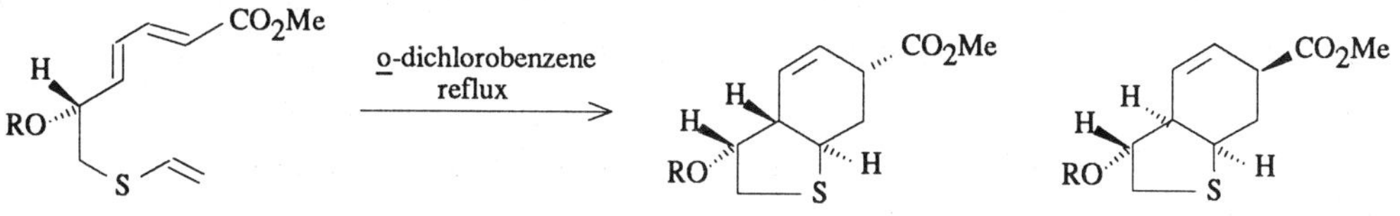

o-dichlorobenzene reflux

(59) (60) (61) (98%; 1:1)

R = t-$BuPh_2Si$

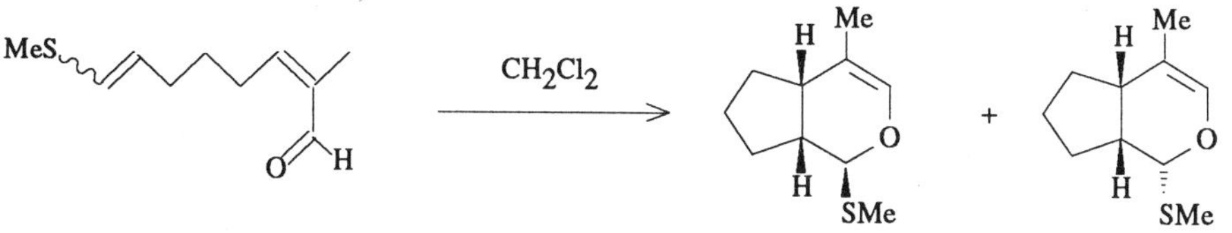

$CH_2Cl_2$

(62) (63) (64)

| | | |
|---|---|---|
| 50°C, 19h, | 35% | 96 : 4 |
| $BF_3.Et_2O$, -78°C, 10min., | 93% | 75 : 25 |

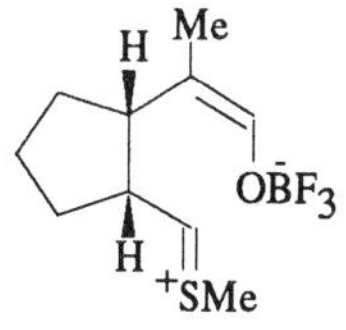

(65)

*Scheme 22*

A reaction of this type was employed in a stereospecific synthesis of (+)-nepetalactone (68) from the ketene dithioacetal (*Z*)-enal (66). Cyclisation catalysed by boron trifluoride etherate gave (67) as a single diastereomer in 55 per cent yield. Hydrolysis then gave (68).

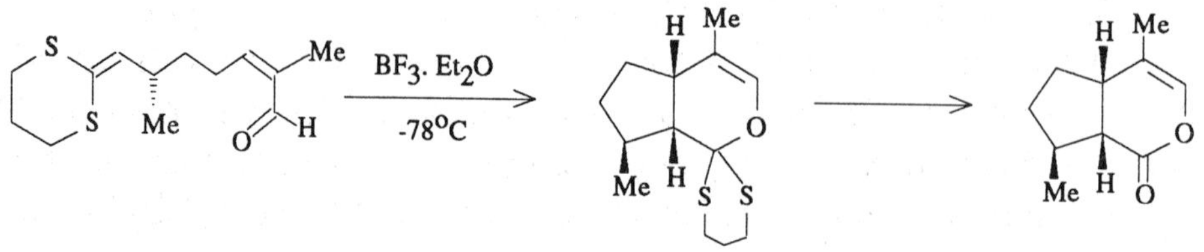

*Scheme 23*

## Diastereoselection

Substituents in the connecting chain may influence the facial selectivity of the cycloadditions as well as the *endo*:*exo* selectivity. Because of non-bonded interactions one approach to the transition state may be more favoured than others and this seems to be particularly the case when a substituent is in the allylic position to the diene system. Thus, in a synthesis of the antibiotic indanomycin the triene (69), heated in toluene, gave the diastereomer (70), with five contiguous chiral centres, in almost quantitative yield.[29] Diastereoselection in favour of (70) is attributed to the preferential formation of the *endo* transition state (71). This transition state is favoured over the alternative (72), which would lead to a product epimeric at C-6, because of steric interaction between the ethyl substituent and the hydrogen at C-3 of the diene system in (72).

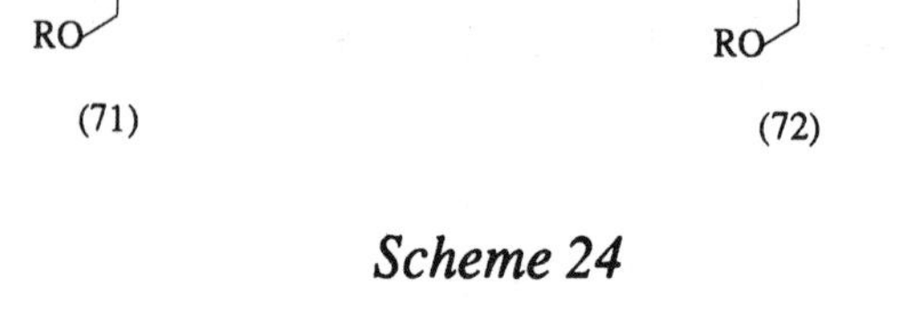

*Scheme 24*

In agreement, if the hydrogen atom at C-3 of the diene is replaced by a sterically more demanding substituent even higher diastereoselectivity can be obtained. Thus, the di-ester (73,R=H), when heated in toluene, gave a mixture of the two diastereomers (74,R=H) and (75,R=H), but with the trimethylsilyl derivative (73,R=$SiMe_3$) only (74,R=$Me_3Si$) was obtained.[30] Formation of (74,R=$Me_3Si$) takes place through the favoured transition state (76) in which steric interference between the ethyl and trimethylsilyl substituents is minimised.

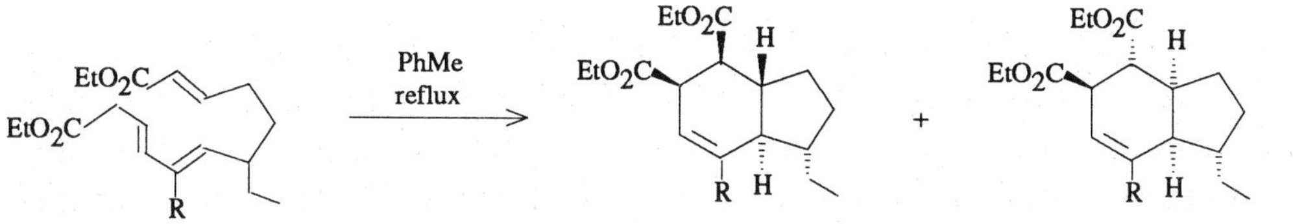

(73) $\xrightarrow{\text{PhMe, reflux}}$ (74) + (75)

R = H ratio 4 : 1
R = $Me_3Si$ (74) only (85% yield)

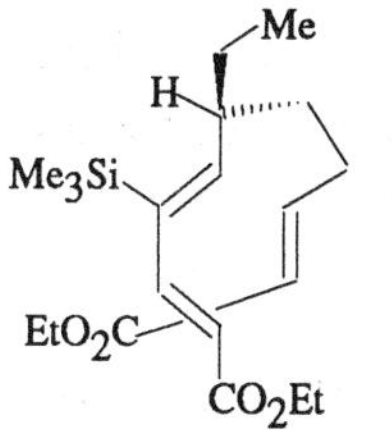

(76)

*Scheme 25*

If the starting material is optically active an enantioselective synthesis of the adduct becomes possible. Thus, in a synthesis of the tetracyclic ketone (79), an intermediate for the synthesis of the antibiotic (+)-ikarugamycin, the optically pure triene (77) gave the bicyclic ester (78) as the major stereoisomer by way of the preferred transition state (80).[31]

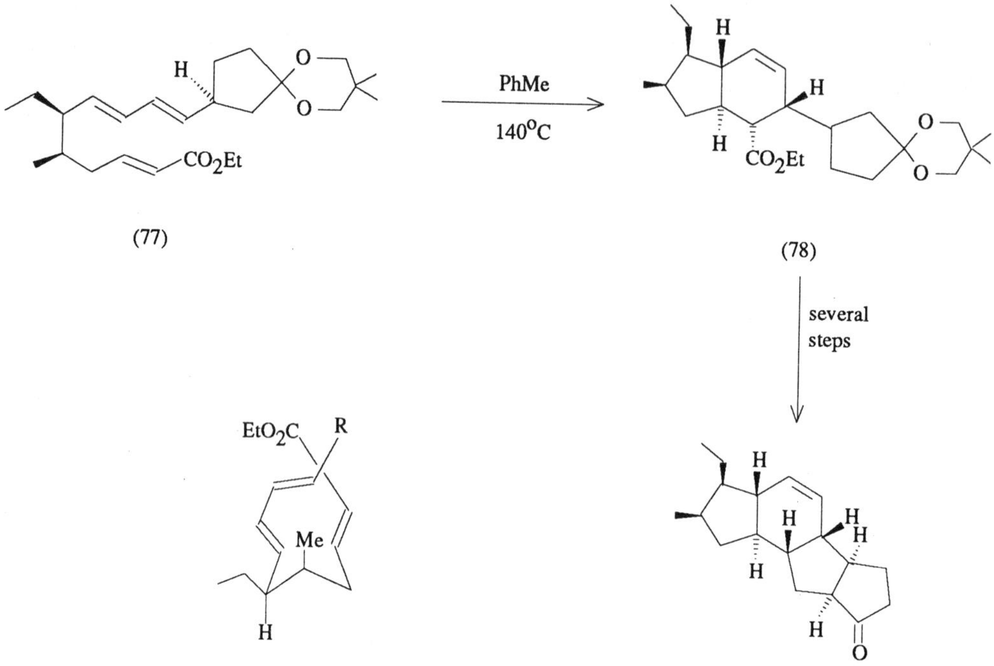

(77)

PhMe

140°C

(78)

several steps

(80)

(79)

*Scheme 26*

The stereochemical advantage of using a (*Z*)-diene in Diels-Alder reactions is illustrated again in the cyclisation of the diene (81) which gave the optically pure product (82) in 95 per cent yield as a single stereoisomer by heating in toluene solution. Under the same conditions the isomeric (*E,E*)-diene did not cyclise.[32] A related reaction was employed in an approach to cytochalasin.[33]

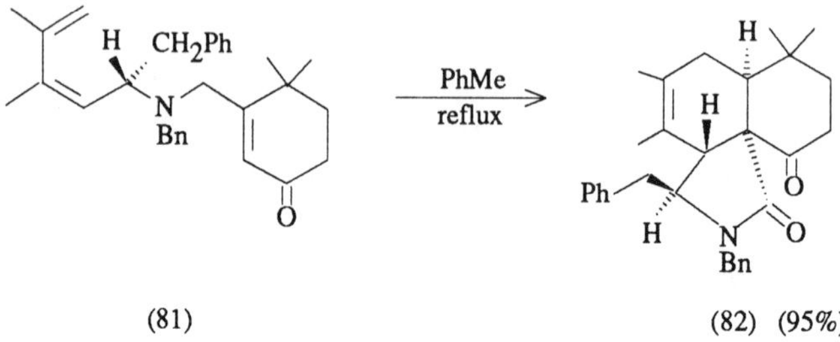

PhMe reflux

(81)

(82) (95%)

*Scheme 27*

Diastereoselectivities in Intramolecular Diels-Alder Reactions of Chiral Amides

RCO (83) $\xrightarrow[CH_2Cl_2,\ -30^{o}C]{Me_2AlCl}$ RCO H (84) + RCO H (85)

| R = | | | (84) | (85) |
|---|---|---|---|---|
| (a) | oxazolidinone N–, H, Ph | ratio | 95 | 5 |
| (b) | oxazolidinone N–, $C_6H_{11}$, H | ratio | 3 | 97 |

RCO (86) $\xrightarrow[CH_2Cl_2,\ -30^{o}C]{Me_2AlCl}$ ROC H (87) + ROC H (88)

| R = | | | (87) | (88) |
|---|---|---|---|---|
| (a) | oxazolidinone N–, H, Ph | ratio | 97 | 3 |
| (b) | oxazolidinone N–, Ph, Me | ratio | 9 | 91 |

(87a) $\xrightarrow{LiOBn}$ $BnO_2C$ H (+)-(89) $BnO_2C$ H (-)-(89) $\xleftarrow{LiOBn}$ (88b)

High diastereofacial selectivities have also been realised in intramolecular cycloaddition reactions of trienes bearing chiral auxiliaries. Particularly good results have been obtained with imides prepared from 1,3,8-decatrienoic acids and 1,3,9-undecatrienoic acids and chiral oxazolidones as shown in the Table.[34] The reactions catalysed by $Me_2AlCl$ or $Et_2AlCl$ took place particularly easily and were highly *endo* and facially selective. Either diastereomer (84/85) or (87/88) could be obtained at will, depending on the choice of chiral auxiliary. The crystalline adducts were readily purified by chromatography and on non-destructive transesterification with lithium benzyloxide or lithium hydroperoxide[35] regenerated the valuable chiral auxiliary, and liberated the optically pure hydrindene or octalin.

These reactions are believed to take place by facially selective addition to the enone double bond in a chelated complex of form (91).

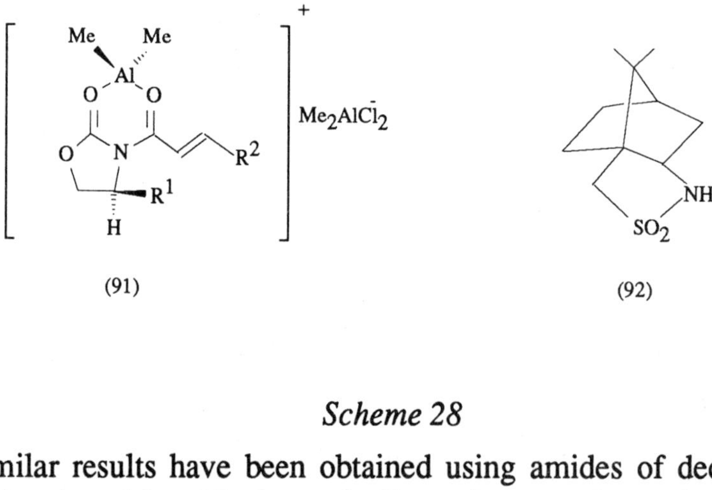

(91)

(92)

*Scheme 28*

Very similar results have been obtained using amides of decatrienoic and undecatrienoic acids with the optically active camphor sultam (92). The sultam is readily available in both antipodal forms, is easily regenerated from the adducts and the generally crystalline adducts are easily purified.[36]

**Synthesis of Terpenoids**

Intramolecular Diels-Alder reactions are now being increasingly employed in the synthesis of natural products where they frequently lead to elegant, economical and stereocontrolled construction of polycyclic ring systems. In many cases, however, the synthesis of the triene required for the Diels-Alder cyclisation is itself a considerable synthetic undertaking, and this sometimes overshadows the elegance of the cyclisation step.

An early example of its application in the sesquiterpene series is seen in the synthesis of the eudesmane hydrocarbon (±)-selena-3,7(11)-diene (97)[37] The eudesmane skeleton, containing a *trans*-fused hydronaphthalene nucleus with an angular methyl substituent can be conveniently constructed by intramolecular Diels-Alder reaction of an appropriate 1,3,9-decatriene. Control of *trans* ring fusion is often difficult in other types of synthesis of eudesmane sesquiterpenes owing to the small energy difference between the *cis*- and *trans*-fused ring systems in this series. Thus, the triene (93) heated in toluene at 200°C, cyclised smoothly to give, after hydrolysis, a mixture of the *trans*-axial (94) and *trans*-equatorial (95) products. Oxidation of the mixture afforded the ketone (96) which was easily converted into the selenadiene (97).

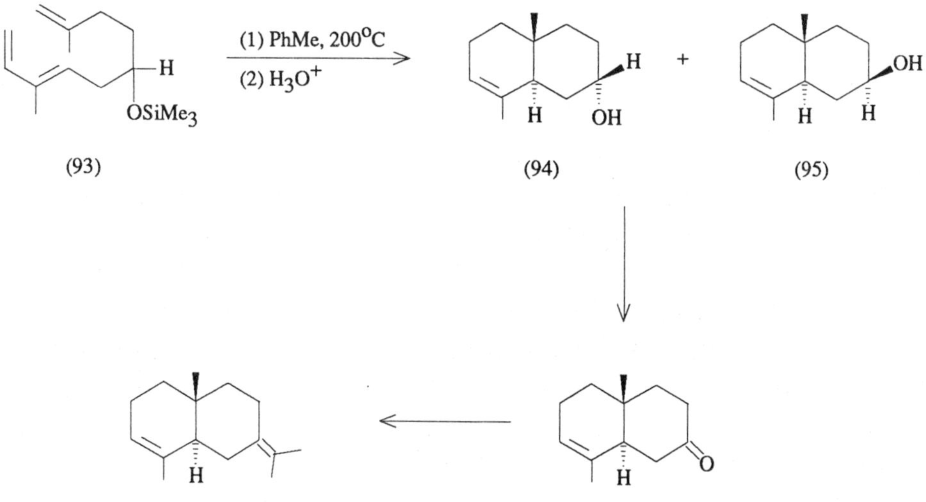

(93) —(1) PhMe, 200°C; (2) $H_3O^+$→ (94) + (95) → (96) → (97)

*Scheme 29*

Eremophilane and valencane sesquiterpenes have also been obtained using intramolecular Diels-Alder reactions to set up the ring structures.[38]

An interesting situation was encountered during cyclisation of the triene (98), in experiments aimed at the diterpene forskolin. The main product of the reaction, in addition to the *endo* adduct (99), was the isomer (100) in which the stereochemical integrity of the double bond of the dienophile appears, very unusually, to have been lost during the cycloaddition. In fact, (100) arises by epimerisation at the α-position to the lactone carbonyl group in the favoured *exo* adduct (101).[39]

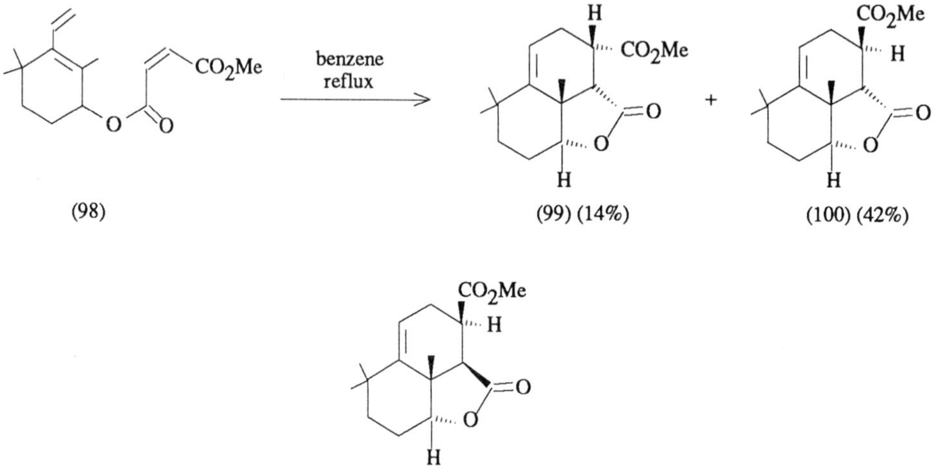

benzene reflux

(98) (99) (14%) (100) (42%)

(101)

*Scheme 30*

Intramolecular cycloaddition to cyclohexadienes and cyclopentadienes provides an elegant route to complex bridged-ring sesquiterpenes. Thus, the cyclohexadiene derivative (102), on heating at 280°C in decalin in the presence of catalytic potassium t-butoxide gave the tricyclic alcohol (103) exclusively in 60 per cent yield, by way of transition state (105) with an equatorial methyl substituent in the connecting chain. In the alternative (106) there is some non-bonded interaction between the two methyl substituents. Hydrogenation of (103) gave (±)-patchouli alcohol.[40] In the absence of potassium t-butoxide only the benzene derivative (107) was obtained. It is suggested that the action of the butoxide may be to activate the diene system by the neighbouring ionised hydroxyl group.

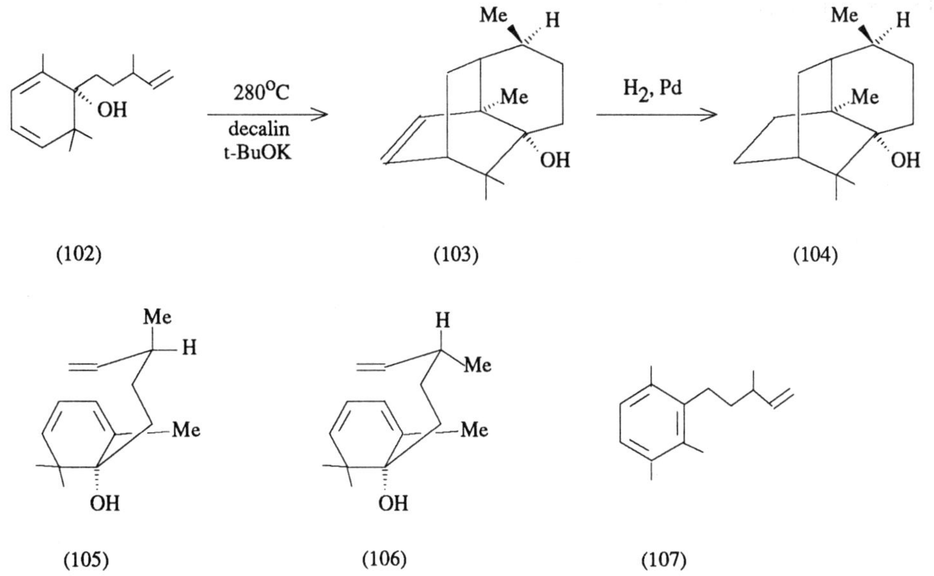
OH
280°C
decalin
t-BuOK
Me
H
Me
OH
$H_2$, Pd
(102)
(103)
(104)
(105)
(106)
(107)

*Scheme 31*

The triene (108) cyclised more readily than (102) because of the activating effect of the carbonyl group. The product (109) was converted in several more steps into (±)-pupukeanone.[41] (110)

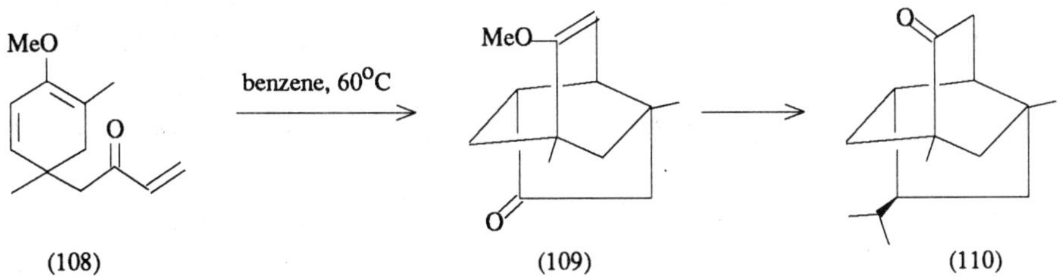
MeO
O
benzene, 60°C
(108)
(109)
(110)

*Scheme 32*

The thermal intramolecular [4+2]cycloaddition of alkenylcyclopentadienes (111) gives one or other of the tricyclic adducts (112) and (113) depending on the length of the connecting chain. A chain length of 3 or 4 atoms leads specifically to adducts of type (112) *via* the 1-cyclopentadienylalkene, whereas a connecting chain of two atoms gives exclusively adducts of type (113) *via* the 5-cyclopentadienylalkene. In the latter case, despite the low equilibrium concentration of the 5-cyclopentadienyl isomer, cycloadducts (113) are favoured because of the highly strained transition states for alternative modes of cyclisation.[42]

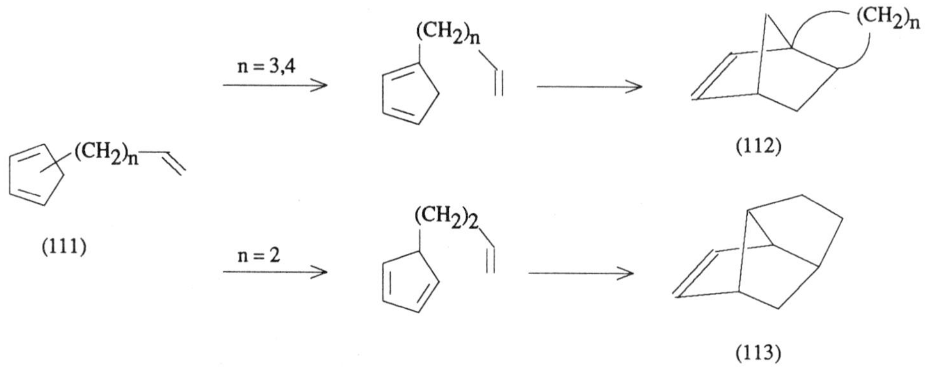

$(CH_2)_n$ (111) n = 3,4 $(CH_2)_n$ (112) n = 2 $(CH_2)_2$ (113)

*Scheme 33*

A reaction of the former type formed the key step in a synthesis of (±)-cedrene.[43] The latter procedure was exploited in a synthesis of (±)-sativene (117). In this synthesis and others like it, the six-membered bridging ring of the sesquiterpene cannot be set up directly in the Diels-Alder step because of the requirement that the connecting chain contain only two atoms (cf.113), and it was developed after the cycloaddition by ring expansion.[44] Thus, treatment of the cyclopentenone (114) with chlorotrimethylsilane, triethylamine and anhydrous zinc chloride in refluxing toluene gave directly, after aqueous work-up, the tricyclic keto-ester (116) in 94 per cent yield by way of the cyclopentadienyl derivative (115).

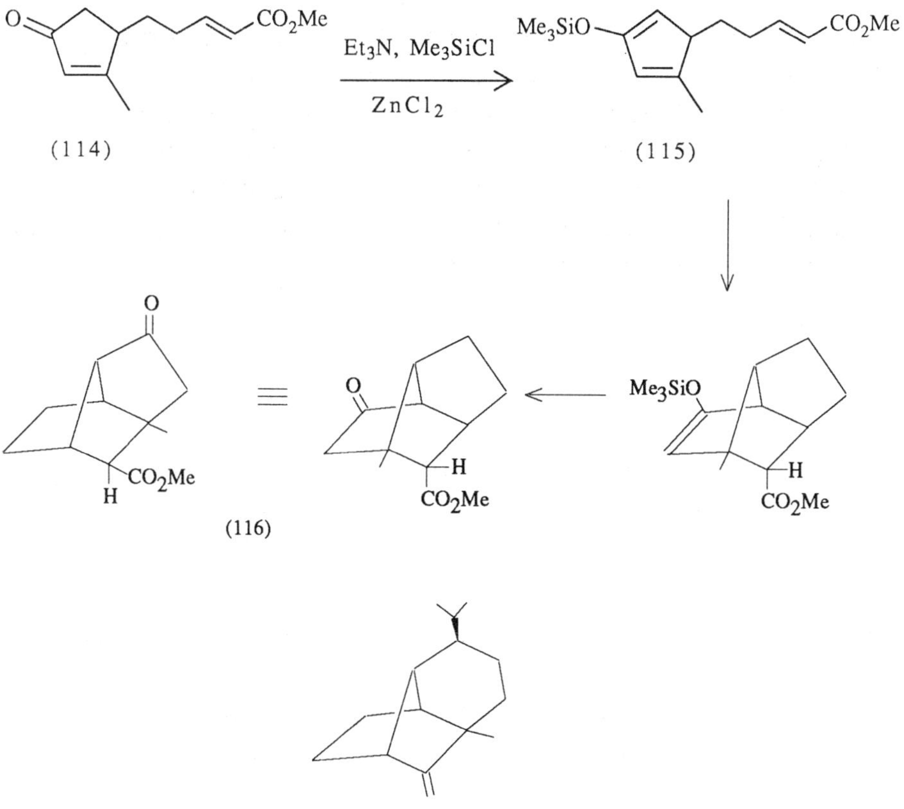

$CO_2Me$ $\xrightarrow[ZnCl_2]{Et_3N,\ Me_3SiCl}$ $Me_3SiO$ $CO_2Me$

(114) (115)

O O $Me_3SiO$

$CO_2Me$ H $\equiv$ H $CO_2Me$ $\leftarrow$ H $CO_2Me$

(116)

(117)

*Scheme 34*

Vinylallenes, although not frequently employed, have some advantages as dienes in intramolecular Diels-Alder reactions. Because of the rigidity of the allene system the number of possible conformations of the transition state may be restricted, leading to greater stereoselectivity in the cycloadditions, and steric interactions which hinder attainment of the s-*cis*-conformation in (*Z*)-dienes is greatly diminished in the allenes. Unfortunately, they also readily undergo 1,5-hydrogen shifts to form 1,3,5-trienes which sometimes results in lowering of the yield of the desired cycloadditon product.[45] (Scheme 35)

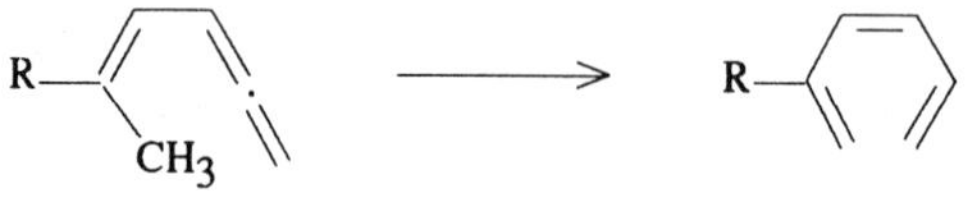

R— CH3 ⟶ R—

*Scheme 35*

Thus, in experiments aimed at the hexahydronaphthalene fragment of the hypocholesterolemic agent compaction (121) the allene (118) readily formed the adduct (119) when heated in benzene at 150°C, by way of the accessible *exo* transition state shown.[46]

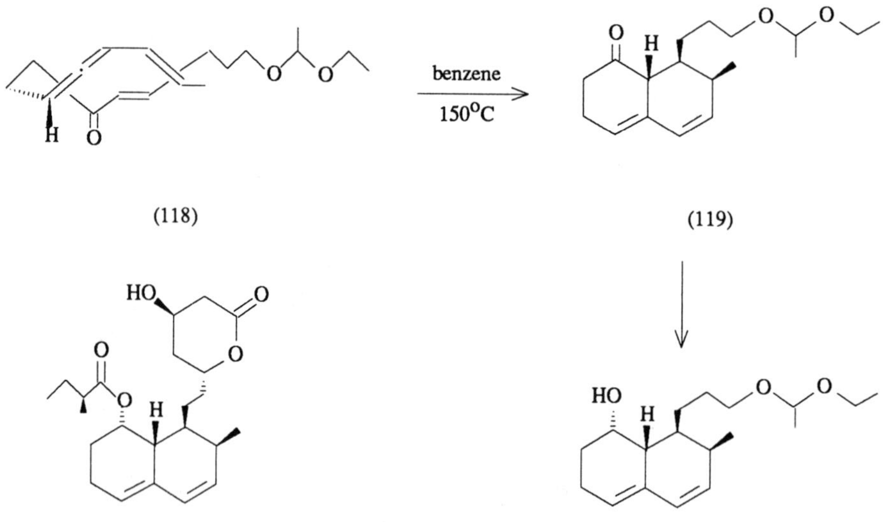

benzene
150°C

(118) (119)

(121) (120)

*Scheme 36*

3-Silyloxyvinylallenes have been used with advantage in some cases, for the silyl enol ethers formed after intramolecular cycloaddition yield cyclic ketones on hydrolysis. Thus, cyclisation of (122) catalysed by diethylaluminium chloride gave the adduct (123) as a mixture of stereoisomers which were used to prepare the eremophilane sesquiterpene (±)-dehydrofukinone (124). The isomeric allene (125) on cyclisation gave (126) which was converted into selena-4(14),7(11)dien-8-one (127).[47]

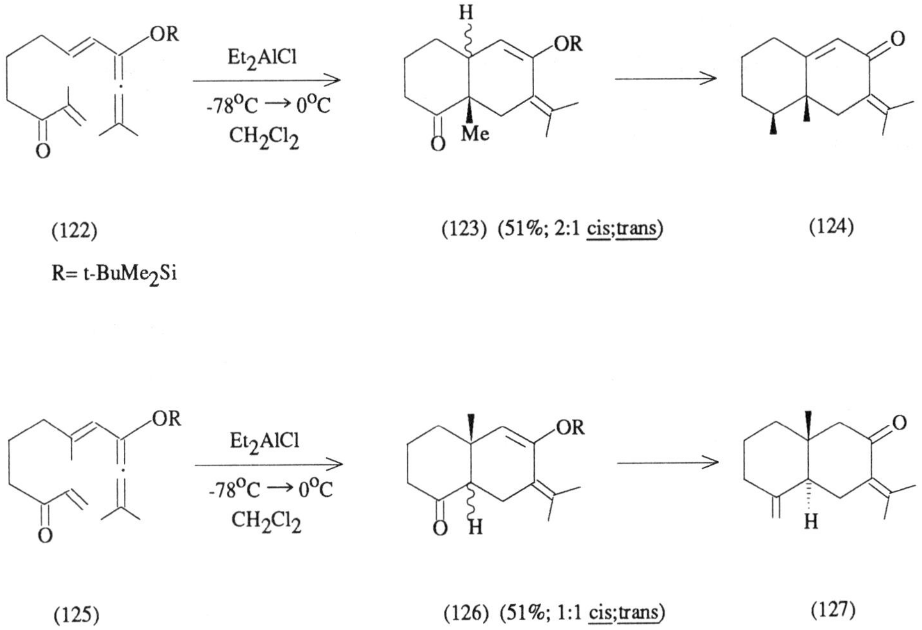

$Et_2AlCl$, -78°C → 0°C, $CH_2Cl_2$

(122) R= t-BuMe$_2$Si    (123) (51%; 2:1 cis;trans)    (124)

$Et_2AlCl$, -78°C → 0°C, $CH_2Cl_2$

(125)    (126) (51%; 1:1 cis;trans)    (127)

*Scheme 37*

Allenic esters have also been employed as dienophiles and have advantages over αβ-unsaturated esters for this.[48]

An interesting example of chirality transfer is found in a neat synthesis of the sesquiterpene (+)-sterpurene (131). The chiral vinylallene (129), generated *in situ* by sulphenate-sulphoxide rearrangement from the chiral propargyl alcohol (128), gave the cyclised product (130) in 70 per cent yield at room temperature, as a mixture of sulphoxide diastereomers. Several more steps led to (+)-sterpurene.[49]

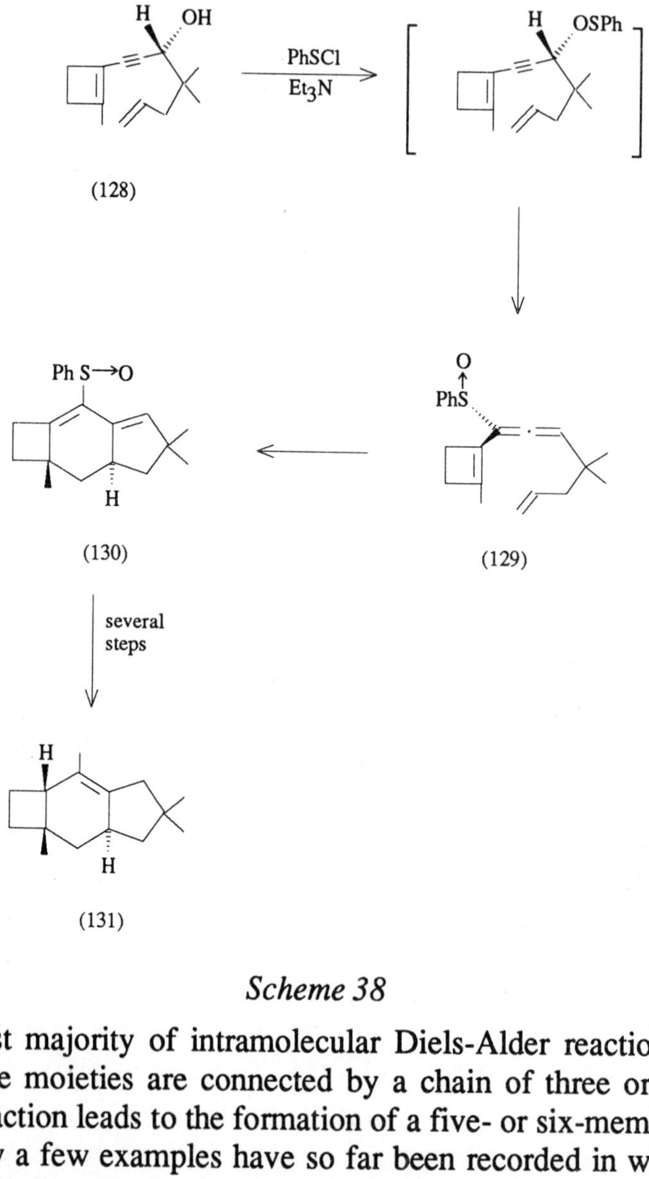

H OH

PhSCl
$Et_3N$

H OSPh

(128)

Ph S→O

H

O
↑
PhS

(130)

(129)

several
steps

H

H

(131)

*Scheme 38*

In the vast majority of intramolecular Diels-Alder reactions the diene and dienophile moieties are connected by a chain of three or four carbon atoms, and reaction leads to the formation of a five- or six-membered alicyclic ring. Only a few examples have so far been recorded in which a larger alicyclic ring is formed. In one successful application, used in an approach to the taxane ring system, catalysed intramolecular cyclisation of the triene (132) at room temperature gave the ketone (133) as a single isomer in 72 per cent yield,[50] a rare application in natural product synthesis of an intramolecular Diels-Alder reaction using a 2-substituted diene.

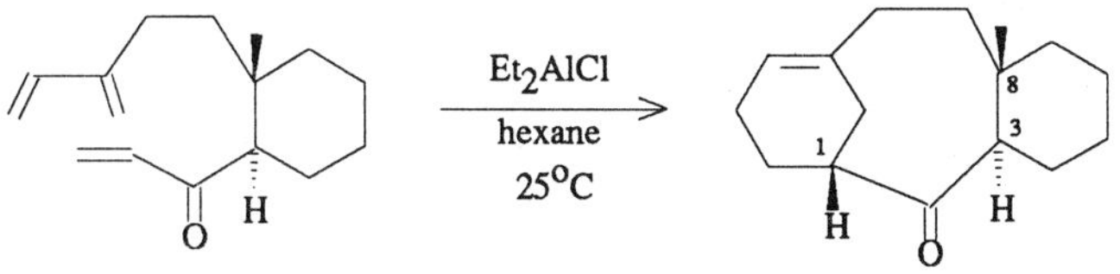

$Et_2AlCl$
hexane
25°C

(132) (133)

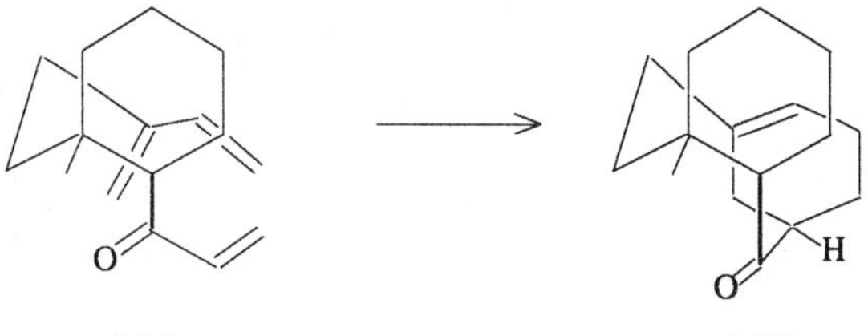

(134) (135)

*Scheme 39*

The tricyclic (133) has the ring system and the relative configuration at C-1, C-3 and C-8 of the taxane diterpenoids. Formation of (133) is believed to take place by way of a transition state of the form (134) leading to the stable boat-chair form of the cyclo-octane product (135).

Again, in a synthesis of the fungal metabolite cytochalasin H (136) the 11-membered ring was formed during the course of an intramolecular Diels-Alder step. Heating the triene-pyrrolinone (137) in toluene gave the adduct (138) as the desired stereoisomer, formed by *endo* addition of the triene to the less-hindered α-face of the pyrrolinone. The cyclisation was highly regio- and stereo-selective; only very small amounts of other products were detected.[51] A similar approach was used in the synthesis of cytochalasin G.[52]

## Synthesis of Alkaloids

With nitrogen in the connecting chain nitrogen heterocycles are obtained. Both dienylamides and enamino amides have proved useful in this way. Thus in a key step in the synthesis of (–)-pumiliotoxin-C the diene-amide (139), derived from (*S*)-norvaline, gave the *cis*-hydroquinoline (141) in 60 per cent yield with only minor amounts of stereoisomeric adducts. In this cyclisation the chiral centre of (139) controls the stereochemistry of the three developing chiral centres, presumably by way of the boat transition state (140) in which the propyl substituent adopts an equatorial orientation.[53]

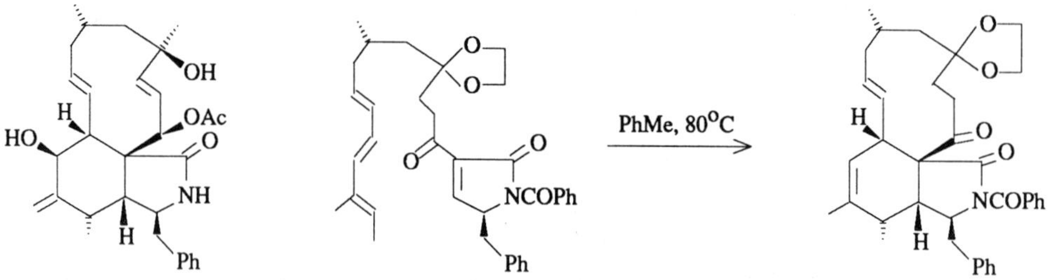

(136) (137) PhMe, 80°C (138) (58%)

*Scheme 40*

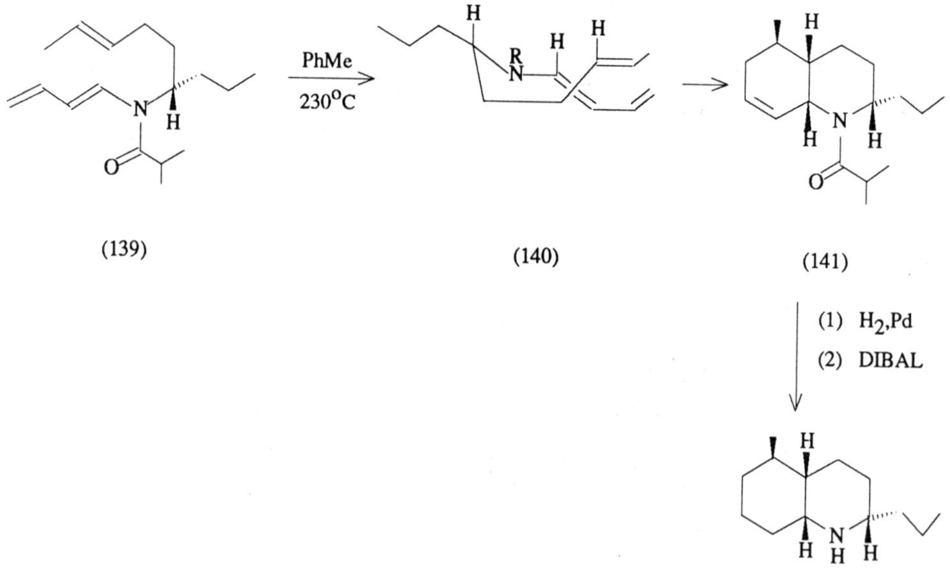

PhMe 230°C

(139) (140) (141)

(1) $H_2$,Pd
(2) DIBAL

(142)

*Scheme 41*

In an approach to the galanthane ring system (as 145) common to many *Amaryllidaceae* alkaloids the dienamide (143) in boiling chlorobenzene gave the single diastereomer (144). Only the *exo* transition state, leading to the observed *trans*-fused product, is reasonably accessible here.[54]

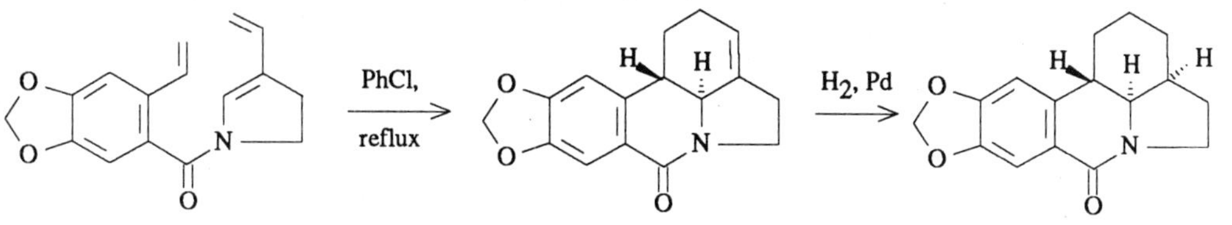

PhCl, reflux

$H_2$, Pd

(143) (144) (52%) (145)

*Scheme 42*

A dienamide (146) generated *in situ* was also employed in the synthesis of *cis*-dihydrolycoricidine (147).[55]

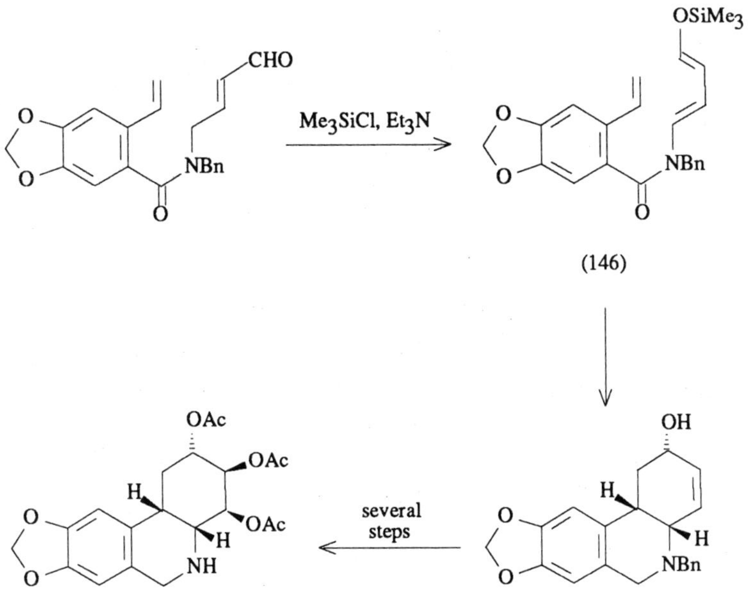

$Me_3SiCl$, $Et_3N$

(146)

several steps

(147) (60%; 3:1)

*Scheme 43*

Annellated hydroindoles and hydroquinolines have also been obtained from endocyclic enamido dienes like (148).[56]

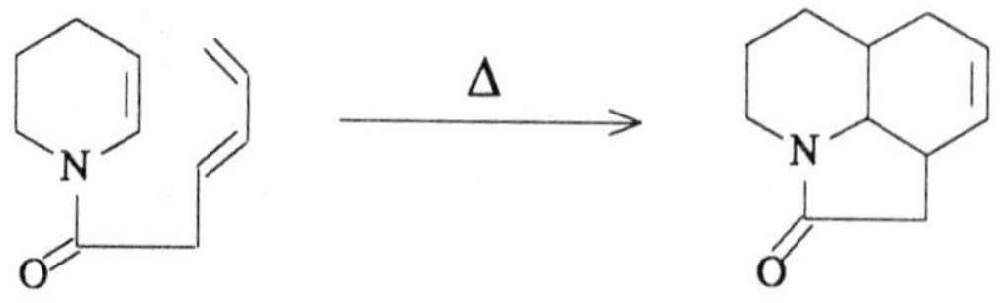
N
O
Δ
N
O

(148)

*Scheme 44*

It was found expedient in practice to use substituted dihydrothiophene dioxides as butadiene equivalents in preparing the enamido dienes for these reactions. Thus, the sulphone (149), on passage through a hot tube at 600°C, formed the enamido diene (150) by cheletropic expulsion of sulphur dioxide, and thence the lulolidine derivative (151) a key intermediate in a synthesis of the *Aspidosperma* alkaloid aspidospermine.

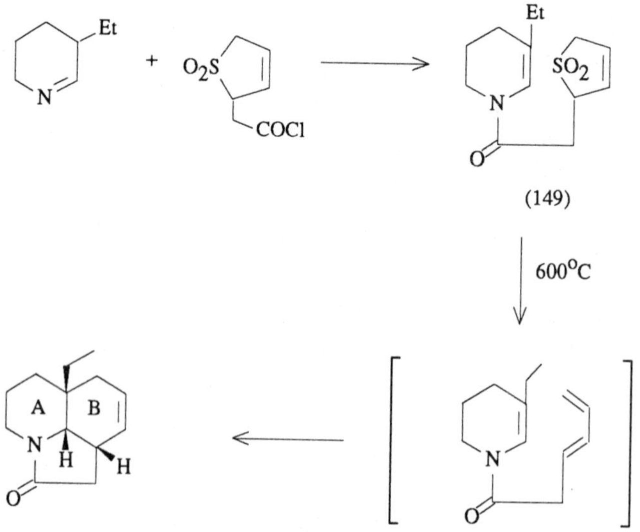
Et
N
+
O2S
COCl
Et
SO2
N
O
(149)
600°C
A
B
N
H
H
O
(151)
N
O
(150)

*Scheme 45*

The single B/C *cis*-fused cycloadduct (151) arises by way of the *endo* transition state, which is less strained than the corresponding *exo* transition state. With the analogue (152), which has one more carbon atom in the connecting chain, a mixture of the B/C *cis*- and *trans*-fused hydrojulolidines (153) and (154) was obtained. Because of the longer chain an *exo* transition state is now accessible.

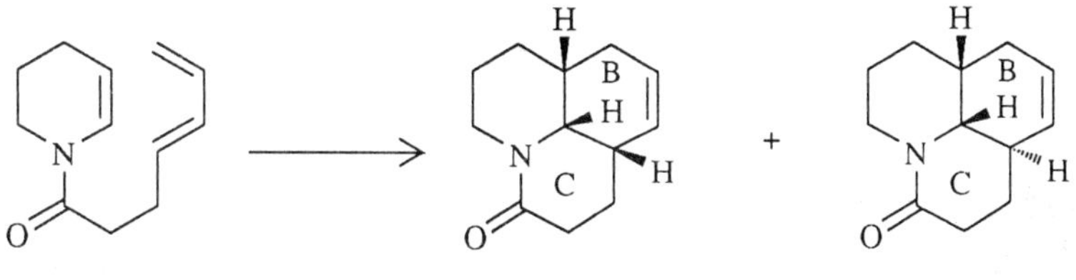

(152) (153) (154) (2:1)

*Scheme 46*

Hydroisoquinolines also have been obtained by intramolecular cycloadditions using both dienylamides and enamides. Thus, the dienylamide (155) in which the α-pyrone serves as a butadiene equivalent gave the cycloadduct (156) in 93 per cent yield in refluxing xylene. This was subsequently converted into the fully functionalised D/E ring fragment (157) in a synthesis of (±)-reserpine.[57] Hydroindoles have been made by a similar procedure.[58]

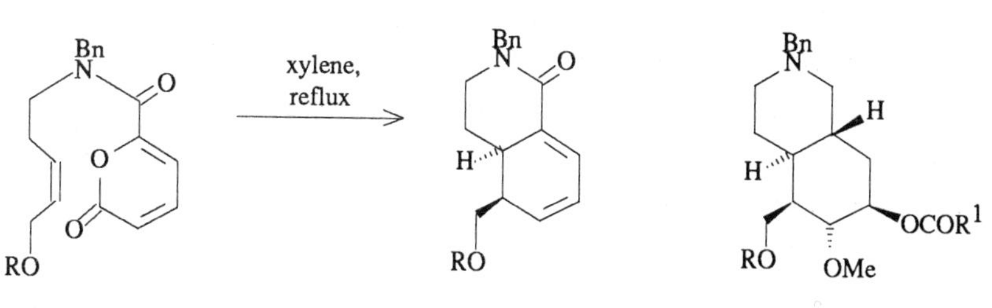

xylene, reflux

(155) (156) (93%) (157)

*Scheme 47*

In a different approach to hydroisoquinolines the ene-amides (158, R=H,$CH_3$) gave mainly the *cis*-fused cycloadducts (159) in toluene at 80-85°C. In contrast, the isomeric secondary amide (161,R=H) did not cyclise at 275°C and cyclisation of the tertiary amide (161,R= $CH_3$) required a temperature of 185°C and gave a much smaller proportion of the *cis*-fused hydroquinoline than (158,R= $CH_3$).[59]

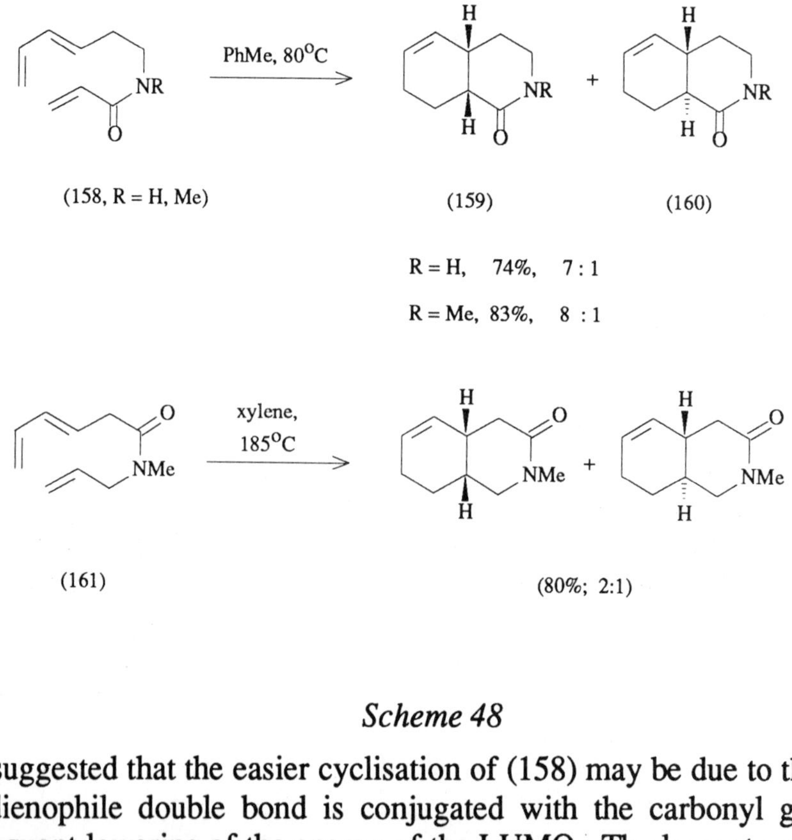

PhMe, 80°C

(158, R = H, Me) (159) (160)

R = H, 74%, 7 : 1

R = Me, 83%, 8 : 1

xylene, 185°C

(161) (80%; 2:1)

*Scheme 48*

It is suggested that the easier cyclisation of (158) may be due to the fact that the dienophile double bond is conjugated with the carbonyl group, with consequent lowering of the energy of the LUMO. The lower temperature of reaction then allows the interplay of secondary orbital interactions favouring an *endo* transition state and a preponderance of the *cis*-fused product. In (161) there is no conjugation of the dienophile double bond to compensate for possible disruption of amide resonance in the transition state, a higher temperature is required for cyclisation and secondary orbital interactions are less important.

A reaction of this kind formed the key step in a novel and expeditious route to the *cis*-D/E ring fragment of heteroyohimbine and corynantheoid alkaloids by intramolecular cycloaddition of an ene-amide to an αβ-unsaturated aldehyde. The aldehyde (162) readily cyclised at 190°C in xylene to a separable mixture of the adducts (163) and (164).[60] The *cis*-fused isomer was converted into (165), a known precursor of tetrahydroalstonine and other heteroyohimboid bases.

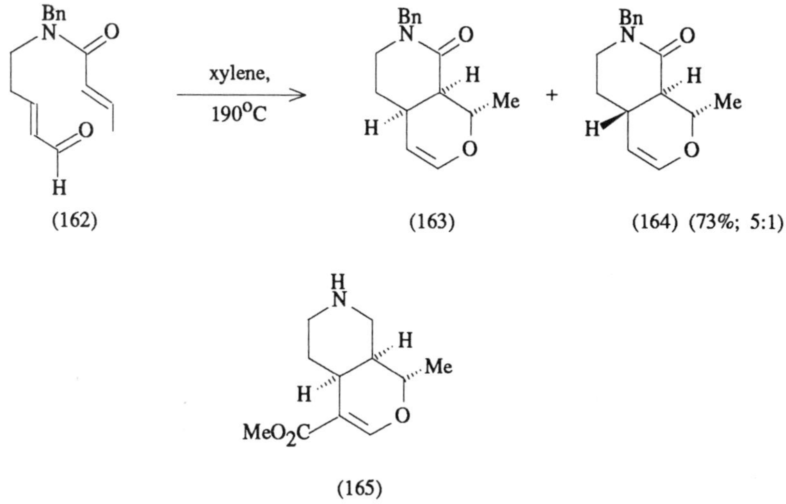

xylene, 190°C

(162) (163) (164) (73%; 5:1)

(165)

*Scheme 49*

Trienic esters on occasion can undergo intramolecular Diels-Alder reactions to give bicyclic lactones but these reactions are frequently more difficult than those of the corresponding amides, possibly due to loss of ester orbital overlap in the transition state conformation required for cyclisation.[61] Thus, the trienic ester (166), in contrast to the related amide (158) required vigorous conditions for cyclisation, and the isomer (167) did not cyclise at temperatures up to 275°C.[62]

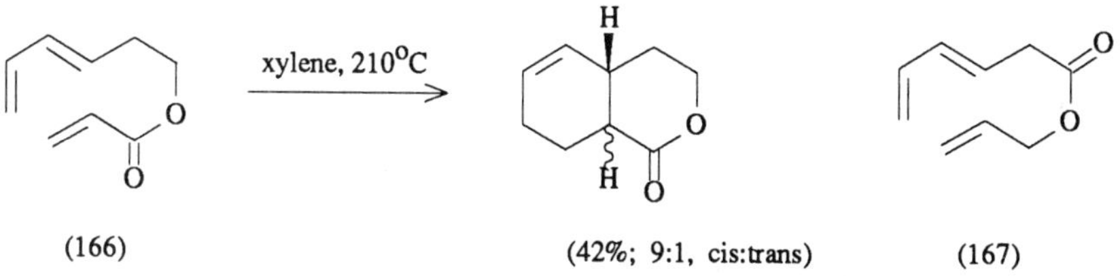

xylene, 210°C

(166) (42%; 9:1, cis:trans) (167)

*Scheme 50*

## Heterodienes and Heterodienophiles

### αβ-Unsaturated Carbonyl compounds; synthesis of Cannabinoids

Diels-Alder reactions of heterodienes and heterodienophiles are well-known and have been employed in the synthesis of heterocyclic compounds. For example, pyran derivatives can be prepared by reaction of unsaturated carbonyl compounds with enol ethers or alkenes. The intramolecular version of this reaction is particularly valuable with educts of the type (168), which are easily prepared by condensation of aldehydes with cyclic 1,3-dicarbonyl compounds. Formed *in situ* they cyclise readily to *cis*- or *trans*-fused polycyclic dihydropyrans (169).

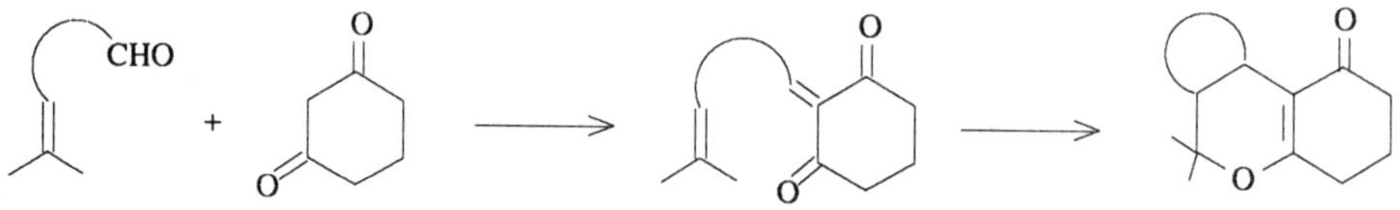

CHO + O O → O O → O O

(168) (169)

*Scheme 51*

Asymmetric induction can be effected by a stereogenic centre in the chain or in the dicarbonyl compounds[63] and this has been exploited in the synthesis of a variety of natural products, including cannabinoids,[64] iridoids[65] and indole alkaloids.[66] Thus, reaction of 1,3-cyclohexadione with (*R*)-citronellal in the presence of sodium methoxide gave the enolate (170). On acidification and chromatography this formed the tricyclic adduct (172) directly in 33 per cent yield, presumably by way of the αβ-unsaturated ketone (171). No other stereoisomer was detected. A chair-like conformation for the transition state (173) with an equatorially disposed methyl substituent seems mostly likely.

Having the (*R*)-configuration at C6a and the *trans* fusion of rings A and B, compound (172) corresponds to the parent skeleton of tetrahydrocannabinol[67] and the sequence was employed in the synthesis of (+)- and (–)-hexahydrocannabinol (174) from (*S*)- and (*R*)-citronellal respectively and 5-n-pentyl-1,3-cyclohexadione.[64]

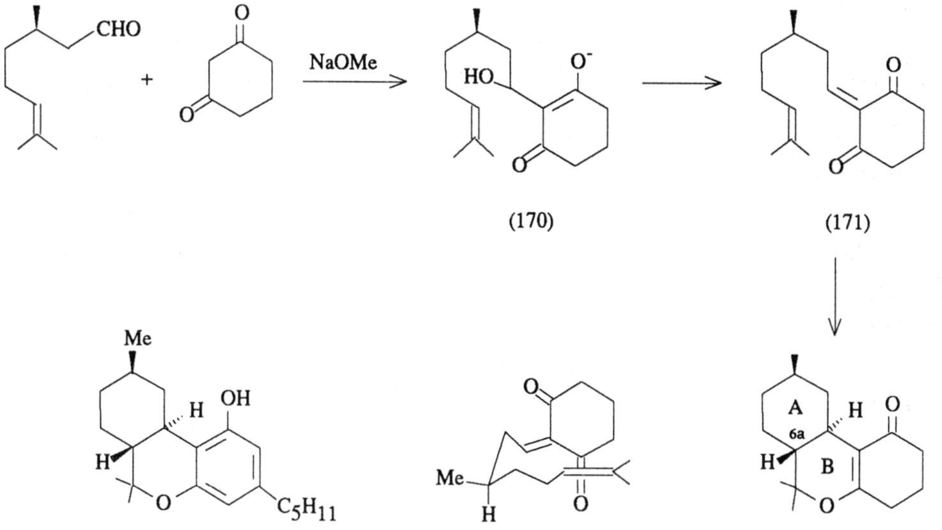

CHO + NaOMe → HO O⁻ (170) → (171) → (172) A 6a H B O; (173) Me H; (174) Me H OH $C_5H_{11}$

*Scheme 52*

In a different approach, the aromatic adduct (177) was obtained from the alcohol (175) in 87 per cent yield by way of the *ortho*-quinonemethide (176). Since the cyclisation leads to the *trans* ring-fused product the *exo* transition state (178) with an equatorially disposed methyl substituent is suggested.[68]

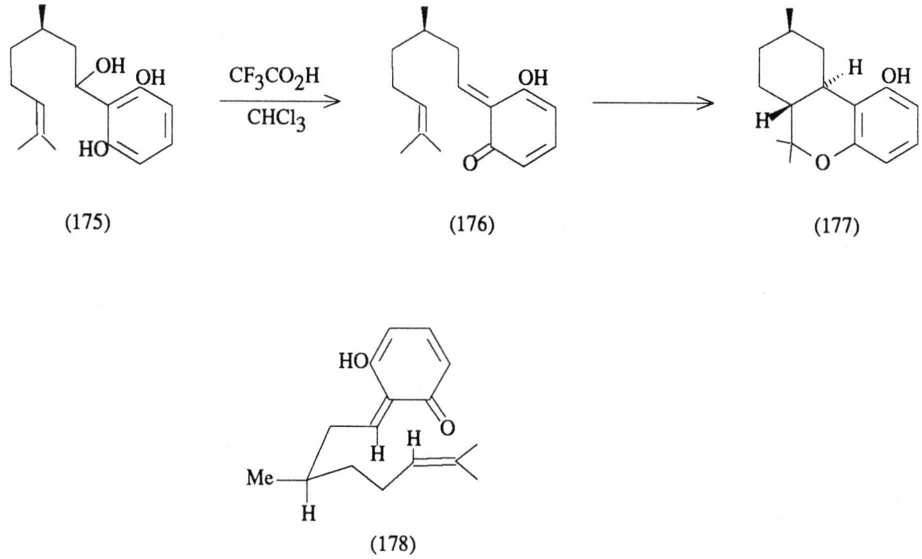

(175) OH OH HO $CF_3CO_2H$ $CHCl_3$ → (176) OH → (177) H OH H O; (178) HO O H H Me H

*Scheme 53*

A related procedure involving an *ortho*-quinonemethide was employed in a synthesis of (+)- and (–)-hexahydrocannabinol (174)[69] and has the advantage over that proceeding from pentyl-1,3-cyclohexadione in that it leads to the aromatic system directly.

In the reactions described above, the chiral centre in citronellal forms an integral part of the product. Enantioselective syntheses in this series have also been effected using the auxiliary chiral dione (180), which is prepared from (–)-ephedrine. Reaction of this dione with the aldehyde (179), for example, catalysed by diethylaluminiun chloride, gave the *cis*-fused adduct (182) with diastereomeric excess of >98 per cent. Successive hydrolysis of the crystallised material with methanolic hydrogen chloride and sodium hydroxide gave the optically pure lactone (183). Using the dione from (+)-ephedrine the enantiomer of (183) was obtained.[70]

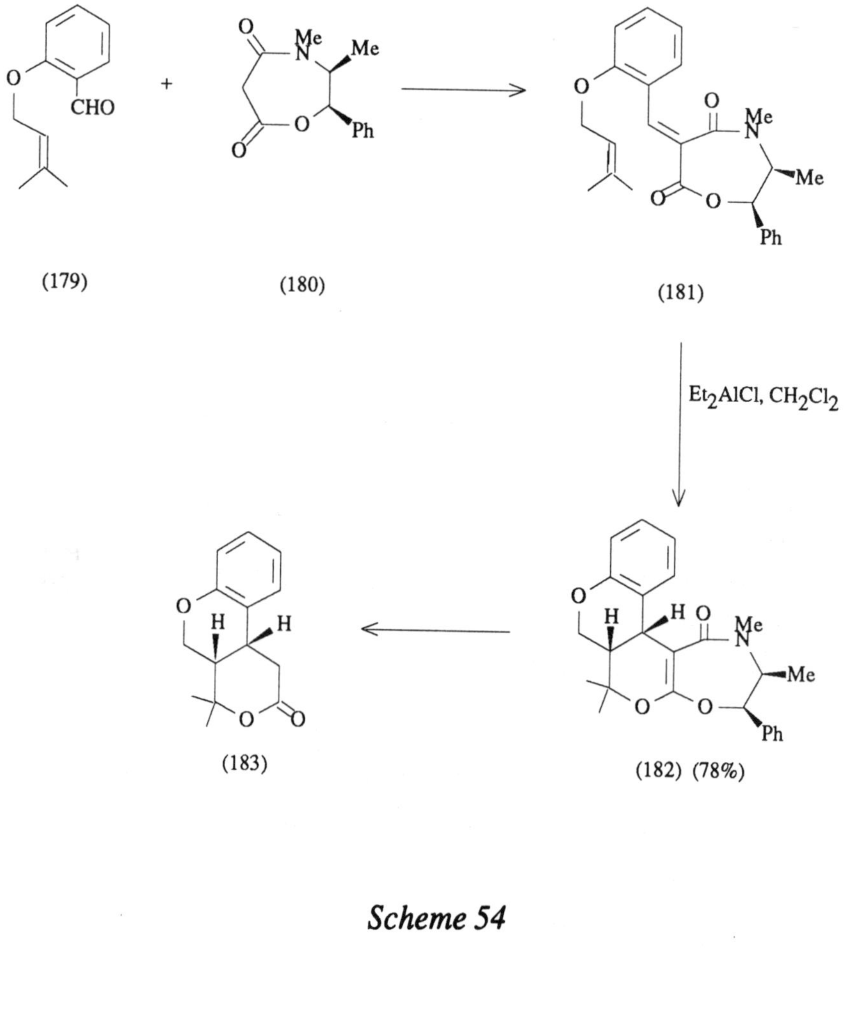

(179) (180) (181)

$Et_2AlCl, CH_2Cl_2$

(183) (182) (78%)

*Scheme 54*

In an excursion in the iridoid field the enone (186), formed transiently from the (*S*)-aldehyde (184) and (185), gave the cyclopenta[c]pyran derivatives (187a) and (187b) in 77 per cent yield in a ratio of 10:1. In several more steps (187a) was converted into the aglycone (188) and thence into deoxyloganin.[71]

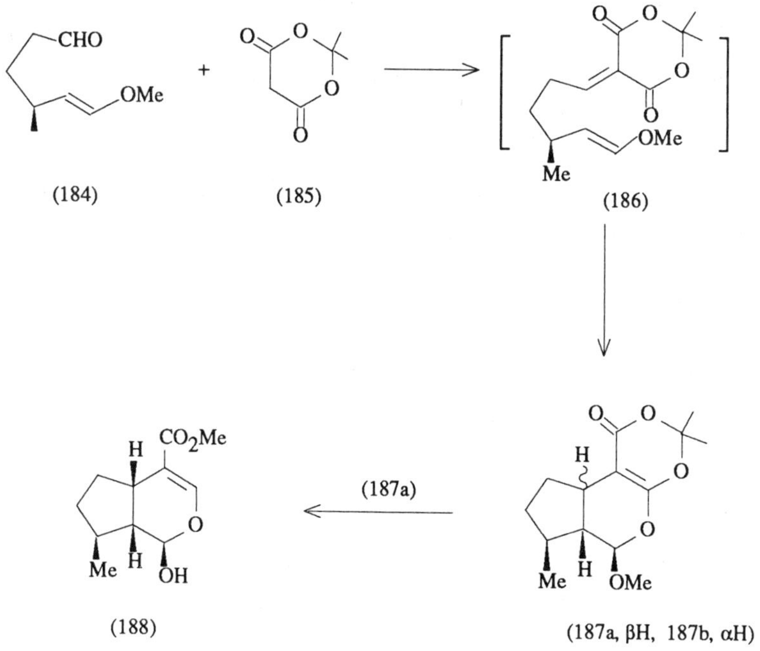

CHO OMe (184) + (185) → (186) → (187a, βH, 187b, αH) → (188) $CO_2Me$ OH

*Scheme 55*

Similarly intramolecular cycloaddition of the enone (189) selectively furnished the *cis* and *trans* ring-fused adducts (190) or (191) depending on the configuration of the alkene. The adduct (190), the stereochemistry of which derives ultimately from that of diethyl L-tartrate, was used in syntheses of (–)-ajmalicine and (–)-tetrahydroalstonine.[72]

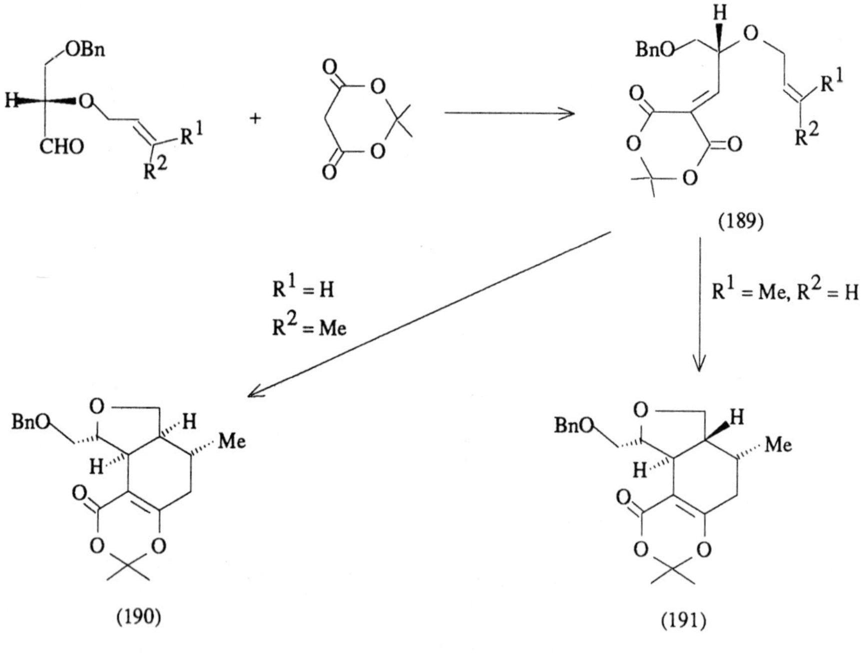

OBn
H O
CHO $R^1$
$R^2$

\+

O O
O O

H
BnO O
O $R^1$
O O
O O $R^2$

(189)

$R^1 = H$
$R^2 = Me$

$R^1 = Me, R^2 = H$

BnO O H
Me
H
O
O O

(190)

BnO O H
Me
H
O
O O

(191)

*Scheme 56*

Intramolecular cycloadditions involving αβ-unsaturated carbonyl compounds as the diene do not always take the expected course. Thus, vapour-phase pyrolysis of the unsaturated aldehyde (192) with a view to obtaining bicyclic dihydropyran (193), a potential intermediate for the synthesis of iridomyrmecin (194), gave a mixture containing considerable amounts of the cyclopentane derivative (195) formed by an intramolecular ene reaction; the catalysed reaction ($BF_3.Et_2O$) gave, not (193) but the bridged bicyclic compound (198). The latter is believed to arise from the (Z)-aldehyde (196) by way of the boat-like transition state (197) which models show allows the best approach of the double bond to the enal.[73]

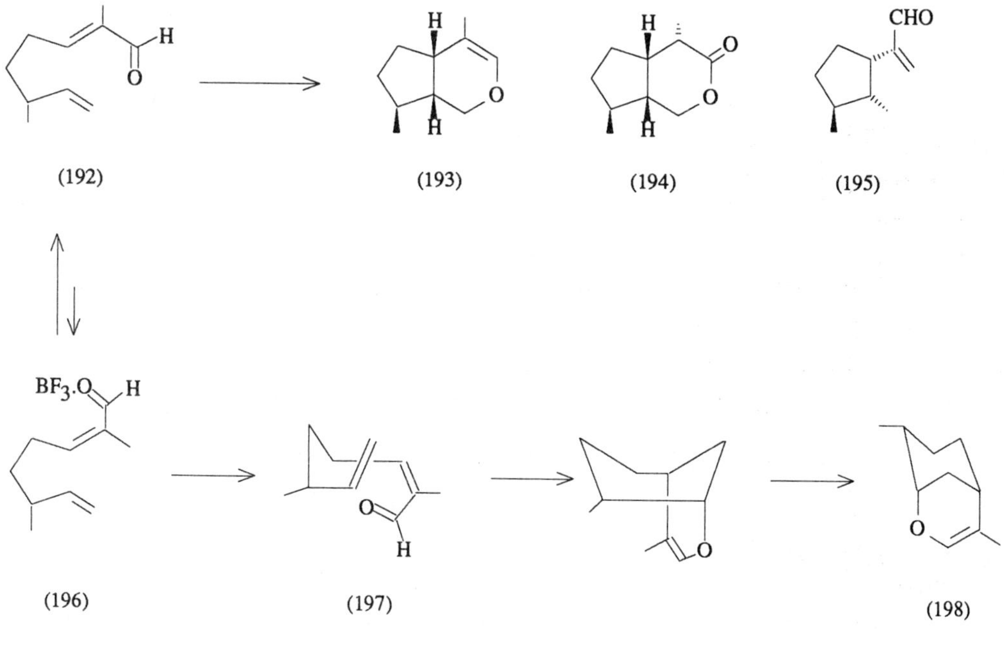

(192) (193) (194) (195)

(196) (197) (198)

*Scheme 57*

## N-Acyl-1-azadienes

Intramolecular Diels-Alder reactions of N-acyl-1-aza-1,3-dienes have been used to prepare hydroquinolizidones and hydroindolizidones, but in general the N-acyl-1-azadienes are less reactive as dienes than αβ-unsaturated ketones. In one procedure the aza-dienes are generated *in situ* by gas phase pyrolysis of N-acyl-O-acetyl-N-allylhydroxylamines, and under the conditions of the reactions cyclise directly to hydroquinolizidones or hydro-indolizidones.[74] A reaction of this kind was employed in a synthesis of the quinolizidine alkaloid (–)-deoxynupharidine.[75] Pyrolysis of the optically active O-acetyl-hydroxamic acid (199) gave a mixture of the diastereomers (201) and (203) by way of the *exo* transition states (200), in which the methyl substituent is equatorial and (202) in which it is axial. The *cis* isomer (201) was converted in several more steps into (–)-deoxynupharidine (204).

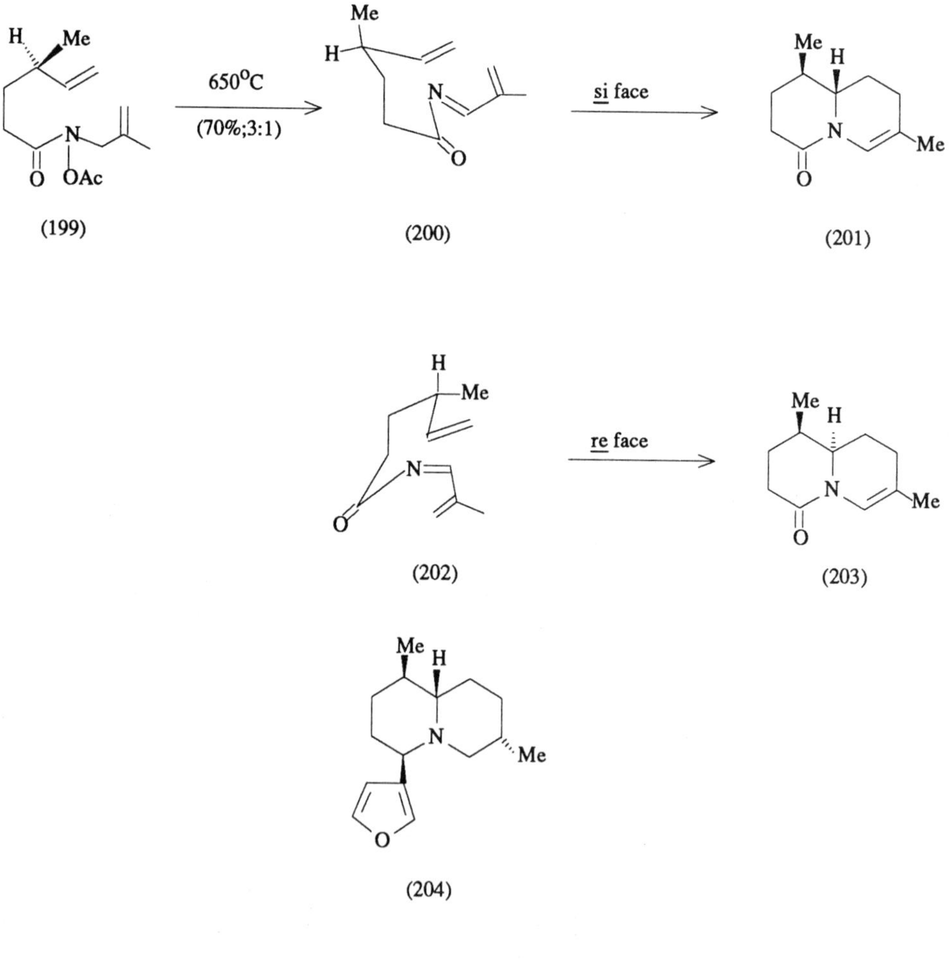

650°C (70%;3:1)

si face

re face

(199) (200) (201) (202) (203) (204)

*Scheme 58*

2-Trimethylsilyloxy-1-azadienes have been obtained also from αβ-unsaturated amides and employed in intramolecular cycloaddition reactions.[76]

**Acylimines; synthesis of nitrogen heterocycles**

Hetero-dienophiles also have been widely used in intramolecular Diels-Alder reactions. Some of the most useful results have been obtained with N-acylimines, which have been converted into piperidines, indolizidines and quinolizidines. These intramolecular reactions are frequently highly stereoselective, in contrast to intermolecular imino Diels-Alder reactions.[77] In many of these reactions the N-acylimine is generated *in situ* by pyrolysis of a suitable N-acetoxymethyl derivative. Thus, pyrolysis of the methylol acetate (205) gave the bicyclic lactam (207) directly by way of the imine (206). Reduction of the double bond and the lactam carbonyl group of (207) gave (±)-δ-coniceine (208).[78]

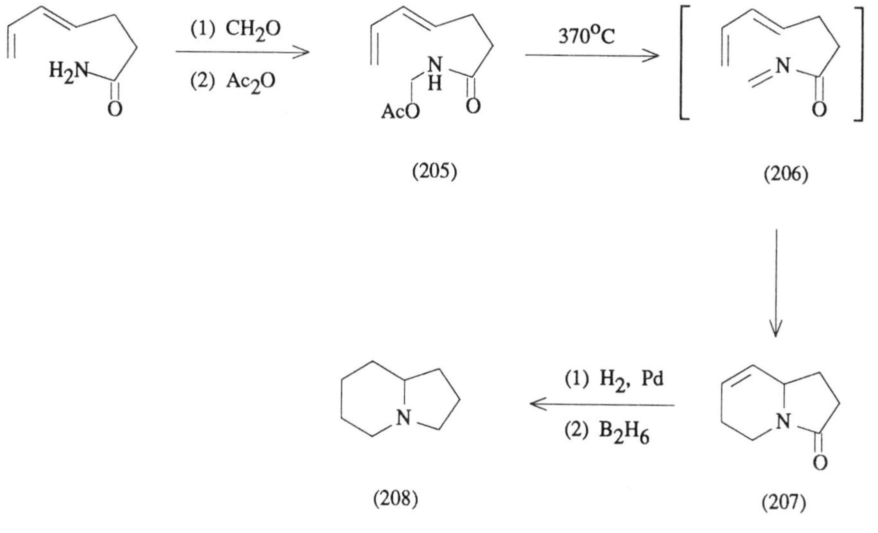

(1) $CH_2O$ (2) $Ac_2O$ ; 370°C ; (1) $H_2$, Pd (2) $B_2H_6$

$H_2N$ ; AcO ; (205) ; (206) ; (207) ; (208)

*Scheme 59*

A similar sequence was used in syntheses of the indolizidine alkaloids elaeokanine A[79] and slaframine[80] and the quinolizidine *epi*-lupinine (212).[81] In the last synthesis the precursor (209), in boiling *ortho*-dichlorobenzene gave the cyclic amide (211) exclusively in 93 per cent yield, by way of the transition state (210) with the carbonyl group in the *endo* orientation and the benzyloxymethyl substituent in the connecting chain adopting a quasi-equatorial position.

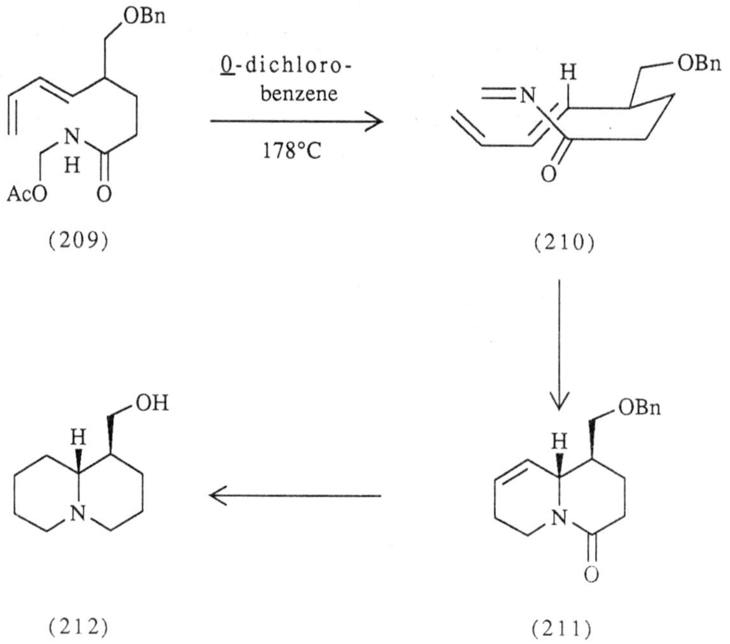

OBn
o-dichloro-
benzene
178°C
(209)
(210)
(211)
(212)

*Scheme 60*

Intramolecular imino Diels-Alder reactions have also been used to make *trans*-2,6-dialkylpiperidines[82] and such a reaction formed the key step in a synthesis of the spermidine alkaloid anhydrocannabisativene.[83] Certain iminium salts, generated *in situ* under Mannich-like conditions undergo inter- and intra-molecular Diels-Alder cycloadditions readily in aqueous solution. Thus, the dienyl aldehyde (213) treated with benzylamine hydrochloride in aqueous ethanol at 70°C gave a mixture of the hydroquinolines (214) and (215) in 63 per cent yield, and the cyclohexadienyl aldehyde (216) and methylamine hydrochloride under the same conditions gave the tricyclic amine (217) in 66 per cent yield; catalytic reduction of (217) gave (±)-dihydrocannivonine (218).[84]

Transient N-acylimines, or the corresponding iminium complexes, can also act as *dienes* in reactions with alkenes to form 5,6-dihydro-1,3-oxazines in a highly stereoselective manner; alkynes also react to give oxazines.[85] Thus, catalysed reaction of the bis-amide (219) with $BF_3.Et_2O$ in dichloromethane at room temperature gave exclusively the *trans*-fused product (221) with four contiguous chiral centres in 75 per cent yield.

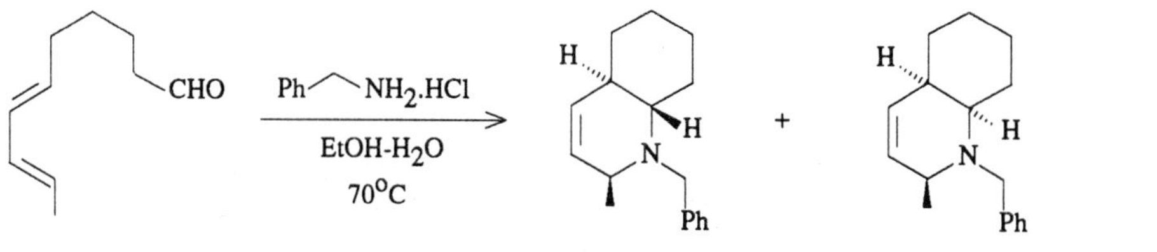

CHO Ph NH$_2$.HCl EtOH-H$_2$O 70°C H H N Ph + H H N Ph

(213) (214) (215) (63%; 2.5:1)

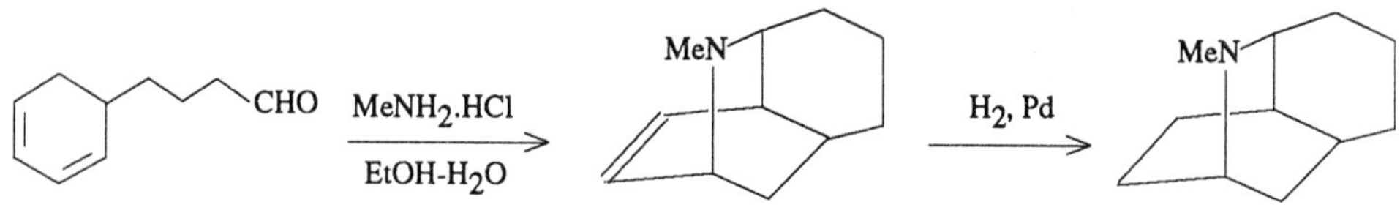

CHO MeNH$_2$.HCl EtOH-H$_2$O MeN H$_2$, Pd MeN

(216) (217) (218)

*Scheme 61*

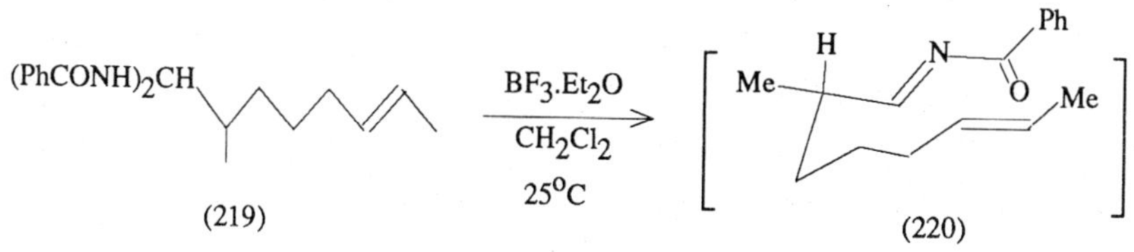

(PhCONH)$_2$CH BF$_3$.Et$_2$O CH$_2$Cl$_2$ 25°C [ H Me N Ph O Me ]

(219) (220)

Me H N Ph O H Me

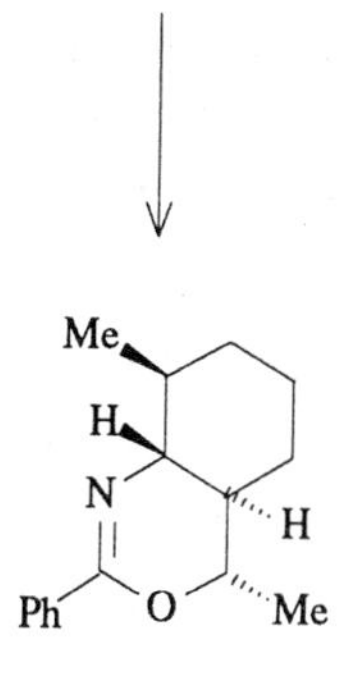

(221) (75%)

*Scheme 62*

The stereochemical results are best accommodated by a transition state of the form (220) in which the catalyst is complexed with the nitrogen lone pair and the methyl substituent on the connecting chain has a quasi equatorial disposition.

In contrast, thermal cyclisation of glyoxylate-derived N-acylimines provides, stereoselectively, *cis*-fused bicyclic dihydro-oxazine $\gamma$-lactones.[86] (Scheme 63)

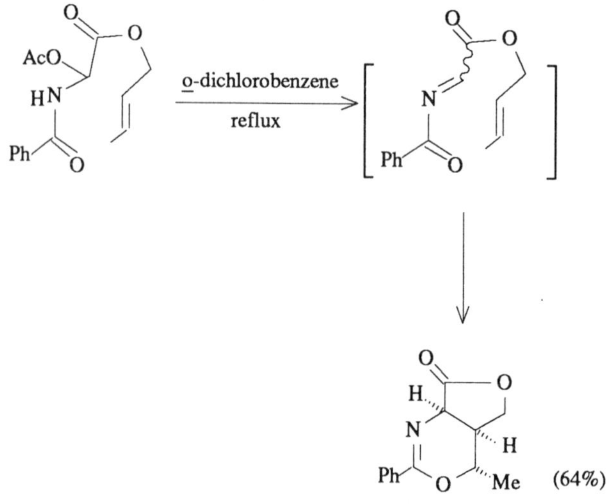

o-dichlorobenzene

reflux

(64%)

*Scheme 63*

## Nitroso compounds

Nitroso compounds have been employed in intramolecular Diels-Alder reactions both as dienes and dienophiles. Thus nitroso-alkenes, generated from silyl derivatives of $\alpha$-chloroketoximes by reaction with fluoride ion cyclise readily to give dihydro-oxazines (e.g.222)[87]

Nitro-alkenes also serve well as $4\pi$ components in highly stereoselective cycloadditions with unactivated alkenes to give nitronates.[88]

Acylnitroso compounds are useful dienophiles and they also undergo highly stereoselective intramolecular cycloadditions. They are frequently formed *in situ* from the corresponding hydroxamic acid, as illustrated in the synthesis of (±)-gephyrotoxin 223AB (227).[89,90] The acylnitroso compound (224), formed *in situ* from hydroxamic acid (223) cyclised spontaneously to the 1,2-oxazine (225) as the sole product in 82 per cent yield.

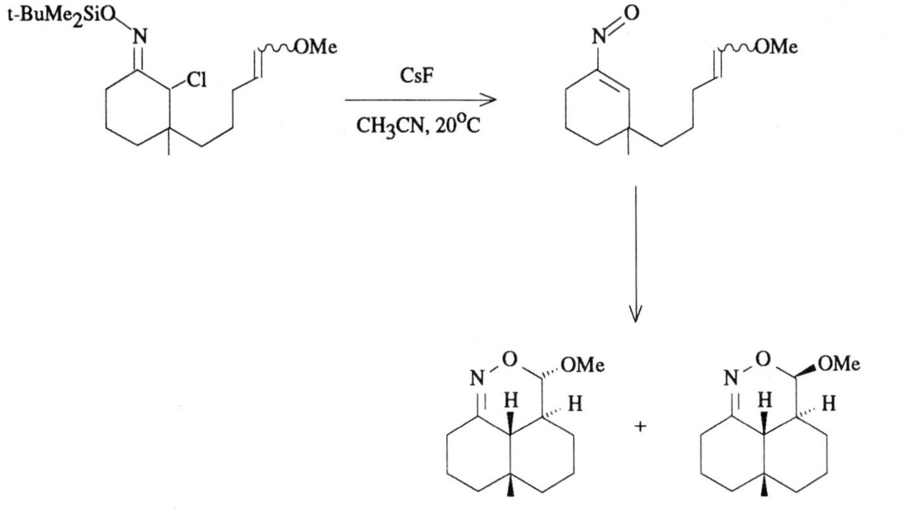

t-BuMe$_2$SiO N Cl OMe

CsF

$CH_3CN$, 20°C

N=O OMe

N O OMe H H + N O OMe H H

(222) (83%; 70:13)

*Scheme 64*

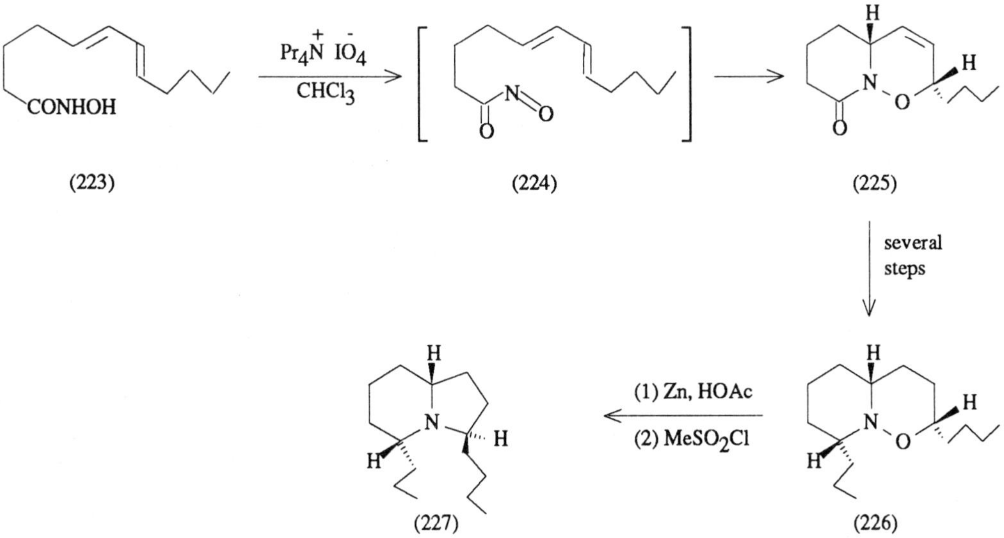

CONHOH

(223)

$Pr_4\overset{+}{N}$ $\overset{-}{IO_4}$

$CHCl_3$

(224)

(225)

several steps

(226)

(1) Zn, HOAc

(2) $MeSO_2Cl$

(227)

*Scheme 65*

## N-Sulphinyl compounds

Another useful group of heterodienophiles are the N-sulphinyl compounds, containing the grouping N=S. By intramolecular Diels-Alder reaction they give dihydrothiazine oxides which can be converted in a stereo-controlled way into unsaturated vicinal amino alcohols, as illustrated in the synthesis of *threo*-sphingosine (232). Reaction of the carbamate (228) with thionyl chloride in pyridine gave directly the dihydrothiazine oxide (230) as a single stereoisomer, by way of the N-sulphinyl carbamate (229). Reaction with phenylmagnesium bromide followed by heating the allylic sulphoxide so formed with trimethyl phosphite converted adduct (230) stereo-specifically into the (*E*)-*threo*-carbamate (231) which gave racemic *threo*-sphingosine (232) on hydrolysis.

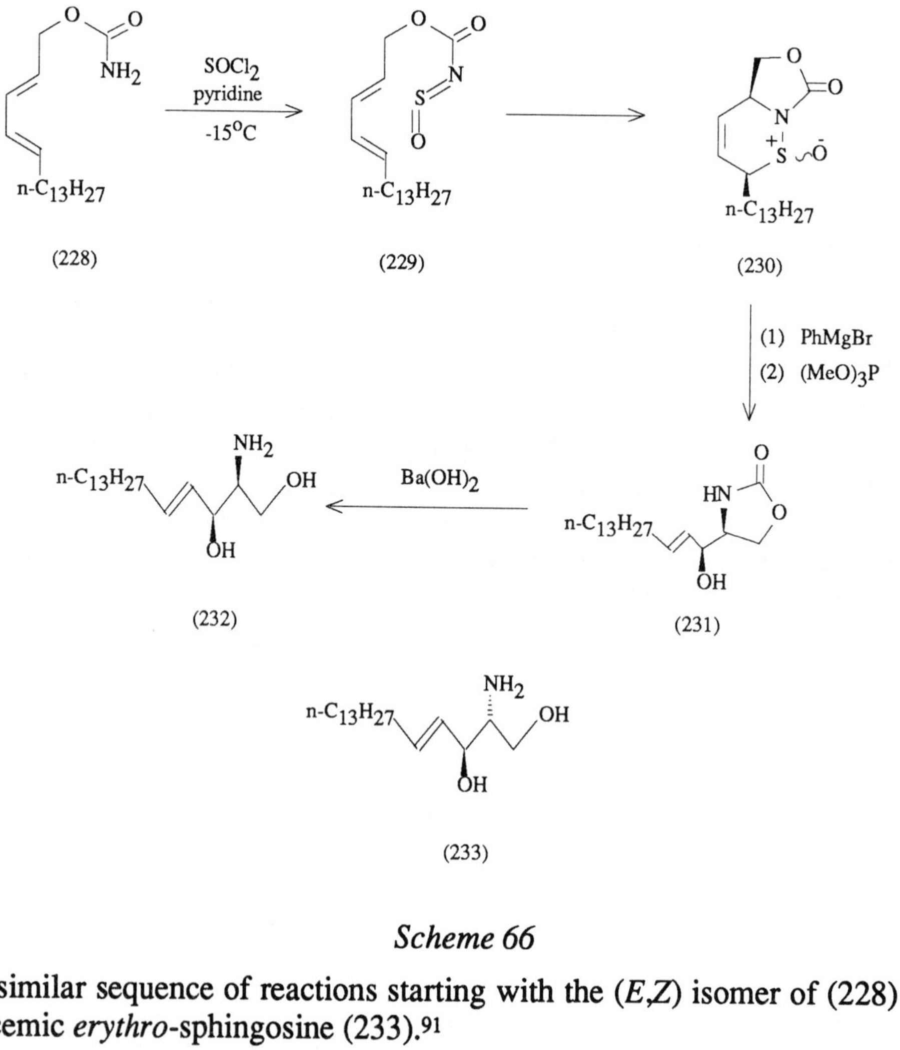

$NH_2$ n-$C_{13}H_{27}$ (228) —$SOCl_2$, pyridine, -15°C→ n-$C_{13}H_{27}$ (229) → n-$C_{13}H_{27}$ (230)

(1) PhMgBr
(2) $(MeO)_3P$

n-$C_{13}H_{27}$ OH (231) —$Ba(OH)_2$→ n-$C_{13}H_{27}$ $NH_2$ OH OH (232)

n-$C_{13}H_{27}$ $NH_2$ OH OH (233)

*Scheme 66*

A similar sequence of reactions starting with the (*E,Z*) isomer of (228) gave racemic *erythro*-sphingosine (233).[91]

This procedure has been exploited in syntheses of 5-epidesosamine (234), the C-5 epimer of the natural amino sugar desosamine[92] and of 3-deoxy-3-methylaminoarabinopyranoside (235), a component of some aminoglycoside antibiotics.[93]

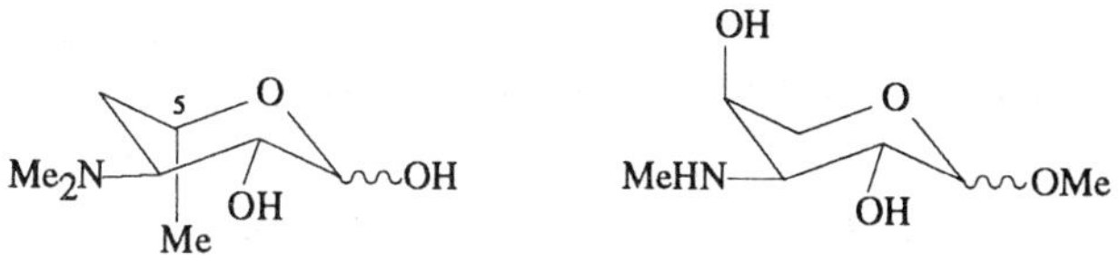

(234) (235)

*Scheme 67*

## ortho-Quinodimethanes; synthesis of steroids and lignanes

Intramolecular Diels-Alder reactions of *ortho*-quinodimethanes (*ortho*-xylylenes) have also been widely employed in the synthesis of natural products, particularly in the steroid and alkaloid fields.[94] These reactions lead smoothly to polycyclic ring systems, generally with high regio- and stereo-selectivity. For the intramolecular reaction there are, in principle, four possible transition states, two *exo* (236) and (238) and two *endo* (240) and (242), but it is found in practice that reactions give fused-ring products exclusively, although it is not always easy to predict whether the *exo* or *endo* mode of reaction will be preferred.

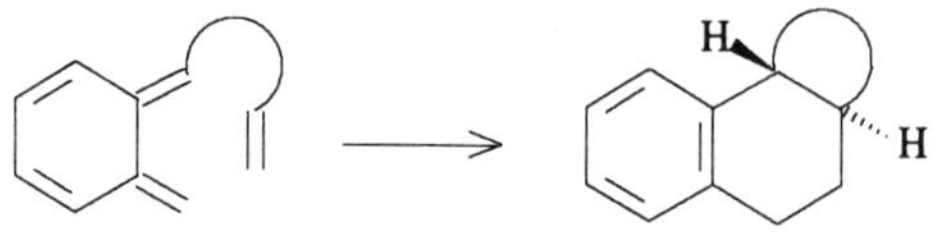

(236) (237)

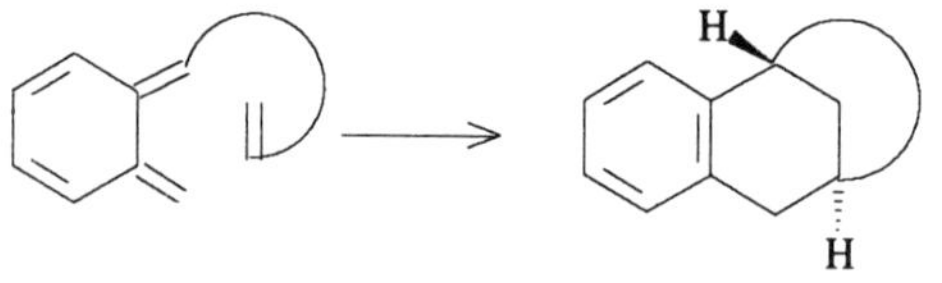

(238) (239)

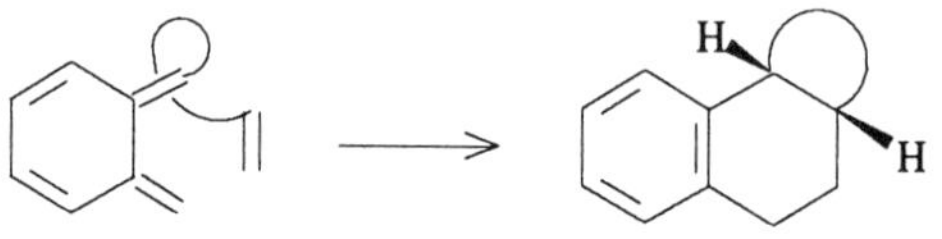

(240) (241)

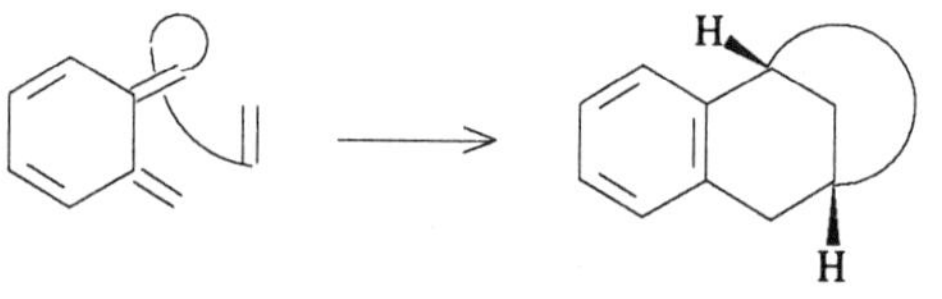

(242) (243)

*Scheme 68*

*Ortho*-quinodimethanes are reactive dienes and even unreactive dienophiles such as isolated carbon-carbon double bonds, acetylenes, aldehydes, nitriles and oxime ethers form cycloadducts.[95] They are generally prepared *in situ*, using one of the procedures described on p.37, for example by thermolysis of benzocyclobutenes. Thus, the optically active benzocyclobutene (244), heated in *ortho*-dichlorobenzene gave the cycloaddition product (246), with the typical steroid *trans,anti,trans* ring fusion, *via* the *exo* transition state (245). Cleavage of the t-butyl and methyl ethers then gave (+)-estradiol (247) of 97 per cent optical purity.[96] A similar route was used to make estrone methyl ether.[97]

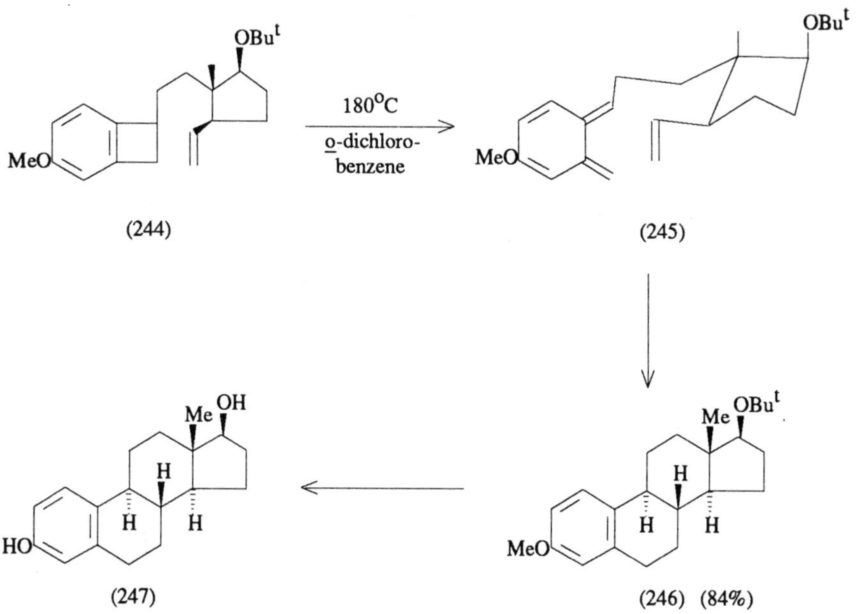

OBu$^t$ OBu$^t$

180°C

o-dichloro-benzene

MeO MeO

(244) (245)

Me OH Me OBu$^t$

H H

H H H H

HO MeO

(247) (246) (84%)

*Scheme 69*

Using the optically active *cis*-disubstituted cyclohexane derivative (248) the optically active *cis,anti,trans* D-homosteroid (250) was obtained and was converted in several more steps into (+)-chenodeoxycholic acid (251).[98]

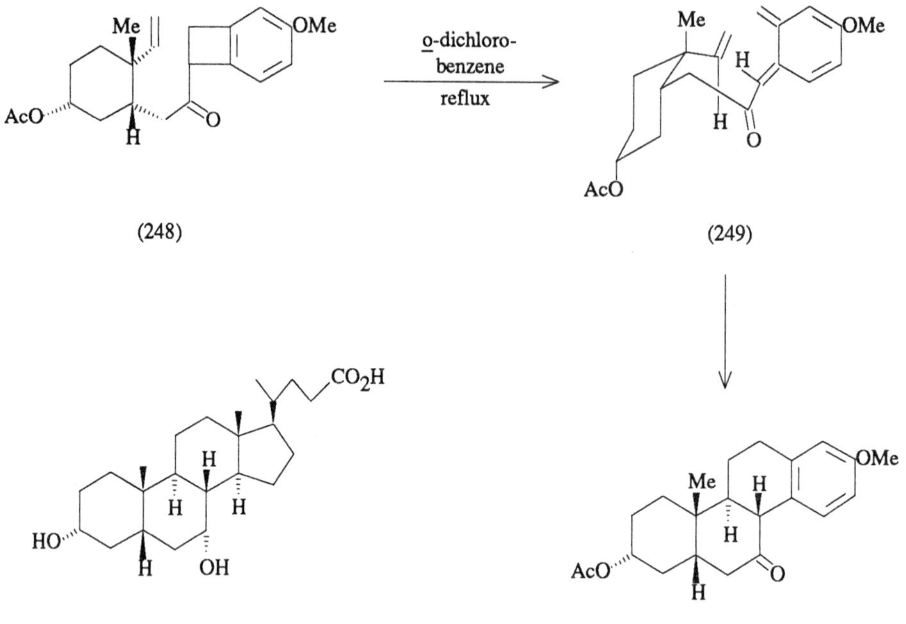

(248) (249) (250) (251)

*Scheme 70*

The factors which affect the preferences for *endo* or *exo* cycloaddition in the reactions of *ortho*-quinodimethanes are not always clearly perceptible, and the balance may be quite markedly affected by modification of the chain connecting the diene and dienophile.[99]

Although the benzocyclobutene route to *ortho*-quinodimethanes has been widely employed, it suffers from the disadvantage that the benzocyclobutenes themselves are frequently difficult to make. A useful advance is provided by the discovery that a variety of benzocyclobutenes can be obtained by co-cyclisation of αω-diynes with substituted acetylenes in the presence of the cobalt(I) catalyst $\eta^5$-cyclopentadienyldicarbonyl cobalt.[100] For example, reaction of the diyne (252), incorporating a potential dienophile, and bistrimethylsilylacetylene gave mainly the *trans*-fused product (254) by way of the initially formed benzocyclobutene. (253).

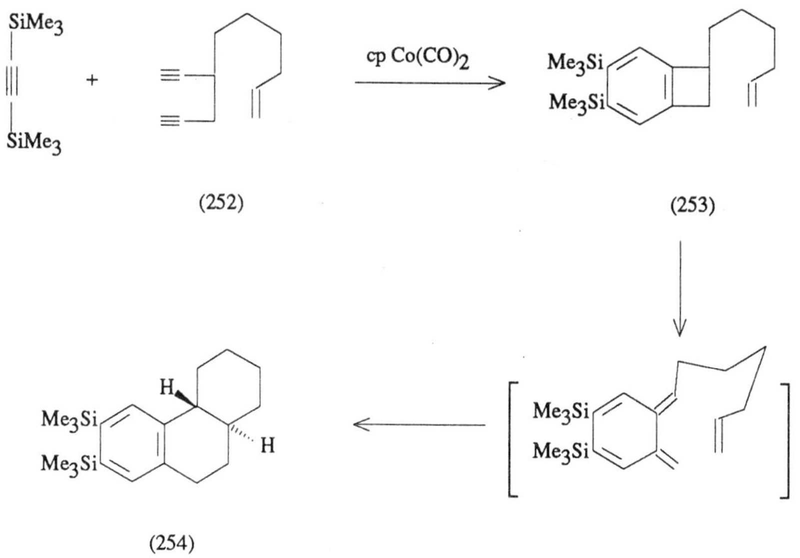

SiMe3 + cp Co(CO)2 → Me3Si Me3Si

(252) (253)

Me3Si Me3Si H H ← [Me3Si Me3Si]

(254)

*Scheme 71*

*Ortho*-quinodimethanes can also be obtained from 1-substituted 1,3-dihydroisothianaphthene-2,2-dioxides; these eliminate sulphur dioxide on pyrolysis to give mainly the (*E*)-dienes (as 256) which cyclise to polycyclic compounds. The required 1-alkenyl sulphones are readily obtained by alkylation of 1,3-dihydroisothionaphthene-2,2-dioxide itself.[101] Thus, thermolysis of the sulphone (255) in boiling trichlorobenzene gave the *trans,anti, trans*-tetracycle (257) in 64 per cent yield, which was subsequently converted into (+)-estradiol.[102]

A disadvantage of the procedures employing benzocyclobutenes or dihydroisothianaphthene sulphones is the high temperatures required to generate the *ortho*-quinodimethane. Milder procedures make use of elimination reactions from suitable *ortho* disubstituted benzene derivatives, for example from benzyltrimethylsilanes as in the conversion of (259) into (260),[103] and of (261) into (262).[104]

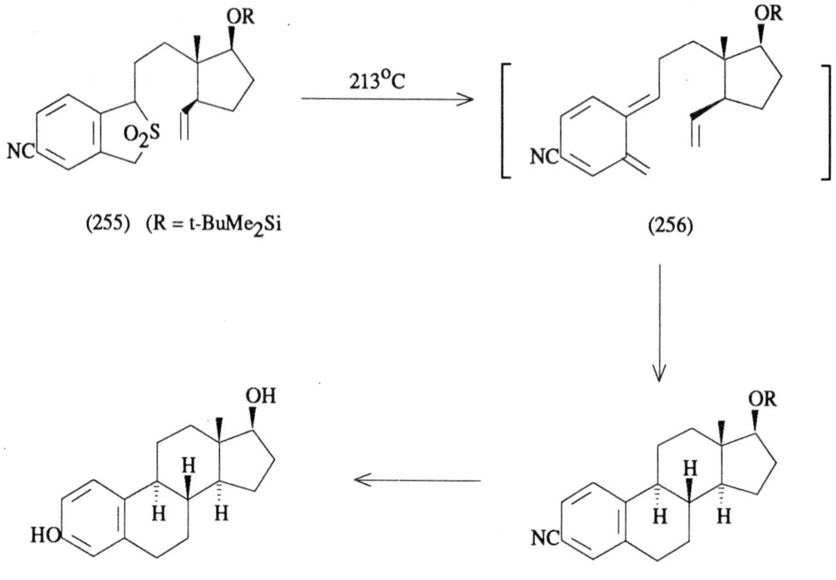

OR
213°C
NC
$O_2S$
OR
NC

(255) (R = t-BuMe$_2$Si

(256)

OH
H
HO
H
H
OR
H
NC
H
H

(258)

(257)

*Scheme 72*

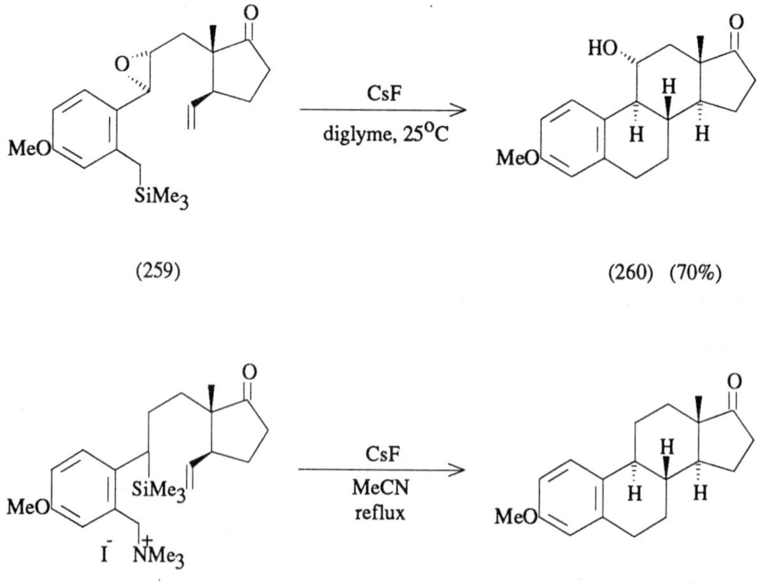

O
O
MeO
SiMe$_3$
CsF
diglyme, 25°C
HO
O
H
MeO
H
H

(259)

(260) (70%)

O
MeO
SiMe$_3$
$I^-$ $\overset{+}{N}Me_3$
CsF
MeCN
reflux
O
H
MeO
H
H

(261)

(262) (86%)

*Scheme 73*

Intramolecular cycloaddition to give polycyclic compounds from *ortho*-quinodimethanes generated by 1,4-elimination from methyl (*ortho*-methylbenzyl) ethers and esters has also been employed.[105] Thus, reaction of the indole derivative (263) with triethylamine in boiling chlorobenzene gave the *trans*-fused octahydropyridocarbazole (265) almost exclusively. Assuming the formation of an indole *ortho*-quinodimethane intermediate, the stereochemistry of the product implies reaction by way of the *exo* transition state (264).[106]

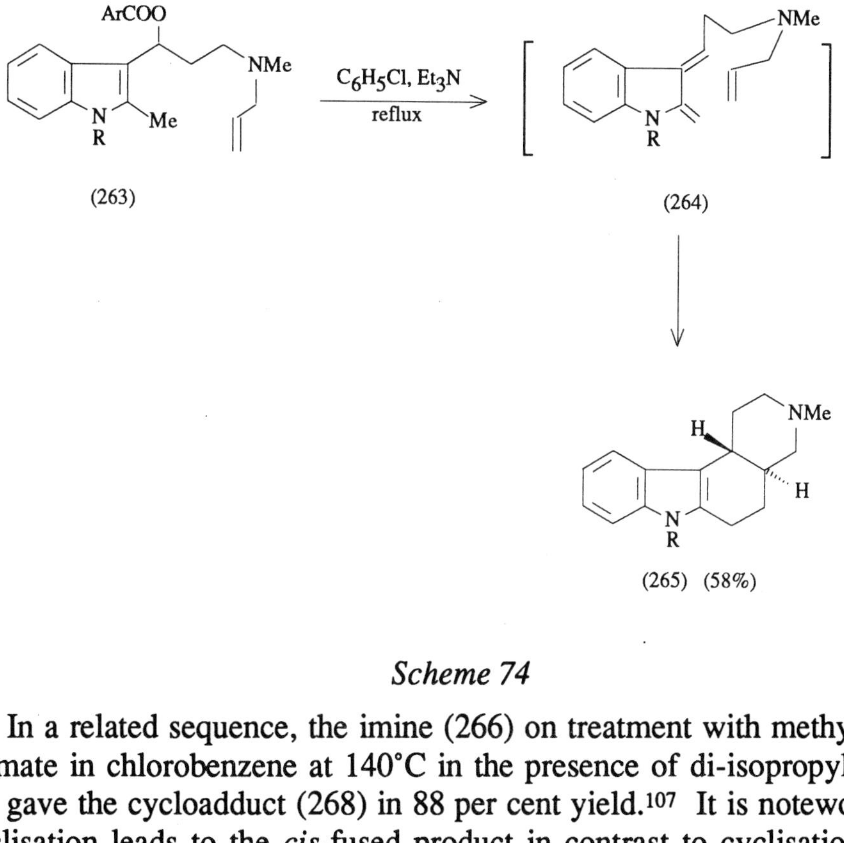

ArCOO NMe N R Me (263)

$C_6H_5Cl$, $Et_3N$ reflux

NMe N R (264)

H NMe H N R (265) (58%)

*Scheme 74*

In a related sequence, the imine (266) on treatment with methyl chloroformate in chlorobenzene at 140°C in the presence of di-isopropylethylamine gave the cycloadduct (268) in 88 per cent yield.[107] It is noteworthy that cyclisation leads to the *cis*-fused product in contrast to cyclisations in the steroid series where the *trans*-fused products predominated. It appears that the (*Z*)-quinodimethane is preferred in the indole series and that cyclisation takes place by way of a transition state of the form (267).

A sequence of this type formed an important step in the synthesis of *Aspidosperma* and other indole alkaloids.[108]

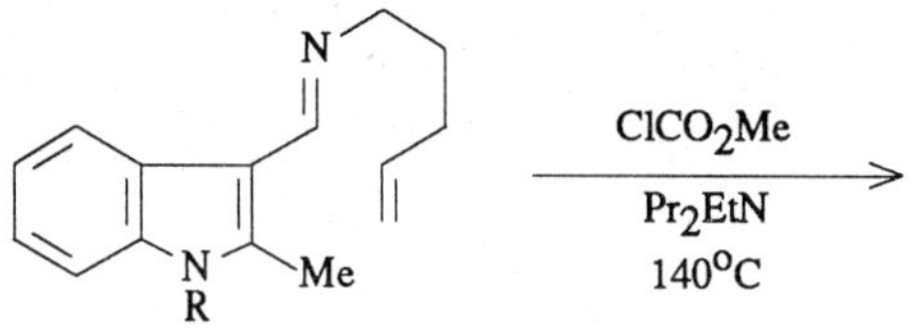

(266) 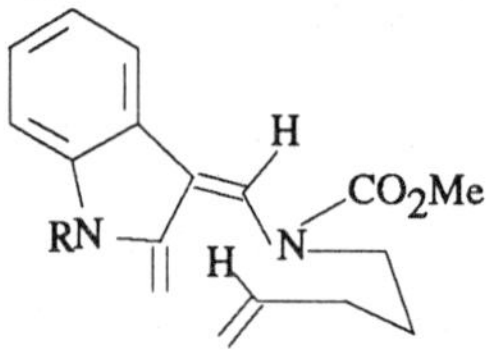 (267) 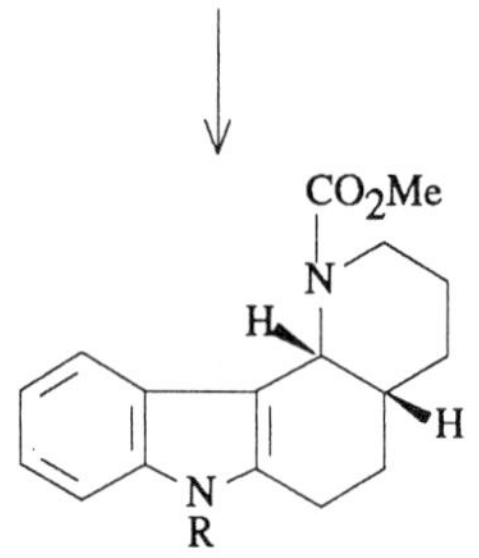 (268)

*Scheme 75*

Hetero-*ortho*-quinonemethides have also been employed. Thus, reaction of the benzylammonium salt (269) with caesium fluoride in boiling acetonitrile gave the benzoquinolizidine (271) by way of the intermediate (270).[109]

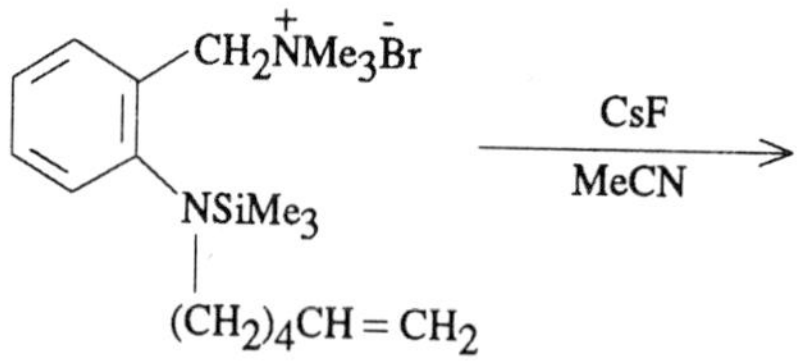

(269) 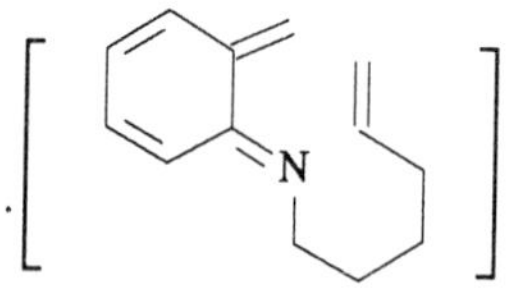 (270) 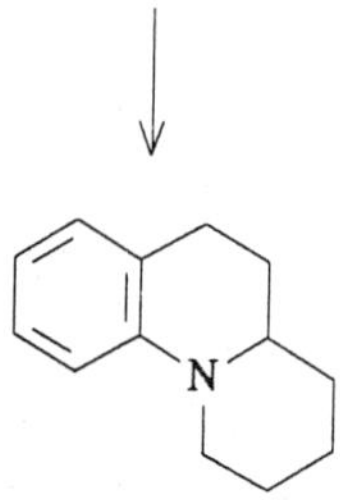 (271) (58%)

*Scheme 76*

Again, in a beautifully conceived synthesis, the *trans*-1-propenylphenol (272) on oxidation with palladium dichloride in aqueous methanol gave the lignan carpanone (274) directly in 46 per cent yield *via* the *ortho*-quinone-methide (273) formed by phenolic oxidation of (272). The product has five contiguous chiral centres, but no other product was detected in the reaction. This approach to the synthesis of the lignan may be related to its biosynthesis.[110]

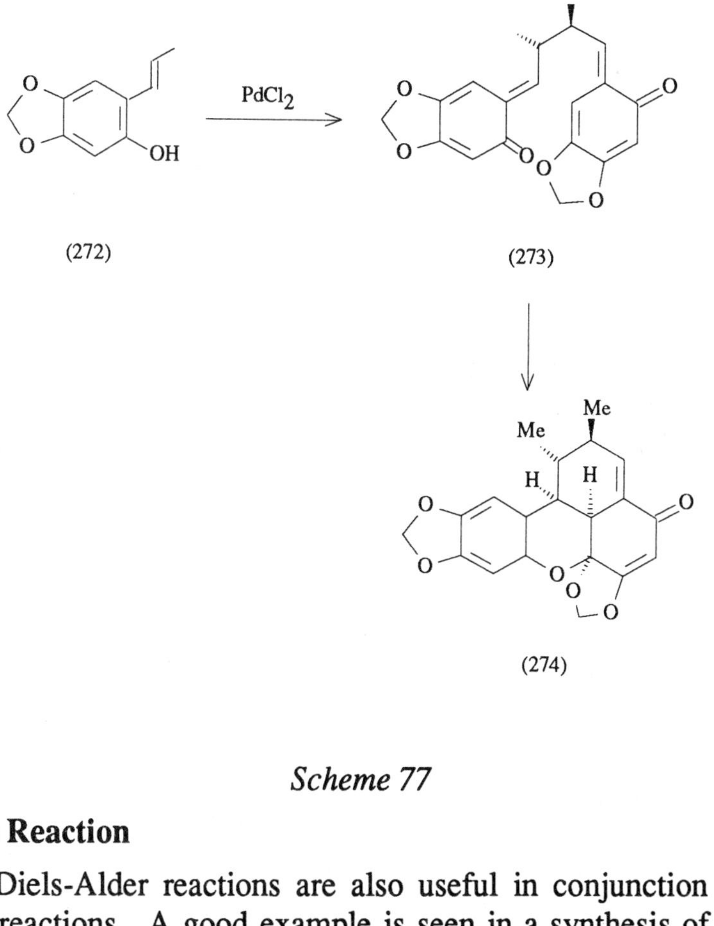

$PdCl_2$

(272)

(273)

Me

Me

H

H

(274)

*Scheme 77*

## The Retro Reaction

Retro Diels-Alder reactions are also useful in conjunction with intramolecular reactions. A good example is seen in a synthesis of the sesquiterpene (±)-paniculide (278). The oxazole (275) on pyrolysis in ethylbenzene afforded the methoxyfuran (276) in 94 per cent yield. Reduction of the carbonyl group and hydrolysis of the enol ether then gave the butenolide (277) which was converted into paniculide in several more steps.[111] Intramolecular addition in related thiazoles was not so easy.[112]

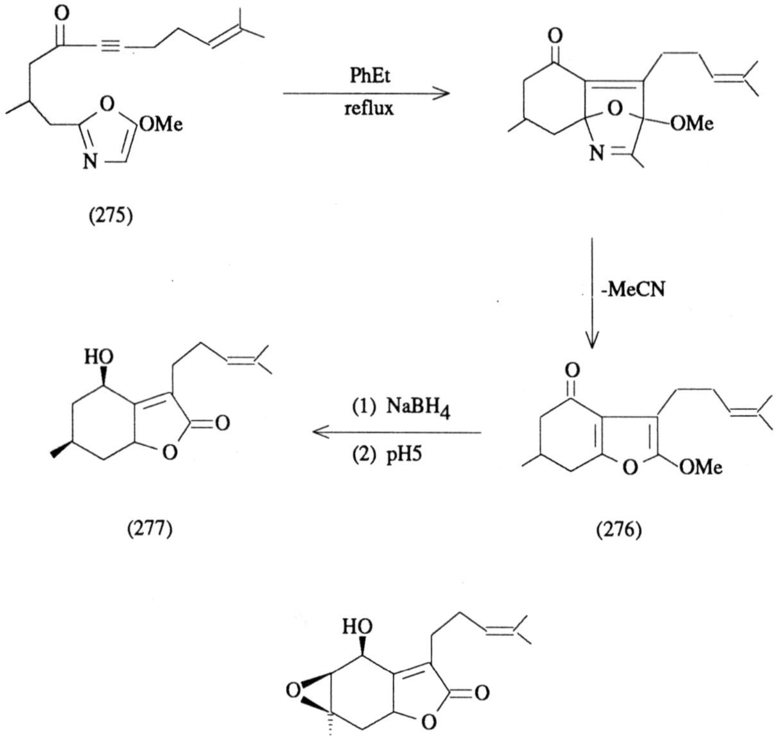

PhEt
reflux

(275)

-MeCN

(1) $NaBH_4$
(2) pH5

(277)

(276)

(278)

*Scheme 78*

**References**


1 K.J. Shea, S. Wise, L.D. Burke, P.D. Davis, J.W. Gilman and A.C. Greeley, *J. Am. Chem. Soc.*, 1982, **104**, 5708; K.J. Shea and E. Wada, *Ibid.*, 5715.

2 K.J. Shea and J.W. Gilman, *Tetrahedron Lett.*, 1983, **24**, 657.

3 K.J. Shea, W.M. Fruscella, R.C. Carr, L.D. Burke and D.K. Cooper, *J. Am. Chem. Soc.*, 1987, **109**, 447.

4 Reviews: G. Brieger and J.N. Bennett, *Chem. Rev.*, 1980, **80**, 63; A.G. Fallis, *Canad. J. Chem.*, 1984, **62**, 183; R. Craig, *Chem. Soc. Rev.*, 1987, **16**, 187; E. Ciganek, *Organic Reactions*, 1984, **32**, 1.

5 H.O. House and T. Cronin, *J. Org. Chem.*, 1965, **30**, 1061.

6 E.J. Corey and M. Petrzilka, *Tetrahedron Lett.*, 1975, 2537; see also J.D. White and B.G. Sheldon, *J. Org. Chem.*, 1981, **46**, 2273.

7 cf. E. Ciganek, *Organic Reactions*, 1984, **32**, 1; D. Craig, *Chem. Soc. Rev.*, 1987, **16**, 187; K.A. Parker and T. Iqbal, *J. Org. Chem.*, 1987, **52**, 4369.

8 W.R. Roush, A.I. Ko and H.R. Gillis, *J. Org. Chem.*, 1980, **45**, 4264; W.R. Roush, H.R. Gillis and A.I. Ko, *J. Am. Chem. Soc.*, 1982, **104**, 2269.

9 W.R. Roush, *J. Am. Chem. Soc.*, 1980, **102**, 1390.

10 Tse-Chong Wu and K.N. Houk, *Tetrahedron Lett.*, 1985, **26**, 2293.

11 W. Oppolzer, C. Fehr and J. Warneke, *Helv. Chim. Acta*, 1977, **60**, 48.

12 R.K. Boeckmann and T.R. Alessi, *J. Am. Chem. Soc.*, 1982, **104**, 3216; R.K. Boeckmann and Soo Sung Ko, *J. Am. Chem. Soc.*, 1980, **102**, 7146; R.K. Boeckmann and Soo Sung Ko, *J. Am. Chem. Soc.*, 1982, **104**, 1033.

13 M. Yoshioka, H. Nakai and M. Ohno, *J. Am. Chem. Soc.*, 1984, **106**, 1133.

14 cf. D. Craig, *Chem. Soc. Rev.*, 1987, **16**, 187; J.A. Marshall, B.G. Shearer and S.L. Crooks, *J. Org. Chem.*, 1987, **52**, 1236.

15 W.R. Roush and S.E. Hall, *J. Am. Chem. Soc.*, 1981, **103**, 5200.

16 W.R. Roush and H.R. Gillis, *J. Org. Chem.*, 1982, **47**, 4825.

17 S.R. Wilson and D.T. Mao, *J. Am. Chem. Soc.*, 1978, **100**, 6289.

18 R.L. Funk and W.E. Zeller, *J. Org. Chem.*, 1982, **47**, 180.

19 W.R. Roush and J.W. Coe, *Tetrahedron Lett.*, 1987, **28**, 931.

20 W.R. Roush and S.E. Hall, *J. Am. Chem. Soc.*, 1981, **103**, 5200.

21 W.R. Roush and H.R. Gillis, *J. Org. Chem.*, 1982, **47**, 4825.

22 W. Oppolzer and R.L. Snowden, *Tetrahedron Lett.*, 1976, 4187; W. Oppolzer, R.L. Snowden and D.P. Simmons, *Helv. Chim. Acta*, 1981, **64**, 2002.

23 G. Stork, G. Clark and C.S. Shiner, *J. Am. Chem. Soc.*, 1981, **103**, 4948.

24 S. Wattanasin, F.G. Kathawala and R.K. Boeckman, *J. Org. Chem.*, 1985, **50**, 3810.

25 R.K. Boeckman and D.M. Demko, *J. Org. Chem.*, 1982, **47**, 1789.

26 cf. D. Craig, *Chem. Soc. Rev.*, 1987, **16**, 187.

27 D.R. Williams and R.D. Gaston, *Tetrahedron Lett.*, 1986, **27**, 1485.

28 S.E. Denmark and J.A. Sternberg, *J. Am. Chem. Soc.*, 1986, **108**, 8277.

29 M.P. Edwards, S.V. Ley and S.G. Lister, *Tetrahedron Lett.*, 1981, **22**, 361; M.P. Edwards, S.V. Ley, S.G. Lister, B.P. Palmer and D.J. Wil-

liams, *J. Org. Chem.*, 1984, **49**, 3503; K.C. Nicolaou and R.L. Magolda, *J. Org. Chem*, 1981, **46**, 1506; cf. also W.R. Roush and A.G. Myers, *J. Org. Chem.*, 1981, **46**, 1509.

30 R.K. Boeckman and T.E. Barta, *J. Org. Chem.*, 1985, **50**, 3421.

31 R.K. Boeckman, J.J. Napier, E.W. Thomas and A.I. Sato, *J. Org. Chem.*, 1983, **48**, 4152; see also M.J. Kurth, D.H. Burns and M.J. O'Brien, *J. Org. Chem.*, 1984, **49**, 731.

32 S.G. Pyne, M.J. Hensel, S.R. Byrn, A.T. McKenzie and P.L. Fuchs, *J. Am. Chem. Soc.*, 1980, **102**, 5960; S.G. Pyne, M.J. Hensel and P.L. Fuchs, *J. Am. Chem. Soc.*, 1982, **104**, 5719.

33 S.G. Pyne, D.C. Spellmeyer, S. Chen and P.L. Fuchs, *J. Am. Chem. Soc.*, 1982, **104**, 5728.

34 D.A. Evans, K.T. Chapman and J. Bisaha, *J. Am. Chem. Soc.*, 1988, **110**, 1238.

35 D.A. Evans, T.C. Britton and J.A. Ellman, *Tetrahedron Lett.*, 1987, **28**, 6141.

36 W. Oppolzer, C. Chapius and G. Bernardinelli, *Helv. Chim. Acta*, 1984, **67**, 1397.

37 S.R. Wilson and D.T. Mao, *J. Am. Chem. Soc.*, 1978, **100**, 6289.

38 cf. F. Näf, R. Decorzant and W. Thommen, *Helv. Chim. Acta*, 1982, **65**, 2212.

39 P. Magnus, C. Walker, P.R. Jenkins and K.A. Menear, *Tetrahedron Lett.*, 1986, **27**, 651.

40 F. Näf and G. Ohloff, *Helv. Chim. Acta*, 1974, **57**, 1868.

41 G.A. Schiehsen and J.D. White, *J. Org. Chem.*, 1980, **45**, 1864; see also H. Yamamoto and H.L. Sham, *J. Am. Chem. Soc.*, 1979, **101**, 1609.

42 R.L. Snowden, *Tetrahedron Lett.*, 1981, **22**, 97,101.

43 E.G. Breithalle and A.G. Fallis, *J. Org. Chem.*, 1978, **43**, 1964.

44 R.L. Snowden, *Tetrahedron*, 1986, **42**, 3277.

45 B.B. Snider and B.W. Burbaum, *J. Org. Chem.*, 1983, **48**, 4370.

46 E.A. Deutsch and B.B. Snider, *J. Org. Chem.*, 1982, **47**, 2682.

47 H.J. Reich, E.K. Eisenhardt, R.E. Olson and M.J. Kelly, *J. Am. Chem. Soc.*, 1986, **108**, 7791.

48 H.M. Saxton, J.K. Sutherland and C. Whaley, *J. Chem. Soc. Chem. Commun.*, 1987, 1449.

49 R.A. Gibbs and W.H. Okamura, *J. Am. Chem. Soc.*, 1988, **100**, 4062.

50 P.A. Brown and P.R. Jenkins, *J. Chem. Soc. Perkin 1*, 1986, 1303; R.V. Bonnert and P.R. Jenkins, *J. Chem. Soc. Chem. Commun.*, 1987, 1540; see also K. Sakan and B.M. Craven, *J. Am. Chem. Soc.*, 1983, **105**, 3732; K.J. Shea and P.D. Davis, *Angew. Chem. internat. edn.*, 1983, **22**, 419; K.J. Shea and C.D. Haffner, *Tetrahedron Lett.*, 1988, **29**, 1367.

51 E.J. Thomas and J.W.F. Whitehead, *J. Chem. Soc. Chem. Commun.*, 1986, 724,727; R. Sauter, E.J. Thomas and J.P. Watts, *J. Chem. Soc. Chem. Commun.*, 1986, 1449.

52 H. Dyke, R. Sauter, P. Steel and E.J. Thomas, *J. Chem. Soc. Chem. Commun.*, 1986, 1447.

53 W. Oppolzer and E. Flashkamp, *Helv. Chim. Acta*, 1977, **60**, 204.

54 G. Stork and D.J. Morgans, *J. Am. Chem. Soc.*, 1979, **101**, 7110.

55 G.E. Keck, E. Boden and U. Sonnewald, *Tetrahedron Lett.*, 1981, **22**, 2615.

56 S.F. Martin, S.R. Desai, G.W. Phillips and A.C. Miller, *J. Am. Chem. Soc.*, 1980, **102**, 3294.

57 S.F. Martin, H. Rueger, S.A. Williamson and S. Grzejszczak, *J. Am. Chem. Soc.*, 1987, **109**, 6124; see also K.J. Shea and J.S. Svoboda, *Tetrahedron Lett.*, 1986, **27**, 4837.

58 M. Noguchi, S. Kakimoto, H. Kawakami and S. Kajigaeshi, *Bull. Chem. Soc. Japan*, 1986, **59**, 1355.

59 S.F. Martin, S.A. Williamson, R.P. Gist and K.M. Smith, *J. Org. Chem.*, 1983, **48**, 5170.

60 S.F. Martin, B. Benage, S.A. Williamson and S.P. Brown, *Tetrahedron*, 1986, **42**, 2903.

61 cf. R.K. Boeckman and D.M. Demke, *J. Org. Chem.*, 1982, **47**, 1789.

62 S.F. Martin, S.A. Williamson, R.P. Gist and K.M. Smith, *J. Org. Chem.*, 1983, **48**, 5170.

63 cf. L-F. Tietze, T. Brumby, M. Pretor and G. Remberg, *J. Org. Chem.*, 1988, **53**, 810.

64 for example L-F. Tietze, G.V. Kiedrowski and B. Berger, *Angew. Chem. internat. edn.*, 1982, **21**,221.

65 J.J. Talley, *J. Org. Chem.*, 1985, **50**, 1695.

66 J.P. Marino and S.L. Dax, *J. Org. Chem.*, 1984, **49**, 3671; L-F. Tietze, *Angew. Chem. internat. edn.*, 1983, **22**, 828.

67 L-F. Tietze, G.V. Kiedrowski, K. Harms, W. Clegg and G. Sheldrick, *Angew. Chem. internat. edn.*, 1980, **19**, 134; L-F. Tietze and G.V. Kiedrowski, *Tetrahedron Lett.*, 1981, **22**, 219.

68 J.J. Talley, *J. Org. Chem.*, 1985, **50**, 1695.

69 J.P. Marino and S.L. Dax, *J. Org. Chem.*, 1984, **49**, 3671.

70 L-F. Tietze, S. Brand and T. Pfeiffer, *Angew. Chem. internat. edn.*, 1985, **24**, 784.

71 L-F. Tietze, H. Denzer, X. Holdgrün and M. Neumann, *Angew. Chem. internat. edn.*, 1987, **26**, 1295.

72 S. Takano, S. Satoh and K. Ogasawara, *J. Chem. Soc. Chem. Commun.*, 1988, 59.

73 B.B. Snider and J.V. Duncia, *J. Org. Chem.*, 1980, **45**, 3461.

74 Y-S. Cheng, F.W. Fowler and A.T. Lupo, *J. Am. Chem. Soc.*, 1981, **103**, 2090; Y-S. Cheng, A.T. Lupo and F.W. Fowler, *J. Am. Chem. Soc.*, 1983, **105**, 7696.

75 Y-C. Hwang and F.W. Fowler, *J. Org. Chem.*, 1985, **50**, 2719.

76 M. Ihara, T. Kirihara, A. Kawaguchi, K. Fukumoto and T. Kametani, *Tetrahedron Lett.*, 1984, **25**, 4541.

77 S.M. Weinreb, *Acc. Chem. Res.*, 1985, **18**, 16.

78 S.M. Weinreb, N.A. Khatri and H. Shringarpure, *J. Am. Chem. Soc.*, 1979, **10**, 5073.

79 N.A. Khatri, H.F. Schmitthener, J. Schringarpure and S.M. Weinreb, *J. Am. Chem. Soc.*, 1981, **103**, 6387.

80 R.A. Gobao, M.L. Bremner and S.M. Weinreb, *J. Am. Chem. Soc.*, 1982, **104**, 7065.

81 M.L. Bremner and S.M. Weinreb, *Tetrahedron Lett.*, 1983, **24**, 261; M.L. Bremner, N.A. Khatri and S.M. Weinreb, *J. Org. Chem.*, 1983, **48**, 3661.

82 B. Noder, T.R. Bailey, R.W. Franck and S.M. Weinreb, *J. Am. Chem. Soc.*, 1981, **103**, 7573.

83 T.R. Bailey, R.S. Garigipati, J.A. Morton and S.M. Weimreb, *J. Am. Chem. Soc.*, 1984, **106**, 3240.

84 S.D. Larsen and P.A. Grieco, *J. Am. Chem. Soc.*, 1985, **107**, 1768; P.A. Grieco and S.D. Larsen, *J. Org. Chem.*, 1986, **51**, 3553.

85 P.M. Scola and S.M. Weinreb, *J. Org. Chem.*, 1986, **51**, 3248.

86 M.J. Melnick and S.M. Weinreb, *J. Org. Chem.*, 1988, **53**, 850.

87 S.E. Denmark, M.S. Dappen and J.A. Sternberg, *J. Org. Chem.*, 1984, **49**, 4741.

88 S.E. Denmark, M.S. Dappen and C.J. Cramer, *J. Am. Chem. Soc.*, 1986, **108**, 1306.

89 I. Iida, Y. Watanabe and C. Kibayashi, *J. Am. Chem. Soc.*, 1985, **107**, 5534.

90 G.E. Keck, *Tetrahedron Lett.*, 1978, 4767; G.E. Keck and D.G. Nickell, *J. Am. Chem. Soc.*, 1980, **102**, 3632.

91 R.S. Garigipati, A.J. Freyer, R.R. Whittle and S.M. Weinreb, *J. Am. Chem. Soc.*, 1984, **106**, 7861.

92 S.W. Remiszewski, R.R. Whittle and S.M. Weinreb, *J. Org. Chem.*, 1984, **49**, 3243.

93 S.W. Remiszewski, T.R. Stouch and S.M. Weinreb, *Tetrahedron*, 1985, **41**, 1173.

94 W. Oppolzer, *Synthesis*, 1978, 793; T. Kametani, *Pure Appl. Chem.*, 1979, **51**, 747; R.L. Funk and K.P.C. Vollhardt, *Chem. Soc. Rev.*, 1980, **9**, 41; J.L. Charlton and M.M. Alauddin, *Tetrahedron*, 1987, **43**, 2873.

95 W. Oppolzer, *Angew. Chem. internat. edn.*, 1972, **11**, 1031.

96 T. Kametani, H. Matsumoto, H. Nemoto and K. Fukumoto, *J. Am. Chem. Soc.*, 1978, **100**, 6218.

97 D.F. Taber, K. Raman and M.D. Gaul, *J. Org. Chem.*, 1987, **52**, 28.

98 T. Kametani, K. Suzuki and H. Nemoto, *J. Org. Chem.*, 1982, **47**, 2331.

99 W. Oppolzer, *Tetrahedron Lett.*, 1974, 1001; *J. Am. Chem. Soc.*, 1971, **93**, 3833.

100 cf. R.L. Funk and K.P.C. Vollhardt, *Chem. Soc. Rev.*, 1980, **9**, 41; K.P.C. Vollhardt, *Angew Chem. internat. edn.*, 1984, **23**, 539.

101 W. Oppolzer, D.A. Roberts and T.G.C. Bird, *Helv. Chim. Acta*, 1979, **62**, 2107.

102 W. Oppolzer and D.A. Roberts, *Helv. Chim. Acta*, 1980, **63**, 1703; cf. also K.C. Nicolaou, W.E. Barnetti and P. Ma, *J. Org. Chem.*, 1980, **45**, 1463.

103 S. Djuric, T. Sakar and P. Magnus, *J. Am. Chem. Soc.*, 1980, **102**, 6885.

104 Y. Ito, M. Nakatsuka and T. Saegusa, *J. Am. Chem. Soc.*, 1981, **103**, 476.

105 cf. T. Tuschka, K. Naito and B. Rickborn, *J. Org. Chem.*, 1983, **48**, 70.

106 M. Herslöf and A.R. Martin, *Tetrahedron Lett.*, 1987, **28**, 3423.

107 T. Gallagher and P. Magnus, *Tetrahedron*, 1981, **37**, 3889.

108 P. Magnus, T. Gallagher and P. Pappalardo, *Acc. Chem. Res.*, 1984, **17**, 35; T. Gallagher, P. Magnus and J. Huffman, *J. Am. Chem. Soc.*, 1982, **104**, 1140; M. Laidlaw, P.M. Cairns and P. Magnus, *J. Chem. Soc. Chem. Commun.*, 1986, 1756; P. Magnus and P.M. Cairns, *J. Am. Chem. Soc.*, 1986, **108**, 217.

109 Y. Ito, S. Miyata, M. Nakatsuka and T. Saegusa, *J. Am. Chem. Soc.*, 1981, **103**, 5250.

110 O.L. Chapman, M.R. Engel, J.P. Springer and J.C. Clardy, *J. Am. Chem. Soc.*, 1971, **93**, 6696; for other approaches to the synthesis of lignans using intramolecular Diels-Alder reactions see D.I. MacDonald and T. Durst, *J. Org. Chem.*, 1986, **51**, 4749; T.L. Holmes and R. Stevenson, *Tetrahedron Lett.*, 1970, 199.

111 P.A. Jacobi, C.S.R. Kaczmarek and U.E. Udodong, *Tetrahedron*, 1987, **43**, 5475.

112 P.A. Jacobi, M. Egbertson, R.F. Frechette, C.K. Miao and K.T. Weiss, *Tetrahedron*, 1988, **44**, 3327.

# 4 MISCELLANEOUS [4+2] CYCLOADDITIONS

## Cycloadditon Reactions with Allyl Cations and Allyl Anions

The Diels-Alder reactions discussed above are concerted [4+2]π cycloadditions involving six π-electrons and lead readily to the formation of six-membered rings. The possibility of analogous six π-electron cycloadditions with allyl anions and allyl cations which would give five- and seven-membered rings respectively was predicted[1] and examples of both processes have been observed, although the synthetic scope of the anion reaction is as yet limited (Scheme 1).

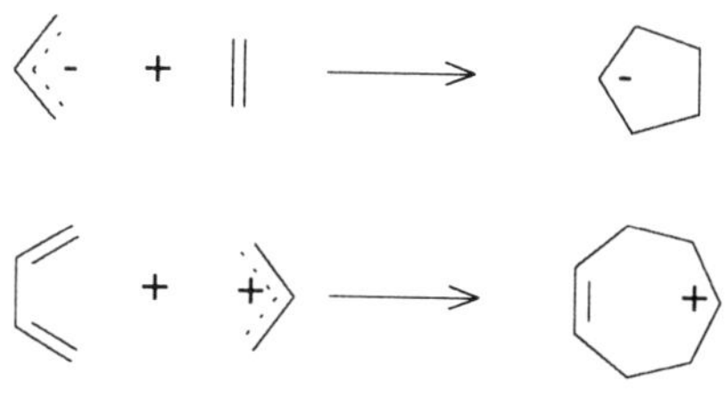

*Scheme 1*

In the anion series the best results have been obtained with the 2-aza-allyl anion which adds to a variety of alkenes to form pyrrolidine derivatives. The all-carbon allyl anions themselves seem to undergo cycloaddition to carbon-carbon double bonds only when an electron-attracting group is attached at C-2 of the allyl system, probably because the allylic delocalised negative charge becomes localised on that carbon atom in the cyclopentyl anion produced. Thus, 2-phenylpropene treated successively with lithium diisopropylamide and *trans*-stilbene, gave the cyclopentane (1) in 41 per cent yield.[2]

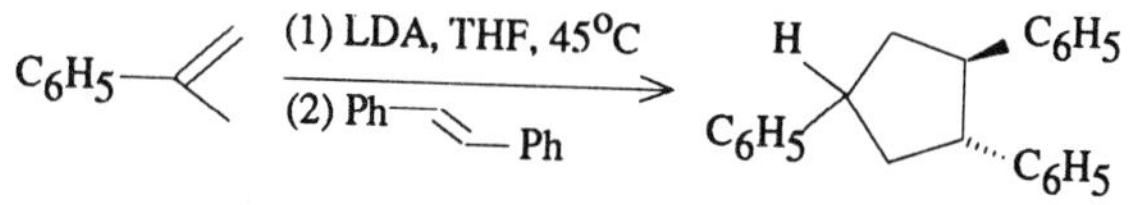

$C_6H_5$ — (1) LDA, THF, 45°C (2) Ph—=—Ph → H, $C_6H_5$, $C_6H_5$, $C_6H_5$ (1)

*Scheme 2*

2-Aza-allyl anions (3) are readily obtained by the action of lithium di-isopropylamide on azomethines of the type (2) where either or both of $R^1$

and $R^2$ is an aryl group. They readily undergo intermolecular cycloaddition reactions with activated olefins, allenes and acetylenes, usually in a highly regioselective manner.[3] Anions bearing more strongly electron-attracting groups are generally too unreactive to be useful in cycloadditions. Thus, reaction of 1,1-diphenyl-2-aza-allyl lithium with styrene gave the sterically less favoured isomer (4) almost exclusively in 85 per cent yield. An alternative procedure which has the advantage that it allows the generation of aza-allyl anions which do not bear stabilising aryl substituents, proceeds from N-(trialkylstannylmethyl)imines by transmetallation with an alkyl-lithium.[4] The reagent (5), for example, with butyl-lithium and *trans*-stilbene, gave the pyrrolidine (6).

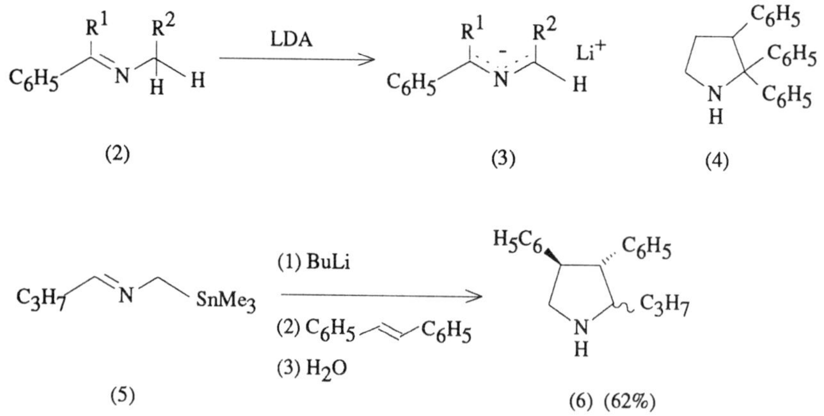

*Scheme 3*

It is uncertain whether these cycloadditions are concerted [π4s+π2s] reactions proceeding *via* a transition state (7), or whether they are stepwise and take place through an intermediate like (8), where rotation about the bonds *a* and *b* is hindered by co-ordination of lithium to the carbon-nitrogen double bond or to the nitrogen.

*Scheme 4*

Intramolecular reactions take place easily, to give pyrrolidine derivatives.[5] Unactivated olefins react readily as anionophiles in intramolecular reactions, and anions bearing only one aryl substituent can be used. Thus, the imine (9) gave a mixture of the two bicyclic pyrrolidines (10) and (11) in 63 per cent yield on treatment with lithium diisopropylamide followed by protonolysis

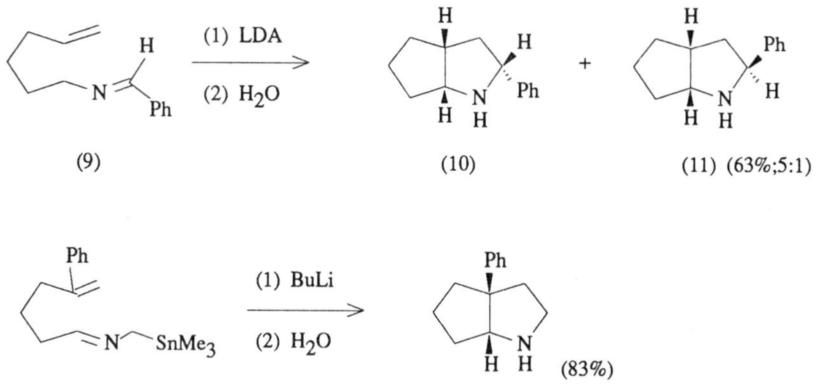

(1) LDA (2) $H_2O$

(9) (10) (11) (63%;5:1)

(1) BuLi (2) $H_2O$ (83%)

*Scheme 5*

Concerted reaction by way of the W-form of the aza-allyl anion (12) rationalises the formation of *cis*-fused bicyclic pyrrolidines and the favoured stereochemistry at C-2.

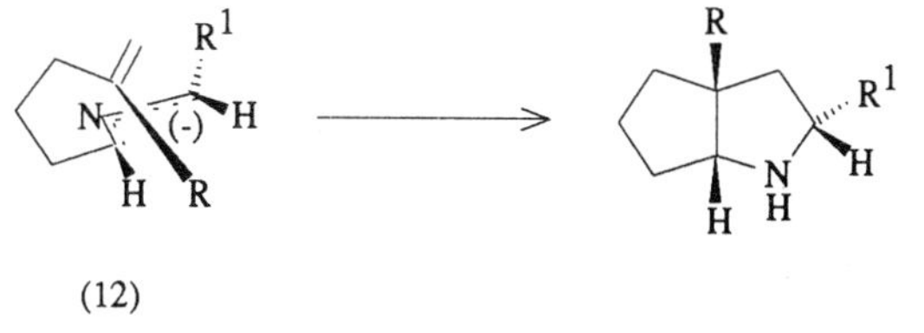

(12)

*Scheme 6*

Cycloaddition of allyl cations to conjugated dienes provides a route to seven-membered carbocycles,[6] and is being increasingly used in synthesis. Several methods may be used to generate the allyl cations. They can be obtained, for example, from an allyl iodide and silver trifluoroacetate or from an allyl alcohol by way of the corresponding trifluoroacetate and zinc chloride. Generated in the presence of a diene they form seven-membered cycloadducts in variable yield. Reaction proceeds best with cyclic dienes. Thus, cyclohexadiene and methylallyl cation gave 3-methylbicyclo[3,2,2]-nona-2,6-diene (13) in 30 per cent yield, and in an intramolecular example the carbon skeleton of the zizaene sesquiterpenes was formed directly from the trifluoroacetate (14) by way of the allyl cation (15).[7] In this sequence the trimethylsilyl group serves both to stabilise the allyl cation and as a trigger for the formation of the exocyclic methylene group.

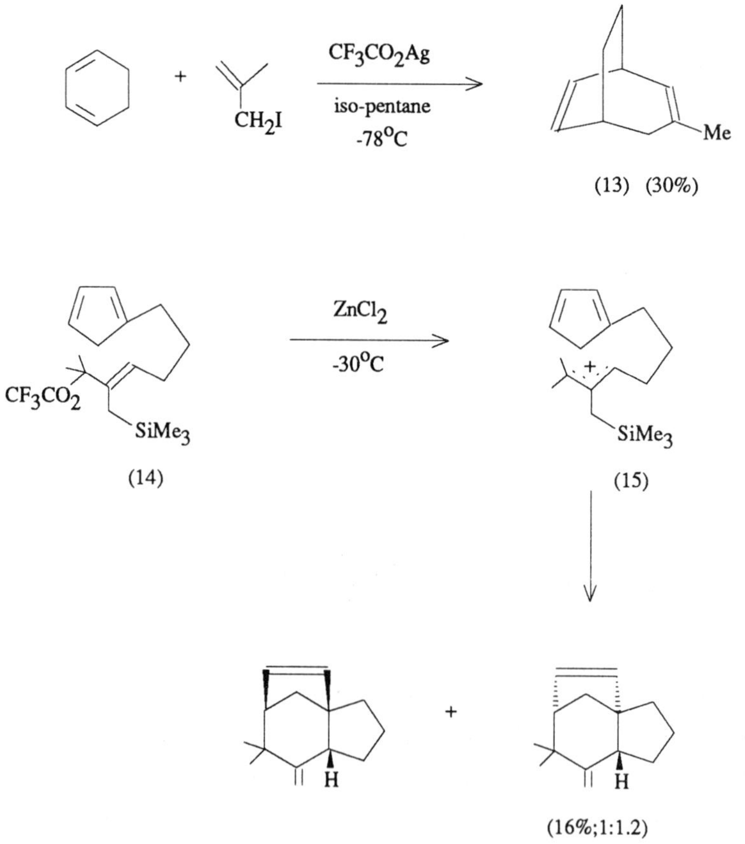

$CH_2I$ ; $CF_3CO_2Ag$, iso-pentane, -78°C ; Me

(13) (30%)

$CF_3CO_2$ ; $SiMe_3$ ; (14) ; $ZnCl_2$, -30°C ; $SiMe_3$ ; (15)

H ; H

(16%;1:1.2)

*Scheme 7*

## Oxyallyl Cations

Synthetically, 2-oxyallyl cations have been the most useful.[8] They are readily obtained by a variety of methods, for example from α,α′-dibromoketones by the action of reducing agents such as zinc-copper couple or iodide ion, or by reaction with di-iron enneacarbonyl, $Fe_2(CO)_9$,[9] or from α-halogeno-trimethylsilyl enol ethers and Lewis acids such as silver trifluoroacetate or zinc chloride.[10] (Scheme 8)

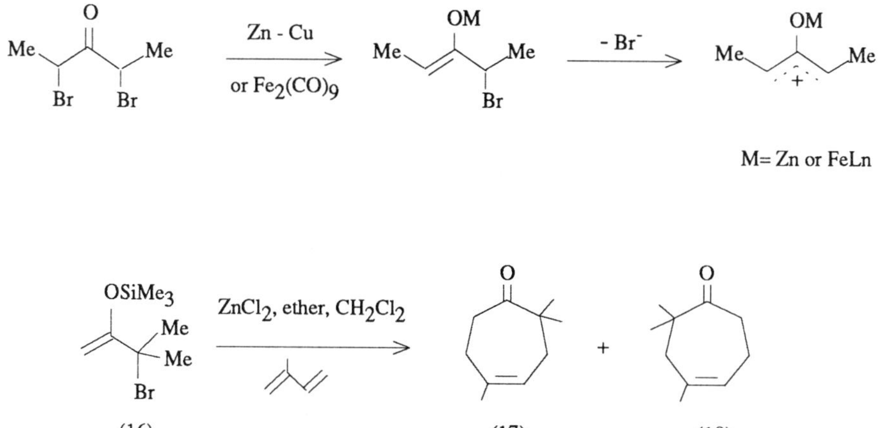

Zn - Cu or $Fe_2(CO)_9$; $- Br^-$

M= Zn or FeLn

$ZnCl_2$, ether, $CH_2Cl_2$

(16) (17) (18)

*Scheme 8*

The zinc enolates are less electrophilic than the iron enolates, because of the greater covalency of the metal-oxygen bond in the iron compounds. Cycloaddition reactions with poorly nucleophilic dienes thus proceed best with diiron enneacarbonyl as reductant. Formation of oxyallyl cations from α,α′-dibromoketones and zinc chloride is promoted by ultrasound and some difficultly accessible adducts have been made available this way.[11]

Generated in the presence of acyclic dienes and some cyclic dienes the oxyallyls readily form seven-membered ring ketones.[12] Thus, reaction of the α-bromosilylenol ether (16) with isoprene in the presence of zinc chloride afforded the naturally-occurring monoterpenoid karahanaenone (17) and the regioisomer (18),[13] and reaction of α,α,α′,α′-tetrabromoacetone and 3-isopropylfuran in the presence of $Fe_2(CO)_9$, followed by debromination of the initial adduct with zinc-copper couple gave the 8-oxabicyclo[3,2,1]oct-6-en-3-one (19), (α,α′-dibromoacetone itself does not form an oxyallyl cation).

By further reaction this was converted into the tropone nezukone (20) and a number of other naturally-occurring tropones and tropolones were prepared in the same way.[14]

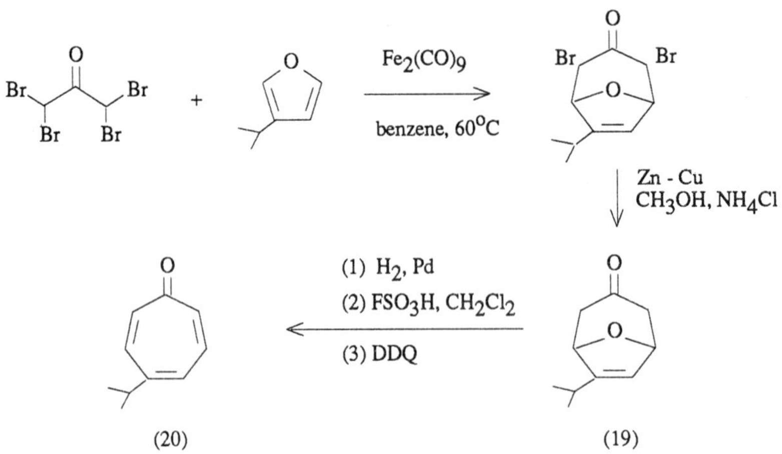

$Fe_2(CO)_9$, benzene, 60°C

Zn - Cu, $CH_3OH$, $NH_4Cl$

(1) $H_2$, Pd
(2) $FSO_3H$, $CH_2Cl_2$
(3) DDQ

(20) (19)

*Scheme 9*

Oxyallyls are being increasingly used in the synthesis of natural products[15] and in this area adducts formed from furans and pyrrole derivatives have been particularly useful. The more electrophilic oxyallyl cations react with N-alkypyrroles to form substitution products, but under the appropriate conditions, using sodium iodide and α,α′-dibromoketones to generate the oxyallyls, cycloadducts can be obtained from N-alkylpyrroles, and have been used to prepare analogues of cocaine[16]. With more electrophilic oxyallyls generated from α-halogenoketones with $Fe_2(CO)_9$ or zinc-copper couple, pyrroles carrying an electron-withdrawing group on the nitrogen must be used. Thus, reaction of N-carbomethoxypyrrole with tetrabromoacetone in the presence of $Fe_2(CO)_9$ gave a mixture of the tropane-like adducts (21) in 70 per cent yield. Debromination with zinc-copper couple and reduction with DIBAL led almost completely to the α-alcohol (22) which served as a common intermediate for the synthesis of a number of tropane alkaloids.[17]

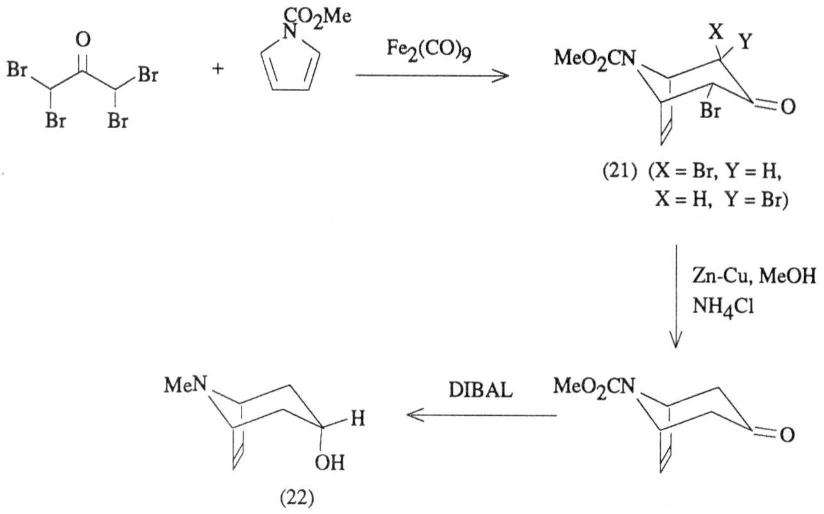

*Scheme 10*

Cycloaddition of oxyallyls to furan derivatives has also been employed in the synthesis of natural products. The initial adducts are 8-oxabicyclo-[3,2,1]oct-6-en-3-ones which can be converted into cycloheptanones or substituted tetrahydrofurans by ring cleavage at the oxygen bridge or at the ketone group. Thus, in a synthesis of nonactic acid (27), the adduct (24), obtained from furan and the oxyallyl (23), gave the lactone (25) after reduction of the double bond and Baeyer-Villiger oxidation. Methanolysis of (25) then gave the *cis*-disubstituted tetrahydrofuran (26) as a single stereo-isomer, which was converted in several more steps into nonactic acid (27).[18] The stereochemistry of the tetrahydrofuran (26) is established at the highly selective cycloaddition step involving the oxyallyl cation in the W-conformation (23) and a preferred boat-like (*endo*) transition state.[19] Baeyer-Villiger oxidation of 8-oxabicyclo[3,2,1]octan-3-ones, as in this example, and opening of the resulting lactone is one of the best methods available for preparing *cis*-2,5-disubstituted tetrahydrofurans. A similar procedure was used to make (28), which served as a precursor for a stereo-selective synthesis of lilac alcohol (29).[8a]

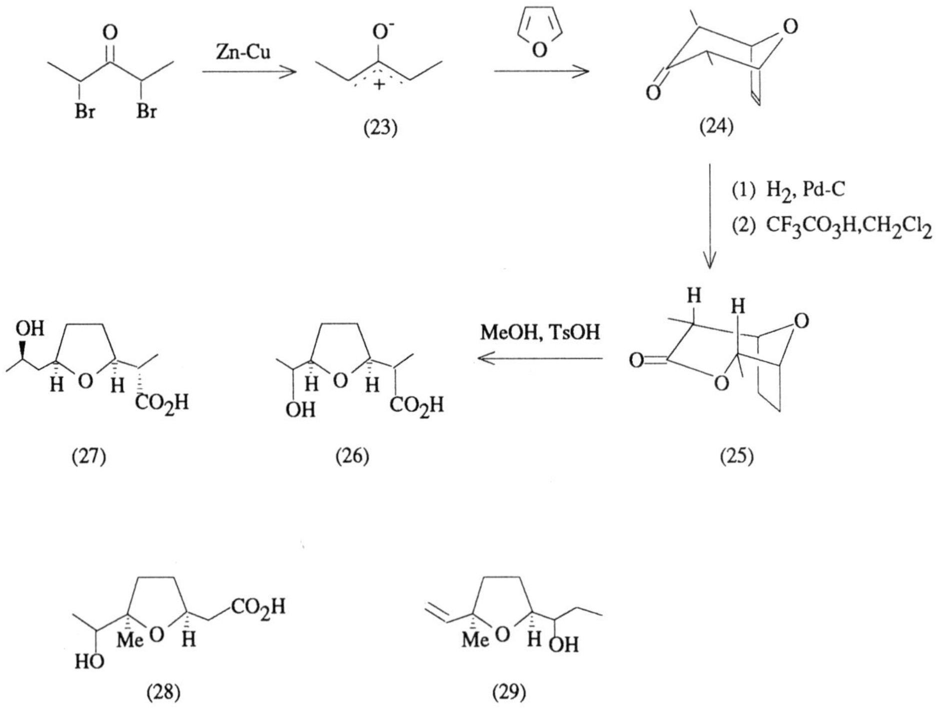

Zn-Cu

(23)

(24)

(1) $H_2$, Pd-C

(2) $CF_3CO_3H, CH_2Cl_2$

MeOH, TsOH

(27)

(26)

(25)

(28)

(29)

*Scheme 11*

In a different area, stereoselective perhydroxylation of the double bond in 8-oxabicyclo[3,2,1]octenones affords products which have been used in the synthesis of C-nucleosides.[20]

Acyclic compounds can also be prepared stereoselectively. Thus, the triol (30), a fragment of the antibiotic rifamycin S, was obtained from the adduct (24), again employing the Baeyer-Villiger procedure to prepare the bicyclo-octane for ring-opening.[21]

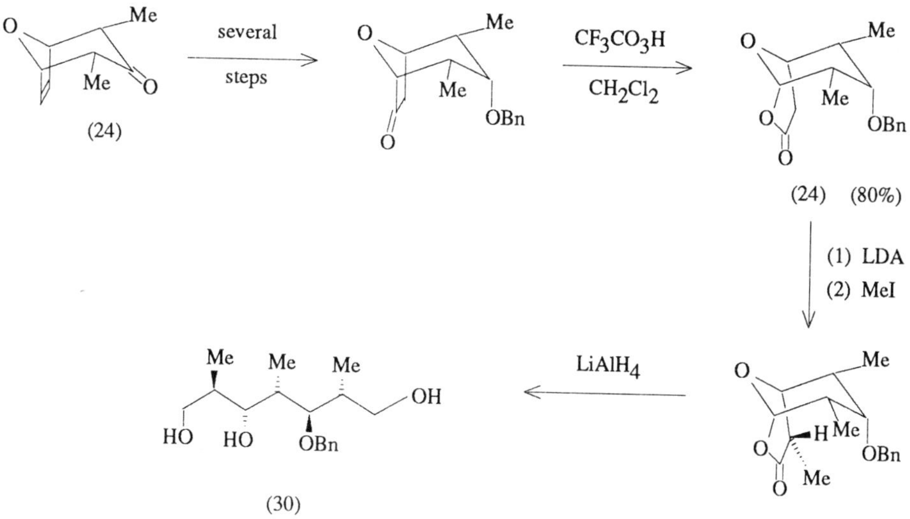

*Scheme 12*

Cleavage at the carbonyl group of the 8-oxabicyclo[3,2,1]octanones has also been effected by Beckmann rearrangement of the derived oximes, and the products used to make analogues of muscarine.[22]

Cleavage of the ether bridge in oxabicyclo[3,2,1]octanones to give seven-membered ring ketones can be effected with acid or Lewis acids, as in the syntheses of nezukone (20), but other procedures can be employed. Thus, the adduct (32) formed exclusively from the 2-substituted furan (31), was converted in several steps into the oxabicyclo-octane (33). Baeyer-Villiger oxidation then led to the acetal (34) which, on hydrolysis, gave specifically the cycloheptanone derivative (35). Again the stereochemistry of the product is determined by the stereoselectivity of the cycloaddition reaction.[23]

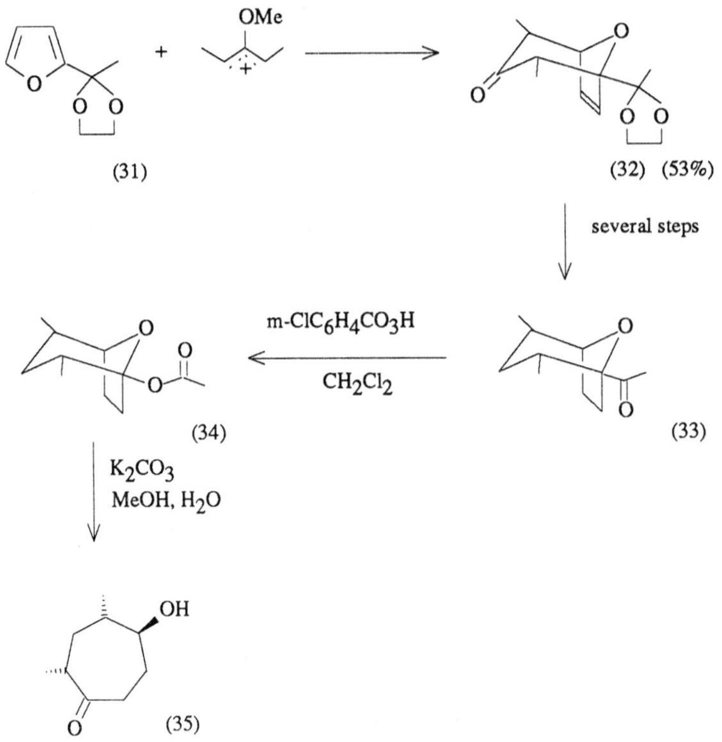

OMe

(31) (32) (53%)

several steps

m-$ClC_6H_4CO_3H$

$CH_2Cl_2$

(34) (33)

$K_2CO_3$
MeOH, $H_2O$

OH

(35)

*Scheme 13*

While many allyl cations, including the 2-methoxyallyl cation, afford oxygen-bridged seven-membered rings on reaction with furan, not all do so. 2-Methylallyl cation and allyl cation itself instead give rise to products of electrophilic substitution rather than cycloaddition products. Extensive work on the mode of formation of the cycloadducts has shown that the reactions may take a concerted and also a stepwise course depending on the nature of the diene and the allyl cation, the reaction conditions and even the solvent.[8a,10] Thus, the oxyallyl species (36), generated in the W-configuration from 2,4-dibromopentan-3-one, reacted with cyclopentadiene stereoselectively to give only the diequatorial-adduct (37) and the diaxial-adduct (38) in proportions depending on the mode of formation of the cations. None of the axial-equatorial isomer was detected, strongly suggesting that these reactions are concerted, taking place through *endo* and *exo* transition states.

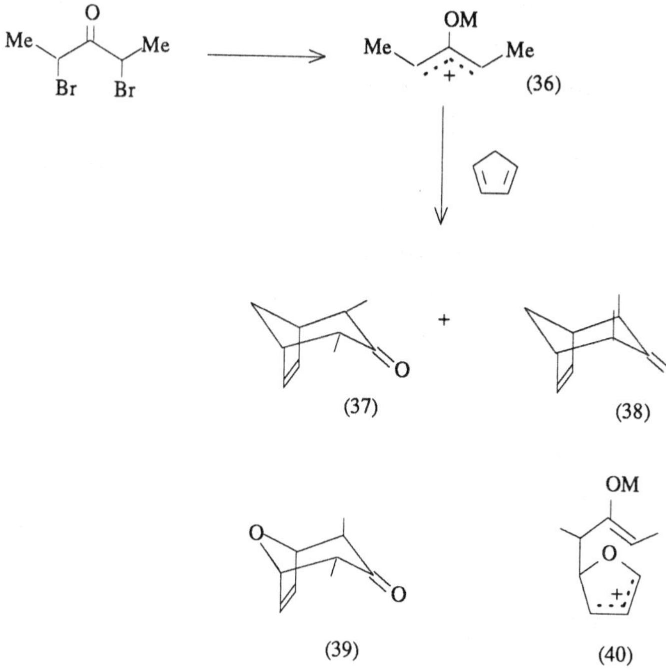

Me Br O Br Me → OM Me Me (36)

(37) + (38)

(39) (40) OM

*Scheme 14*

On the other hand furan, under the same conditions, gave mixtures containing variable proportions of the axial-equatorial isomer (39), formed in a stepwise addition by way of (40).

Oxyallyl cations also react with certain olefins to form five-membered rings but these are stepwise reactions. The preferred orientation of addition is that one that affords the maximum stabilisation of the zwitterionic intermediate.[24]

## Pentadienyl Cations

In principle, seven-membered rings might also be formed by a [$2\pi$+$4\pi$]-electron cycloaddition of pentadienyl cations to olefins.[1] A spectacular realisation of this is seen in the thermal conversion of perezone (41) into the isomeric pipitzols (42),[25] which has been shown indeed to be a concerted reaction.[26]

Reactions of this kind have formed the key step in several elegant synthesis of natural products containing the bicyclo[3,2,1]octane ring system. In these the cyclopentadienyl cation is formed *in situ* from a *para*-quinone monoacetal. Thus, the sesquiterpene gymnomitral (45) was readily obtained from the adduct (44) formed in the catalysed reaction of 1,2-dimethylcyclopentene with the hemiacetal (43).[27]

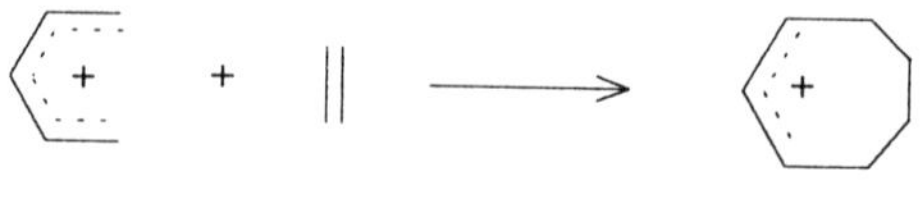

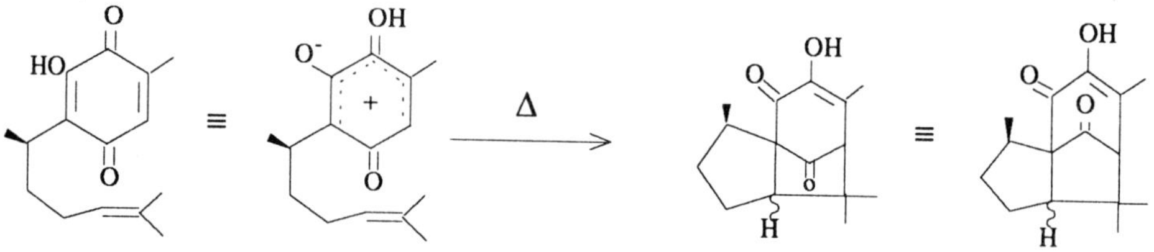

(41) (42)

*Scheme 15*

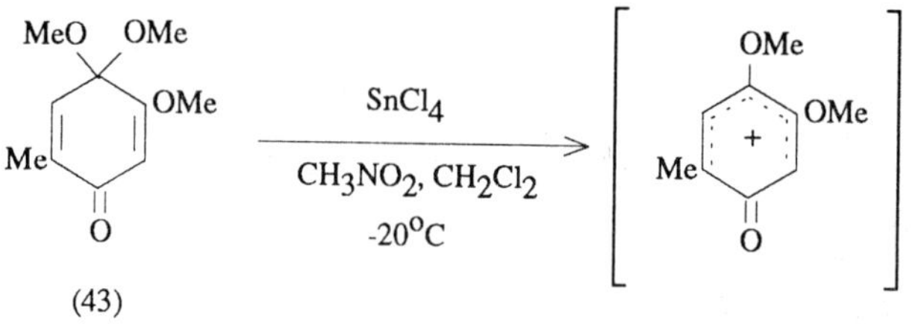

SnCl$_4$ CH$_3$NO$_2$, CH$_2$Cl$_2$ -20°C

(43) (44) (45)

several steps

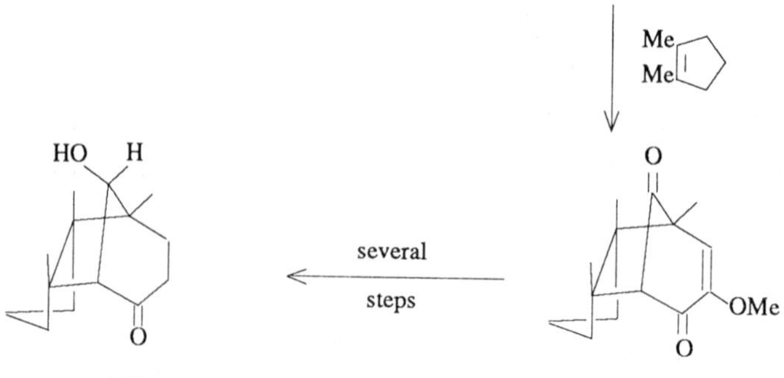

*Scheme 16*

In an intramolecular example, the adducts (47) and (48) were obtained in 80 per cent yield by way of the pentadienyl cation generated from the phenol (46) by anodic oxidation. Further transformations of these products gave (49) and thence (±)-8,14-cedranoxide (50).[28]

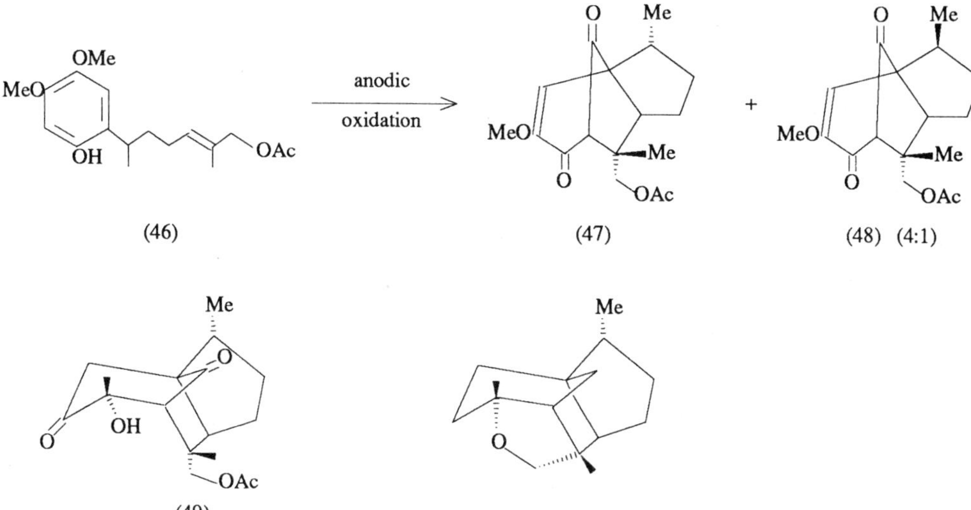

*Scheme 17*

## Cycloadditions with Trimethylenemethane

A useful route to five-membered rings is provided by the cycloaddition of Pd(0) complexes of trimethylenemethane to olefins bearing electron-attracting substituents.[29] The complexes are prepared *in situ* from 2-trimethylsilylmethylallyl acetate, and react with a range of electron-deficient olefins to form methyleneclyclopentanes (Scheme 18). Simple alkyl-substituted olefins, and electron-rich olefins like enol ethers and enamines do not give adducts.

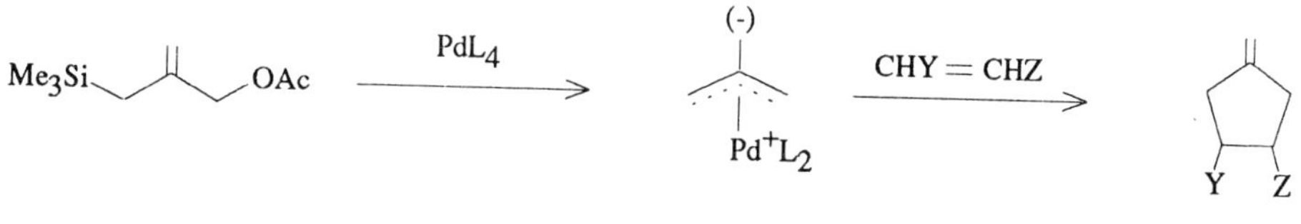

*Scheme 18*

Thus, methyl methacrylate gave the methylenecyclopentane derivative (51), and cyclopentyl phenyl sulphone formed the bicyclic product (52).

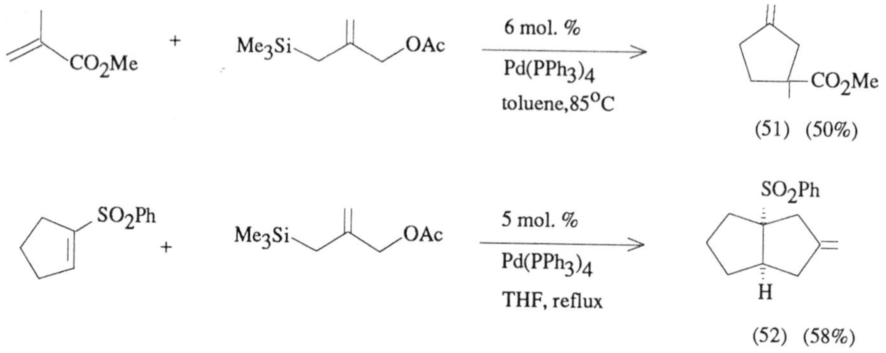

$CO_2Me$ + $Me_3Si$ OAc → 6 mol. % $Pd(PPh_3)_4$, toluene, 85°C → $CO_2Me$ (51) (50%)

$SO_2Ph$ + $Me_3Si$ OAc → 5 mol. % $Pd(PPh_3)_4$, THF, reflux → $SO_2Ph$, H (52) (58%)

*Scheme 19*

For (*E*)-olefins the reactions are highly stereoselective, but (*Z*)-olefins frequently give mixtures of stereoisomers, sometimes because of partial isomerisation of the olefin before reaction.

These reactions are believed to take place in a stepwise fashion by initial addition of the nucleophilic species (53) to the double bond, followed by cyclisation. Competitive bond rotation in the intermediate (54) before ring closure results in partial loss of stereochemical integrity in some cases.

With complexes bearing an alkyl substituent on the trimethylenemethane moiety equilibration takes place faster than ring closure to give a complex in which the negative charge resides on the carbon bearing the substituent. Thus, the two isomeric precursors (55) and (56) both gave the same product (58) on reaction with cyclopentenone, by way of the complex (57).[30]

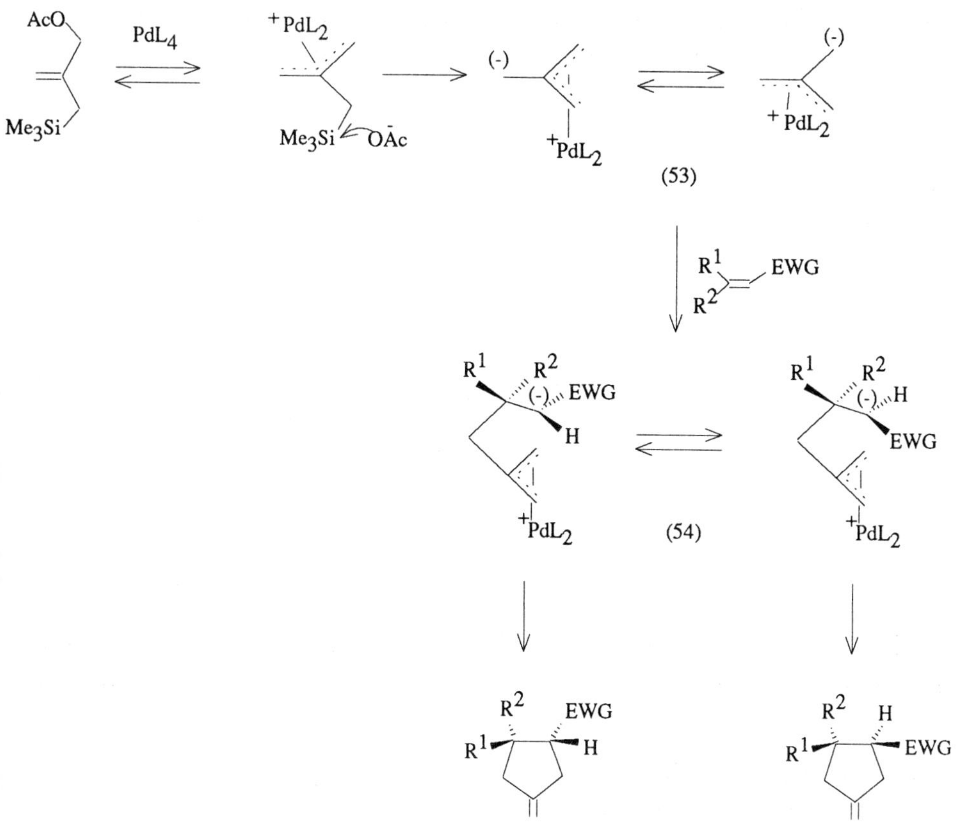
AcO
Me3Si
PdL4
+PdL2
Me3Si OAc
(-)
+PdL2
(-)
+ PdL2
(53)
R1 EWG
R2
R1 R2
(-) EWG
H
+PdL2
(54)
R1 R2
(-) H
EWG
+PdL2
R2 EWG
R1 H
R2 H
R1 EWG

*Scheme 20*

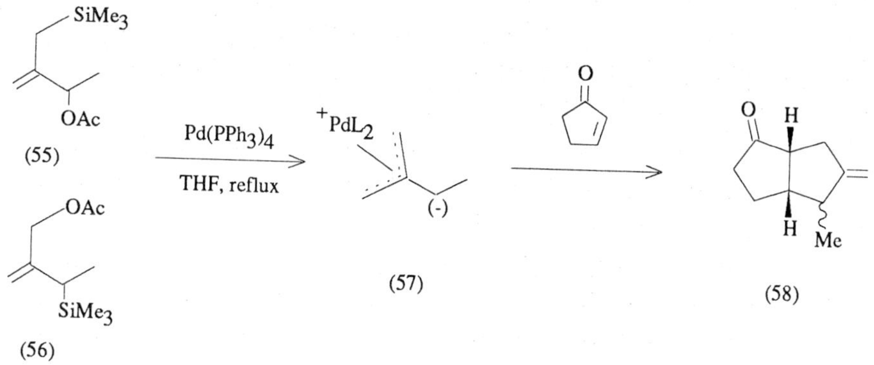
SiMe3
OAc
(55)
OAc
SiMe3
(56)
Pd(PPh3)4
THF, reflux
+PdL2
(-)
(57)
O
O
H
H
Me
(58)

*Scheme 21*

These reactions have been employed in the synthesis of a number of natural products containing five-membered rings.[29a] Thus, in a synthesis of (+)-brefeldin A (62) the five-membered ring was formed stereoselectively by cycloaddition of the Pd(0) complex of trimethylenemethane to the optically active acrylate (59), giving a mixture of the two diastereoisomeric adducts (60) and (61) in 87 per cent yield.[31]

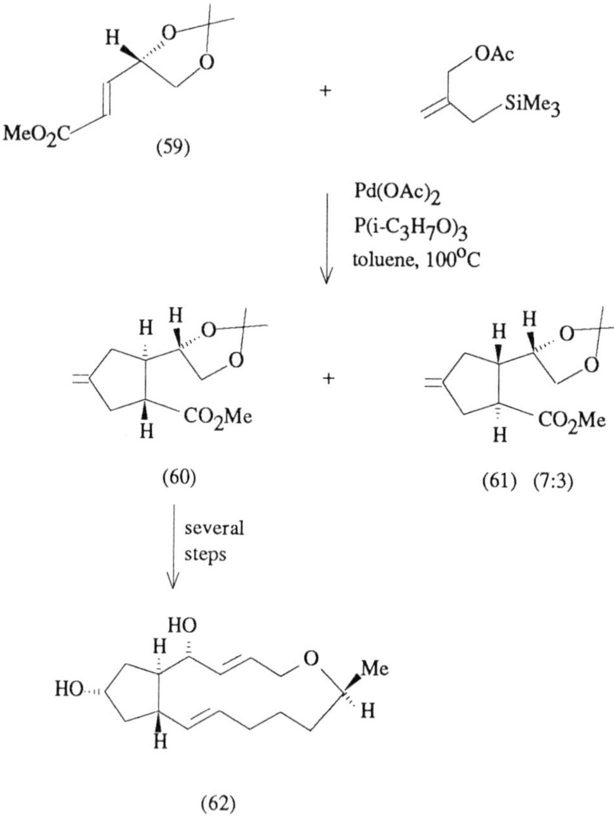

H
O
O
MeO2C
(59)
+
OAc
SiMe3
Pd(OAc)2
P(i-C3H7O)3
toluene, 100°C
H
H
O
O
CO2Me
H
(60)
+
H
H
O
O
CO2Me
H
(61) (7:3)
several
steps
HO
H
O
Me
HO
H
H
(62)

*Scheme 22*

Intramolecular reactions have also been effected[32], and derivatives of tetrahydrofuran and of pyrrolidine have been obtained by cycloaddition to the carbonyl group of aldehydes and ketones, and to imines respectively.[33] Stepwise [4+3] cycloaddition of trimethylenemethane to dialkylidenecyclopentanes has been employed to prepare hydroazulenes.[34]

In general, attempts to trap trimethylenemethane itself with olefins have not been successful because of intramolecular closure to methylenecyclopropane. But the related diyls (64), obtained by thermal or photolytic extrusion of nitrogen from the diazenes (63) can be trapped by electron-deficient olefins to give fused-ring bicyclo[3,3,0]octanes (65) or bridged-ring 7-alkylidenebicyclo[2,2,1]heptanes (66).[35] Under the appropriate experimental conditions the fused-ring products can be obtained preferentially.

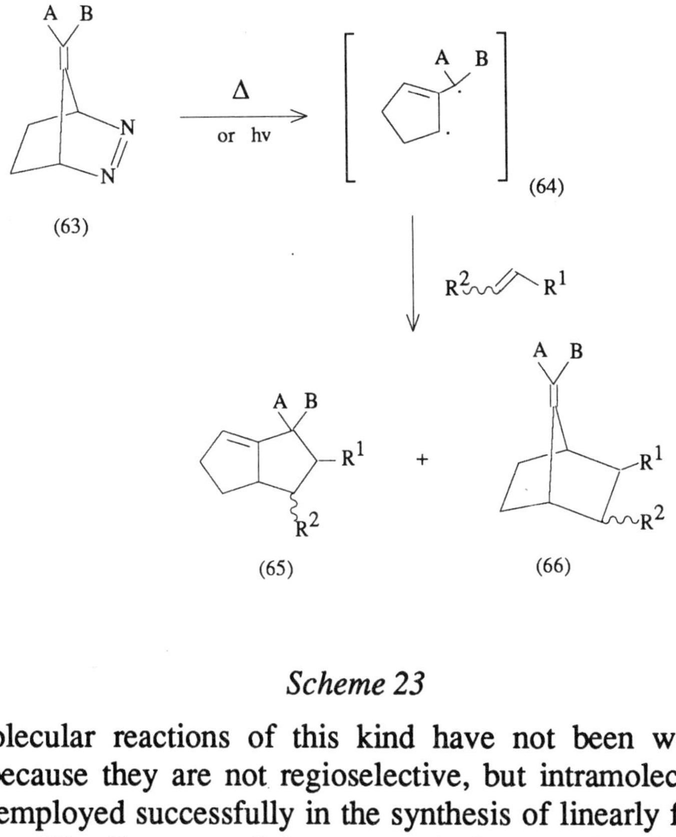

A B
N
N
Δ
or hv
A B
(63)
(64)
$R^2$ $R^1$
A B
A B
$R^1$
+
$R^1$
$R^2$
$R^2$
(65)
(66)

*Scheme 23*

Intermolecular reactions of this kind have not been widely used in synthesis because they are not regioselective, but intramolecular reactions have been employed successfully in the synthesis of linearly fused tricyclopentanes (as 67). Because of the geometrical constraints imposed by the connecting chain these reactions are regioselective and frequently stereoselective as well. Furthermore, they generally give rise to the *cis,anti* ring fusion of the tricyclopentanes commonly found in natural products of this series.[36]

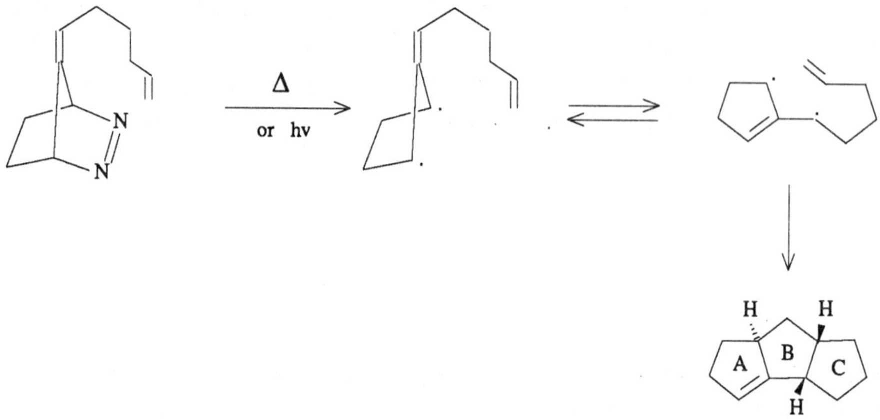
Δ
or hv
H
H
A
B
C
H

(67)

*Scheme 24*

Thus, the diazene (68) in boiling acetonitrile solution gave the two cyclisation products (69) and (70) in 85 per cent yield in the ratio of about 9:1 in favour of the desired *cis,anti* ring-fused isomer (69) which was subsequently converted into hirsutene (71). In both cyclisation products the stereochemistry about the diylophile double bond was maintained.[37] This is a general feature of these reactions.

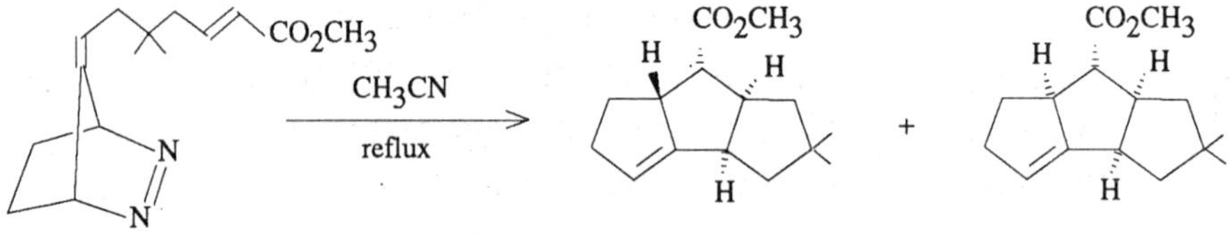
CO2CH3
CH3CN
reflux
N
N
CO2CH3
H
H
H
+
CO2CH3
H
H
H

(68) (69) (70) (9:1)

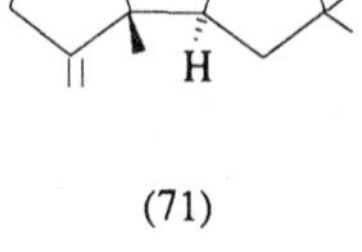
H
H
H

(71)

*Scheme 25*

These reactions are regarded as taking place preferentially by way of the *endo* transition state (72) in which the connecting chain adopts a pseudo chair conformation, rather than the *exo* form (73) which would lead to the *cis,syn* ring-fused product. In general, however, secondary orbital interactions appear to be less important than conformational factors and non-bonded interactions in controlling the course of these reactions.[36]

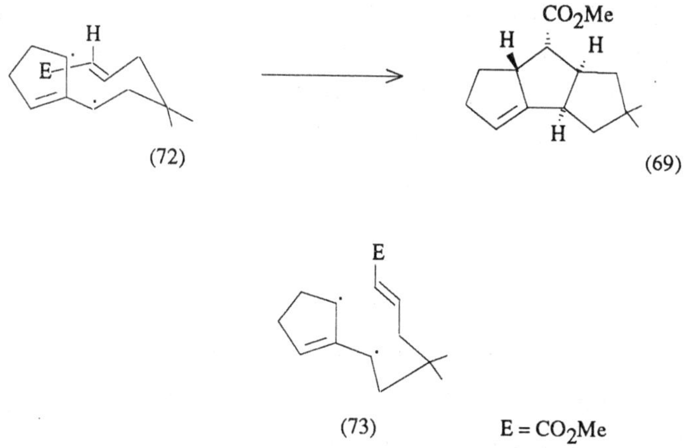

(72) (69) (73) E = $CO_2Me$

*Scheme 26*

Intramolecular reactions of this kind can sometimes be effected with non-activated double bonds.[38] With a chiral centre in the acyclic chain of the diyl *cis,anti*-tricyclopentanes containing an additional chiral centre can be set up with high selectivity. Thus, the diazene (74) gave preponderantly the product (76) with four chiral centres, by way of the preferred transition state (75). The stereochemical course of the cyclisation was dependent on the temperature: photolytic decomposition of (74) in methanol at – 60°C gave (76) almost exclusively. This product was converted into the enone (77) which was used to make (±)-coriolin and (±)-hypnophilin.[39]

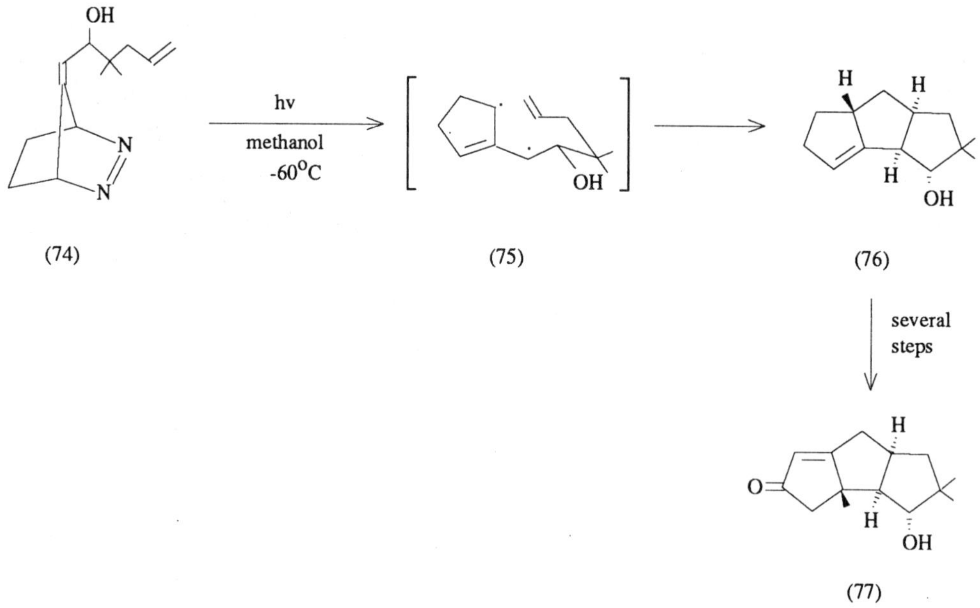

*Scheme 27*

There has been little systematic work on the mechanism of these cyclisations. Such evidence as there is suggests that they take place by a stepwise pathway.[36]

## Meta-Photocycloaddition to Benzene Derivatives

Another route to complex polycyclic systems containing five-membered rings is based on the *meta*-photocycloaddition of benzene derivatives to olefins, which takes place by irradiation of the mixture in an inert solvent at 254nm.[40]

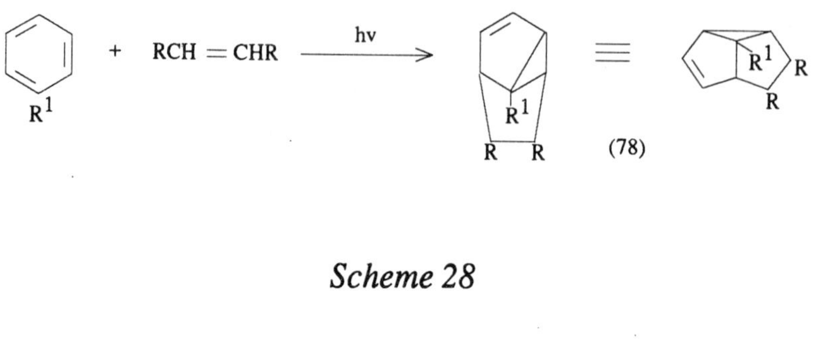

*Scheme 28*

The reactions generally involve the singlet excited state of the aromatic component, and the stereochemistry of the olefins is preserved in the products, supporting the supposition that the reactions are concerted. With substituted benzenes the reactions are highly regioselective: addition of the alkene takes place at the two carbon atoms of the ring *ortho* to the strongest electron-donating substituent. Anisole, therefore, gives mainly the adduct (78, $R^1$=OMe).[41] The regiochemistry of the additions has been rationalised on the basis of frontier orbital overlap in exciplex intermediates.[42]

By controlled cleavage of bonds in the three-membered ring the photo-adducts can be converted into cycloheptenes, bicyclo[3,2,1]octanes and bicyclo[3,3,0]octanes (Scheme 29), ring systems commonly found in natural products. A number of naturally-occurring tricycloundecanes have been expeditiously synthesised in this way.

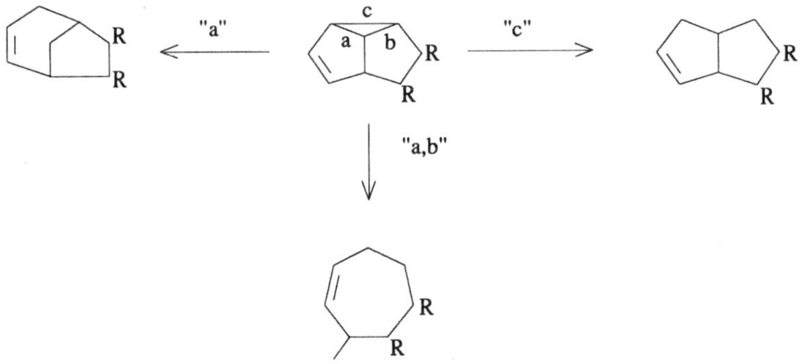

*Scheme 29*

Both inter- and intra-molecular cycloadditions have been exploited. Thus, in a synthesis of modhephene (81) the propellane skeleton (79) was assembled in one step by photoaddition of vinyl acetate to indane. Conversion of (79) into the ketone (80) followed by further transformations including α-alkylation and conjugate addition of cuprate, with opening of the cyclopropane ring, then afforded modhephene (81).[43]

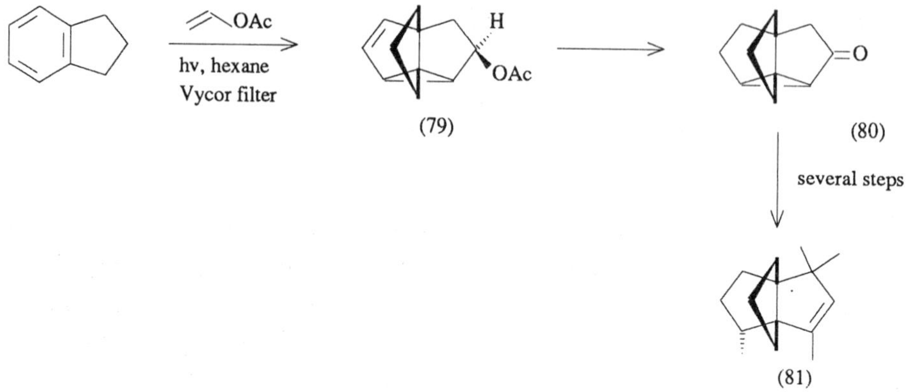

OAc
hv, hexane
Vycor filter
H
OAc
(79)
O
(80)
several steps
(81)

*Scheme 30*

The synthesis of (±)-isocomene (86)[44] employed intramolecular cyclisation of the alkenylbezene (82). Although twenty-four different *meta*-cycloadducts are possible here, geometric constraints and mechanistic factors ensured that only the two products (83) and (84) were obtained. Thermolysis of (83) led directly to dehydroisocomene (85) by a 1,5-hydrogen shift. The stereochemistry of (83) and (84) is consistent with an exciplex intermediate of the form (87) in which the methyl group in the acyclic chain is as far removed as possible from the ring substituent.

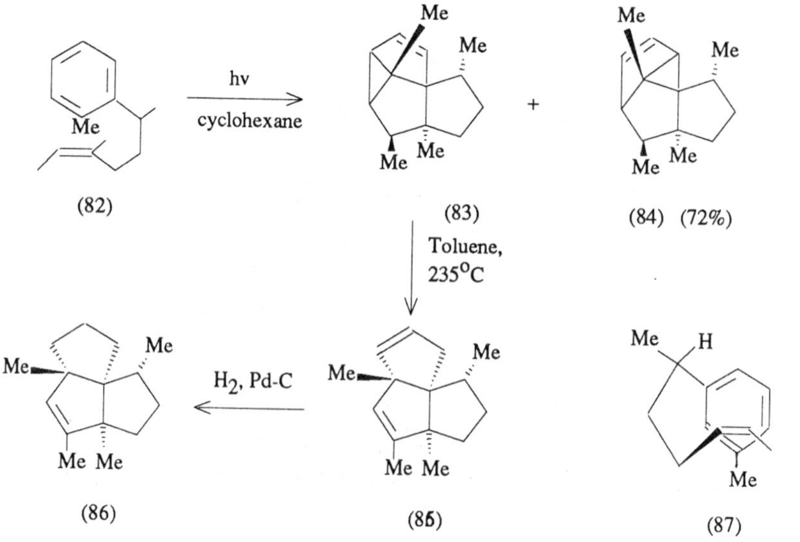

Me
(82)
hv
cyclohexane
Me Me Me Me
(83)
+
Me Me Me Me
(84) (72%)
Toluene, 235°C
Me Me Me Me
(86)
$H_2$, Pd-C
Me Me Me Me
(85)
Me H Me
(87)

*Scheme 31*

Cedrene,[45] hirsutene,[46] the pseudoguaianolide rudmollin,[47] coriolin,[48] sulphinene[49] and isoiridomyrmecin[50] were synthesised by similar methods by appropriate control of cyclopropane ring cleavage in the initial photo-adducts.

## Cobalt-catalysed Trimerisation of Acetylenes

The cycloaddition reactions described above result in the generation of ring compounds by the formation of two new sigma bonds. Metal-catalysed trimerisation of acetylenes to benzene derivatives is well-known and can be realised with many transition metals, as well as Ziegler-type catalysts. One of the most useful reactions of this kind is catalysed by (dicarbonyl $\eta^5$-cyclopentadienyl)cobalt, $CpCo(CO)_2$, and leads to the formation of annelated benzene derivatives from αω-diynes and acetylenes with the formation of three new bonds by [2+2+2] cycloaddition (Scheme 32).[51]

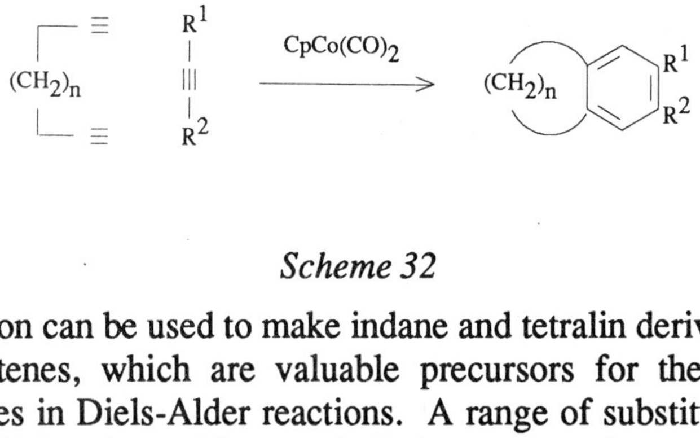

$(CH_2)_n$ diyne + $R^1-C\equiv C-R^2$ $\xrightarrow{CpCo(CO)_2}$ $(CH_2)_n$-annelated benzene bearing $R^1$, $R^2$

*Scheme 32*

The reaction can be used to make indane and tetralin derivatives and also benzocyclobutenes, which are valuable precursors for the generation of *ortho*-xylylenes in Diels-Alder reactions. A range of substituents (R=alkyl, $CH_2OH$, $CO_2R$) is tolerated but particularly good results are obtained when $R^1$ and $R^2$ are bulky trimethylsilyl substituents. These hinder the co-cyclisation of the acetylenic component, and the trimethylsilyl substituents in the benzenes produced serve as excellent leaving groups in electrophilic substitution reactions.

Formation of benzene derivatives is believed to take place by way of cobaltacyclopentadienes of the form (88) and (89) which subsequently undergo Diels-Alder addition of the alkyne with expulsion of cobalt and generation of a benzene ring.

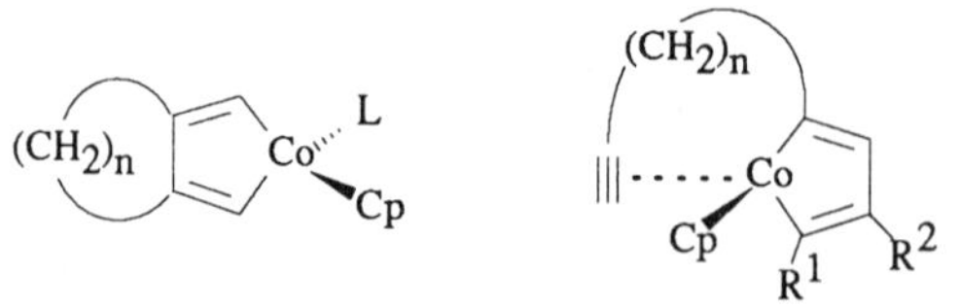

(CH2)n Co L Cp

(88)

(CH2)n Co Cp R1 R2

(89)

*Scheme 33*

Thus, 1,5-hexadiyne and the acetylene (90) gave the benzocyclobutene (91) in 53 per cent yield, which, after electrophilic substitution of the trimethylsilyl group by bromine, was used to make the highly strained benzene derivative (92).[52] In an intramolecular example, leading to (±)-oestrone (96), the diyne (93) and bistrimethylsilylacetylene gave the tetracyclic compound (95) stereospecifically by intramolecular Diels-Alder addition to the *ortho*-xylylene (94).[53]

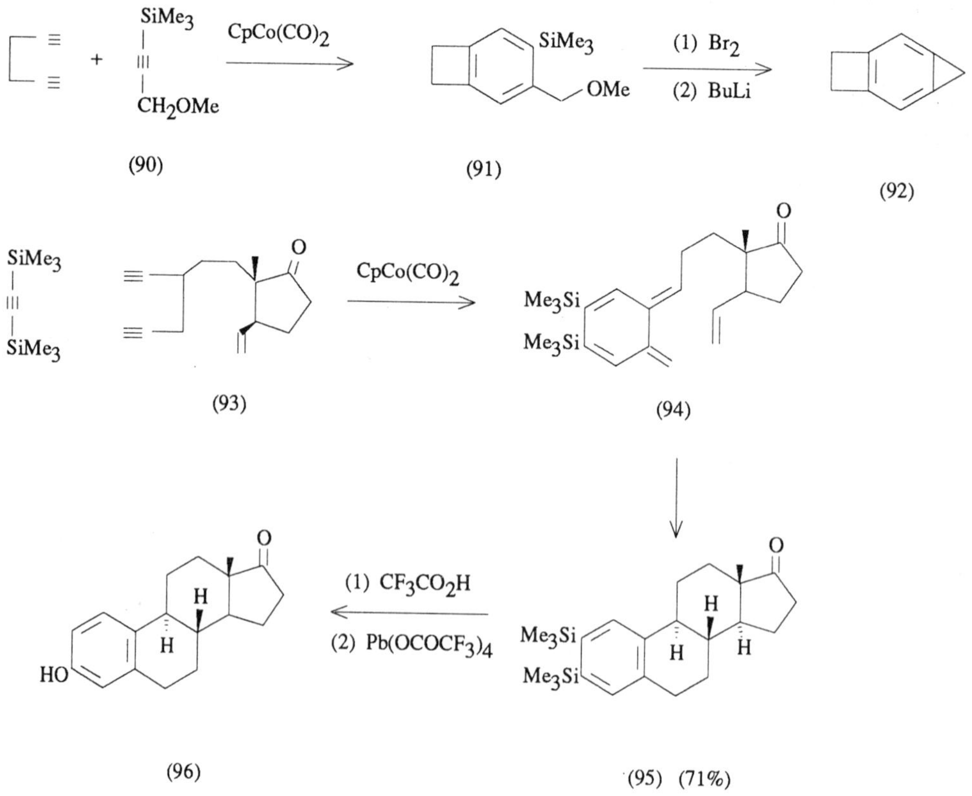

SiMe3 CH2OMe CpCo(CO)2 SiMe3 OMe (1) Br2 (2) BuLi

(90) (91) (92)

SiMe3 SiMe3 O CpCo(CO)2 Me3Si Me3Si O

(93) (94)

HO H H O (1) $CF_3CO_2H$ (2) $Pb(OCOCF_3)_4$ Me3Si Me3Si H H H O

(96) (95) (71%)

*Scheme 34*

In a related approach all four rings of the B-ring aromatic steroid (97) were assembled in one step from the open-chain precursor (96) with complete control of the stereochemistry of the C-D ring junction.[54]

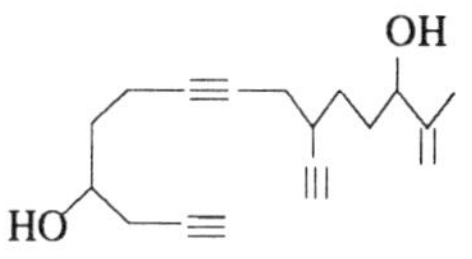

OH
HO
OH
C D
H
HO

(96) 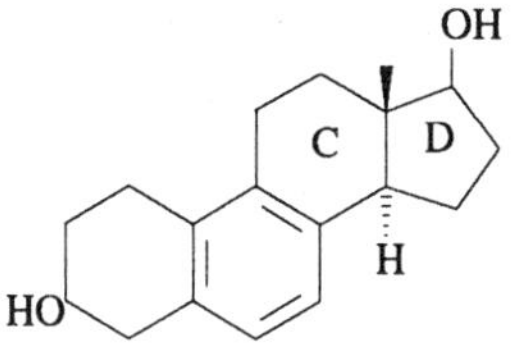 (97)

*Scheme 35*

Intramolecular (2+2+2) cycloadditions of ene-diynes have also been employed to assemble polycyclic structures. The inital products are cyclohexadiene-cobalt complexes from which the polycycle is easily liberated by oxidative demetallation.[55] These reactions also have been exploited in the synthesis of aromatic steroids.[56]

Heterocyclic compounds can also be made. Thus, nitriles and terminal diacetylenes give pyridine derivatives,[57] and 5-isocyanato-1-alkynes and acetylenes form annelated 2-pyridones.[58]

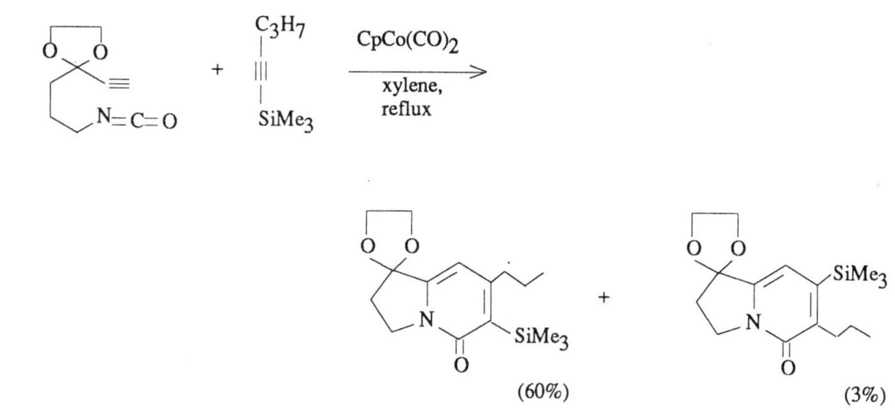

≡ — $SiMe_3$
≡ — $SiMe_3$
Me
C
N
$CpCo(CO)_2$
xylene, reflux
$SiMe_3$
Me
N
$SiMe_3$
(76%)

O O
N=C=O
+
$C_3H_7$
$SiMe_3$
$CpCo(CO)_2$
xylene, reflux

O O
N
O
$SiMe_3$
(60%)
+
O O
N
O
$SiMe_3$
(3%)

*Scheme 36*

Trimerisation of acetylenes to benzene derivatives has also been conveniently achieved using Wilkinson's catalyst, $RhCl(PPh_3)_3$.[59] Thus, diynes (98) gave the substituted benzenes (99) in good yield at 0-80°C. Intramolecular reactions take place readily. In a key step in the synthesis of the sesquiterpene calomelanolactone the pentasubstituted benzene (101) was obtained in 86 per cent yield from the triyne (100) at 25°C.[60]

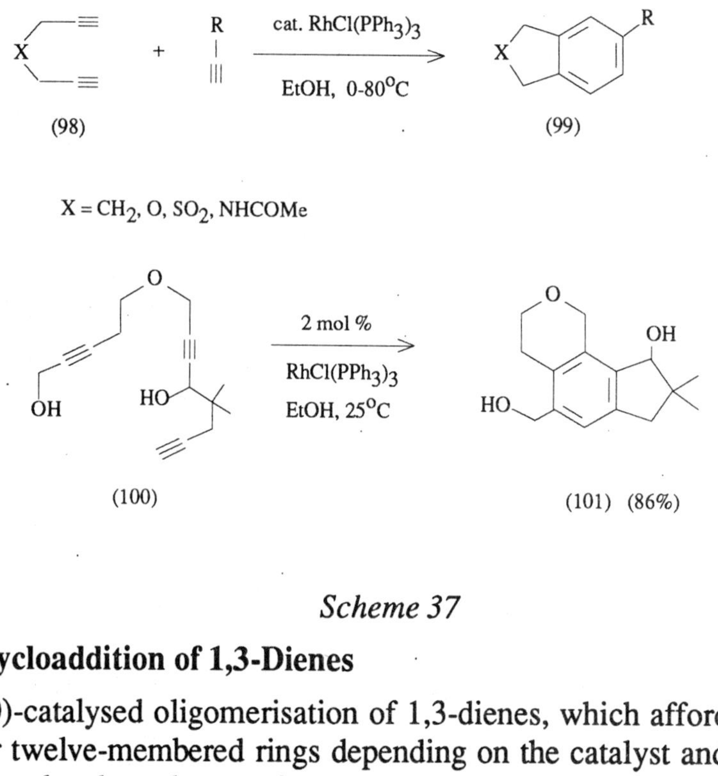

cat. $RhCl(PPh_3)_3$

EtOH, 0-80°C

(98) (99)

X = $CH_2$, O, $SO_2$, NHCOMe

2 mol %

$RhCl(PPh_3)_3$

EtOH, 25°C

(100) (101) (86%)

*Scheme 37*

## [4+4]-Cycloaddition of 1,3-Dienes

Nickel(0)-catalysed oligomerisation of 1,3-dienes, which affords four-, six-, eight- or twelve-membered rings depending on the catalyst and the reaction conditions, has been known for many years but has not been employed in complex syntheses, due mainly to the lack of control of regio- and stereochemistry, and the general observation that substituted dienes are much less reactive than butadiene. It has been found, however that intramolecular Nickel(0)-catalysed [4+4] dimerisation of 1,3-dienes under the appropriate conditions provides an efficient method for the synthesis of polycyclic compounds containing eight-membered rings.[61] For example, the bis-diene (102), treated with $Ni(COD)_2$ and triphenylphosphine in toluene at 60°C gave a mixture of the cyclo-octadienes (103) and (104) in which the *cis*-isomer, the product of a formal *endo* addition, predominated by a factor of 19:1.[61a]

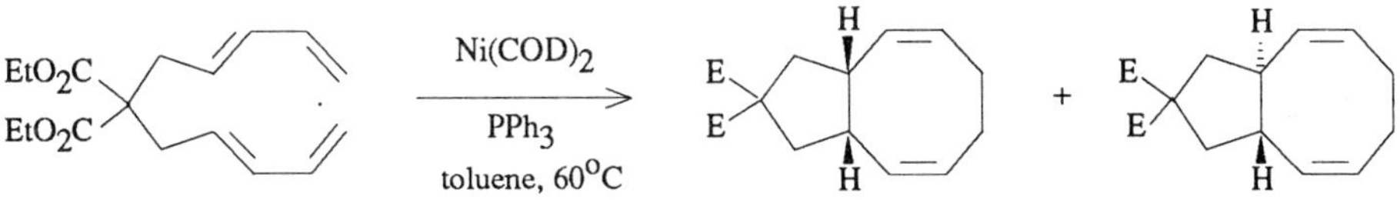

$EtO_2C$ $EtO_2C$ (102) —Ni(COD)$_2$, PPh$_3$, toluene, 60°C→ (103) + (104) (70%:19:1)

*Scheme 38*

In contrast bis-dienes in which the two diene units are joined by a chain of four carbon atoms give predominantly the corresponding *trans*-fused products. Thus, the bis-diene (105) gave the *trans*-fused cyclo-octadiene (106) in 92 per cent yield with greater than 97 per cent diastereoselectivity.[61b] Bridged bicyclic compounds can be formed as well, for example (108). The diastereoselection of these cyclisations is believed to be determined by the kinetically controlled collapse of intermediate organo-nickel complexes.[61c]

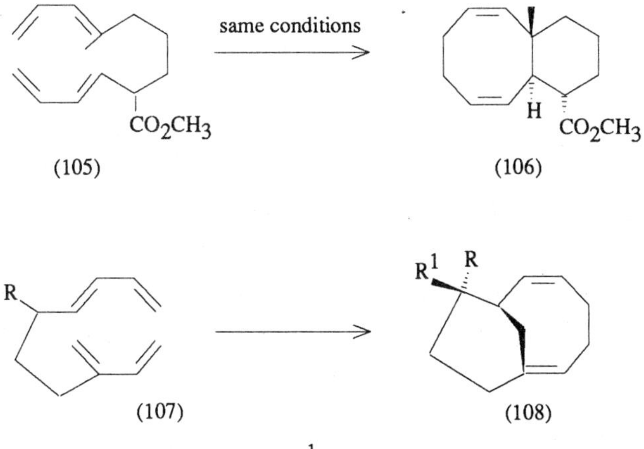

(105) $CO_2CH_3$ —same conditions→ (106) $CO_2CH_3$

(107) → (108)

(R or $R^1$ = t-BuMe$_2$SiO)

*Scheme 39*

The reaction has been employed in an asymmetric synthesis of the sesquiterpene lactone (+)-asteriscanolide (111) starting from the optically active bis-diene (109).[62]

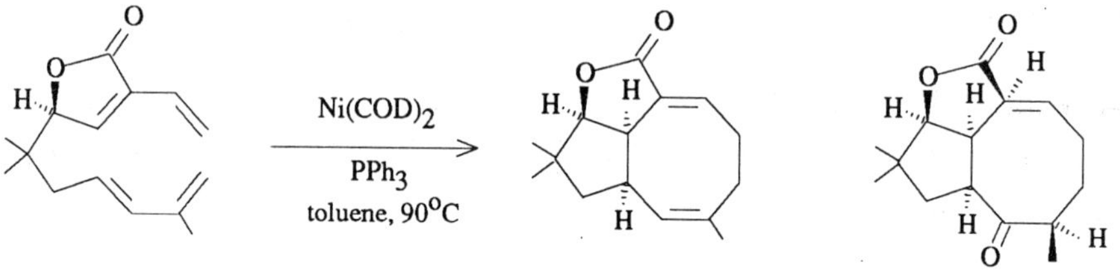

Ni(COD)$_2$

PPh$_3$
toluene, 90°C

(109)

(110) (67%)

(111)

*Scheme 40*

Cyclisation proceeded well with excellent stereoselectivity furnishing (110) in 67 per cent yield.

**References.**

1 R.B. Woodward and R. Hoffman, "The Conservation of Orbital Symmetry", 1970, Academic Press, New York.

2 A. Eidenschink and T. Kauffmann, *Angew. Chem. internat. edn.*, 1972, **11**, 292; see also J.P. Marino and W.B. Mesbergen, *J. Am. Chem. Soc.*, 1974, **96**, 4050.

3 T. Kauffmann, *Angew. Chem. internat. edn.*, 1974, **13**, 627.

4 W.H. Pearson, D.P. Szura and W.G. Harter, *Tetrahedron Lett.*, 1988, **29**, 761.

5 W.H. Pearson, M.A. Walters and K.D. Oswell, *J. Am. Chem. Soc.*, 1986, *108*, 2769.

6 H.M.R. Hoffmann, *Angew. Chem. internat. edn.*, 1973, **12**, 819; 1984, **23**, 1.

7 H.M.R. Hoffmann, R. Henning and O.R. Lalko, *Angew. Chem. internat. edn.*, 1982, *21*, 442.

8 (a)H.M.R. Hoffmann, *Angew. Chem. internat. edn.*, 1984, **23**, 1; (b)J. Mann, *Tetrahedron*, 1986, **42**, 4611.

9 R. Noyori and Y. Hayakawa, *Organic Reactions*, 1983, **29**, 163.

10 N. Shimizu, M. Tanaka and Y. Tsuno, *J. Am. Chem. Soc.*, 1982, **104**, 1330.

11 N.N. Joshi and H.M.R. Hoffmann, *Tetrahedron Lett.*, 1986, **27**, 687.

12 T. Takaya, S. Makino, Y. Hayakawa and R. Noyori, *J. Am. Chem. Soc.*, 1978, **100**, 1765.

13 H. Sakurai, A. Shirahata and A. Hosomi, *Angew. Chem. internat. edn.*, 1979, **18**, 163.

14 H. Takaya, Y. Hayakawa, S. Makino and R. Noyori, *J. Am. Chem. Soc.*, 1978, **100**, 1778.

15 J. Mann, *Tetrahedron*, 1986, **42**, 4611.

16 A.P. Cowling and J. Mann, *J. Chem. Soc. Perkin 1*, 1978, 1564; G. Fierz, R. Chidgey and H.M.R. Hoffmann, *Angew. Chem. internat. edn.*, 1974, **13**, 410.

17 Y. Hayakawa, Y. Baba, S. Makino and R. Noyori, *J. Am. Chem. Soc.*, 1978, **100**, 1786; R. Noyori and Y. Hayakawa, *Tetrahedron*, 1985, **41**, 5879.

18 M.J. Arco, M.H. Trammell and J.D. White, *J. Org. Chem.*, 1976, **41**, 2075.

19 H.M.R. Hoffmann, K.F. Clemens and R.H. Smithers, *J. Am. Chem. Soc.*, 1972, **94**, 3940.

20 cf. R. Noyori and Y. Hayakawa, *Tetrahedron*, 1985, **41**, 5879.

21 A.V. Rama Rao, J.S. Yadav and V. Vidyasagar, *J. Chem. Soc. Chem. Commun.*, 1985, 55.

22 A.P. Cowling, J. Mann and A.A. Usmani, *J. Chem. Soc. Perkin 1*, 1981, 2116.

23 J.D. White and Y. Fukuyama, *J. Am. Chem. Soc.*, 1979, **101**, 226.

24 R. Noyori, F. Shimizu, K. Fukuta, H. Takaya and Y. Hayakawa, *J. Am. Chem. Soc.*, 1977, **99**, 5196; Y. Hayakawa, K. Yokoyama and R. Noyori, *J. Am. Chem. Soc.*, 1978, **100**, 1791, 1799; R. Chidgey and H.M.R. Hoffmann, *Tetrahedron Lett.*, 1977, 2633.

25 F. Walls, J. Padilla, P. Joseph-Nathan, F. Géral, M. Escobar and J. Romo, *Tetrahedron*, 1966, **22**, 2387; P. Joseph-Nathan, M.E. Gambay and R.L. Santillan, *J. Org. Chem.*, 1987, **52**, 759.

26 P. Joseph-Nathan, V. Mendoza and E. Garcia, *Tetrahedron*, 1977, **33**, 1573.

27 G. Büchi and Ping-Sun Chu, *J. Am. Chem. Soc.*, 1979, **101**, 6767; 1981, **103**, 2718; see also G. Büchi and Ching-Pong Mak, *J. Am. Chem. Soc.*, 1977, **99**, 8073. For another example see S.V. Mortlock, J.K. Seckington and E.J. Thomas, *J. Chem. Soc. Perkin 1*, 1988, 2305.

28 Y. Shizuri, Y. Okuno, H. Shigemori and S. Yamamura, *Tetrahedron*

*Lett.*, 1987, **28**, 6661.

29 (a)B.M. Trost, *Angew. Chem. internat. edn.*, 1986, **25**, 1; *Chem. Soc. Rev.*, 1982, **11**, 141; (b)B.M. Trost and D.M.T. Chan, *J. Am. Chem. Soc.*, 1983, **105**, 2315, 2326.

30 B.M. Trost and D.M.T. Chan, *J. Am. Chem. Soc.*, 1981, **103**, 5972.

31 B.M. Trost, J. Lynch, P. Renaut and D.H. Steinman, *J. Am. Chem. Soc.*, 1986, **108**, 2843; B.M. Trost and S.G. Mignani, *Tetrahedron Lett.*, 1986, **27**, 4137.

32 B.M. Trost and D.M.T. Chan, *J. Am. Chem. Soc.*, 1982, **104**, 3733.

33 B.M. Trost and P.J. Bonk, *J. Am. Chem. Soc.*, 1985, **107**, 1778.

34 B.M. Trost and D.T. MacPherson, *J. Am. Chem. Soc.*, 1987, **109**, 3483.

35 J.A. Berson, *Acc. Chem. Res.*, 1978, **11**, 446.

36 cf. R.D. Little, *Chem. Rev.*, 1986, **86**, 875.

37 R.D. Little and G.W. Muller, *J. Am. Chem. Soc.*, 1981, **103**, 2744.

38 R.D. Little, R.G. Higby and K.D. Moeller, *J. Org. Chem.*, 1983, **46**, 3139; R.D. Little and G.L. Carroll, *Tetrahedron Lett.*, 1981, **22**, 4309.

39 L. Van Hijfte, R.D. Little, J.L. Petersen and K.D. Moeller, *J. Org. Chem.*, 1987, **52**, 4647.

40 cf. A. Gilbert, *Pure Appl. Chem.*, 1980, **52**, 2669; *Photochemistry*, 1984. **15**, 291.

41 cf. R. Srinivasan, V.Y. Merritt and G. Subrahmanyan, *Tetrahedron Lett.*, 1974, 2715.

42 K.N. Houk, *Pure Appl. Chem.*, 1982, **54**, 1632.

43 P.A. Wender and G.B. Dreyer, *J. Am. Chem. Soc.*, 1982, **104**, 5805.

44 P.A. Wender and G.B. Dreyer, *Tetrahedron*, 1981, **37**, 4445.

45 P.A. Wender and J.J. Howbert, *J. Am. Chem. Soc.*, 1981, **103**, 688.

46 P.A. Wender and J.J. Howbert, *Tetrahedron Lett.*, 1982, **23**, 3983.

47 P.A. Wender and K. Fisher, *Tetrahedron Lett.*, 1986, **27**, 1857.

48 P.A. Wender and J.J. Howbert, *Tetrahedron Lett.*, 1983, **24**, 5325.

49 P.A. Wender and R.J. Teransky, *Tetrahedron Lett.*, 1985, **26** 2625; P.A. Wender and S.K. Singh, *loc.cit.*, 5987.

50 P.A. Wender and G.B. Dreyer, *Tetrahedron Lett.*, 1983, **24**, 4543.

51 K.P.C. Vollhardt, *Angew. Chem. internat. edn.*, 1084, **23**, 539.

52 C.J. Saward and K.P.C. Vollhardt, *Tetrahedron Lett.*, 1975, 4539.

53 R.L. Funk and K.P.C. Vollhardt, *J. Am. Chem. Soc.*, 1977, **99**, 5483; 1979, **101**, 215; 1980, **102**, 5253.

54 S.H. Lecker, N.H. Nguyen and K.P.C. Vollhardt, *J. Am. Chem. Soc.*, 1986, **108**, 856.

55 cf. K.P.C. Vollhardt, *Angew. Chem. internat. edn.*, 1984, **23**, 539; E.D. Sternberg and K.P.C. Vollhardt, *J. Org. Chem.*, 1984, **49**, 1564, 1574.

56 E.D. Sternberg and K.P.C. Vollhardt, *J. Org. Chem.*, 1984, **49**, 1574; 1982, **47**, 3447. J.C. Clinet, E. Duñach and K.P.C. Vollhardt, *J. Am. Chem. Soc.*, 1983, **105**, 6710.

57 cf. C.A. Parnell and K.P.C. Vollhardt, *Tetrahedron*, 1985, **41**, 5791; R.E. Geiger, M. Lalonde, H. Stoller and K. Schleich, *Helv. Chim. Acta*, 1984, **67**, 1274.

58 cf. R.A. Earl and K.P.C. Vollhardt, *J. Am. Chem. Soc.*, 1983, **105**, 6991.

59 R. Grigg, R. Scott and P. Stevenson, *Tetrahedron Lett.*, 1982, **23**, 2691; R. Grigg, P. Stevenson and T. Worakun, *Tetrahedron*, 1988, **44**, 4967.

60 S.J. Nelson and P.J. Stevenson, *Tetrahedron Lett.*, 1988, **29**, 813.

61 (a)P.A. Wender and N.C. Ihle, *J. Am. Chem. Soc.*, 1986, **108**, 4678; (b)P.A. Wender and M.L. Snapper, *Tetrahedron Lett.,* 1987, **28**, 2221; (c)P.A. Wender and N.C. Ihle, *Tetrahedron Lett.*, 1987, **28**, 2451.

62 P.A. Wender, N.C. Ihle and C.R.D. Correia, *J. Am. Chem. Soc.*, 1988, **110**, 5904.

# 5 THE ENE REACTION

In the ene reaction an olefin bearing an allylic hydrogen atom (the ene) reacts with a compound containing an activated multiple bond (the enophile) to give an adduct, with formation of a new σ-bond between unsaturated centres of the ene and enophile and migration of the allylic hydrogen to the other terminus of the enophile multiple bond.

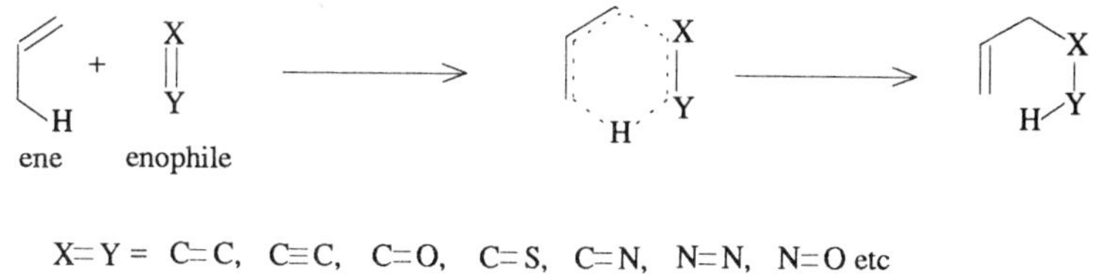

ene enophile

X=Y = C=C, C≡C, C=O, C=S, C=N, N=N, N=O etc

*Scheme 1*

Although, strictly speaking, not a *cyclo*addition reaction, its formal resemblance[1] to the Diels-Alder reaction justifies its discussion here.

In the ene reaction the two electrons of the allylic C–H sigma bond take the place of two π-electrons of the diene in the Diels-Alder reaction. The activation energy is thus greater and higher temperatures are generally required than in most Diels-Alder reactions. This formerly precluded the widespread use of the ene reation in complex synthesis, but it has now been found that many ene reactions are catalysed by Lewis acids and then proceed under mild conditions, often with improved stereoselectivity. A number of catalysts have been used, $AlCl_3$,$SnCl_4$,$TiCl_4$ alkylaluminium halides; the last act as proton scavengers as well as Lewis acids, with the advantage that unwanted proton-catalysed side reactions are prevented.[2] Thus, the uncatalysed reaction of methyl acrylate with 2-methyl-2-propene requires a temperature of 230°C, but with ethylaluminium dichloride as catalyst good yields of adducts are obtained from reactive olefins at 25°C. Most ene reactions involve interaction of the HOMO of the ene component with the LUMO of the enophile and, as in the Diels-Alder reaction, the catalysts exert their effect by lowering the energy of the LUMO of the enophile.

Most "all-carbon" ene reactions are believed to proceed by a concerted mechanism through a six-membered cyclic transition state in which the new C–C bond is frequently more highly developed than the C–H bond. Stepwise radical or zwitterionic pathways may intervene in particular cases.[3] Hetero-ene reactions are more likely to be stepwise.[4] Lewis acid-catalysed ene reactions can be stepwise with a zwitterionic intermediate, or concerted, with a polar transition state, depending on the reaction components and the catalyst employed.[5]

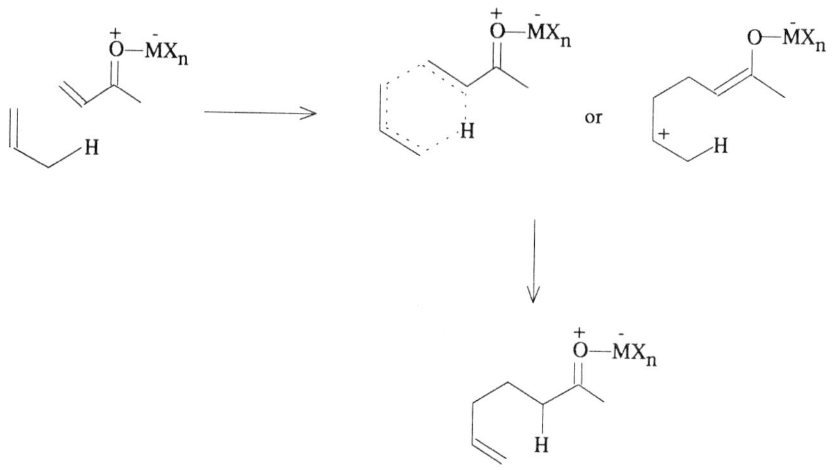

*Scheme 2*

The course of thermal and catalysed reactions need not be the same. Ene components with at least one disubstituted olefinic carbon atom are more reactive than monosubstituted or 1,2-disubstituted olefins in catalysed reactions because of the development of positive charge at the central carbon atom of the ene. Thermal ene reactions, on the other hand, are strongly influenced by steric factors, and the steric accessibility of the double bond and the allylic hydrogen atom of the ene now become more important. In general, a primary allylic hydrogen is abstracted more readily than a secondary and much faster than a tertiary one.[6] Thus, thermal ene reaction of dimethyl oxomalonate (2) with the diene (1) takes place mainly at the monosubstituted double bond to give (3), but in the catalysed reaction the isomer (4) is the main product.[5]

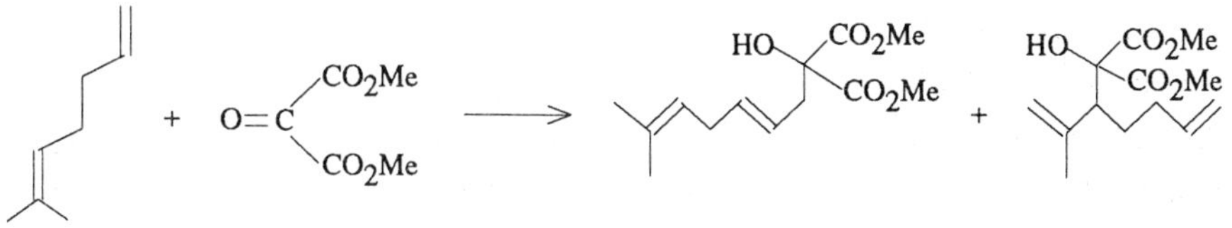

(1) (2) (3) (4)

| | (3) | (4) |
|---|---|---|
| 180°C, 48h | 69% | 6% |
| $SnCl_4$, 0°C, 5min. | 1% | 39% |

*Scheme 3*

Like the Diels-Alder reaction the ene reaction is reversible at high temperatures, and numerous fragmentations of the type

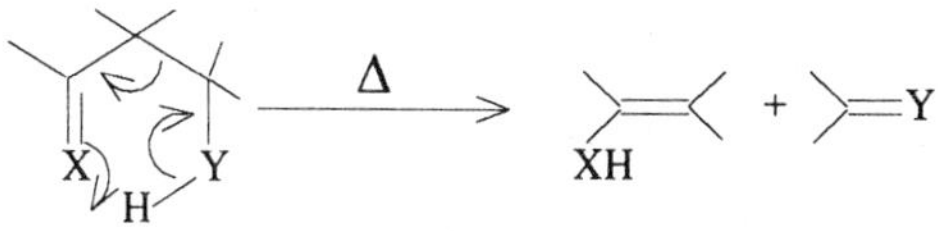

*Scheme 4*

have been observed. Retro-ene reactions like this can sometimes play a valuable part in syntheses.[6,7]

Again like the Diels-Alder reaction the ene reaction is highly stereoselective in many cases, showing *cis* addition and a preference for the formation of *endo* products. Since the ene reaction does not lead to cyclic products a distinction between *endo* and *exo* transition states is not always so easily made as in the Diels-Alder reaction, but many ene reactions do appear to take place preferentially by way of a transition state in which the two $\pi$-systems overlap as much as possible.[8] Thus, in the reaction of maleic anhydride with *cis*-2-butene the adduct (5) was obtained almost exclusively by way of the *endo* transition state (7). With *trans*-2-butene the stereoselectivity was less, but the *endo* product (6) still predominated. The poorer selectivity in the reaction with *trans*-2-butene is attributed to the non-bonding repulsion between the non-reacting methyl substituent and the opposing carbonyl group in the *endo* transition state.[9]

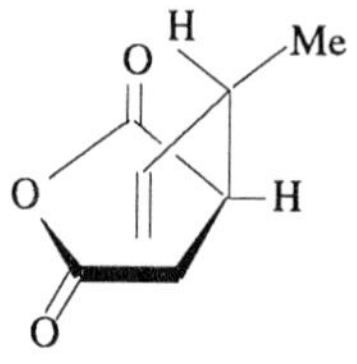

(5) 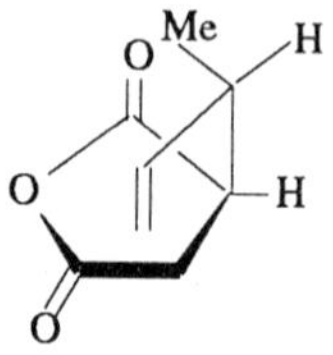 (6)

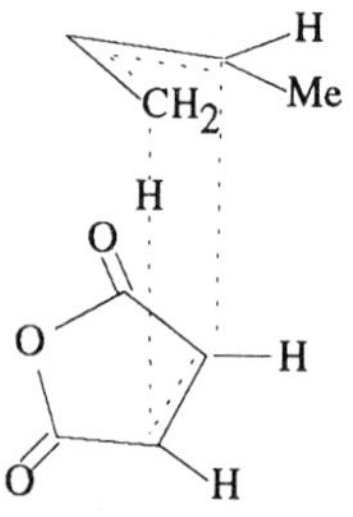

(7) 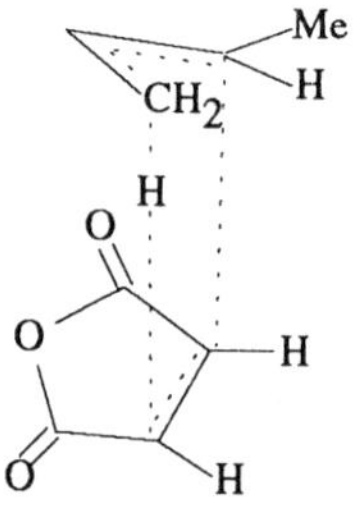 (8)

*Scheme 5*

*Endo* addition in ene reactions is favoured by orbital symmetry relationships, but the preference is not so marked as in the Diels-Alder reaction and may easily be overcome by other factors. In many intramolecular reactions, for example, steric constraints enforce an *exo* transition state.[10]

## Hetero-ene Reactions

There appear to be very few ene reactions involving hetero-ene components, but several hetero-enophiles have been employed, including carbonyl compounds, thio-carbonyl compounds, imines, nitroso compounds, hydrazones and azo compounds. Carbonyl compounds have probably been most widely used. Although their reactions with olefins could apparently give rise to either homoallylic alcohols or allyl ethers, the alcohols are formed exclusively due to the greater gain in bond energy in this mode of reaction.

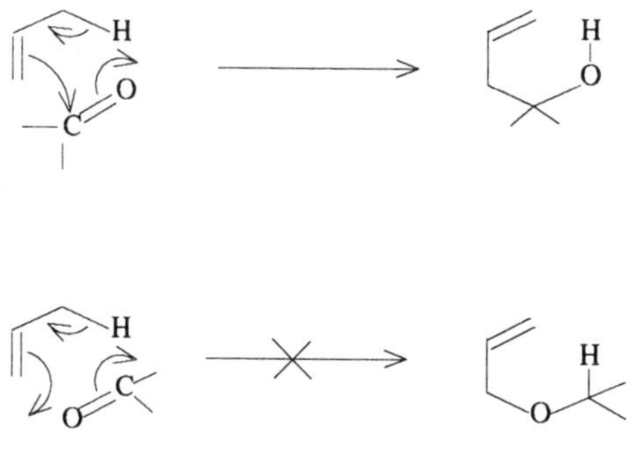

*Scheme 6*

Purely thermal ene reactions of carbonyl compounds are only effective with formaldehyde and other reactive aldehydes such as chloral or glyoxylate esters, but reactions catalysed by dimethylaluminium chloride[11] or, better, ethylaluminium dichloride[12] take place easily and provide a convenient route to homoallylic alcohols from a range of aliphatic and aromatic aldehydes under mild conditions. Thus, reaction of 9-decenoic acid with acetaldehyde in presence of ethylaluminium dichloride gave 11-hydroxy-8-dodecenoic acid in 66 per cent yield as a 4:1 mixture of (*E*) and (*Z*) isomers.

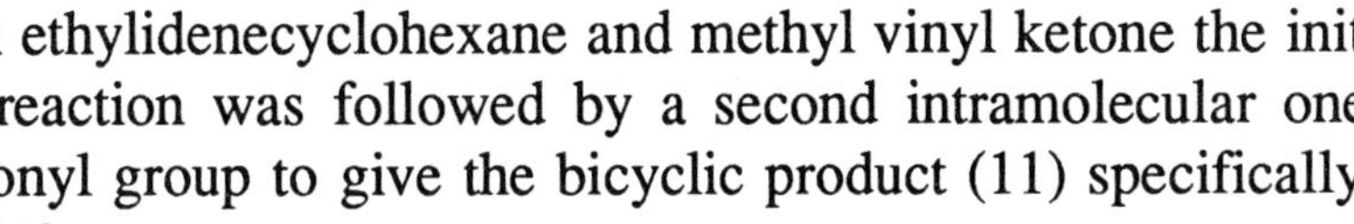

$(CH_2)_6CO_2H$ —MeCHO, $EtAlCl_2$, heptane, 0°C→ OH $(CH_2)_6CO_2H$

*Scheme 7*

With ethylidenecyclohexane and methyl vinyl ketone the initial 'all-carbon' ene reaction was followed by a second intramolecular one involving the carbonyl group to give the bicyclic product (11) specifically in 61 per cent yield.[13]

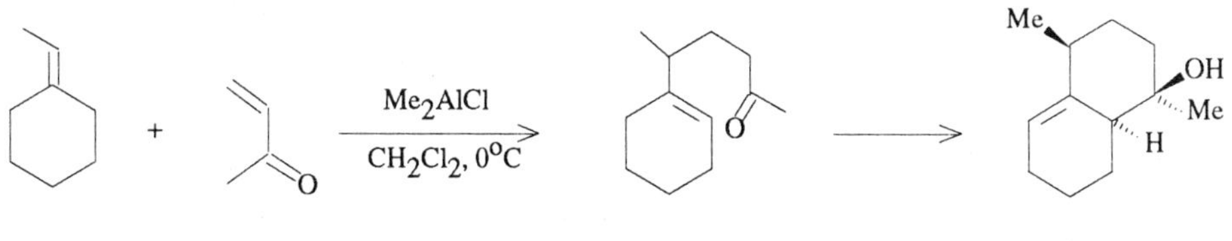

+ —$Me_2AlCl$, $CH_2Cl_2$, 0°C→ → Me, OH, Me, H (11)

*Scheme 8*

Ene reactions of glyoxylates give α-hydroxy esters and these reactions also are often highly stereoselective. Thus, thermal reaction of methyl glyoxylate with *cis*-2-butene gave a 7.4:1 mixture of the epimeric methyl esters (12) and (13) in 54 per cent yield. Hydrogenation of the main product (12) gave racemic methyl isoleucate (14) thus establishing the stereochemistry. This isomer results from the *endo* transition state (15) while the minor isomer derives from the *exo* transition state (16). The inherent preference for an *endo* transition state and the steric repulsion between the ester group and the non-reacting methyl substituent in the *exo* transition state result in a large preference for the formation of (12) from *cis*-2-butene. With *trans*-2-butene the reaction was less selective (12:13=0.57:1). Here, steric repulsion between methyl and ester groups in the *endo* transition state (18) results in lower selectivity for the *endo* isomer (13).[14]

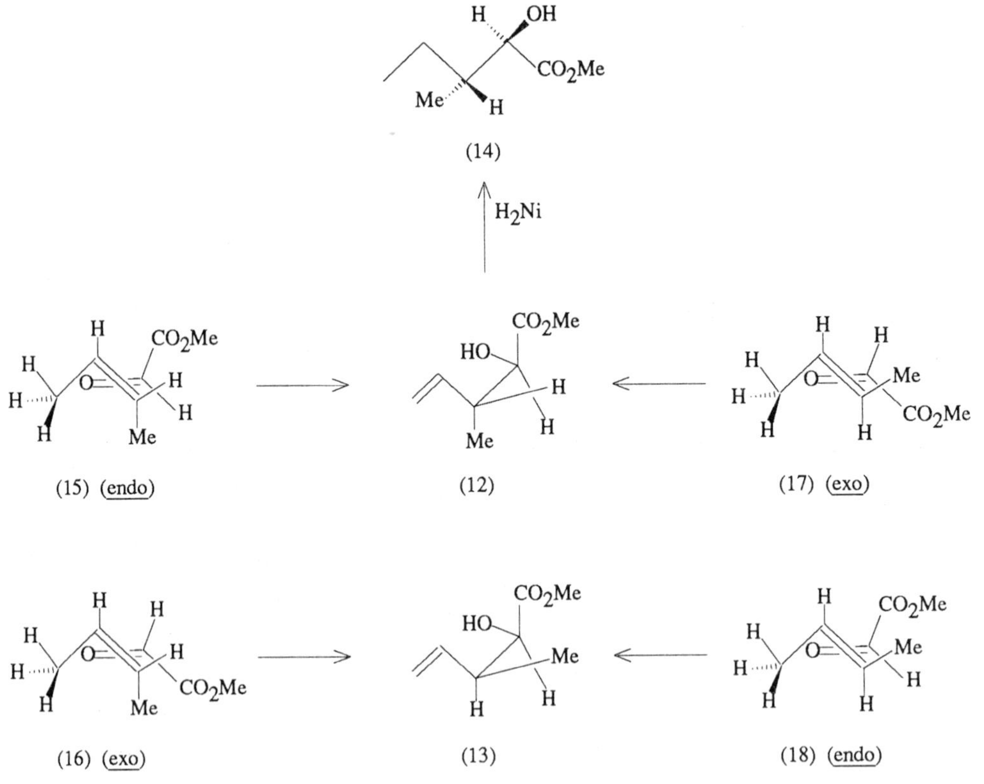

(14)

$H_2$Ni

(15) (endo) (12) (17) (exo)

(16) (exo) (13) (18) (endo)

*Scheme 9*

Intramolecular reactions take place readily even with compounds containing a normally unreactive enophile. For example, the unsaturated aldehyde (19) gave the cyclopentane derivative (20) in 93 per cent yield,[15] and the axial alcohol (22) was the exclusive product from the catalysed cyclisation of (21).[16] These catalysed reactions of aldehydes and ketones are believed to proceed through zwitterionic intermediates.[16,17]

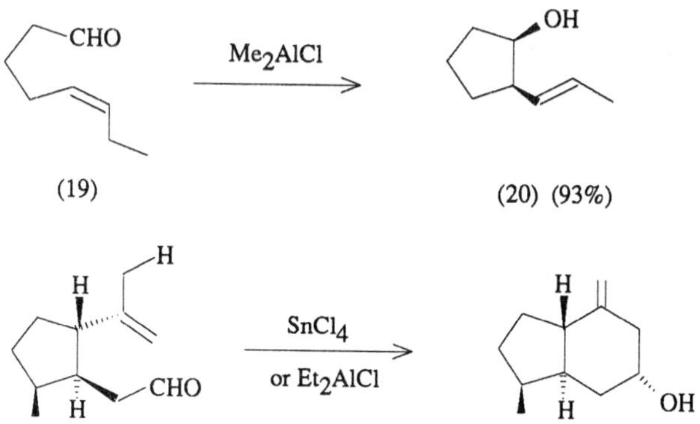

*Scheme 10*

Intramolecular thermal ene reactions of unsaturated thio-aldehydes similarly lead to homoallylic mercaptans. Thus, the prenyl ester (23) of thioxoacetic acid, liberated *in situ* by thermolysis of its adduct with 9,10-dimethylanthracene, gave the mercaptan (24).[18]

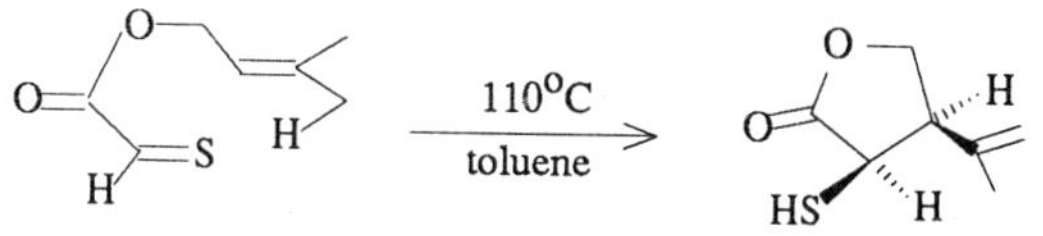

*Scheme 11*

Intermolecular reactions with thioaldehydes, on the other hand, yield substantial amounts of the alternative allyl sulphides.[19]

The ene reaction of imines also could apparently take two different directions, with formation of either a new C–C or C–N bond. Consideration of the bond energies involved suggests that neither course would be as favourable as that in all-carbon ene reactions and, in agreement, simple non-activated imines do not take part in the reaction. Some activated imines do, however. Thus butyl N-(*p*-tolylsulphonyl)imino-acetate (25) reacts with olefins under thermal or catalytic conditions, giving adducts which are readily converted into γδ-unsaturated amino acids.[20] The adduct (26), for example, was obtained in 90 per cent yield, almost entirely as the (*E*)-isomer, by way of the transition state (27) in which steric interaction is minimised. Selective formation of the (*E*)-isomers in these reactions is taken as evidence for a concerted pathway but other evidence suggests a non-concerted two-step mechanism for ene reactions with hetero-enophiles of this kind.[21]

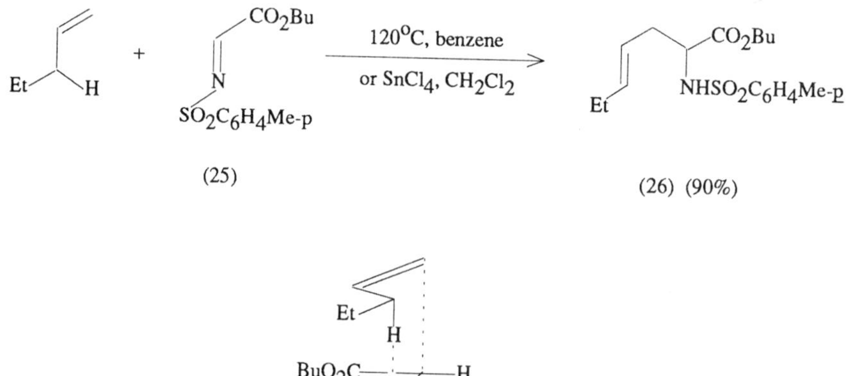

Et H + $CO_2Bu$ N $SO_2C_6H_4Me$-p (25) → 120°C, benzene or $SnCl_4$, $CH_2Cl_2$ → $CO_2Bu$ $NHSO_2C_6H_4Me$-p Et (26) (90%)

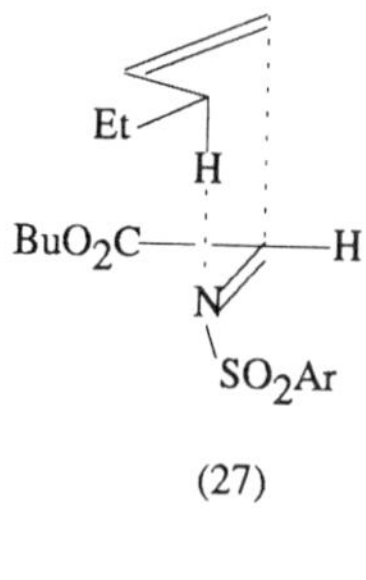

Et H $BuO_2C$ H N $SO_2Ar$ (27)

*Scheme 12*

Some acylimines also react, although only under fairly vigorous conditions. Thus, the acylimine (29) generated *in situ* by flash vacuum pyrolysis of the acetate (28), gave the cyclic amide (30) in 40 per cent yield. In this intramolecular example the reaction takes the alternative course with formation of a new C–N bond rather than a C–C bond.[22]

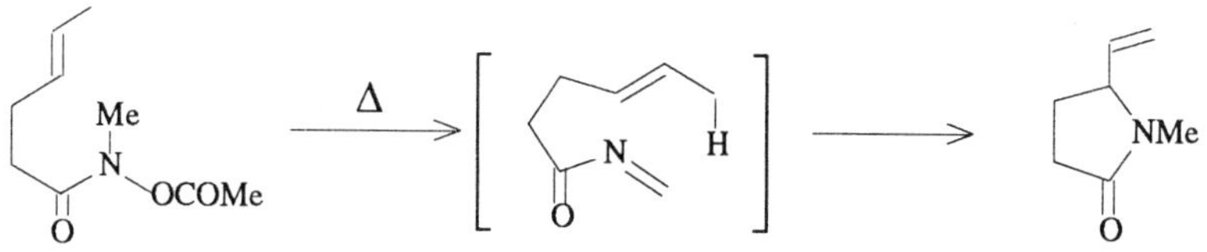

(28) (29) (30) (40%)

*Scheme 13*

Activated nitroso compounds form another group of useful enophiles, although they are less reactive here than in Diels-Alder additions. Nitrosoarenes react with certain mono-olefins to give allylic hydroxylamines, and C-nitrosocarbonyl compounds, released thermally from their 9,10-dimethylanthracene adducts, react with olefins to form N-allylhydroxamic acids. Since these are readily reduced to the corresponding amides and amines the sequence provides a method for the allylic amination of olefins. Here again, although the enophile is unsymmetrical only one mode of addition is observed, reaction always leading to the formation of a C–N bond and generation of an N–OH compound.[23] Thus, nitrosocarbonylmethane (31), liberated *in situ* from its Diels-Alder adduct in the presence of cyclohexene gave the adduct (32) in 85 per cent yield.

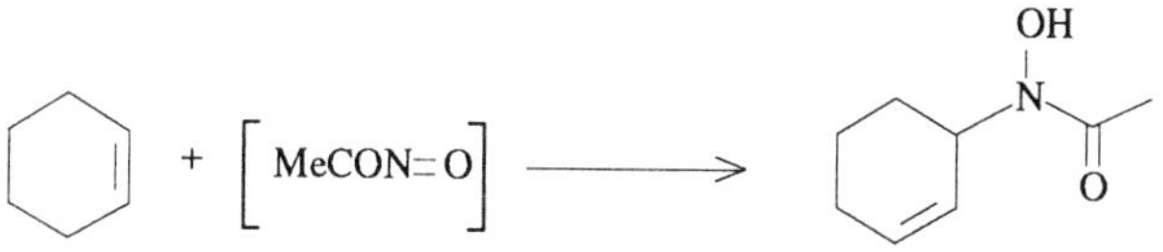

+ [MeCON=O] →

(31) (32) (85%)

*Scheme 14*

Intramolecular reactions take place readily. Oxidation of the hydroxamic acid (33) in the presence of 9,10-dimethylanthracene afforded the Diels-Alder adduct of the nitroso-ketone (34) which on thermal release in boiling toluene gave the adduct (35) in quantitative yield. This was converted into the amine (36) and thence into (±)-crinane (37).[24] (±)-Mesembrine was synthesised by a similar procedure.[25]

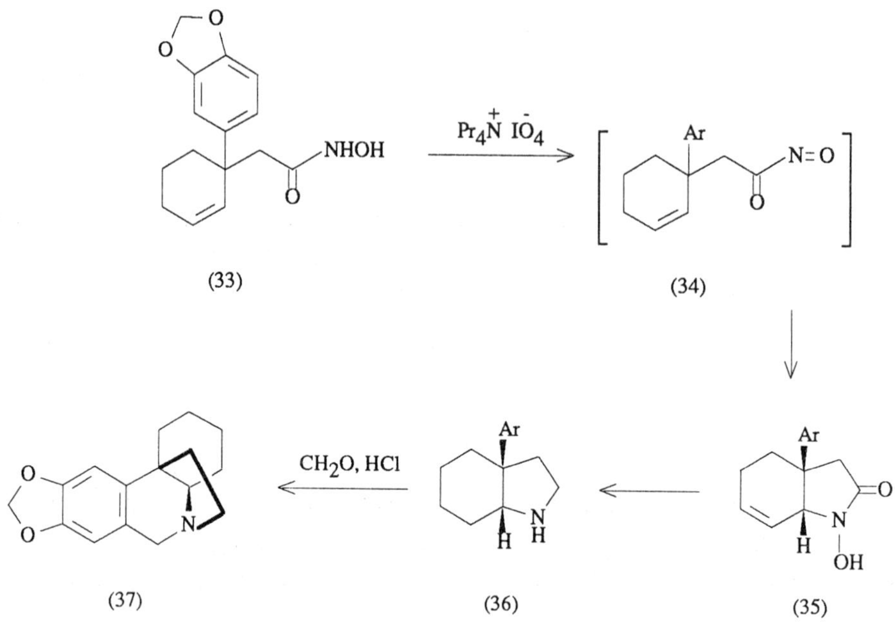

*Scheme 15*

Allylic amination can also be effected by reaction of olefins with the powerful enophiles bis(N-*p*-toluenesulphonyl)selenodiimide[26] or bis(N-*p*-toluenesulphonyl)sulphodiimide,[27] aza analogues of selenium dioxide and sulphur dioxide. Thus, bis(*p*-toluenesulphonyl)sulphodiimide reacted instantly with methylenecyclohexane in dichloromethane to give the single product (39) which was smoothly cleaved to the corresponding sulphonamide (40) with trimethyl phosphite. This sequence was applied in a neat synthesis of (±)-gabaculine.[28] Reaction takes the same course as allylic oxidation of olefins with selenium dioxide;[29] initial ene reaction to (38) is followed by a [2,3]-sigmatropic rearrangement to (39).

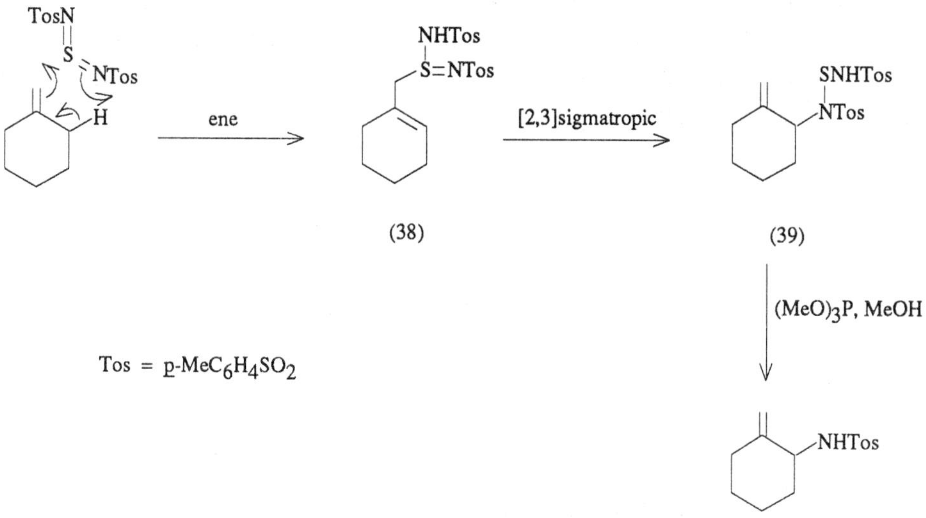

TosN S NTos H — ene → (38) NHTos S=NTos — [2,3]sigmatropic → (39) SNHTos NTos — (MeO)$_3$P, MeOH → (40) (84%) NHTos

Tos = p-$MeC_6H_4SO_2$

*Scheme 16*

Ene reactions with the related mono-oxo compound, N-sulphinylsulphonamide, are reversible under mild conditions, and this can be exploited to introduce deuterium or tritium specifically into the allylic position of olefins.[30]

Diethyl azodicarboxylate also reacts readily with a variety of olefins,[6] and thermal ene reactions of t-butylhydrazones with methyl acrylate and acrylonitrile forms the key step in a useful synthesis of β-keto esters and γ-keto nitriles.[31]

Oxygen also is a valuable dienophile and can be used in the allylic oxidation of olefins by photosensitised oxygenation.[32] The first products of the reactions are allylic hydroperoxides formed in an ene reaction with oxygen, and these are easily reduced to the corresponding allylic alcohol. α-Pinene, for example, is converted into *trans*-pinocarveyl hydroperoxide (41) which on reduction gives *trans*-pinocarveol (42).

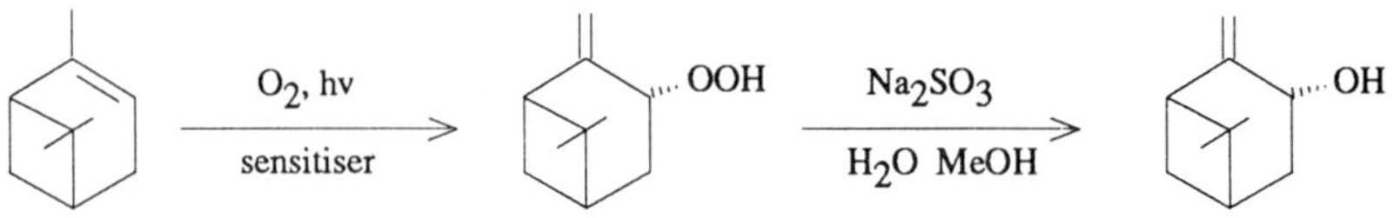

$O_2$, hv / sensitiser → (41) OOH; $Na_2SO_3$ / $H_2O$ MeOH → (42) OH

*Scheme 17*

The reactive species in these reactions is believed to be singlet oxygen and the reaction is generally held to proceed by a concerted mechanism analogous to that of the ene reaction.

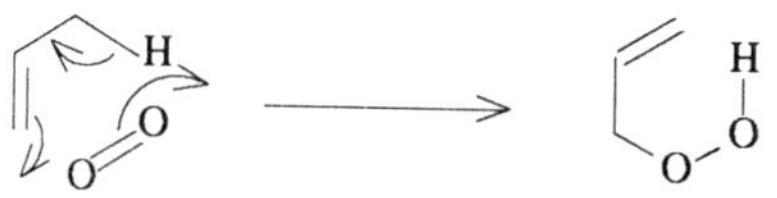

*Scheme 18*

## Intramolecular Ene Reactions

Intramolecular ene reactions are now being widely employed in synthesis. They generally take place more easily than the corresponding intermolecular reactions, particularly when they give rise to five-membered rings.[33] Because of the constraints imposed on the transition state by the chain linking the ene and enophile components, the intramolecular reactions are frequently highly stereoselective. Thus, kinetically controlled thermal cyclisation of the *cis* 1,6-diene (43) gave preponderantly the *cis* disubstituted cyclopentane (44) by way of the *exo* transition state (47); the *endo* transition state here is highly strained. Cyclisation of the corresponding *trans* diene again gave mainly the *cis*-cyclopentane derivative (44), this time clearly by way of the *endo* transition state. At higher temperatures the reactions are reversible, and the *trans* isomer (45) accumulates. This example again illustrates the fact that intramolecular ene reactions can frequently be effected with a non-activated enophile double bond, in contrast to most intermolecular reactions.

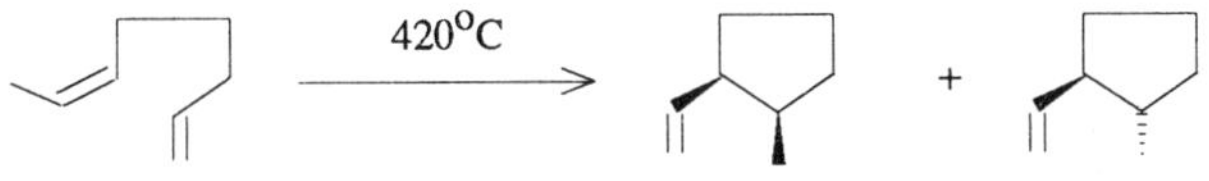

420°C

(43) (44) + (45) (14:1)

(46) (endo) (47) (exo)

*Scheme 19*

Acetylenic bonds also take part readily in intramolecular ene reactions. Thus, dehydrolinalool (48) was converted into (49) in quantitative yield at 200°C[34] and in the di-yne (50) both the ene and enophile are triple bonds.[35]

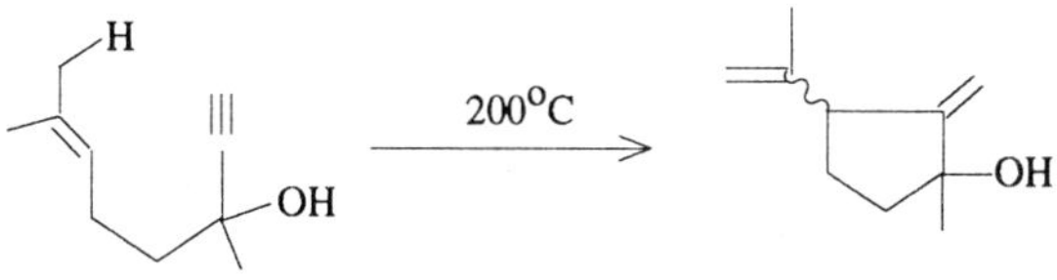

200°C

(48) (49)

550°C ene [1,5] H-shift

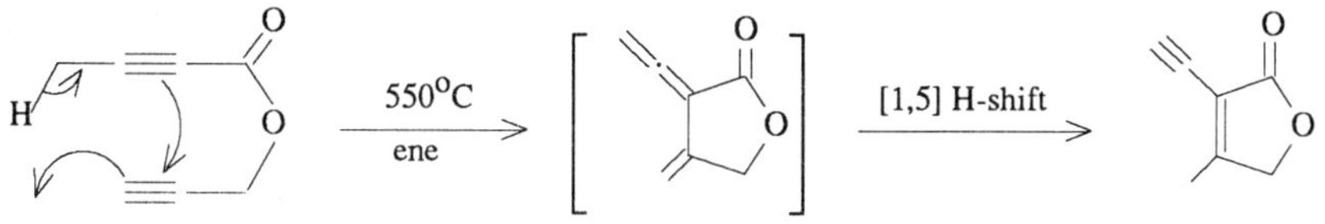

(50) (80%)

*Scheme 20*

The general rule with 1,6-dienes like (43), where the enophile is linked to the terminal olefinic carbon of the allyl group, appears to be that the two nearest carbon atoms of the ene and the enophile combine to form a five-membered ring. The same applies to the corresponding 1,7-dienes which cyclise to form six-membered rings, although generally not so easily as the 1,6-dienes; higher temperatures are required and yields are lower.

The formation of six-membered rings can sometimes be facilitated under catalytic conditions with appropriate location of an activating group in the enophile. Thus, even the 1,6-diene (51) gave the cyclohexane derivative (53) exclusively in 82 per cent yield on catalysed cyclisation with ethylaluminium dichloride, by way of the boat-like transition state (52).[36]

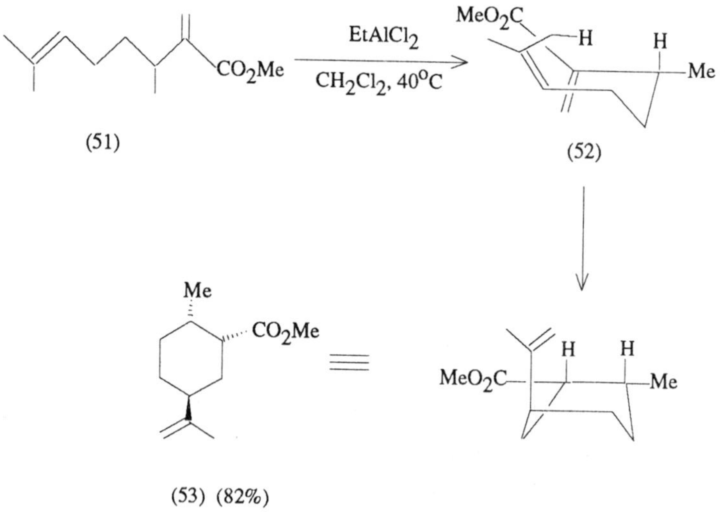

$CO_2Me$ (51) — $EtAlCl_2$, $CH_2Cl_2$, 40°C → $MeO_2C$, H, H, Me (52)

Me, $CO_2Me$ (53) (82%) ≡ $MeO_2C$, H, H, Me

*Scheme 21*

The exclusive formation of (53) is a consequence of the electronic effects of the Lewis acid catalyst and the location of the activating ester group. The expected cyclopentane derivative (54) is not observed here because the position of the ester group precludes Lewis acid catalysis of this reaction.

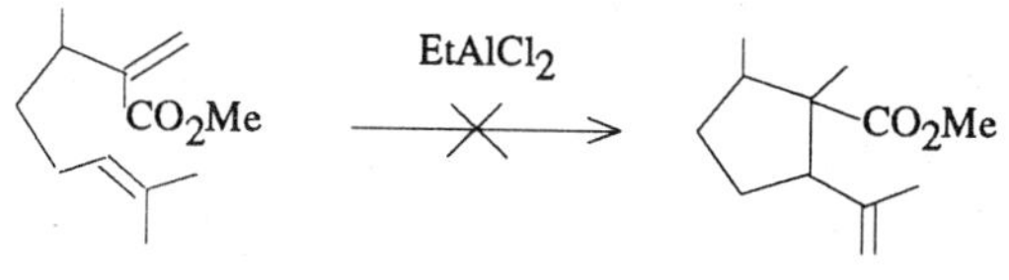
EtAlCl2
CO2Me
CO2Me

(51) (54)

*Scheme 22*

Again, the 1,7-dienes (55), where $R^1$ and $R^2$ are electron-withdrawing groups, gave the *trans* disubstituted cyclohexanes (57) almost exclusively on thermal or, better, catalysed cyclisation, by way of the transition state (56).[37]

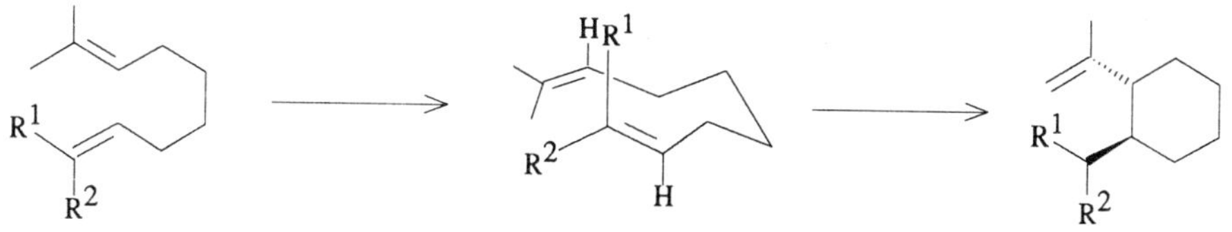
R1
R2
HR1
R2
H
R1
R2

(55) (56) (57)

*Scheme 23*

Thus, the diene (58) obtained from (*R*)-citronellal and dimethyl malonate, cyclised in the presence of catalytic ferric chloride on alumina to give the cyclohexane (59) almost exclusively in 92 per cent yield.[38]

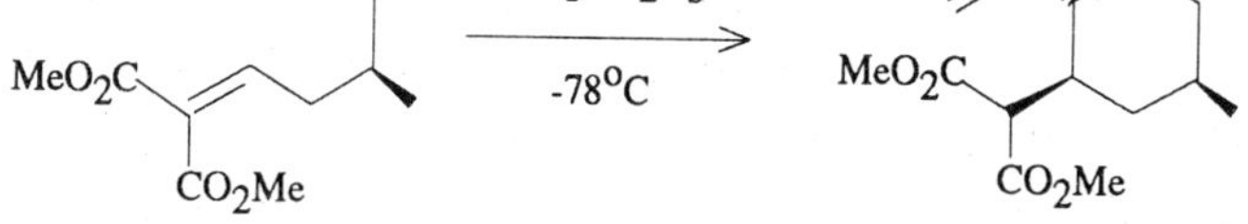
MeO2C
CO2Me
FeCl3-Al2O3
-78°C
MeO2C
CO2Me

(58) (59)

*Scheme 24*

Related to these reactions of 1,6- and 1,7-dienes is the thermal cyclisation of olefinic ketones, which leads, in appropriate cases, to cyclopentanones, cyclohexanones or cycloalkyl ketones in high yield, by an intramolecular ene reaction of the corresponding enols.[39] The sequence is of wide application and can be used to make fused-ring, spiro and bridged-ring compounds. The dienyl ketone (60), for example, on pyrolysis at 350°C gave the bicyclic ketone (63) by two successive enol ene reactions.

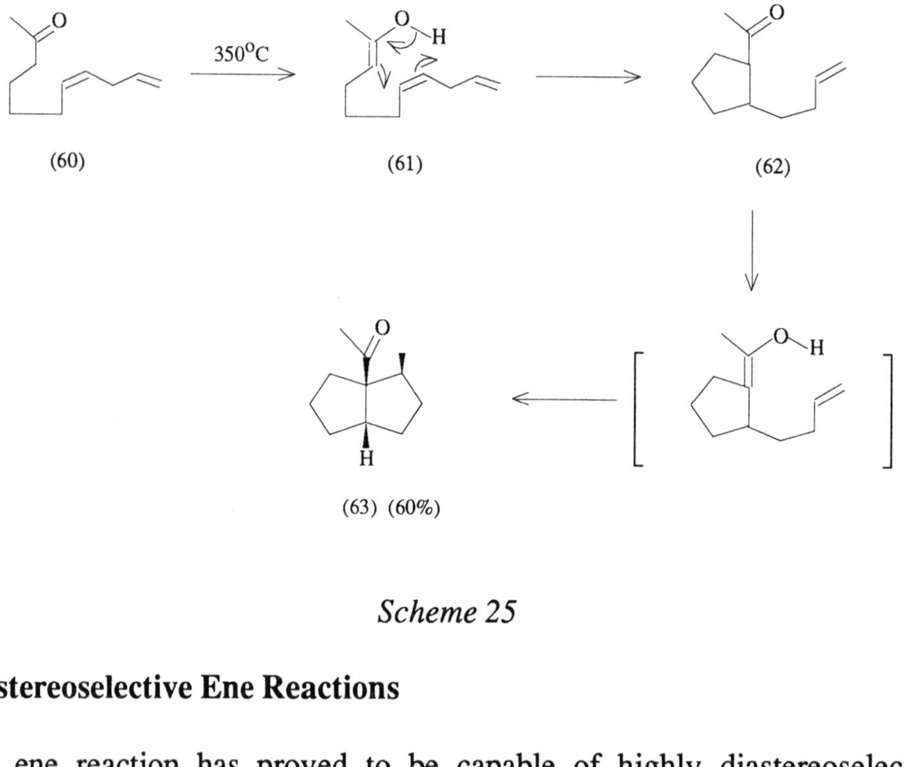

350°C

(60) (61) (62)

(63) (60%)

*Scheme 25*

**Diastereoselective Ene Reactions**

The ene reaction has proved to be capable of highly diastereoselective transformations both where a chiral centre in either the ene or enophile is finally incorporated in an optically active target molecule, or where one or other component carries an optically active auxiliary group which is removed after the diasteroselective ene reaction has been effected. Thus, in a synthesis of (–)-α-kainic acid (66), the optically active diene (64), derived ultimately from (*S*)-glutamic acid, cyclised to the pyrrolidine derivative (65) with almost complete diastereoselectivity[40] and, in an intermolecular example, catalysed reaction of the pregnadiene (67) and methyl acrylate gave the adduct (68), with the natural (*S*)-configuration at C-20 in 85 per cent yield.[41]

The stereocontrol in the latter reaction is attributed to virtual exclusive attack of the enophile from the less hindered side of (67).

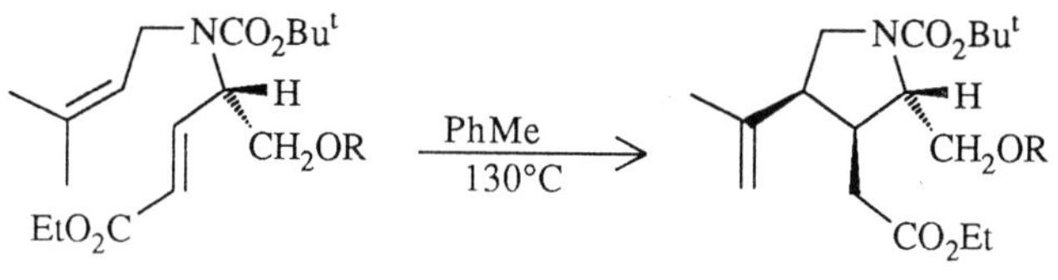

$NCO_2Bu^t$ H $CH_2OR$ $EtO_2C$ PhMe 130°C $NCO_2Bu^t$ H $CH_2OR$ $CO_2Et$

(64)
R = t-$BuMe_2Si$

(65) (74%)

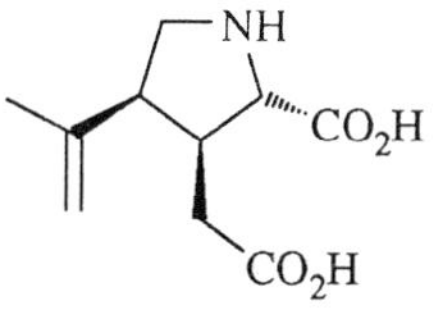

NH $CO_2H$ $CO_2H$

(66)

AcO $CO_2Me$ $CH_2Cl_2$ $EtAlCl_2$ 20 $CO_2Me$ $H_2$, Pd $CO_2Me$

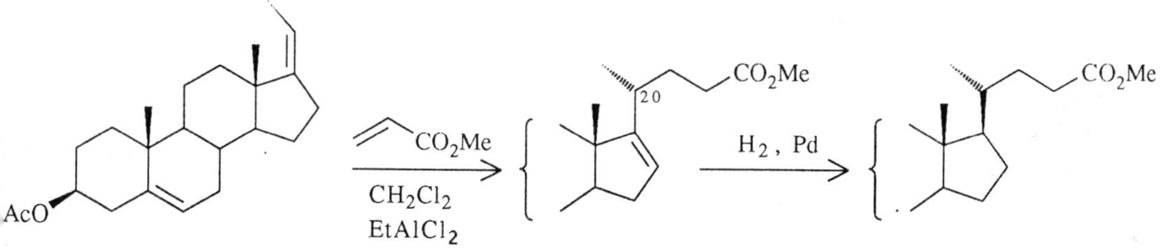

(67) (68)

*Scheme 26*

In reactions involving the use of chiral auxiliaries, excellent results have been obtained with glyoxylate esters of 8-phenylmenthol[42] and the readily available (+)- and (–)-forms of *trans*-2-phenylcyclohexanol.[43] Thus, reaction of 1-hexene with 8-phenylmenthyl glyoxylate (69) in the presence of stannic chloride gave the ene adduct (70) in 92 per cent yield and with greater than 99 per cent diastereoselectivity. The absolute sense of the stereochemistry in this and other ene reactions of (69) is the same as that obtained in nucleophilic additions to the glyoxylate and arises by selective *si* attack on the aldehyde carbonyl group of the ester in the syn conformation shown. Equally good results were obtained in reactions with phenylmenthyl esters of α-keto acids, generating tertiary alcohols.[44]

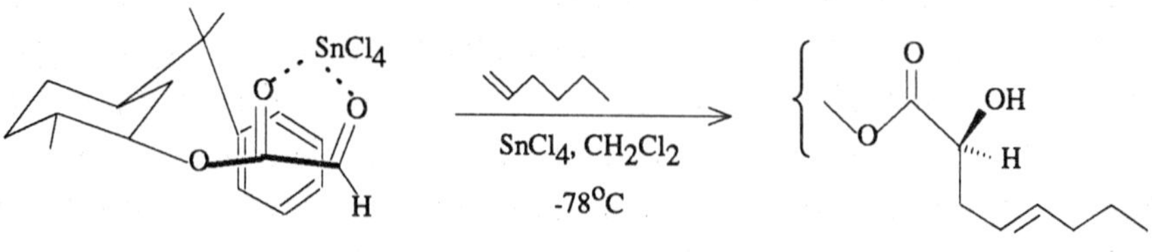

$SnCl_4$ ; $SnCl_4$, $CH_2Cl_2$ ; -78°C ; O ; OH ; H

(69)

(70) (92%; >99% de)

*Scheme 27*

Reaction of the glyoxylate with 1,2-disubstituted olefins generates two new chiral centres, with high stereoselectivity at each centre. With *trans*-2-butene the *anti* isomer (71) was the predominant product, but with *cis*-2-butene the reaction was less selective because of partial isomerisation of the olefin during reaction. 1-Trimethylsilyl-*cis*-2-butene, however, gave a high proportion of the *syn* alcohol (72). These products are masked aldols (through oxidative cleavage of the olefinic bond) and this sequence of reactions provides a stereoselective route to some aldols which might otherwise be difficultly accessible. The stereochemical results could be explained by a concerted addition, but there is evidence that these reactions in fact follow a two-step ionic pathway.[42]

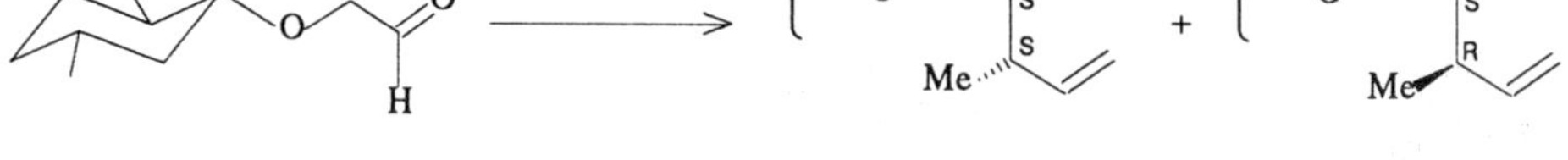

Ph ; O ; H ; OH ; S ; S ; Me ; + ; R

(71) (72)

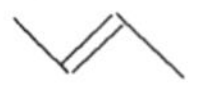

, $SnCl_4$, $CH_2Cl_2$, -78°C; (85%; 15:1)

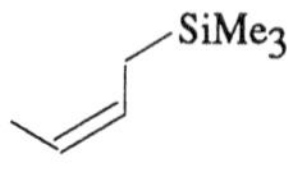

$SiMe_3$

, $SnCl_4$, $CH_2Cl_2$, -78°C; (100%; 1:15)

*Scheme 28*

Reaction of optically active glyoxylates with olefins which are themselves chiral as well as leading to stereocontrolled formation of new chiral centres, can also effect kinetic resolution of the racemic olefin. Thus, the racemic diene (73) on reaction with 8-phenylmenthyl glyoxylate gave the two adducts (74) and (75) in a ratio of 8:1. The main product (74) was reduced to the diol (77), with recovery of the valuable auxiliary, and thence to (76), with inversion at the secondary alcohol. Several further steps, involving selective cleavage of the two olefinic bonds, gave (–)-xylomollin (78).[45]

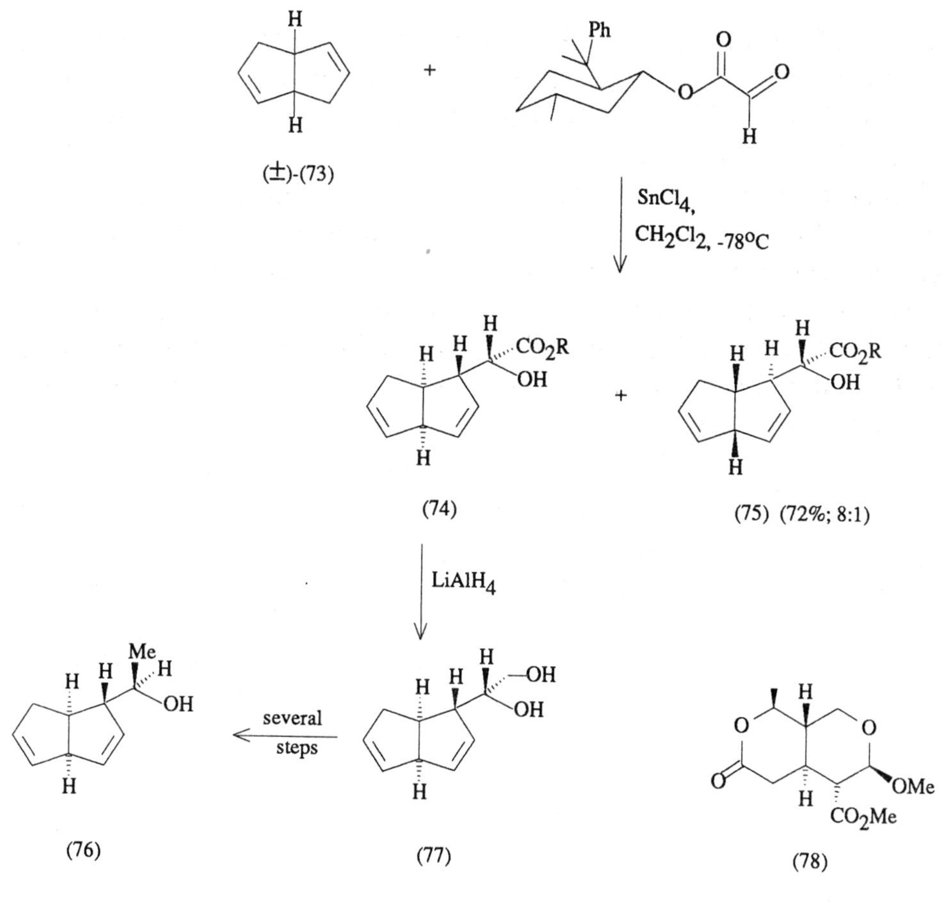

*Scheme 29*

Kinetic selection was also realised between the two olefinic bonds in the *meso* diene (79). Reaction with the glyoxylate ester of (*R*)-*trans*-2-phenylcyclohexanol gave the adduct (80) in a remarkable 81 per cent yield. With the glyoxylate ester of (*S*)-2-phenylcyclohexanol the epimer (81) was obtained. Reduction of (80) with lithium aluminium hydride furnished the diol (83) which was used to make optically pure (–)-specionin (82), with excellent stereocontrol at each step by the bicyclo[3,3,0]octane framework.[46]

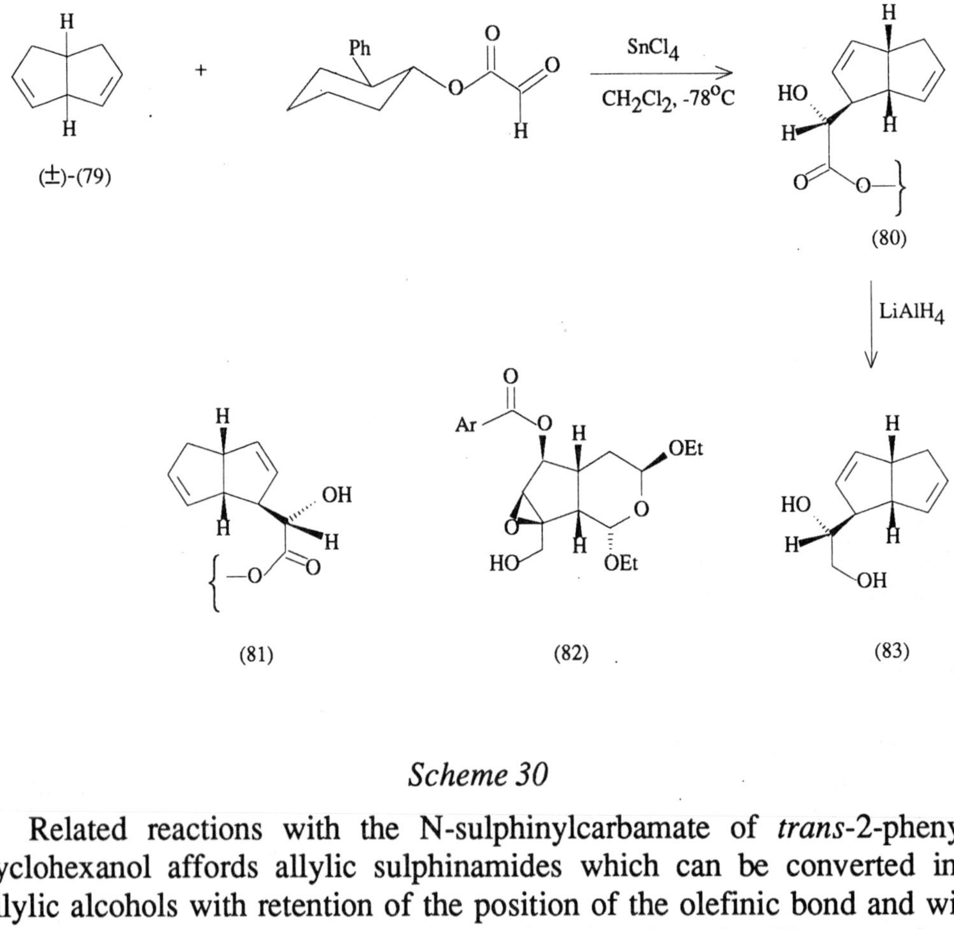

H H (±)-(79) + Ph O O O H $SnCl_4$ $CH_2Cl_2$, -78°C HO H H H O O (80)

$LiAlH_4$

H H OH H O O (81)

O Ar O H OEt O O H HO OEt (82)

H HO H H OH (83)

*Scheme 30*

Related reactions with the N-sulphinylcarbamate of *trans*-2-phenylcyclohexanol affords allylic sulphinamides which can be converted into allylic alcohols with retention of the position of the olefinic bond and with high levels of controllable absolute stereochemistry.[47] Thus, catalysed reaction of the olefin (84) with the sulphinylcarbamate (85) gave the adduct (86) which by N-ethylation followed by reaction with phenylmagnesium bromide afforded the allylic sulphoxide (87). Sequential sulphoxide-sulphenate rearrangement and sulphur cleavage then gave the allylic alcohol (88) in 56 per cent overall yield and of 91 per cent enantiomeric purity.

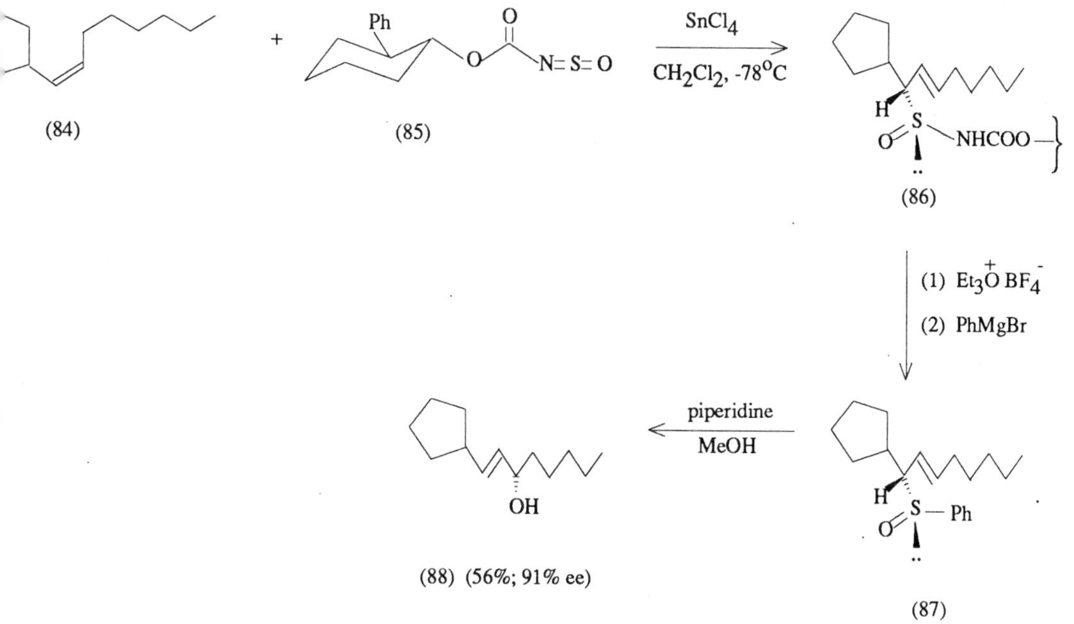

Ph O O N=S=O

(84) + (85) $\xrightarrow[CH_2Cl_2,\ -78^\circ C]{SnCl_4}$ (86)

H S O NHCOO

(1) $Et_3\overset{+}{O}\,BF_4^-$
(2) PhMgBr

(87) H S — Ph O

piperidine / MeOH

(88) (56%; 91% ee) OH

*Scheme 31*

There appear to be few examples of diastereoselective ene reactions with chiral ene components or with ene components bearing chiral auxiliary groups. Some success has been achieved in reactions catalysed by chiral organoaluminium reagents.[48]

## Magnesio-ene Reactions

The scope of the ene reaction has been extended by the discovery that allylic Grignard reagents take part in the reaction under mild conditions by migration of magnesium and formation of a new carbon-magnesium bond.[49]

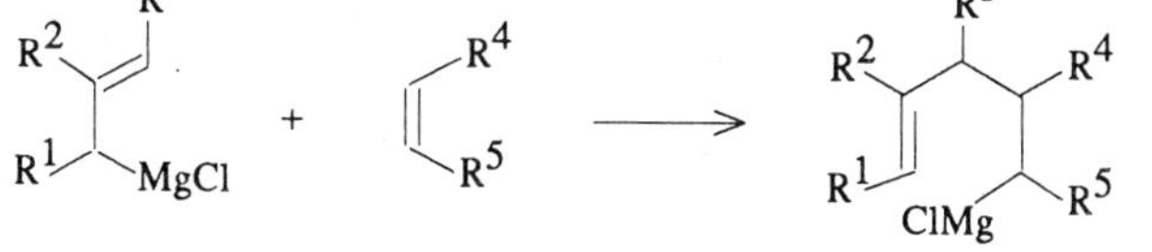

$R^1$ $R^2$ $R^3$ MgCl + $R^4$ $R^5$ ⟶ $R^1$ $R^2$ $R^3$ $R^4$ $R^5$ ClMg

*Scheme 32*

Intramolecular reactions of this kind have been exploited in the synthesis of a number of natural products. Two modes of reaction have been distinguished in which the enophile is linked at either end of the double bond of the allyl group, to give products (89) or (90), usually with high regio- and stereoselectivity.[50]

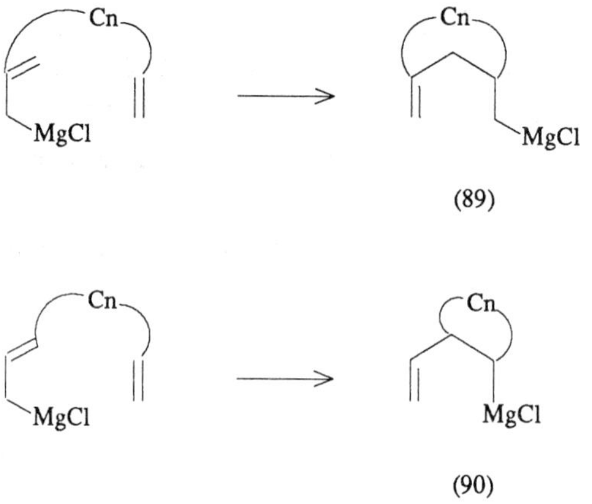

Cn
MgCl
Cn
MgCl
(89)

Cn
MgCl
Cn
MgCl
(90)

*Scheme 33*

Thus, in a reaction of the first type, the Grignard reagent from (91), on heating at 130°C, gave the 'ene' product (92) and thence, with oxygen, the alcohol (93). This key intermediate was converted in several more steps into the sesquiterpene alcohol (±)-chokol-A (94).[51]

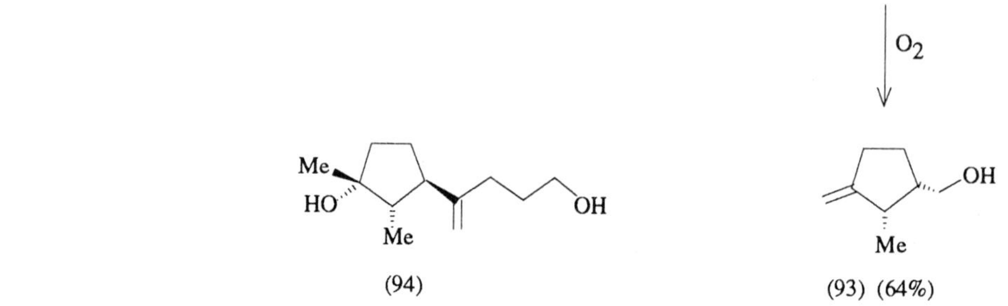

Cl
(91)
Mg, THF
-78°C
ClMg
H
H
Me
ClMg
H
H
Me
(92)
$O_2$
Me
HO
Me
OH
(94)
OH
Me
(93) (64%)

*Scheme 34*

Alternatively, cyclisation of (95), in which the enophile is linked to the terminal carbon of the allylic Grignard reagent, formed the key step in a synthesis of (+)-(α)-skytanthine (98) and (+)-iridomyrmecin (99). The initial 'ene' product (96) was converted directly into the alcohol (97) and thence in several more steps into the two target molecules.[52] The highly diastereoselective ene cyclisation is believed to proceed by way of the favoured transition state (95).

In an interesting development it has been found that 1,3-dienes can act as enophiles in reactions with allylic ethers or acetals in the presence of ligated Fe(0) catalysts, in what are formal [4+4] ene reactions. Thus, the acetal (100) and 2,3-dimethylbutadiene gave very largely the adduct (101). The choice of ligand was critical in determining the sense and degree of diastereoselection in this reaction.[53]

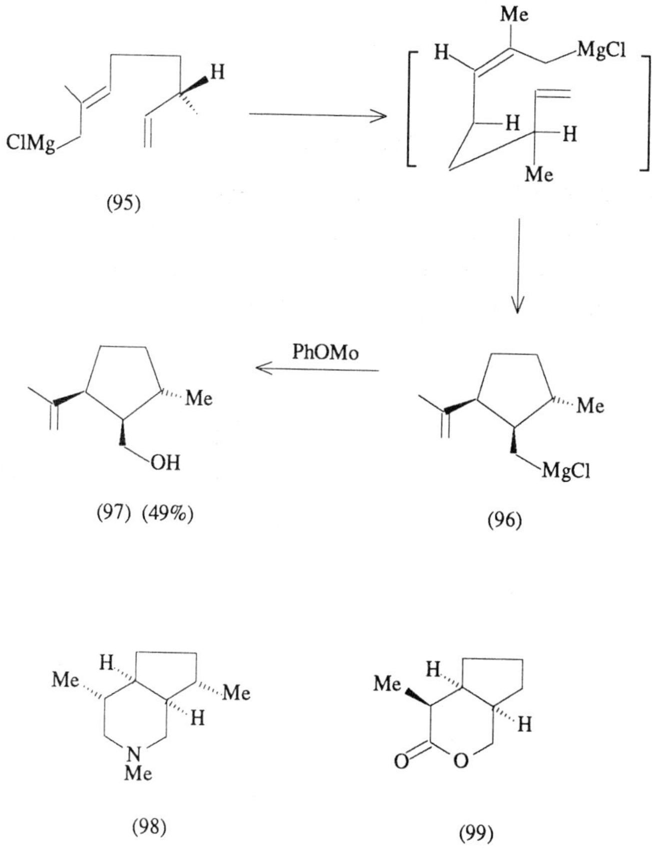

ClMg H (95)

Me H MgCl H H Me

PhOMo

Me OH (97) (49%)

Me MgCl (96)

H Me Me H N Me (98)

Me H H O O (99)

*Scheme 35*

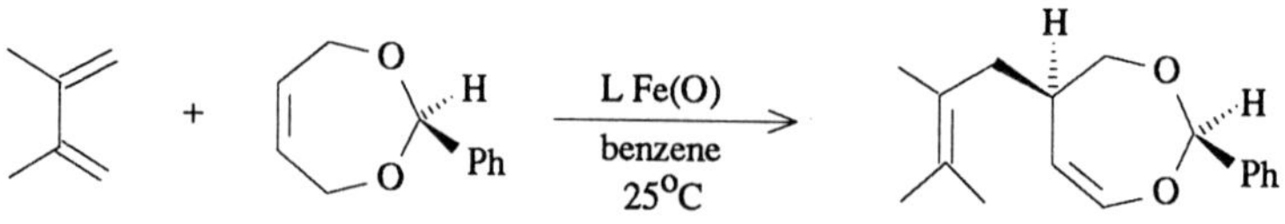

\+ O H Ph O

L Fe(O)
benzene
25°C

H O H Ph O

(100)

(101)

*Scheme 36*

## References


1 H.M.R. Hoffmann, *Angew. Chem. internat. edn.*, 1969, **8**, 556; E.C. Keung and H. Alper, *J. Chem. Educ.*, 1972, **49**, 97.

2 B.B. Snider, *Acc. Chem. Res.*, 1980, **13**, 426; N.H. Andersen, S.W. Hadley, J.D. Kelly and E.R. Bacon, *J. Org. Chem.*, 1985, **50**, 4144; B.B. Snider and E. Ron, *J. Am. Chem. Soc.*, 1985, **107**, 8160; J.V. Duncia, P.T. Lansbury, T. Miller and B.B. Snider, *J. Am. Chem. Soc.*, 1982, **104**, 1930.

3 cf. L.M. Stephenson and M. Orfanopoulos, *J. Org. Chem.*, 1981, **46**, 2200; S.H. Nahm and H.N. Cheng, *J. Org. Chem.*, 1986, **51**, 5093; B.B. Snider and E. Ron, *J. Am. Chem. Soc.*, 1985, **107**, 8160; H.M.R. Hoffmann, *Angew. Chem. internat. edn.*, 1969, **17**, 476.

4 cf. W. Starflinger, G. Kresze and K. Huss, *J. Org. Chem.*, 1986, **51**,37; N.H. Andersen, S.W. Hadley, J.D. Kelly and E.R. Bacon, *J. Org. Chem.*, 1985, **50**, 4144.

5 cf. B.B. Snider, *Acc. Chem. Res.*, 1980, **13**, 426.

6 cf. H.M.R. Hoffmann, *Angew. Chem. internat. edn.*, 1969, **17**, 476.

7 cf. H.M.R. Hoffmann, *Angew. Chem. internat. edn.*, 1969, **8**, 556; W. Oppolzer and V. Snieckus, *Angew. Chem. internat. edn.*, 1978, **17**, 476; W. Oppolzer, *Helv. Chim. Acta*, 1973, **56**, 1812; D. Armesto, W.M. Horspool, R. Pérez-Ossorio and A. Ramos, *J. Org. Chem.*, 1987, **52**, 3378.

8 K. Alder and H. von Brachel, *Annalen*, 1962, **651**, 141.

9 J.A. Berson, R.G. Wahl and H.D. Perlmutter, *J. Am. Chem. Soc.*, 1966, **88**, 187; S.H. Nahm and H.N. Cheng, *J. Org. Chem.*, 1986, **51**, 5093.

10 cf. W. Oppolzer, *Pure Appl. Chem.*, 1981, **53**, 1181.

11 B.B. Snider and D.J. Rodini, *Tetrahedron Lett.*, 1980, **21**, 1815; B.B. Snider, D.J. Rodini, T.C. Kirk and R. Cordova, *J. Am. Chem. Soc.*, 1982, **104**, 555.

12 B.B. Snider and G.B. Phillips, *J. Org. Chem.*, 1983, **48**, 464.

13 B.B. Snider and E.A. Deutsch, *J. Org. Chem.*, 1982, **47**, 745.

14 B.B. Snider and J.W. van Straten, *J. Org. Chem.*, 1979, **44**, 3567.

15 B.B. Snider, D.J. Rodini, M. Karras, T.C. Kirk, E.A. Deutsch, R. Cordova and R.T. Price, *Tetrahedron*, 1981, **37**, 3927.

16 N.H. Andersen, S.W. Hadley, J.D. Kelly and E.R. Bacon, *J. Org. Chem.*, 1985, **50**, 4144.

17 cf. J.K. Whitesell, *Acc. Chem. Res.*, 1985, **18**, 280.

18 S.S-M. Choi and G.W. Kirby, *J. Chem. Soc. Chem. Commun.*, 1988, 177.

19 cf. E. Vedejs, T.H. Eberlein and R.G. Wilde, *J. Org. Chem.*, 1988, **53**, 2220.

20 O. Achmatowicz and M.Peitraszkiewicz, *J. Chem. Soc. Perkin 1*, 1981, 2680.

21 W. Starflinger, G. Kresze and K. Huss, *J. Org. Chem.*, 1986, **51**, 37.

22 J-M. Lin, K. Koch and F.W. Fowler, *J. Org. Chem.*, 1986, **51**, 167.

23 G.W. Kirby *Chem. Soc. Rev.*, 1977, **6**, 1; G.E. Keck, R.R. Webb and J.B. Yates, *Tetrahedron*, 1981, **37**, 4007.

24 G.E. Keck and R.R. Webb, *J. Am. Chem. Soc.*, 1981, **103**, 3173.

25 G.E. Keck and R.R. Webb, *J. Org. Chem.*, 1982, **47**, 1302.

26 K.B. Sharpless, T. Hori, L.K. Truesdale and C.O. Dietrich, *J. Am. Chem. Soc.*, 1976, **98**, 269.

27 K.B. Sharpless and T. Hori, *J. Org. Chem.*, 1976, **41**, 176.

28 S.P. Singer and K.B. Sharpless, *J. Org. Chem.*, 1978, **43**, 1448.

29 D. Arigoni, A. Vasella, K.B. Sharpless and H.P. Jensen, *J. Am. Chem. Soc.*, 1973, **95**, 7917; W-D. Woggon, F. Ruther and H. Egli, *J. Chem. Soc. Chem. Commun.*, 1980, 706.

30 T. Hori, S.P. Singer and K.B. Sharpless, *J. Org. Chem.*, 1978, **43**, 1456.

31 J.E. Baldwin, R.M. Adlington, A.U. Jain, J.N. Kohle and M.W.D. Perry, *Tetrahedron*, 1986, **42**, 4247.

32 K. Gollnick, *Adv. Photochem.*, 1968, **6**, 1; W. Carruthers, *Some Modern Methods of Organic Synthesis*, Cambridge University Press, 3rd edn., 1986, p.392.

33 W. Oppolzer and V. Snieckus, *Angew. Chem. internat. edn.*, 1978, **17**, 476; W. Oppolzer, *Pure and Appl. Chem.*, 1981, **53**, 1181.

34 W. Hoffmann, H. Pasedach, H. Pommer and W. Reif, *Annalen*, 1971, **747**, 60; P. Naegli and R. Kaiser, *Tetrahedron Lett.*, 1972, 2013.

35 V. Bilinski, M. Karpf and A.S. Dreiding, *Helv. Chim. Acta*, 1986, **69**, 1734.

36 B.B. Snider and G.B. Phillips, *J. Org. Chem.*, 1984, **49**, 183.

37 L.F. Tietze and U. Beifuss, *Annalen*, 1988, 321.

38 L.F. Tietze, U. Beifuss, J. Antel and G.M. Sheldrick, *Angew. Chem. internat. edn.*, 1988, **27**, 703.

39 J. Brocard, G. Moinet and J-M. Conia, *Bull. Soc. Chim. France*, 1972, **II**, 1711; G. Mandville, F. Leyendecker and J-M. Conia, *Bull. Soc. Chim. France*, 1973, **II**, 963; F. Leyendecker, J. Drouin and J-M. Conia, *Tetrahedron Lett.*, 1974, 2931; J-M. Conia and P. Le Perchec, *Synthesis*, 1975, 1; W. Oppolzer, *Helv. Chim. Acta*, 1973, **56**, 1812.

40 W. Oppolzer and K. Thirring, *J. Am. Chem. Soc.*, 1982, **104**, 4978.

41 A.D. Batcho, D.E. Berger, S.G. Davoust, P.M. Wovkulich and M.R. Uskokovic, *Helv. Chim. Acta*, 1981, **64**, 1682.

42 J.K. Whitesell, A. Battacharya, C.M. Buchanan, H.H. Chen, D. Deyo, D. James, C-L. Liu and M.A. Minton, *Tetrahedron.*, 1986, **42**, 2993.

43 J.K. Whitesell, H.H. Chen and R.M. Lawrence, *J. Org. Chem.*, 1985, **50**, 4663.

44 J.K. Whitesell, D. Deyo and A. Battacharya, *J. Chem. Soc. Chem. Commun.*, 1983, 802.

45 J.K. Whitesell and M.A. Minton, *J. Am. Chem. Soc.*, 1986, **108**, 6802.

46 J.K. Whitesell and D.E. Allen, *J. Am. Chem. Soc.*, 1988, **110**, 3585.

47 J.K. Whitesell and R.M. Lawrence, *Chimia*, 1986, **40**, 318; J.K. Whitesell and J.F. Carpenter, *J. Am. Chem. Soc.*, 1987, **109**, 2839.

48 K. Maruoka, Y. Hoshino, T. Shirasaka and H. Yamamoto, *Tetrahedron Lett.*, 1988, **29**, 3967.

49 cf. H. Lehmkuhl, *Bull. Soc. Chim. France*, 1981, **II**, 87.

50 W. Oppolzer, R. Pitteloud and H.F. Strauss, *J. Am. Chem. Soc.*, 1982, **104**, 6476; W. Oppolzer and K. Bättig, *Tetrahedron Lett.*, 1982, **23**, 4669.

51 W. Oppolzer and A.F. Cunningham, *Tetrahedron Lett.*, 1986, **27**, 5467.

52 W. Oppolzer and E.J. Jacobsen, *Tetrahedron Lett.*, 1986, **27**, 1141.

53 J.M. Takacs, L.G. Anderson and P.W. Newsome, *J. Am. Chem. Soc.*, 1987, **109**, 2542; J.M. Takacs and L.G., Anderson, *J. Am. Chem. Soc.*, 1987, **109**, 2200.

# 6 1,3-DIPOLAR CYCLOADDITION REACTIONS

1,3-Dipolar cycloadditon reactions, like the Diels-Alder reaction, are [$\pi$4s+$\pi$2s] reactions and proceed through a 6$\pi$-electron 'aromatic' transition state, but they differ from Diels-Alder reactions in that the 4$\pi$-electron component (1) like the allyl anion contains only three atoms, at least one of which is a hetero-atom. Cycloaddition to a double or triple bond leads to a five-membered heterocyclic compound.[1]

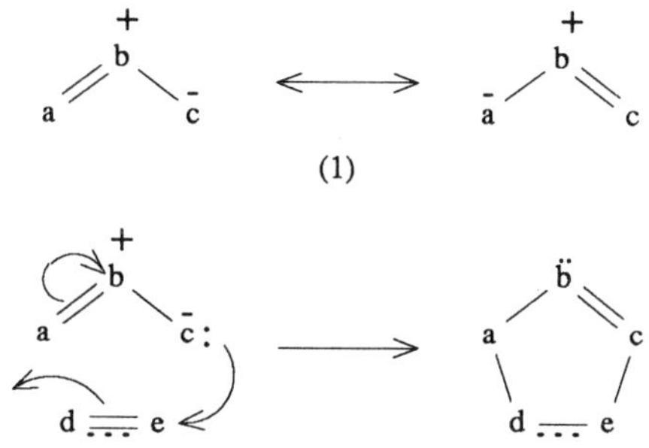

(1)

*Scheme 1*

The 4$\pi$-electron component, called the 1,3-dipole, is of such a nature that the stabilised all octet structure can only be represented by zwitterionic forms in which the positive charge is located on the central atom and the negative charge is distributed over the two terminal atoms. The 2$\pi$-electron component is called the dipolarophile.

A considerable number of 1,3-dipoles containing various combinations of carbon and hetero-atoms is theoretically possible and many have been made and their reactions with dipolarophiles studied. Restricting the permutations to second row elements Huisgen has classified the eighteen possibilities shown in the Table, six of the propargyl-allene type and twelve of the allyl type.[2,3]

Like allyl anions, all 1,3-dipoles have a system of four $\pi$-electrons in three parallel atomic Pz orbitals. Dipoles of the propargyl-allenyl type, which have a triple bond in one canonical form, contain an additional $\pi$-orbital orthogonal to the allyl-anion type molecular orbital; they have a linear structure. Allyl-type 1,3-dipoles are bent.

## Classification of 1,3-Dipoles Containing Carbon, Nitrogen and Oxygen Centres

### Propargyl-Allenyl Type

|  |  | |  |
|---|---|---|---|
| Nitrile ylides | $-C\equiv\overset{+}{N}-\overset{-}{\underset{\cdot\cdot}{C}}<$ | $\longleftrightarrow$ | $-\overset{-}{\underset{\cdot\cdot}{C}}=\overset{+}{N}=C<$ |
| Nitrile Imines | $-C\equiv\overset{+}{N}-\overset{-}{\underset{\cdot\cdot}{N}}<$ | $\longleftrightarrow$ | $-\overset{-}{\underset{\cdot\cdot}{C}}=\overset{+}{N}=N<$ |
| Nitrile Oxides | $-C\equiv\overset{+}{N}-\overset{-}{\underset{\cdot\cdot}{O}}$ | $\longleftrightarrow$ | $-\overset{-}{\underset{\cdot\cdot}{C}}=\overset{+}{N}=O$ |
| Diazoalkanes | $N\equiv\overset{+}{N}-\overset{-}{\underset{\cdot\cdot}{C}}<$ | $\longleftrightarrow$ | $\overset{-}{\underset{\cdot\cdot}{N}}=\overset{+}{N}=C<$ |
| Azides | $N\equiv\overset{+}{N}-\overset{-}{\underset{\cdot\cdot}{N}}-$ | $\longleftrightarrow$ | $\overset{-}{\underset{\cdot\cdot}{N}}=\overset{+}{N}=N-$ |
| Nitrous oxide | $N\equiv\overset{+}{N}-\overset{-}{\underset{\cdot\cdot}{O}}$ | $\longleftrightarrow$ | $\overset{-}{\underset{\cdot\cdot}{N}}=\overset{+}{N}=O$ |

### Allyl Type

|  |  | |  |
|---|---|---|---|
| Azomethine ylides | $>C=\underset{\vert}{\overset{+}{N}}-\overset{-}{\underset{\cdot\cdot}{C}}-$ | $\longleftrightarrow$ | $>\overset{-}{\underset{\cdot\cdot}{C}}-\underset{\vert}{\overset{+}{N}}=C<$ |
| Azomethine imines | $>C=\underset{\vert}{\overset{+}{N}}-\overset{-}{\underset{\cdot\cdot}{N}}-$ | $\longleftrightarrow$ | $>\overset{-}{\underset{\cdot\cdot}{C}}-\underset{\vert}{\overset{+}{N}}=N-$ |
| Nitrones | $>C=\underset{\vert}{\overset{+}{N}}-\overset{-}{\underset{\cdot\cdot}{O}}$ | $\longleftrightarrow$ | $>\overset{-}{\underset{\cdot\cdot}{C}}-\underset{\vert}{\overset{+}{N}}=O$ |
| Azimes | $-N=\underset{\vert}{\overset{+}{N}}-\overset{-}{\underset{\cdot\cdot}{N}}-$ | $\longleftrightarrow$ | $-\overset{-}{\underset{\cdot\cdot}{N}}-\underset{\vert}{\overset{+}{N}}=N-$ |
| Azoxy compounds | $-N=\underset{\vert}{\overset{+}{N}}-\overset{-}{\underset{\cdot\cdot}{O}}$ | $\longleftrightarrow$ | $-\overset{-}{\underset{\cdot\cdot}{N}}-\underset{\vert}{\overset{+}{N}}=O$ |
| Nitro compounds | $O=\underset{\vert}{\overset{+}{N}}-\overset{-}{\underset{\cdot\cdot}{O}}$ | $\longleftrightarrow$ | $\overset{-}{\underset{\cdot\cdot}{O}}-\underset{\vert}{\overset{+}{N}}=O$ |
| Carbonyl ylides | $>C=\overset{+}{\dot{O}}-\overset{-}{\underset{\cdot\cdot}{C}}<$ | $\longleftrightarrow$ | $>\overset{-}{\underset{\cdot\cdot}{C}}-\overset{+}{\dot{O}}=C<$ |
| Carbonyl imines | $>C=\overset{+}{\dot{O}}-\overset{-}{\underset{\cdot\cdot}{N}}-$ | $\longleftrightarrow$ | $>\overset{-}{\underset{\cdot\cdot}{C}}-\overset{+}{\dot{O}}=N-$ |
| Carbonyl oxides | $>C=\overset{+}{\dot{O}}-\overset{-}{\underset{\cdot\cdot}{O}}$ | $\longleftrightarrow$ | $>\overset{-}{\underset{\cdot\cdot}{C}}-\overset{+}{\dot{O}}=O$ |
| Nitrosimines | $-N=\overset{+}{\dot{O}}-\overset{-}{\underset{\cdot\cdot}{N}}-$ | $\longleftrightarrow$ | $-\overset{-}{\underset{\cdot\cdot}{N}}-\overset{+}{\dot{O}}=N-$ |
| Nitrosoxides | $-N=\overset{+}{\dot{O}}-\overset{-}{\underset{\cdot\cdot}{O}}$ | $\longleftrightarrow$ | $-\overset{-}{\underset{\cdot\cdot}{N}}-\overset{+}{\dot{O}}=O$ |
| Ozone | $O=\overset{+}{\dot{O}}-\overset{-}{O}$ | $\longleftrightarrow$ | $\overset{-}{O}-\overset{+}{\dot{O}}=O$ |

(After R.Huisgen, J.Org.Chem., 1976, 41, 403.)

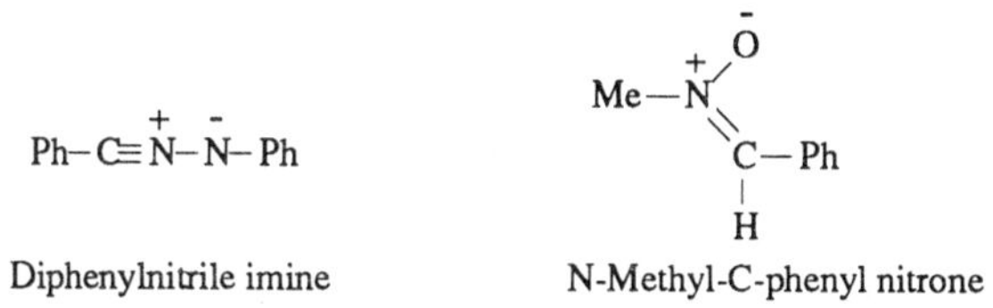

$Ph-C\equiv\overset{+}{N}-\overset{-}{N}-Ph$

Diphenylnitrile imine

$Me-\overset{+}{N}(-\overset{-}{O})=C(H)-Ph$

N-Methyl-C-phenyl nitrone

*Scheme 2*

It is generally agreed that 1,3-dipolar cycloadditions are concerted,[3] like Diels-Alder reactions. This is borne out by their high regio- and stereo-selectivities, among other factors.[2,4] Some 1,3-dipolar cycloaddition reactions are controlled mainly by the HOMO (dipole) - LUMO (dipolarophile) interaction and others by the LUMO (dipole) - HOMO (dipolarophile) interaction. The smaller the energy gap between the controlling orbitals the faster the reaction. The former are accelerated by electron-donating substituents in the dipole and electron-attracting substituents in the dipolarophile, and the latter by electron-attracting substituents in the dipole and electron-donating substituents on the dipolarophile; in each case the energy gap between the controlling orbitals in the two components is diminished. However, most 1,3-dipolar cycloadditions fall into a third class in which both the HOMO (dipole) - LUMO (dipolarophile) and LUMO (dipole) - HOMO (dipolarophile) interactions may be important. The controlling interaction in these cases depends on the nature of the dipolarophile and on the electronic nature of substituents on the dipole. These reactions can be accelerated by both electron-donating and electron-withdrawing substituents in either component.[2] This change in orbital control from HOMO (dipole) to LUMO (dipole) or *vice versa* from one reaction to another of a particular dipole may have consequences for the regioselectivities of the reactions. Regioselectivities are controlled by the magnitudes of the atomic orbital coefficients in the orbitals concerned; the atoms in each component with the largest coefficients interact, but these may not be the same in the HOMO and LUMO of any given compound.[5] For example, reaction of phenyl azide with 1-hexene, which is a LUMO (dipole) - HOMO (dipolarophile) interaction, gives mainly the 1-phenyl-5-butyltriazoline (2), but with methyl acrylate the 1-phenyl-4-carboxylic ester (3) is obtained. The reaction is now HOMO (dipole) - LUMO (dipolarophile) controlled, because of the lowering of the energies of the frontier orbitals of the dipolarophile by the ester group, with a change in the relative magnitudes of the atomic orbital coefficients in the interacting frontier orbitals of the two components.[6] In some reactions electronically preferred orientations may be disfavoured by steric effects.[2]

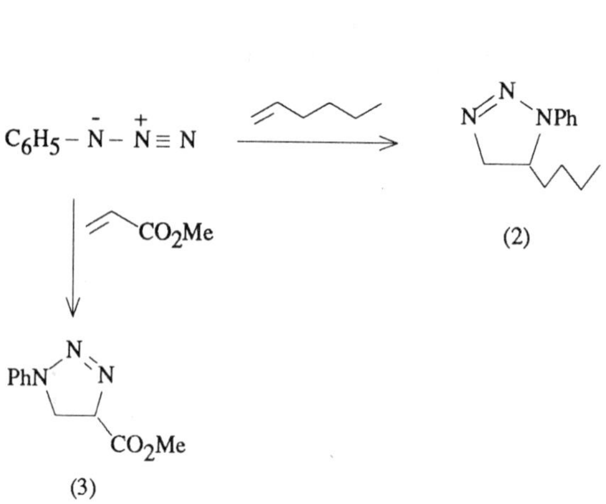

$C_6H_5-\overset{-}{N}-\overset{+}{N}\equiv N$

(2)

$CO_2Me$

PhN

$CO_2Me$

(3)

*Scheme 3*

As well as being regioselective, 1,3-dipolar cycloadditions are highly stereoselective. Numerous examples have shown that the stereochemistry of substituents on the double bond of the dipolarophile is retained in the adducts. Apparent exceptions have been shown to be due to isomerisation either before or after the cycloaddition.[7] Similarly for the dipoles, as long as the cycloadditions are noticeably faster than their possible isomerisation by rotation. Thus, the *cis* azomethine ylide (5), generated stereospecifically from the aziridine (4), gave exclusively the pyrrolidinetetracarboxylate (6) with *cis* ester groups at C-2 and C-5, on reaction with acetylene dicarboxylic ester. The isomeric *trans* azomethine ylide (8) gave only the *trans* pyrrolidine derivative (9).[8] With less reactive dipolarophiles, however, reactions with the *trans* ylide (8) were less selective because of competing isomerisation to the *cis* compound (5).

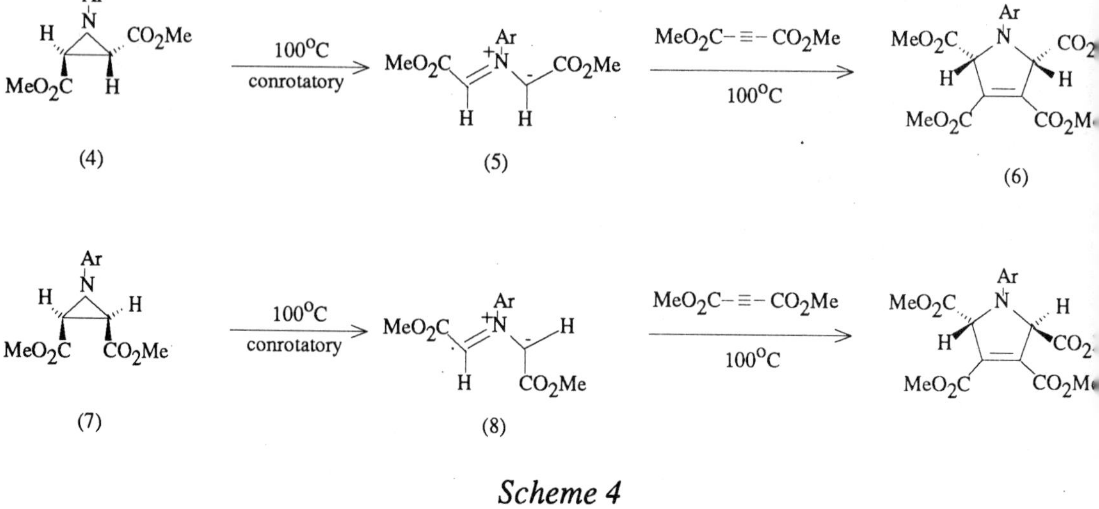

(4) —100°C, conrotatory→ (5) —$MeO_2C-\equiv-CO_2Me$, 100°C→ (6)

(7) —100°C, conrotatory→ (8) —$MeO_2C-\equiv-CO_2Me$, 100°C→

*Scheme 4*

When two chiral centres are created in the cycloaddition, one arising from each reactant, diastereomeric (*cis* and *trans*) products may be formed by way of the *endo* and *exo* transition states, but it is not always easy to predict the stereochemical course of the reactions. The outcome depends on the interplay of two generally opposing forces in the transition state — attractive π-orbital overlap of unsaturated substituents favouring an *endo* transition state, and repulsive van der Waals steric interactions favouring an *exo* transition state. Frequently mixtures of diastereomers are obtained. Thus, in the reaction of N-methyl-C-phenylnitrone (10) with various dipolarophiles to give 5-substituted isoxazolidines (11), the stereochemical results were not consistent, appearing sometimes to favour the *endo* transition state (10a) and sometimes the *exo* (10b) (Scheme 5). This may reflect an interplay of secondary orbital interactions and steric effects in the transition state, and possibly also interconversion of *cis* and *trans* forms of the nitrone before cycloaddition.[9]

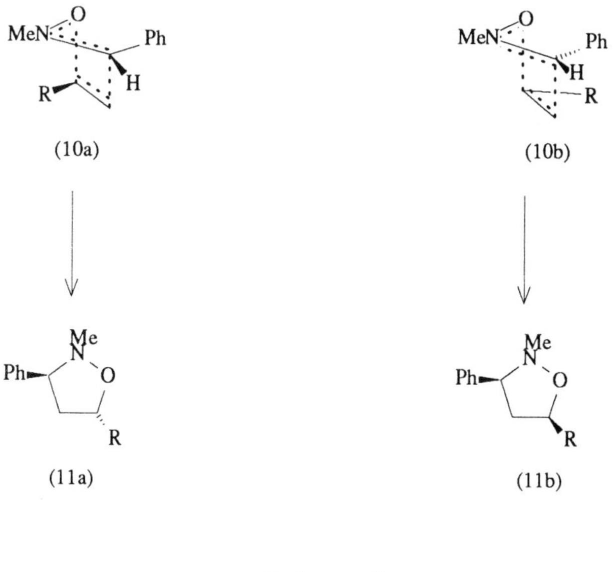

(10a) (10b)

(11a) (11b)

*Scheme 5*

In some cases, however, one effect predominates and a single product is obtained.

In most reactions the 1,3-dipole is not isolated but is generated *in situ* in the presence of the dipolarophile, which is generally an alkene or an alkyne, although this is not essential. Cycloadditions to dipolarophiles containing hetero multiple bonds, such as imines, nitriles and carbonyl compounds, have also been effected, so that a wide variety of five-membered heterocyclic compounds can be made by this general route. Some of the heterocyclic systems produced in this way can be converted easily into other synthetically useful compounds; ring-opening reactions which result in the stereocontrolled formation of acyclic compounds are particularly valuable. Not all the 1,3-dipoles listed in the Table are equally useful in this respect. Nitrones, nitrile oxides, azomethine ylides and, to a lesser extent, azomethine imines, azides and diazoalkanes have been most widely used, and the present chapter will be concerned more with recent work in this area rather than in providing a comprehensive account of the straightforward synthesis of heterocyclic ring systems.

## Azomethine Ylides

Cycloaddition of azomethine ylides to olefinic and acetylenic dipolarophiles leads, respectively, to pyrrolidines and $\Delta^3$-pyrrolines; the latter are easily converted into pyrroles (Scheme 6).

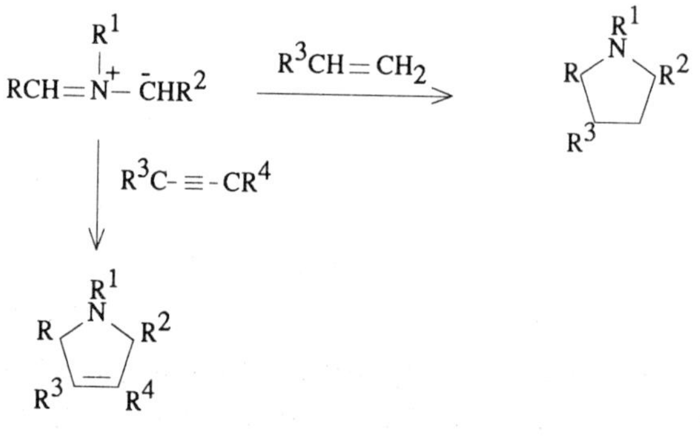

$$RCH{=}\overset{R^1}{\overset{|}{N^+}}{-}\bar{C}HR^2 \xrightarrow{R^3CH=CH_2} \text{pyrrolidine}$$

$$RCH{=}\overset{R^1}{\overset{|}{N^+}}{-}\bar{C}HR^2 \xrightarrow{R^3C\equiv CR^4} \Delta^3\text{-pyrroline}$$

*Scheme 6*

The reactions take place most easily with dipolarophiles bearing electron-withdrawing substituents through a HOMO (dipole) – LUMO (dipolarophile) interaction.[10] Molecular orbital calculations suggest that electron-rich species such as enamines and enol ethers should also react rapidly by LUMO dipole - HOMO dipolarophile interaction,[11] but few reactions of this type with azomethine ylides appear to have been reported.

Azomethine ylides are generated *in situ* and not isolated. Stabilised azomethine ylides with an electron-withdrawing substituent on the carbon atom bearing the negative charge have traditionally been prepared by thermolysis or photolysis of suitably substituted aziridines. Generated in this way in the presence of a dipolarophile they form pyrrolidines or pyrrolines directly. Both the ring-opening of the aziridines and the subsequent cycloaddition are stereospecific, provided that the cycloaddition takes place before bond rotation in the transient azomethine ylide of Scheme 4.[12]

With olefinic dipolarophiles the stereochemistry of substituents on the double bond is maintained in the product of cycloaddition. With mono-substituted olefins the reactions are regioselecctive, giving preferentially 2,4-disubstituted pyrrolidines, for example (12). Reaction of aziridine (13) with the enone (15) in toluene at 175°C gave the pyrrolidines (16) and (17) in 70 per cent yield. Wittig olefination of the main product gave (18) which was converted into the neuroexcitatory amino acid *allo*-kainic acid (19). An attempt to use the same procedure to make kainic acid (20) from the corresponding *cis*-enone was frustrated by its partial isomerisation to the *trans* compound under the reaction conditions.[13]

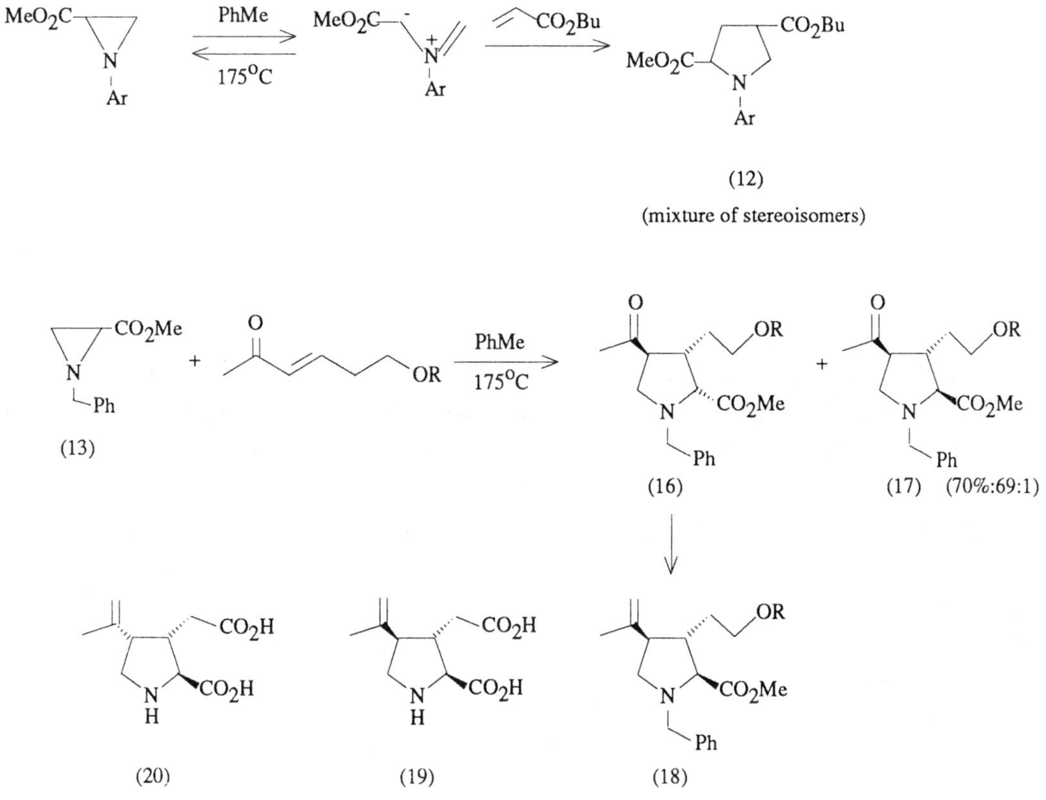

$MeO_2C$ N Ar PhMe 175°C $MeO_2C$ N Ar $CO_2Bu$ $MeO_2C$ N Ar $CO_2Bu$

(12)
(mixture of stereoisomers)

$CO_2Me$ N Ph (13) + O OR PhMe 175°C O OR N $CO_2Me$ Ph (16) + O OR N $CO_2Me$ Ph (17) (70%:69:1)

$CO_2H$ N H $CO_2H$ (20) $CO_2H$ N H $CO_2H$ (19) OR N $CO_2Me$ Ph (18)

*Scheme 7*

Intramolecular reactions take place easily. Thus, in a synthesis of acromelic acid A (25), intramolecular cycloaddition of the azomethine ylide (22), derived from the optically active aziridine (21), gave the trisubstituted pyrrolidine (23) in 73 per cent yield as a single stereoisomer. Further manipulation and epimerisation of the C-2 hydrogen atom gave (24) and thence acromelic acid A (25).[14]

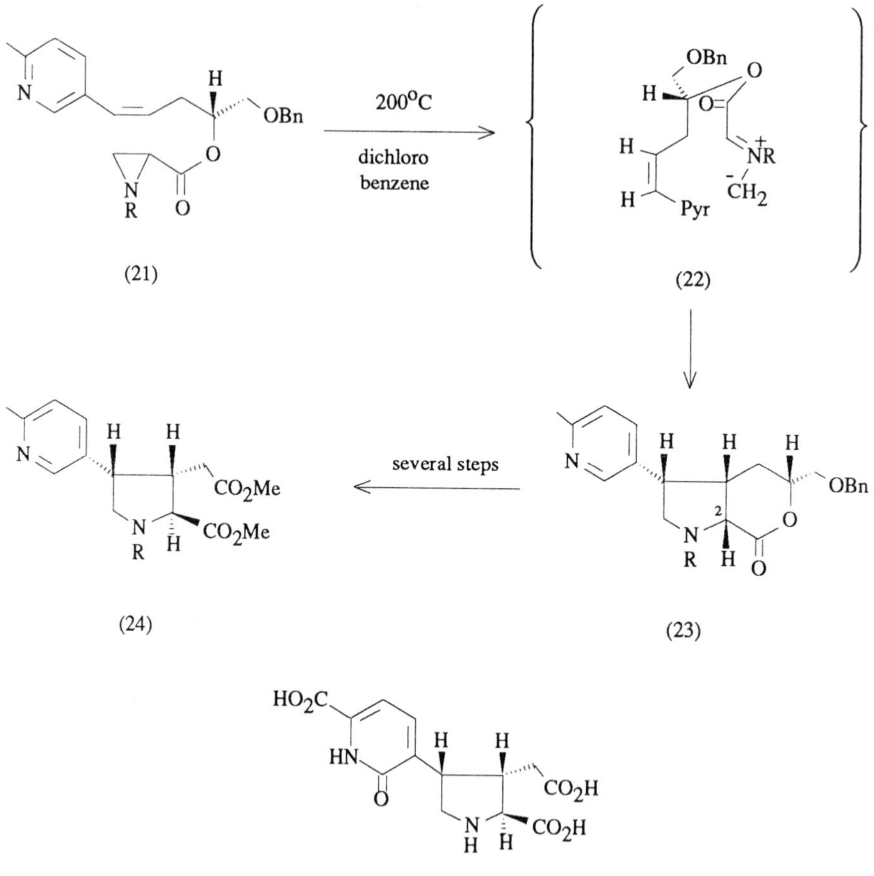

200°C

dichloro benzene

(21)

(22)

several steps

(24)

(23)

(25)

*Scheme 8*

The very high diastereofacial selectivity of the cycloaddition step in this synthesis is in line with the preferential formation of a transition state of the form (22) in which the bulky benzyloxymethyl substituent takes up the least hindered position, leading to the all-*cis* product (23).

A disadvantage of the aziridine route to stabilised azomethine ylides is the high temperatures required. In a valuable alternative, acyl-stabilised azomethine ylides are formed at ordinary temperatures from the tautomeric 4-oxazolines, themselves obtained by reduction of the corresponding oxazolium salts with phenylsilane and caesium fluoride. The resulting 4-oxazolines open spontaneously to the corresponding dipoles (as 28)[15] which can be trapped by suitable dipolarophiles. Thus, treatment of the oxazolium salt (26) with reducing agent in the presence of methyl acrylate gave a mixture of the regioisomeric pyrrolidines (29) and (30) by way of the oxazoline (27) and the dipole (28), with the 2,4-diester (29) in preponderant amount as expected.[16]

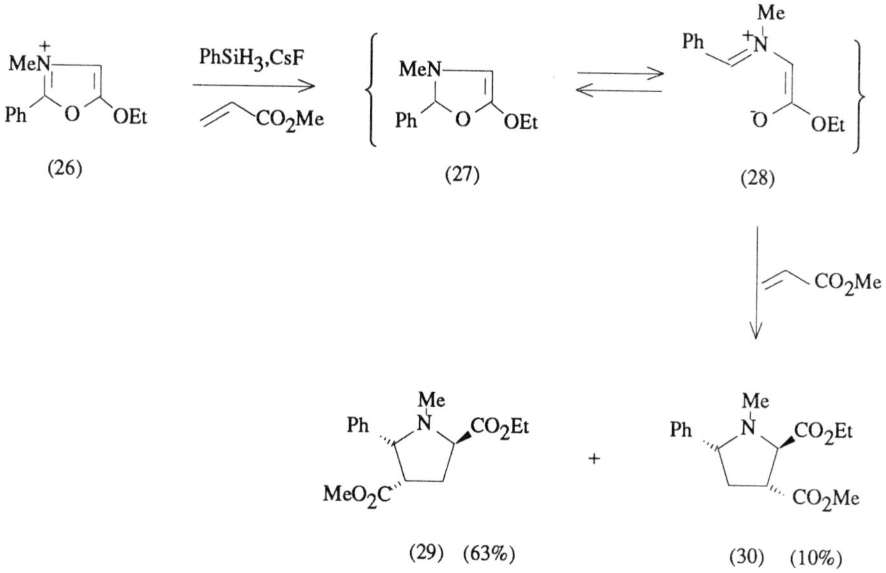

MeN⁺ Ph O OEt (26) — $PhSiH_3$,CsF / $CO_2Me$ → MeN Ph O OEt (27) ⇌ Me Ph N⁺ O⁻ OEt (28)

↓ $CO_2Me$

Me N Ph $CO_2Et$ $MeO_2C$ (29) (63%) + Me N Ph $CO_2Et$ $CO_2Me$ (30) (10%)

*Scheme 9*

This reaction is highly stereoselective, and in this respect the oxazoline route to azomethine ylides is superior to the aziridine route. Although the familiar N-phenylaziridines often give adducts derived by conrotatory ring-opening with high selectivity, analogous N-alkylaziridines frequently form mixtures because of equilibration of the dipole. In addition the oxazoline route makes available a much wider range of acyl-stabilised azomethine ylides.

Azomethine ylides stabilised by ester groups can also be obtained by thermal isomerisation of imines of α-amino acid esters. Thus, the imine (31), heated in toluene in the presence of N-phenylmaleimide gave the adduct (33) stereospecifically in 86 per cent yield, by way of the transient azomethine ylide (32).[17]

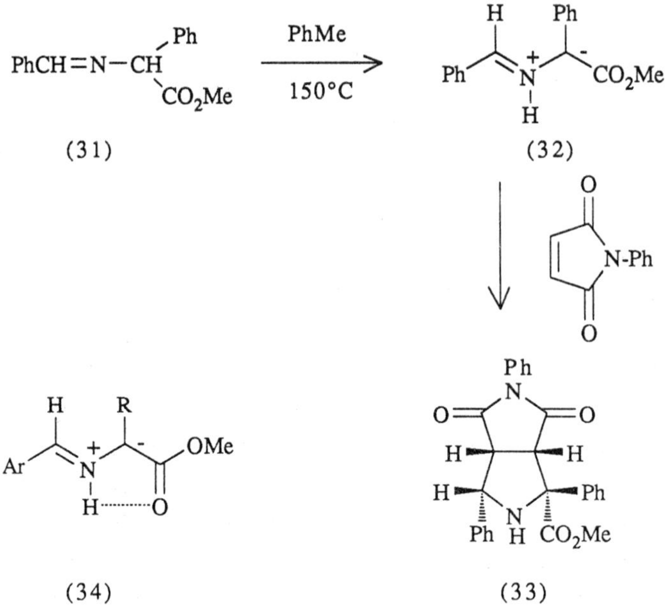

PhCH=N—CH(Ph)CO$_2$Me (31) $\xrightarrow[150°C]{PhMe}$ (32) → (33)

(31) (32) (34) (33)

*Scheme 10*

This is one of a series of reactions involving 1,2-prototropic shifts of the type

$$X{=}Y{-}ZH \rightleftharpoons X{=}\overset{+}{Y}H{-}\overset{-}{Z}$$

*Scheme 11*

which have been widely studied, and provide a route to 1,3-dipoles from imines, oximes and hydrazones.[18] The formation of azomethine ylides from imines is catalysed by Lewis acids and Brönsted acids and this allows many cycloadditions to be effected at room temperature. Dipole formation is generally stereospecific and is believed to involve a hydrogen-bonded form

(34). In some cases stereomutation of the dipole is observed, particularly in slow reactions with unactivated dipolarophiles such as terminal alkenes.

Azomethine ylides generated in this way from imines undergo cycloadditions with a wide range of dipolarophiles, generally *via* an *endo* transition state. An interesting application leads to the formation of dehydroamino acid esters by reaction of imines with ethyl azodicarboxylate.[19] Thus, the imine (35) and ethyl azodicarboxylate gave the dehydro ester derivative (38); there is good evidence for a reaction pathway *via* (36) and (37).

$EtO_2CN{=}NCO_2Et$ + $PhCH{=}N{-}CH(CO_2Me)CHMe_2$ (35) —130°C→ (36)

(36) → (37)

(37) → $EtO_2CN{=}NCO_2Et$ + $PhCH{=}N{-}C(CO_2Me){=}CMe_2$ (38)

*Scheme 12*

Intramolecular reactions take place easily, even with unactivated alkenes which do not react intermolecularly.[20]

A related sequence employs azomethine ylides derived from aldehydes and esters of N-alkylamino acids.[21] Thus, the aldehyde (39), heated in xylene with the ethyl ester of sarcosine gave stereospecifically the tricyclic product (40) in 40 per cent yield. Decarboxylation then led to the *Sceletium* alkaloid (41).

A better procedure, which leads directly to the decarboxylated product, employs the trimethylsilyl ester of sarcosine instead of the ethyl ester. Decarboxylation of the derived cyclic intermediate (42) affords the non-stabilised azomethine ylide (43) which adds readily to the internal double bond.

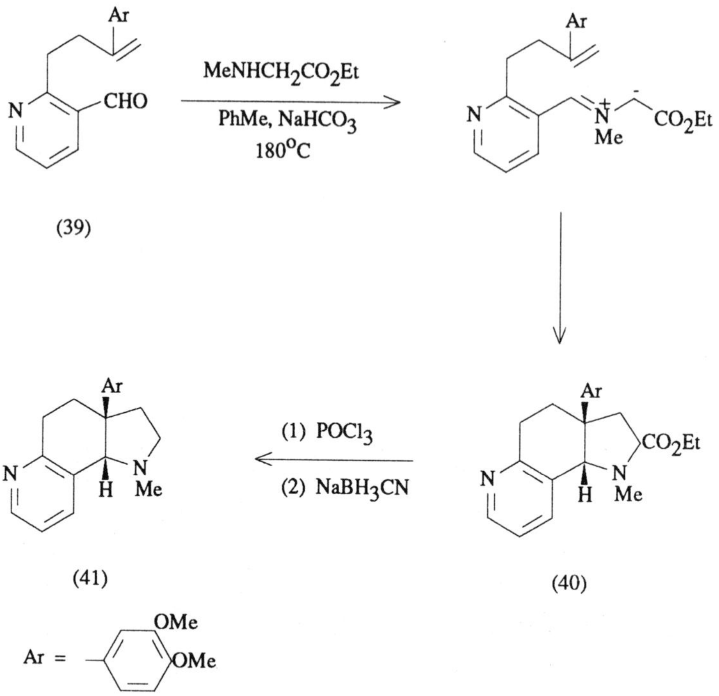

Ar
CHO
N
(39)

$MeNHCH_2CO_2Et$

PhMe, $NaHCO_3$

180°C

Ar
N
$\overset{+}{N}$
Me
$CO_2Et$

Ar
N
N
H Me
(41)

(1) $POCl_3$

(2) $NaBH_3CN$

Ar
$CO_2Et$
N
N
H Me
(40)

Ar = OMe OMe

*Scheme 13*

Ar
N
O
O
:N
Me
(42)

Ar
N
Me $N^+$ $CH_2^-$
(43)

*Scheme 14*

This procedure was employed in a highly stereoselective synthesis of (±)-lycorine (46).[22] Heating the aldehyde (44) in boiling toluene with N-benzylglycine which had been pretreated with hexamethyldisilazane gave the single stereoisomer (45) by way of the transition state (47). This was converted into lycorine by debenzylation and reaction with formaldehyde.

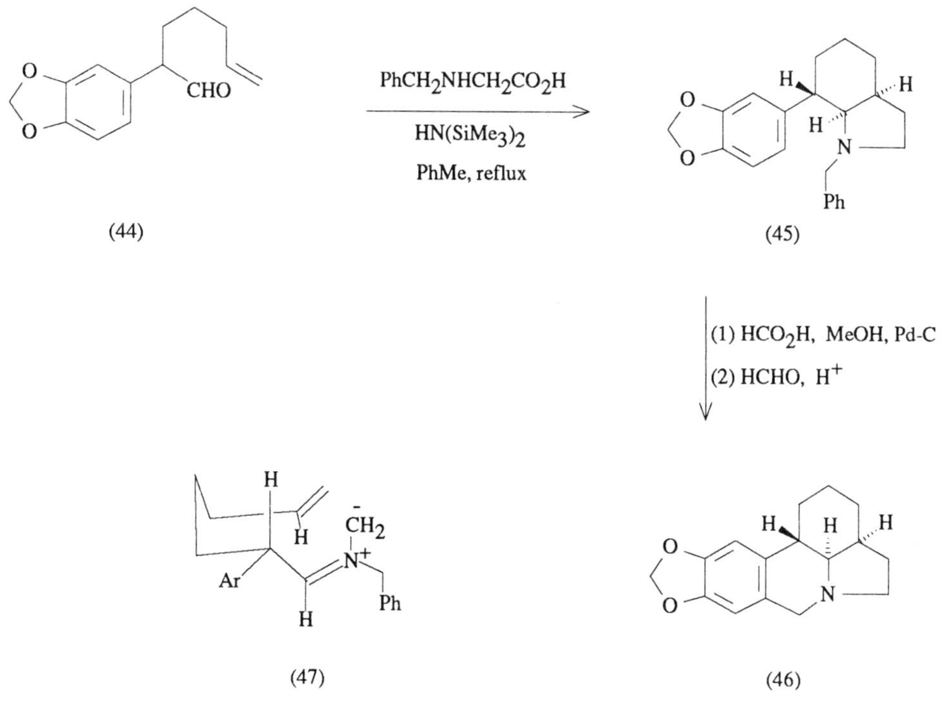

$PhCH_2NHCH_2CO_2H$

$HN(SiMe_3)_2$

PhMe, reflux

(44) (45)

(1) $HCO_2H$, MeOH, Pd-C

(2) HCHO, $H^+$

(47) (46)

*Scheme 15*

*Non-stabilised* azomethine ylides have not been so readily available as the stabilised analogues, but recently a number of methods have been developed for generating them *in situ*, several of them based on the desilylation of trimethylsilylmethyl imminium salts.[23] Thus, the imine (48) and trimethylsilylmethyl triflate in acetonitrile gave the salt (49); without isolation this was treated with caesium fluoride and dimethyl acetylenedicarboxylate and gave the pyrrolidine derivative (51) in 70 per cent yield by way of the azomethine ylide (50).[24]

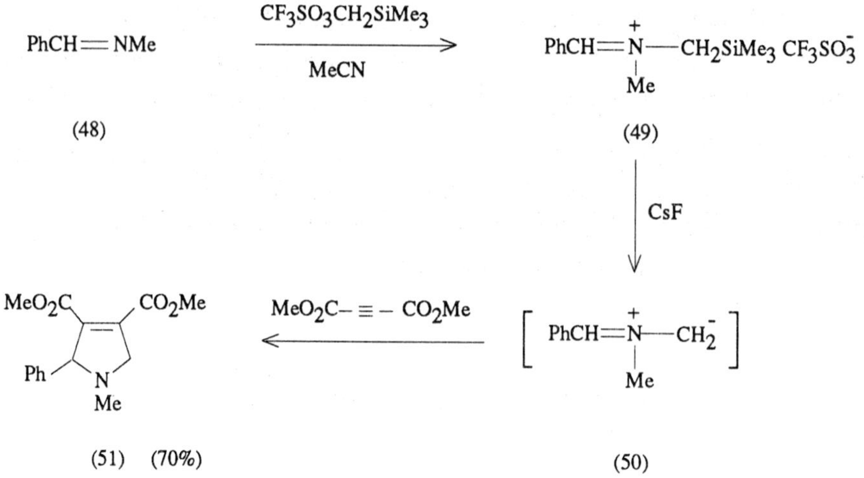

PhCH=NMe (48) $\xrightarrow[\text{MeCN}]{CF_3SO_3CH_2SiMe_3}$ $PhCH=\overset{+}{N}(Me)—CH_2SiMe_3\ CF_3SO_3^-$ (49)

(49) $\xrightarrow{\text{CsF}}$ [ $PhCH=\overset{+}{N}(Me)—CH_2^-$ ] (50)

(50) $\xrightarrow{MeO_2C—\equiv—CO_2Me}$ (51) (70%)

(51): $MeO_2C$, $CO_2Me$, Ph, N, Me

*Scheme 16*

OBn, O, N, $SiMe_3$ $\xrightarrow{CF_3SO_3Me}$ (52): OBn, OMe, $^+N$, $SiMe_3$ $\xrightarrow{\text{CsF}}$ [ OBn, OMe, $^+N$, $CH_2^-$ ]

$\xrightarrow{=\!\!\!\diagup CO_2Me}$ (53): BnO, OMe, $CO_2Me$, N

(53) ⟶ (54) (51%): OBn, $CO_2Me$, N

(54) $\xrightarrow{H_2,\ Pd}$ (55): BnO, H, $CO_2Me$, N

(55) ⟶ (56): HO, H, $CH_2OH$, N

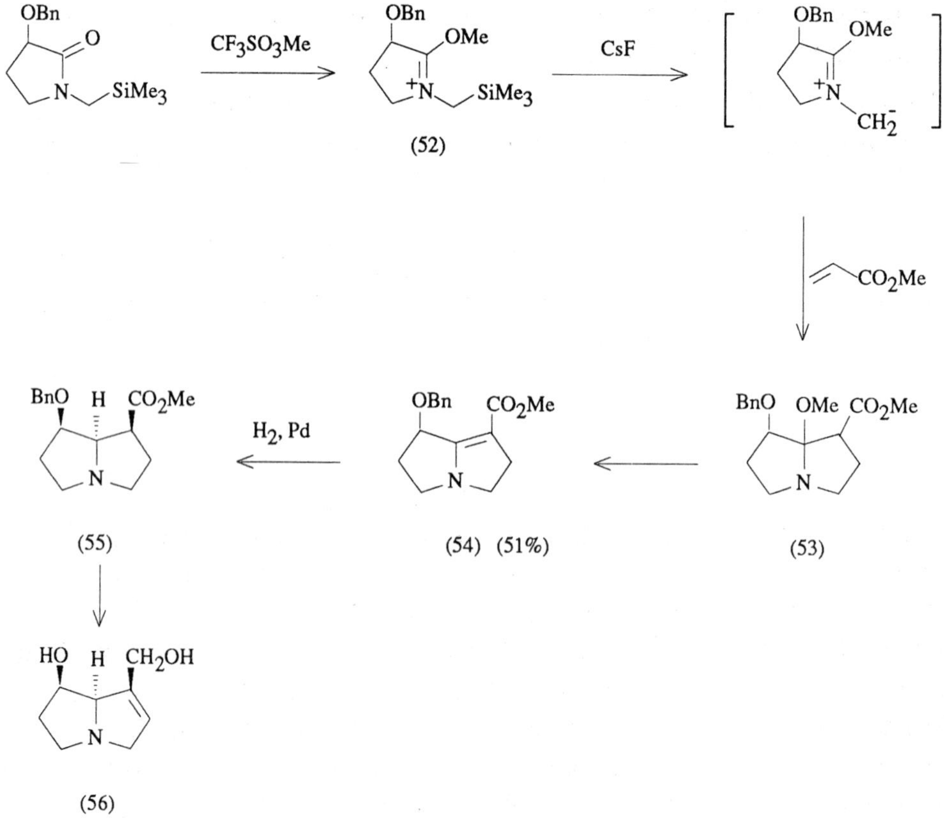

*Scheme 17*

This procedure has been employed in the synthesis of pyrrolizidine bases from non-stabilised imidate ylides prepared from the appropriate lactam.[25] Thus, the crude salt (52) treated with methyl acrylate and caesium fluoride gave (54) directly by loss of methanol from the initial cycloadduct (53). Catalytic reduction of (54) gave (55) specifically, which was transformed by standard reactions into the pyrrolizidine base retronecine (56) (Scheme 17).

The desilylation procedure provides a route to the parent azomethine ylide (58) from the readily accessible silane (57), and this has been employed in the stereoselective synthesis of 3,4-disubstituted pyrrolidines by trapping with activated dipolarophiles.[26]

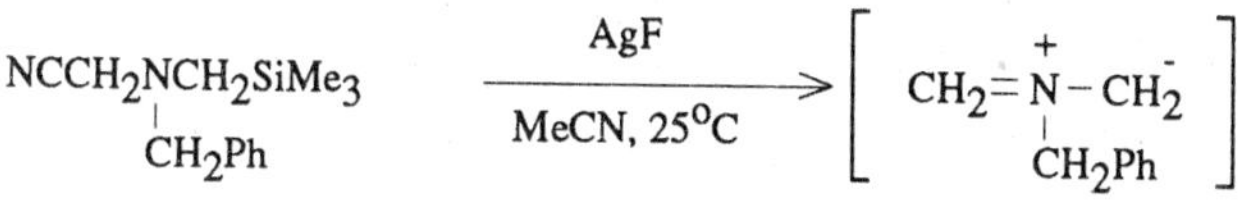

$NCCH_2N(CH_2Ph)CH_2SiMe_3$ (57) $\xrightarrow[\text{MeCN, 25}^{\circ}\text{C}]{\text{AgF}}$ $[CH_2{=}\overset{+}{N}(CH_2Ph){-}\overset{-}{C}H_2]$ (58)

*Scheme 18*

Intramolecular versions of the reaction have been used in the synthesis of physostigmine[27] and erythrinane.[28]

Non-stabilised azomethine ylides are also produced by decarboxylation of imminium ions derived from primary and secondary α-amino acids,[29]

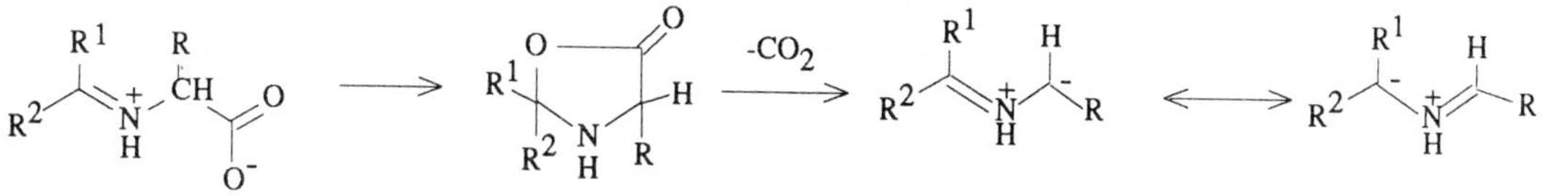

(59) $-CO_2$

*Scheme 19*

Decarboxylation to give the azomethine ylide is believed to occur through an intermediate oxazolidin-5-one (as 59), which loses carbon dioxide in a 1,3-dipolar cycloreversion to generate the ylide stereospecifically.[30] In the presence of a dipolarophile cycloadducts are obtained. Thus alanine heated with benzaldehyde and N-phenylmaleimide in dimethylformamide gave the *endo* and *exo* adducts (60) and (61) along with the stereoisomer (62); the latter arises by addition after bond rotation in the azomethine ylide. In an intramolecular example, the aldehyde (63) heated with sarcosine in dimethylformamide, gave the cycloadduct (64) stereospecifically.[31]

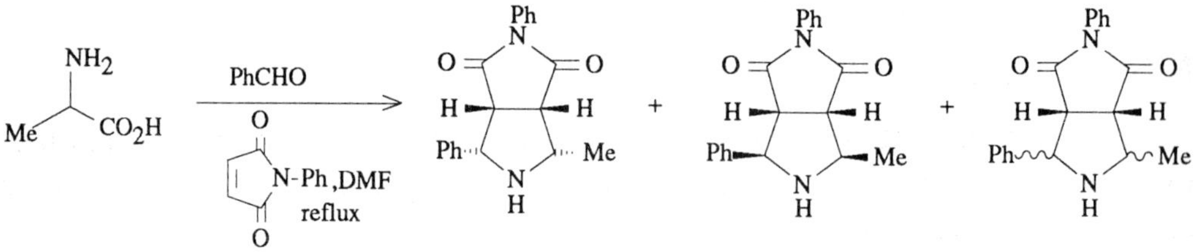

$NH_2$ Me $CO_2H$ → PhCHO, N-Ph-maleimide, DMF, reflux → (60) + (61) + (62) [Ph, N, O, H, Ph, Me, N, H]

(60) (61) (62)

CHO, O → $MeNHCH_2CO_2H$ / DMF, 100°C → MeN, H, H, O

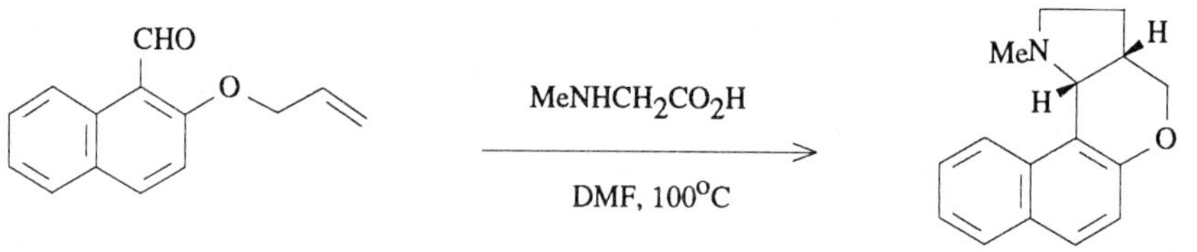

(63) (64) (44%)

*Scheme 20*

In yet another procedure a combination of metal salt and triethylamine effects regio- and stereo-specific, or highly stereoselective cycloaddition of imines of α-amino acid esters to a range of dipolarophiles at room temperature, probably by way of metallo-1,3-dipole formation.[32] (Scheme 21)

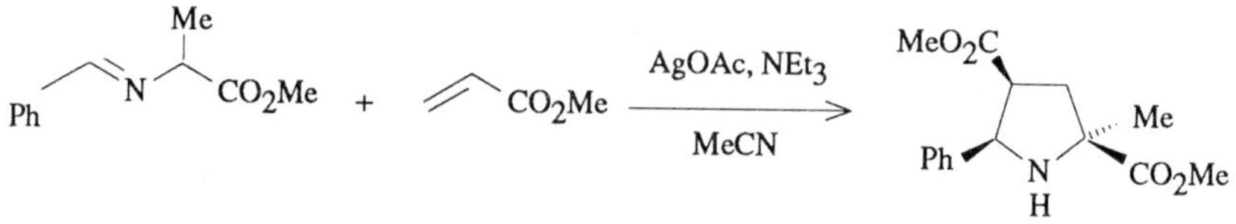

Ph, N, Me, $CO_2Me$ + $CO_2Me$ → AgOAc, $NEt_3$ / MeCN → $MeO_2C$, Ph, N, H, Me, $CO_2Me$

(80%)

*Scheme 21*

$R-C\equiv\overset{+}{N}-\overset{-}{O}$ ⟷ $R-\overset{-}{C}=\overset{+}{N}=O$

$R^1$ (alkene) ↙ ↘ $R-\equiv$

R, N, O, 1, 2, 3, 4, 5, $R^1$ R, N, O, $R^1$

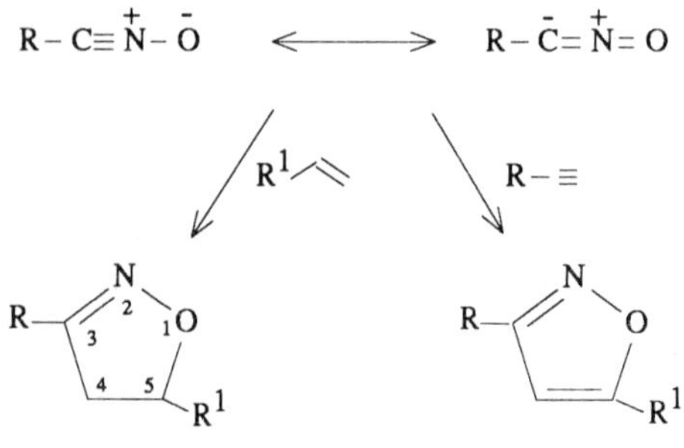

*Scheme 22*

**Nitrile Oxides**

Nitrile oxides are reactive 1,3-dipoles which readily form $\Delta^2$-isoxazolines on reaction with ethylenic dipolarophiles; acetylenic dipolarophiles give isoxazoles (Scheme 22).[33]

With monosubstituted olefins reaction gives exclusively or predominantly the 5-substituted isoxazoline, whatever the nature of the substituent on the dipolarophile. Reactions with electron-rich or conjugated dipolarophiles are controlled by the LUMO (dipole) - HOMO (dipolarophile) interaction, and union of the atoms with the larger coefficients leads to 5-substituted isoxazolines. Electron-deficient dipolarophiles also react rapidly due to the influence of both HOMO (dipole) and LUMO (dipole) interactions; the latter still predominates and the 5-substituted isoxazolines are again the main products.

Isoxazolines with a substituent at C-4, which cannot be obtained directly by 1,3-dipolar cycloaddition, can sometimes be prepared from the preformed isoxazoline by reaction of the derived C-4 carbanion with an appropriate electrophile.[34]

Cycloaddition of nitrile oxides to acetylenic dipolarophiles leads directly to isoxazoles. These also serve as useful building blocks in synthesis through chemical modification and ring cleavage, and they have been employed in the synthesis of natural products.[35] With mono-substituted electron-deficient acetylenic dipolarophiles the HOMO (dipole) - LUMO (dipolarophile) interaction becomes more important and some of the 4-substituted isoxazole is generally produced.[36]

Nitrile oxides are usually prepared *in situ* and not isolated, although stable isolable examples have been obtained.[37] Generated in the presence of a dipolarophile they form the cycloadduct directly, often in high yield. Two general methods have been used for their preparation - dehydrochlorination of hydroximoyl chlorides with triethylamine[38] and dehydration of primary nitro compounds with an aryl isocyanate.[39] They have also been conveniently obtained by oxidation of aldoximes.[40]

Cycloaddition of nitrile oxides to olefinic and acetylenic dipolarophiles is a good method for the preparation of isoxazolines and isoxazoles, but the real value of the reactions in synthesis lies in the access it provides to acyclic compounds by cleavage of the isoxazoline ring. Since the initial cycloadditons are usually highly regio- and stereo-selective, the sequence of reactions - cycloadditon, manipulation of functional groups and ring cleavage of the isoxazoline - provides a valuable stereocontrolled route to a variety of acyclic compounds including $\alpha\beta$-unsaturated ketones, $\beta\gamma$-hydroxyketones and $\gamma$-amino-alcohols[41] (Scheme 23).

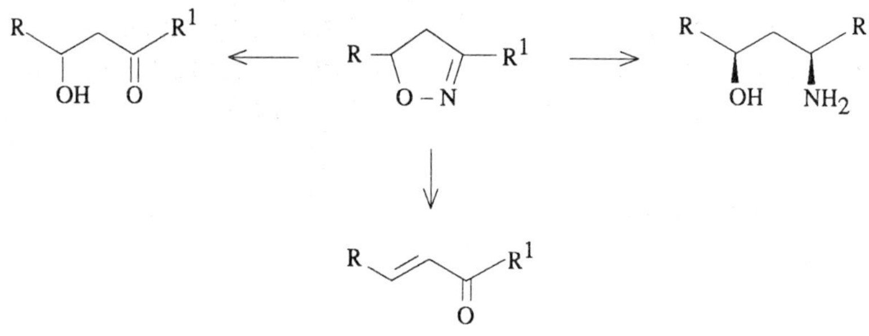

Scheme 23

Thus, decarboxylative ring-opening of the 3-carboxyisoxazoline prepared by cycloaddition of carbethoxyformonitrile oxide to an olefin, leads stereospecifically to a β-hydroxynitrile, formed, in effect, by *cis*-cyanohydroxylation of the double bond of the olefin. *Trans*-2-butene, for example, gave the adduct (66) and thence the hydroxynitrile (67), while *cis*-2-butene gave the isomeric hydroxynitrile (68).[42] The nitrile oxide was generated *in situ* from the readily available chloro-oxime (65). The decarboxylative ring opening is easily effected by heating the carboxylic acid at its melting point.

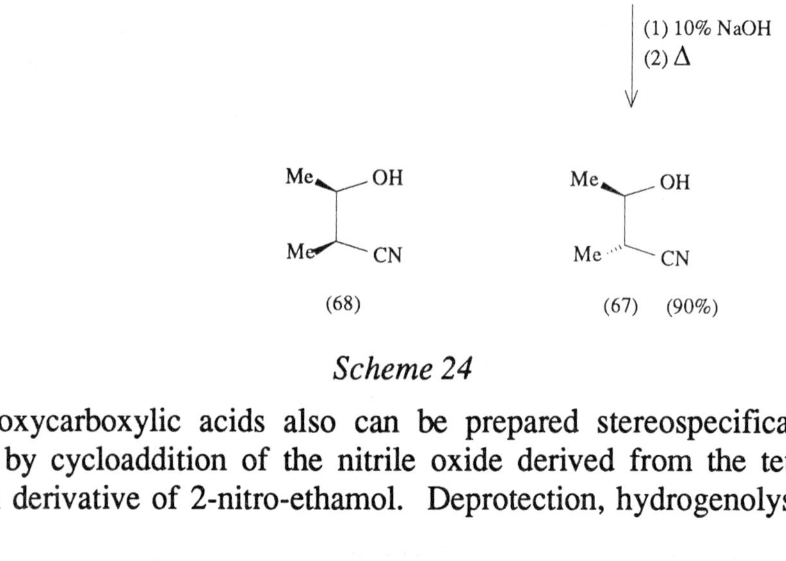

Scheme 24

β-Hydroxycarboxylic acids also can be prepared stereospecifically from olefins by cycloaddition of the nitrile oxide derived from the tetrahydro-pyranyl derivative of 2-nitro-ethamol. Deprotection, hydrogenolysis of the isoxazoline ring (see later) and oxidative cleavage of the α-hydroxyketone obtained yields the stereochemically pure β-hydroxy carboxylic acid (for example (69)) (Scheme 25).[42] In other words, cycloaddition reactions of nitrile oxides can be used to functionalise olefinic bonds stereospecifically.

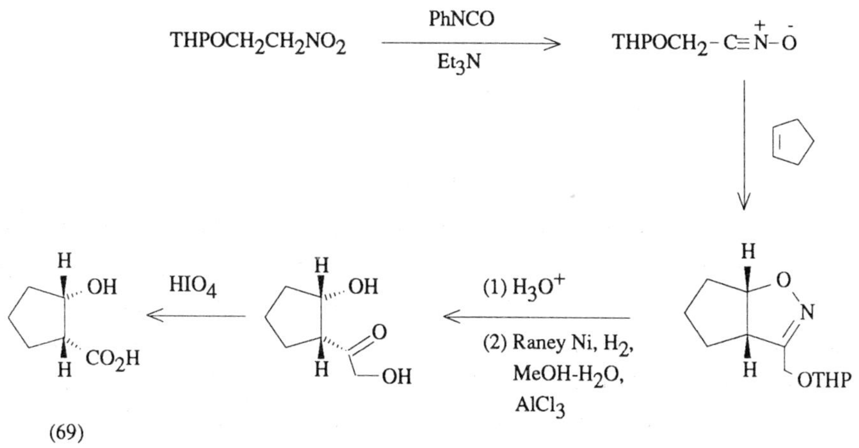

THPOCH$_2$CH$_2$NO$_2$ —(PhNCO, Et$_3$N)→ THPOCH$_2$–C≡N$^+$–O$^-$ —(cyclopentene)→ isoxazoline —(1) H$_3$O$^+$; (2) Raney Ni, H$_2$, MeOH-H$_2$O, AlCl$_3$→ α-hydroxyketone —(HIO$_4$)→ (69)

*Scheme 25*

Cleavage of the isoxazoline to form a β-hydroxy aldehyde has been employed in the synthesis of sugars.[43] Thus, the adduct (70) from formonitrile oxide and butadiene was selectively converted into the diol (71) with osmium tetroxide and hydrogen peroxide. Hydrogenolysis of the isoxazoline with hydrogen and Raney nickel then led smoothly to 2-deoxyribose (73) by way of the β-hydroxy-aldehyde (72).[44]

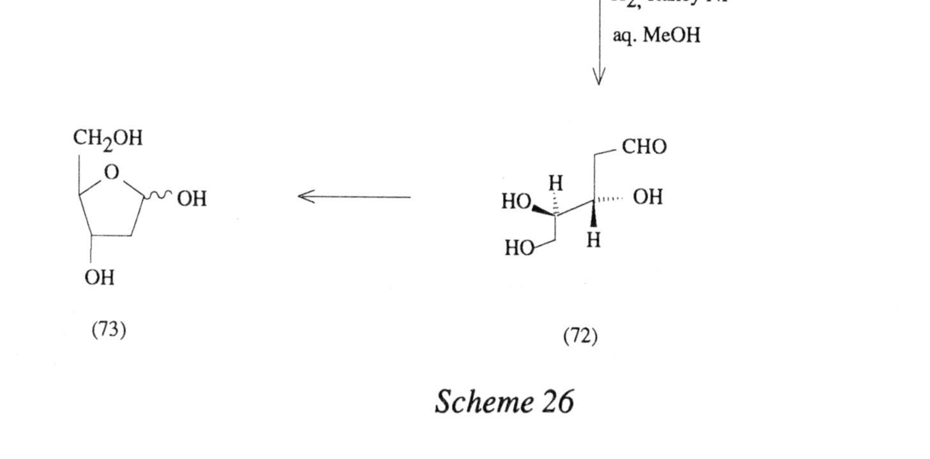

butadiene —(MeNO$_2$, PhNCO)→ (70) —(OsO$_4$, H$_2$O$_2$)→ (71) (80%) + (20%)

(71) —(H$_2$, Raney Ni, aq. MeOH)→ (72) → (73)

*Scheme 26*

Hydrogenolysis of isoxazolines to β-hydroxy aldehydes or ketones, as in the conversion of (71) into (72), takes place by way of the corresponding β-hydroxy imine which is hydrolised to the carbonyl compound under the reaction conditions. A variety of conditions has been used; good results are usually obtained with hydrogen and Raney nickel in buffered aqueous methanol.[45] This sequence, in which a preformed isoxazoline ring serves as the synthetic equivalent of a β-hydroxycarbonyl compound, has been used in the synthesis of a number of natural products containing a β-hydroxy carbonyl group or the derived αβ-unsaturated carbonyl group, and also in a valuable alternative procedure for effecting directed aldol condensations. Thus, generation of the nitrile oxide (74) in the presence of methyl vinyl ketone gave the isoxazoline (75) in 92 per cent yield. The hydroxy-1,4-diketone was then unmasked by reductive hydrolysis and was cyclised to (±)-dihydrocinerolone (76).[46] A number of other γ-hydroxycyclopentenones have been synthesised in the same way.

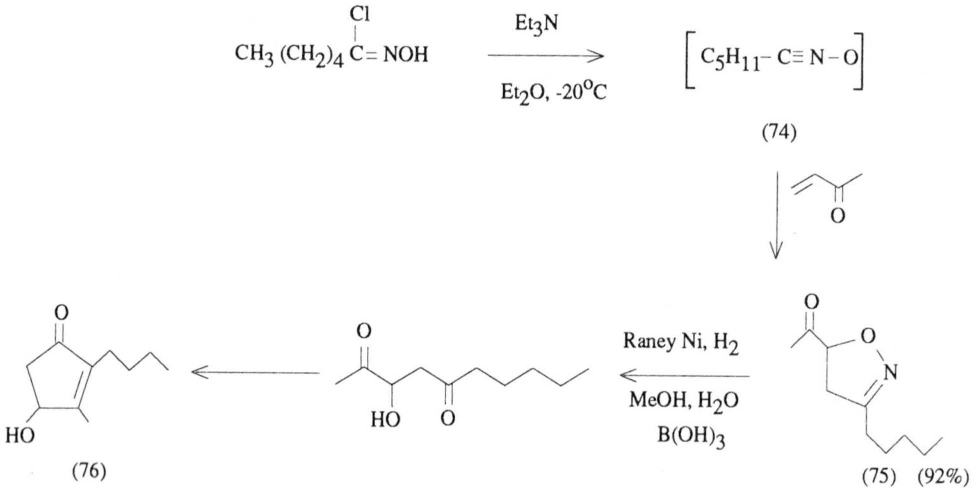

$CH_3(CH_2)_4C(Cl){=}NOH$ → ($Et_3N$, $Et_2O$, -20°C) → $[C_5H_{11}{-}C{\equiv}N{-}O]$ (74)

(74) → (methyl vinyl ketone) → (75) (92%)

(75) → (Raney Ni, $H_2$; MeOH, $H_2O$, $B(OH)_3$) → hydroxy-1,4-diketone → (76)

*Scheme 27*

In an intramolecular example, the ω-nitroalkene (77) was converted into the nitrile oxide (78) which cyclised spontaneously to the fifteen-membered macrocycle (79). Reductive hydrolysis and dehydration of the aldol gave the αβ-unsaturated ketone (80) which was converted into (±)-muscone with lithium dimethylcuprate.[47]

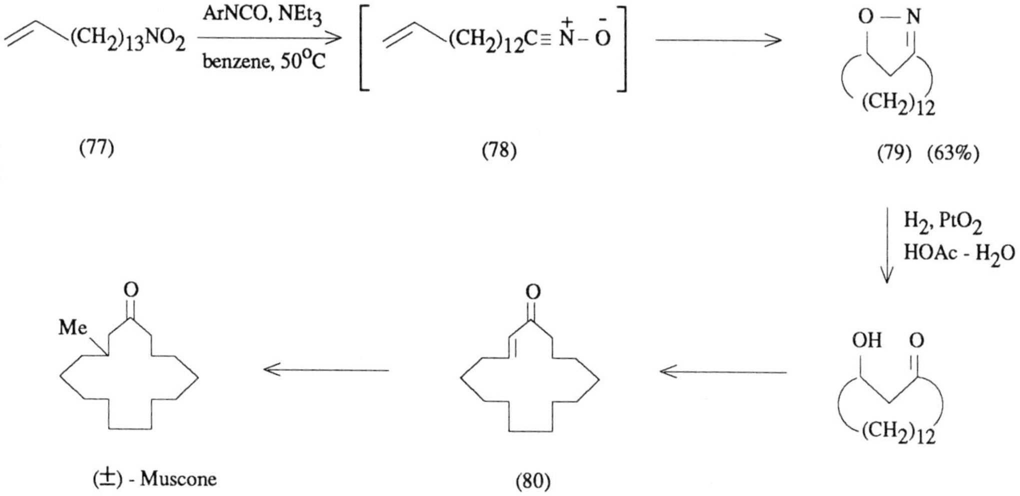

(CH$_2$)$_{13}$NO$_2$ ArNCO, NEt$_3$ benzene, 50°C (CH$_2$)$_{12}$C≡$\overset{+}{N}$–$\overset{-}{O}$ O — N (CH$_2$)$_{12}$

(77) (78) (79) (63%)

H$_2$, PtO$_2$
HOAc - H$_2$O

O Me O OH O (CH$_2$)$_{12}$

(±) - Muscone (80)

*Scheme 28*

$\overset{-}{O}$–$\overset{+}{N}$≡CCO$_2$Et
Et$_2$O, 25°C

O H O H CO$_2$Et H O – N + O H H CO$_2$Et O – N

(81) (82) (83) (71%:4:1)

(1) NaOH, EtOH
(2) H$_3$O$^+$
(3) CH$_2$N$_2$

HO O OH OH 2-Deoxy-D-ribose

(1) H$_3$O$^+$
(2) DIBAL

O H O CO$_2$Me OH

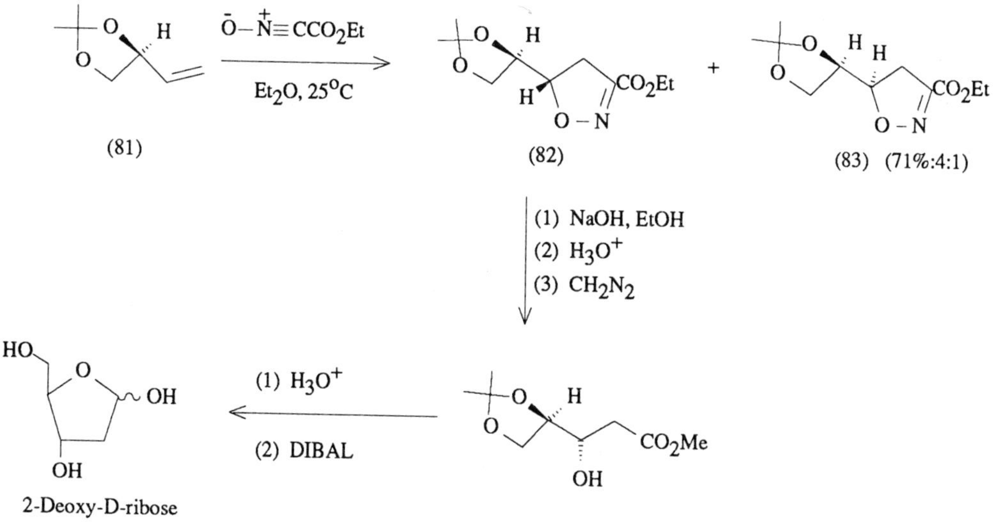

*Scheme 29*

Enantioselective syntheses of aldols have been effected using optically active alkenes as dipolarophiles. These reactions are particularly effective where the alkene contains an allylic alkoxy group. Thus, reaction of (+)-(*S*)-isopropylidene-3-buten-1,2-diol (81) and carbethoxyformonitrile oxide gave a 4:1 mixture of the *erythro* and *threo* isoxazolines (82) and (83) in 71 per cent yield at room temperature. The major, *erythro*, product was isolated by chromatography and in a few additional steps was converted into optically pure 2-deoxy-D-ribose, *via* decarboxylative ring opening of the isoxazoline ring to a β-hydroxynitrile and hydrolysis to the carboxylic acid.[48]

A number of other nitrile oxides reacted with (81) in the same sense and with a similar degree of diastereoselection.[49]

With simple allylic ethers cycloaddition to nitrile oxides again leads preferentially to *erythro* products, but only the t-butyldimethylsilyl ethers give a degree of diastereoselection comparable to that obtained with the cyclic acetals (81). Selectivity increases with increase in the size of the alkyl group attached to the allylic chiral centre.[50] (Scheme 30)

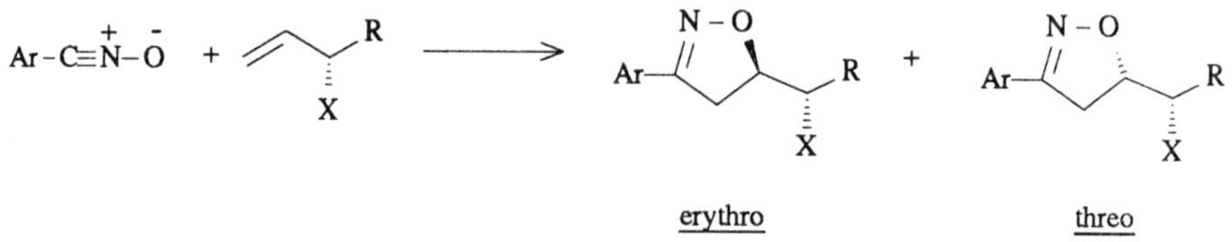

| | erythro | threo |
|---|---|---|
| Ar = Ph, R = Me, X = $OSiMe_3$ | 71 | 29 |
| Ar = p-$NO_2C_6H_4$, R = t-Bu, X = $OSiMe_3$ | >95 | <5 |

*Scheme 30*

Very modest diastereoselection is observed when there is little to distinguish the allylic groups on a steric or electronic basis.[51]

The preferential formation of *erythro* adducts in the cycloaddition of nitrile oxides to allylic ethers and acetals is consistent with addition of the dipole to the less hindered face of the double bond in that conformation of the olefin in which the allylic alkoxy group is in an "inside" position (cf. 84). Approach of the dipole antiperiplanar to the alkyl substituent, as in (85) for addition to an acetal and (86) for addition to an allyl ether, gives the *erythro* product.[52]

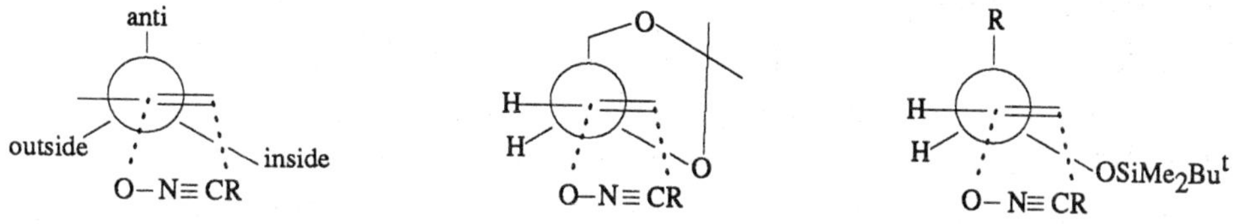

anti
outside
inside
O–N≡CR

O
H
H
O
O–N≡CR

R
H
H
OSiMe$_2$Bu$^t$
O–N≡CR

(84) (85) (86)

*Scheme 31*

Diastereoselection is less in cycloadditions of nitrile oxides to olefins in which the chiral allylic position bears only alkyl groups, although the *erythro* product is still preferred.[53] The best results have been obtained in intramolecular reactions.[54]

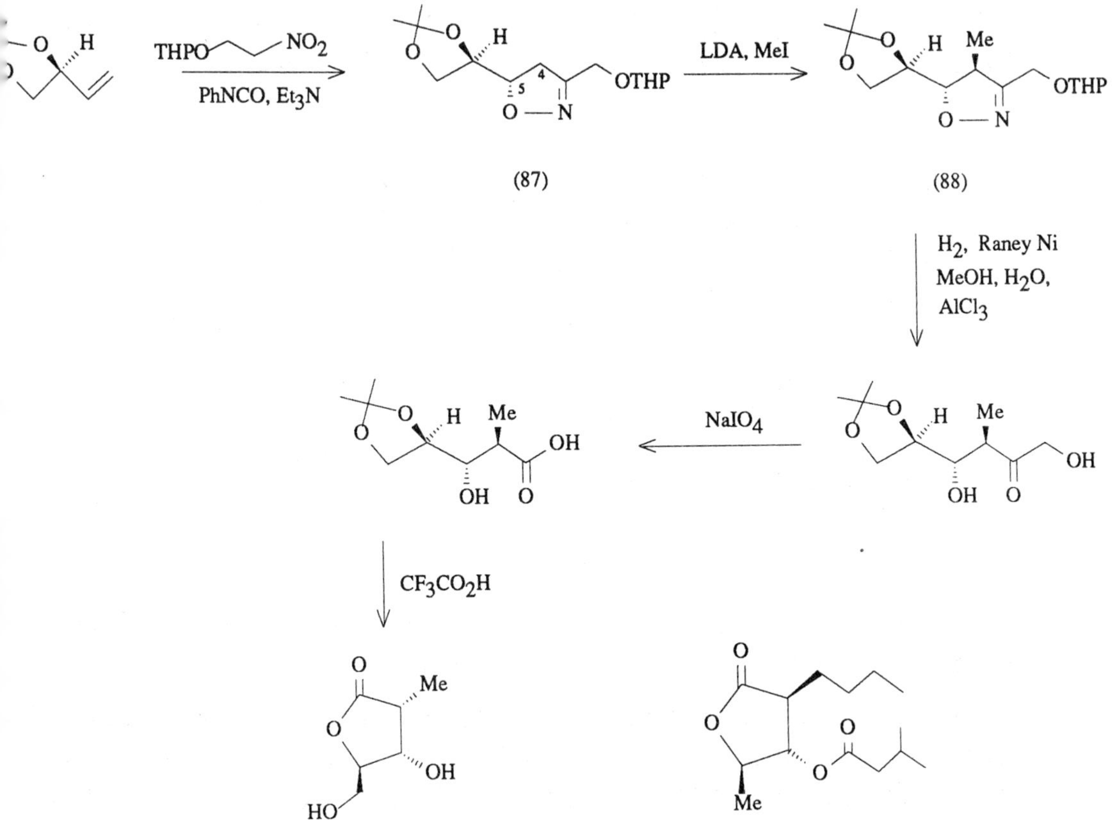

THPO NO$_2$
PhNCO, Et$_3$N
OTHP
LDA, MeI
(87) (88)
H$_2$, Raney Ni
MeOH, H$_2$O,
AlCl$_3$
NaIO$_4$
CF$_3$CO$_2$H
(89) (90)

*Scheme 32*

Attempts to prepare stereochemically pure 4,5-disubstituted isoxazolines by cycloaddition of nitrile oxides to allylic ethers with a substituent on the terminal carbon atom of the double bond have been less successful because of the reduced diastereoselection and uncertian regioselection in reactions with 1,2-disubstituted olefins. However, this difficulty can be circumvented by taking advantage of the regio- and stereo-selective alkylation at C-4 of isoxazolines with an alkoxymethyl substituent at C-5.[55] Thus, the isoxazoline (87) gave the single isomer (88) by methylation *anti* to the C-5 substituent, and thence, in a series of steps, the optically active lactone (89). A similar sequence was used to synthesise (±)-blastmycinone (90) a degradation product of the antibiotic antimycin A3 (Scheme 32).[56]

Attempts to achieve enantioselective syntheses by way of isoxazolines prepared from optically active nitrile oxides have so far been less successful than those using optically active dipolarophiles.[57] Thus, the nitrile oxide (92), derived from the optically active nitro compound (91), formed cycloadducts with several olefins with diastereoselectivities of only 1.5-3.0:1. The main product (93) from the reaction of (92) with *cis*-but-2-ene was converted into the optically pure β-hydroxycarboxylic acid (94), and the sequence provides a useful route to optically active compounds of this class.[58]

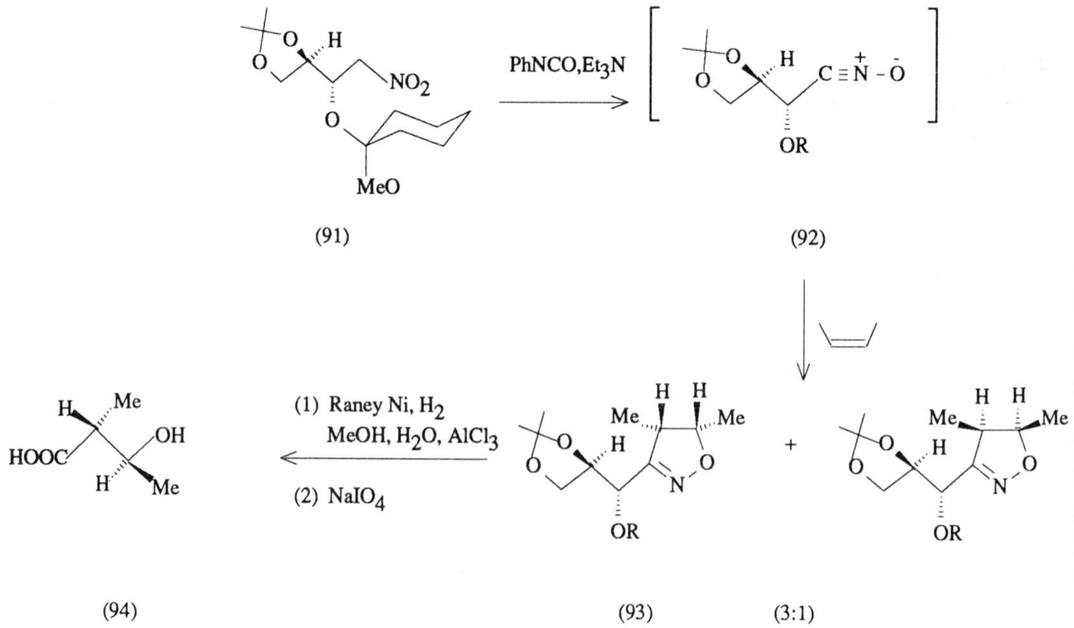

PhNCO,$Et_3N$

(91) (92)

(1) Raney Ni, $H_2$
MeOH, $H_2O$, $AlCl_3$

(2) $NaIO_4$

(94) (93) (3:1)

*Scheme 33*

The two-step sequence of isoxazoline formation followed by ring cleavage to a β-hydroxy ketone provides a valuable stereocontrolled alternative for the preparation of crossed aldols. It has the advantage over the straightforward aldol reaction in that it takes place under mild, more or less neutral conditions, and it complements the normal aldol reaction in that the new carbon-carbon bond in the aldol produced is formed between C-1 and C-2 rather than between C-2 and C-3 (cf.95).

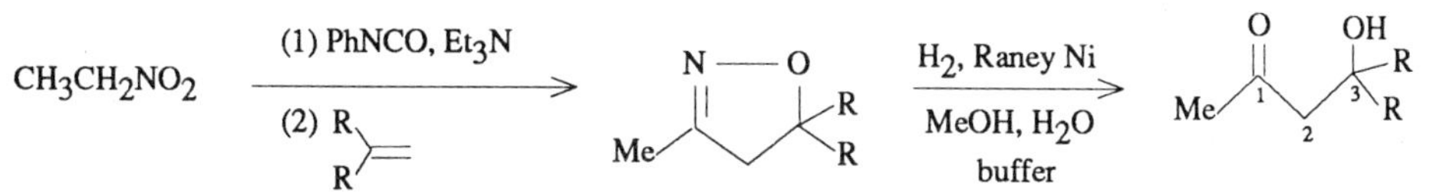

$CH_3CH_2NO_2$ —(1) PhNCO, $Et_3N$; (2) $R_2C{=}CH_2$→ isoxazoline —$H_2$, Raney Ni; MeOH, $H_2O$ buffer→ (95)

*Scheme 34*

Further, since the cycloadditions are usually stereospecific as regards the substituents on the double bond of the dipolarophile, the sequence makes possible the diastereospecific formation of 2,3-dialkylated aldols from 1,2-disubstituted alkenes. In principle, the relative stereochemistry of substituents at C-2 and C-3 of the aldol product depends only on the stereochemistry of the olefin and conditions have been established which avoid epimerisation during the hydrolytic cleavage of the isoxazoline ring. Thus, reaction of the nitrile oxide (96) with *trans*-but-2-ene gave the isoxazoline (97) specifically, and on hydrolytic cleavage this formed the β-hydroxy ketone (98) with less than 2 per cent epimerisation. Under the same conditons *cis*-but-2-ene gave the isomeric β-hydroxy ketone (99).[59]

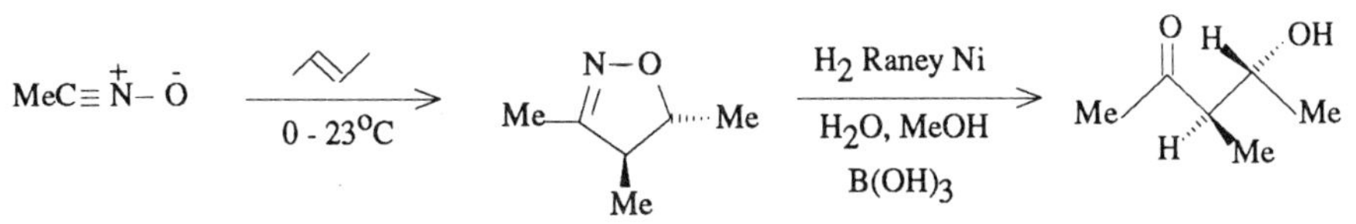

$MeC{\equiv}\overset{+}{N}{-}\overset{-}{O}$ (96) —trans-but-2-ene, 0 - 23°C→ (97) —$H_2$ Raney Ni; $H_2O$, MeOH; $B(OH)_3$→ (98) (83%)

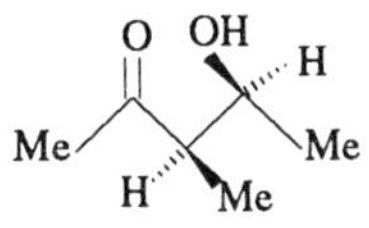

(99)

*Scheme 35*

Intermolecular addition to unsymmetrical 1,2-disubstituted olefins has limitations in synthesis because of the uncertain regioselectivity, but in intramolecular reactions the regiochemistry is controlled by the length of the connecting chain and excellent selectivity is often achieved. Thus, the nitro-olefin (100) gave the cycloadduct (101) specifically in excellent yield and thence the aldol (102). The latter is formally the product of a diastereo-selective directed aldol condensation between two ketones, a transformation difficult to achieve by a straightforward aldol procedure.[59,60]

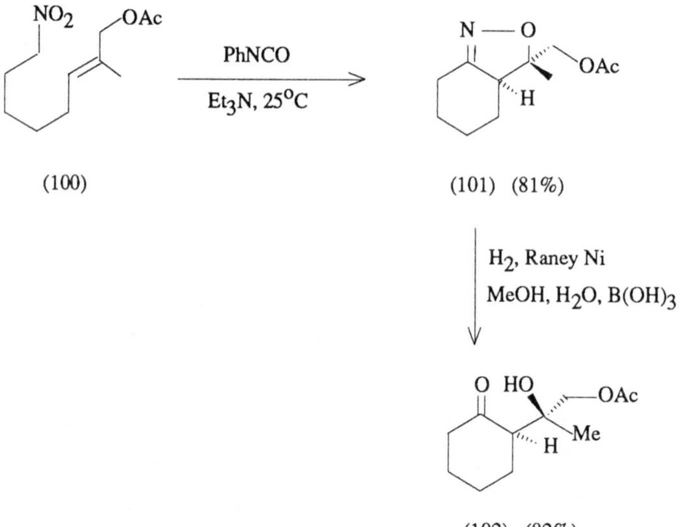

$NO_2$ OAc

PhNCO

$Et_3N$, 25°C

N — O OAc H

(100)

(101) (81%)

$H_2$, Raney Ni

MeOH, $H_2O$, $B(OH)_3$

O HO OAc Me H

(102) (82%)

*Scheme 36*

Aldols containing three contiguous chiral centres can be obtained by alkylation at C-4 of the intermediate isoxazoline. Thus, methylation of the main product (104) from the cycloaddition of benzonitrile oxide to the olefin (103) gave largely (105) and thence the β-hydroxy ketone (106) (Scheme 37).[61]

Another useful conversion of the isoxazolines is their reductive cleavage to 1,3-amino-alcohols. Lithium aluminium hydride is the best reagent, producing 1,3-amino-alcohols in excellent yield with good stereoselectivity (Scheme 38).[62]

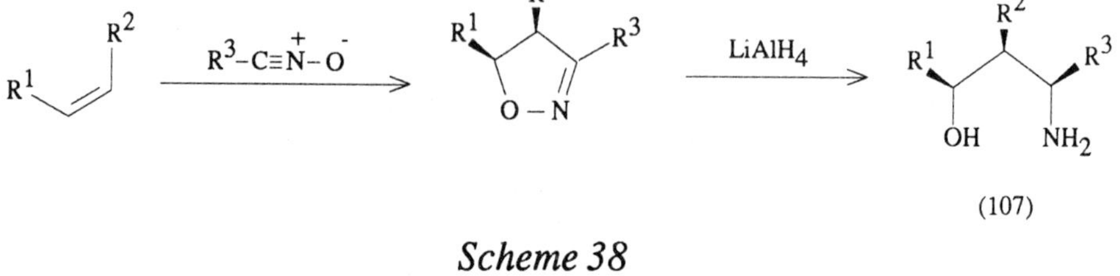

$R^1$ $R^2$

$R^3-C\equiv\overset{+}{N}-\overset{-}{O}$

$R^1$ $R^2$ $R^3$ O – N

$LiAlH_4$

$R^1$ $R^2$ $R^3$ OH $NH_2$

(107)

*Scheme 38*

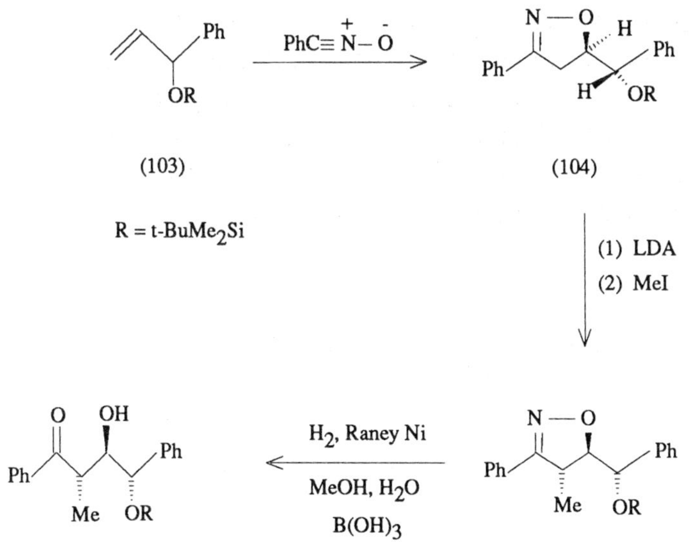

Ph
OR
PhC≡N⁺–O⁻
N—O
H
Ph
Ph
H
OR
R = t-BuMe2Si
(1) LDA
(2) MeI
O
OH
Ph
Ph
Me
OR
H2, Raney Ni
MeOH, H2O
B(OH)3
N—O
Ph
Ph
Me
OR

(103) (104) (105) (106)

*Scheme 37*

With alkyl or other non-coordinating substituents at C-4 or C-5 of the isoxazoline ring addition of hydride takes place *trans* (*anti*) to the substituents to give the *erythro* 1,3-amino-alcohol (as 107). Hydroxyl or hydroxymethyl substituents, on the other hand, direct attack of lithium aluminium hydride to the *syn* face of the C=N double bond to give predominantly the *threo* amino-alcohol (Scheme 39).

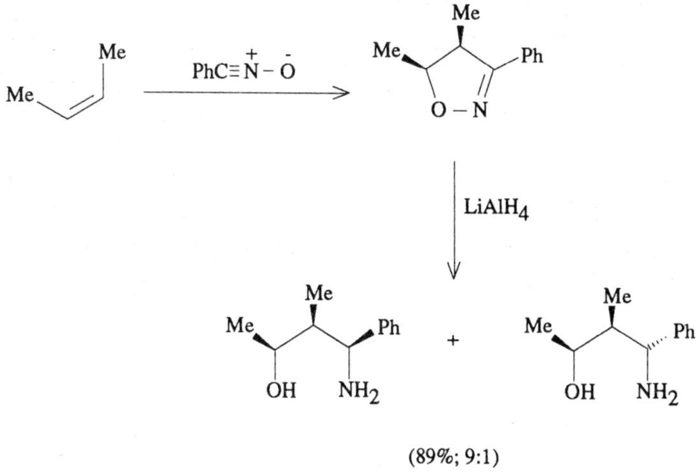

Me
Me
PhC≡N⁺–O⁻
Me
Me
Ph
O – N
LiAlH4
Me
Me
Ph
OH
NH2
+
Me
Me
Ph
OH
NH2
(89%; 9:1)

*Scheme 39*

The formation of 1,3-amino-alcohols by reduction of isoxazolines has formed an important step in the synthesis of γ-hydroxy-α-amino acids,[63] aminopolyols such as phytosphingosine,[64] amino sugars[65] and alkaloids.[66] Thus, in a synthesis of D-lividosamine-N,O,O-triacetate (111) the isoxazoline (108), obtained from reaction of the nitrile oxide derived from the acetal of 2-nitroacetaldehyde with the (*S*)-acetal (81), gave a mixture (*ca.*4:1) of the *syn* and *anti* amino alcohols (109) on reduction with lithium aluminium hydride. Cleavage of the two protecting groups with acid led directly to the two pyranoses (110) from which the pure glycoside (111) was isolated by chromatography of the N,O,O-tri-acetate.[67]

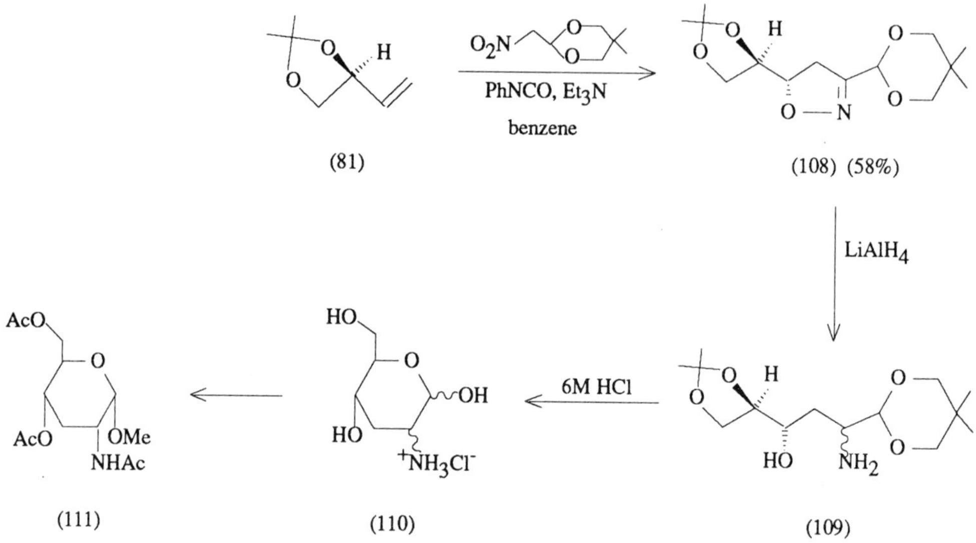

PhNCO, $Et_3N$, benzene; (81); (108) (58%); $LiAlH_4$; 6M HCl; (109); (110); (111)

*Scheme 40*

Cycloadducts from furan and nitrile oxides have been useful for synthesis of aminopolyols.[64] The adduct (112), for example, by cleavage of the dihydrofuran ring and further manipulation gave the isoxazoline (113) which on reduction with lithium aluminium hydride was converted specifically into the protected amino-tetraol (114).[68]

An intramolecular 1,3-cycloaddition followed by reductive cleavage of the isoxazoline ring formed the key step in a synthesis of the ergot alkaloid (+)-paliclavine (116). The cycloaddition step led to formation of the C-5/C-10 bond and set up the correct relative stereochemistry at C-10 and C-10a. Although there was, disappointingly, no diastereoselection in the cycloaddition step, the γ-amino-alcohol system of the alkaloid was readily formed with the correct relative and absolute stereochemistry at three chiral centres by stereoselective reductive cleavage of the isoxazoline ring of the isomer (115). Three new chiral centres are set up in this synthesis and the original one is destroyed.[69]

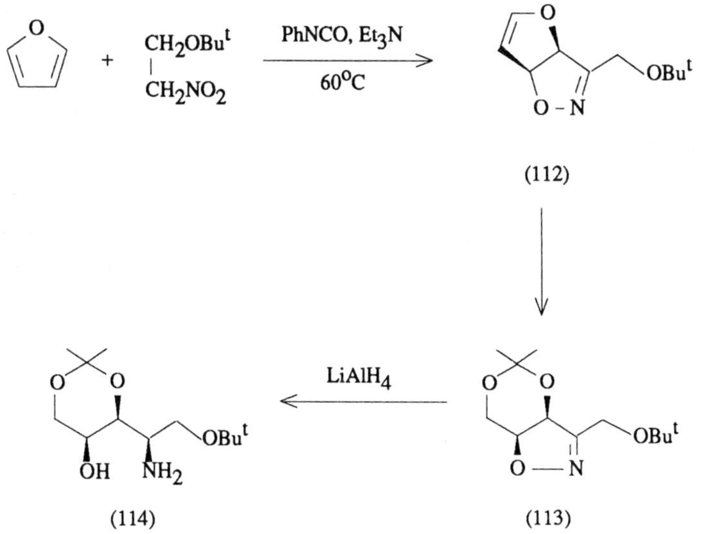
PhNCO, $Et_3N$
60°C
$CH_2OBu^t$
$CH_2NO_2$
OBu$^t$
LiAlH$_4$
(112)
(113)
(114)

*Scheme 41*

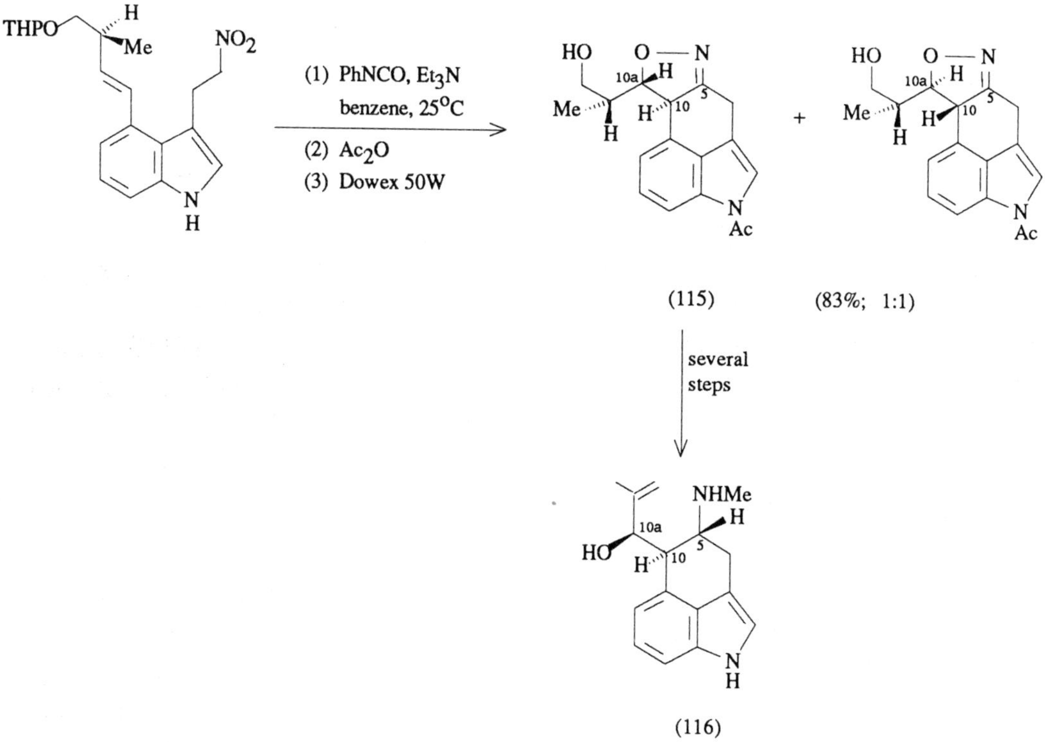
(1) PhNCO, $Et_3N$
benzene, 25°C
(2) $Ac_2O$
(3) Dowex 50W
several steps
(115)
(83%; 1:1)
(116)

*Scheme 42*

## Nitrones

Nitrones are reactive 1,3-dipoles and with olefinic and acetylenic dipolarophiles they form isoxazolidines and $\Delta^2$-isoxazolines respectively.[70]

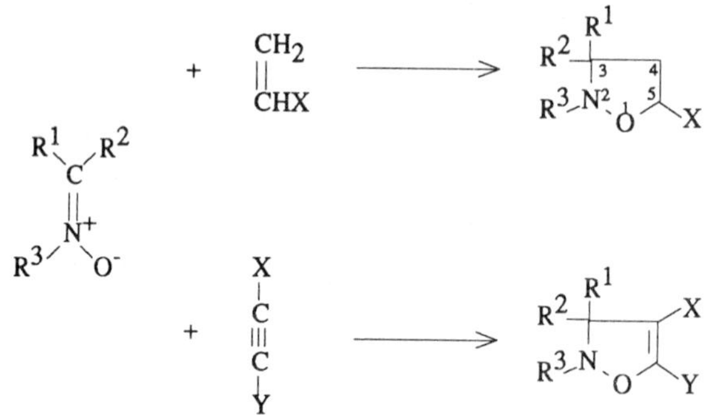

*Scheme 43*

With monosubstituted olefinic dipolarophiles the 5-substituted isoxazolidine is generally formed predominantly; but with olefins bearing strongly electron-withdrawing substituents 4-substituted derivates may also be obtained.[71] Thus N-methyl-C-phenylnitrone reacts with a range of olefins including styrene, simple alkenes such as hex-1-ene, and vinyl ethers to form the 5-substituted isoxazolidines almost exclusively. But with methyl acrylate a mixture of the 5- and 4-substituted products was formed and nitroethylene gave the 4-nitroisoxazolidine exclusively.[72] The situation is not straightforward, however, for acrylonitrile gave the 5-cyano-isoxazolidine, apparently exclusively. An increasing tendency for the formation of the 4-substituted isoxazolidine is found also in reactions of nitrones bearing electron-donating substituents on the nitrone carbon atom.

These effects are accommodated within the molecular orbital description of 1,3-dipolar cycloadditions.[73] Nitrones, like nitrile oxides, have relatively high-lying HOMOs and relatively low-lying LUMOs and in their cycloaddition reactions both the HOMO (dipole) - LUMO (dipolarophile) and LUMO (dipole) - HOMO (dipolarophile) interactions may be important; whichever is predominant depends on the structures of the dipole and the dipolarophile. Reaction rates are increased by electron-donating or electron-withdrawing substituent on either component. Most reactions are LUMO (dipole) controlled and in these the favoured interaction of the atoms on each component with the largest atomic orbital coefficients leads to preferential formation of the 5-substituted isoxazolidine. In reactions involving dipolarophiles bearing electron-withdrawing substituents, however, the HOMO (dipole) - LUMO (dipolarophile) interaction becomes more important and, eventually, predominant, with increasing electron-attracting power of the substituent. The favoured interaction of the atoms on the dipole and dipolarophile with the largest atomic orbital coefficients then leads to the formation of the 4-substituted isoxazolidines.

The stereochemistry of nitrone cycloadditions has been the subejct of several studies.[74] Reactions with olefinic dipolarophiles are stereospecific as regards the disposition of substituents on the olefinic bond, but with acyclic nitrones it is not always easy to predict the relative stereochemistry in the adduct of substituents originating in the dipole and dipolarophile, and the degree of stereoselectivity here may not be high. Reactions appear sometimes to favour an *endo* and sometimes an *exo* transition state due to a complex interplay of secondary orbital interactions and steric effects. There is also the possibility of interconversion of *cis* and *trans* forms of the nitrone under the conditions of the reaction before cycloaddition takes place. Thus, reaction of N-methyl-C-phenylnitrone with acrylonitrile gave mainly the *trans* product (117), but with nitro-ethylene the *cis* isomer (118) predominated.[75]

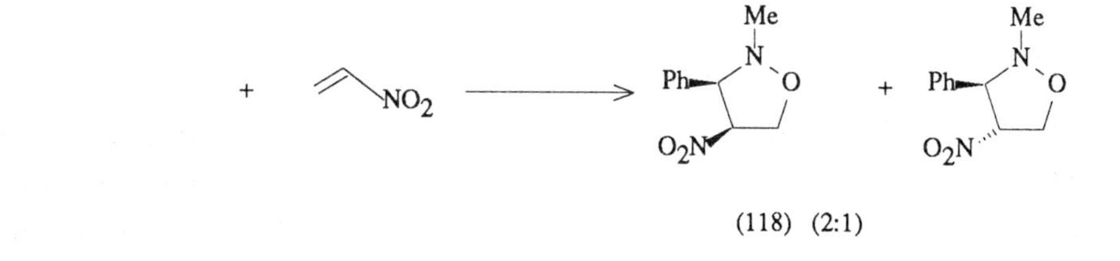

Me Ph N O CN benzene reflux Me N Ph O CN Me N Ph O CN

(117) (3:1)

Me N Ph O $NO_2$ Me N Ph O $O_2N$ Me N Ph O $O_2N$

(118) (2:1)

*Scheme 44*

With cyclic nitrones such as Δ[1]-pyrroline-N-oxide (119), where *cis-trans* isomerisation of the nitrone is ruled out, stereoselectivities are frequently much higher than with acyclic nitrones. Reaction of (119) with 4-phenylbut-1-ene, for example, gave the single adduct (120) in 73 per cent yield by way of the *exo* transition state.[76] In reactions with cyclic nitrones *endo* transition states are disfavoured by steric interaction of substituents on the dipolar-ophile with methylene groups of the ring.

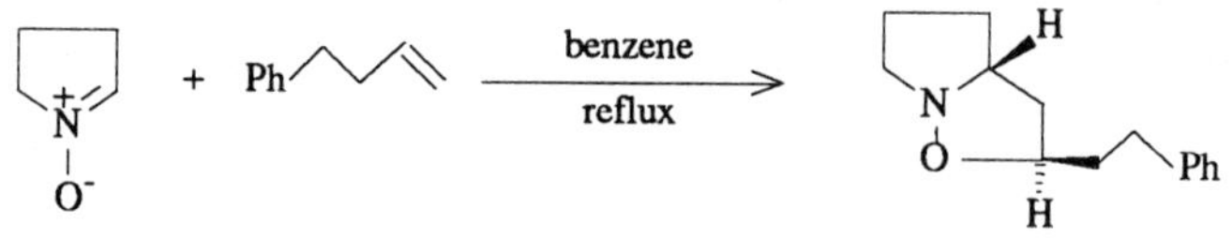

benzene
reflux

(119) (120)

*Scheme 45*

Intramolecular reactions take place easily and are being widely employed in synthesis. With C-alk-4-enylnitrones such as (121) *cis*-fused adducts are formed exclusively in nearly all cases because of constraints in the transition state due to the connecting chain and reactions of this kind have been used in the stereocontrolled synthesis of cyclopentane derivatives. The amino-diol (124), for example, was obtained from the nitrone (122) by way of the cycloadduct (123).[77]

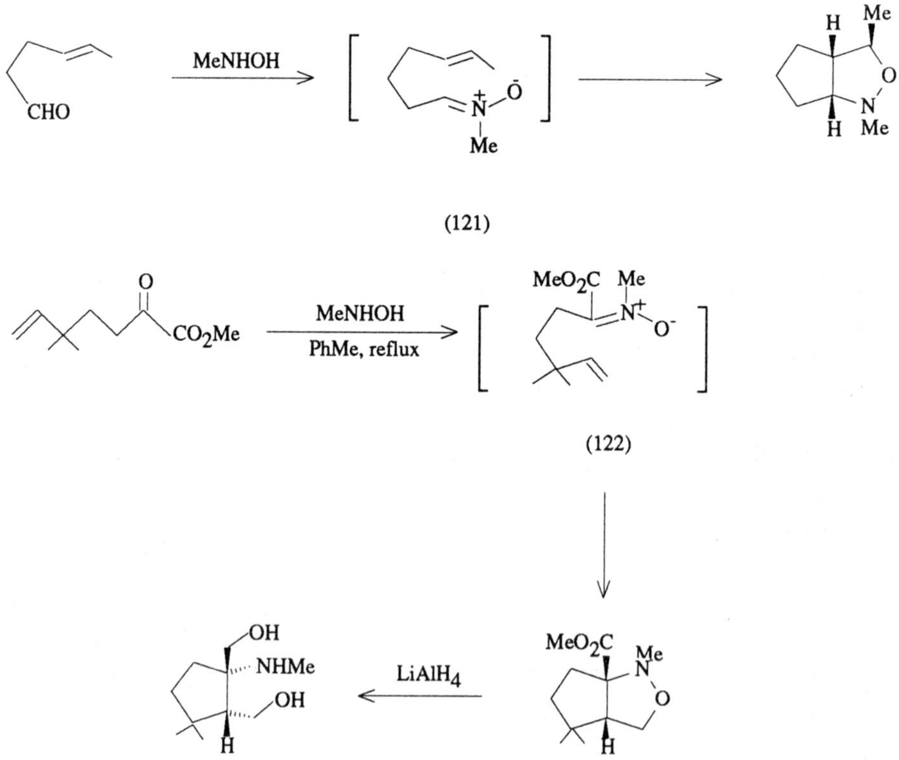

MeNHOH

(121)

MeNHOH
PhMe, reflux

(122)

$LiAlH_4$

(124) (123) (83%)

*Scheme 46*

C-Alk-5-enylnitrones, such as (125), however, because of the greater flexibility allowed by the longer connecting chain, give mixtures of *cis-* and *trans* ring-fused products (126) together with the alternative bridged-ring products (as 127) formed by attack of the oxygen atom of the nitrone at each end of the double bond.[78]

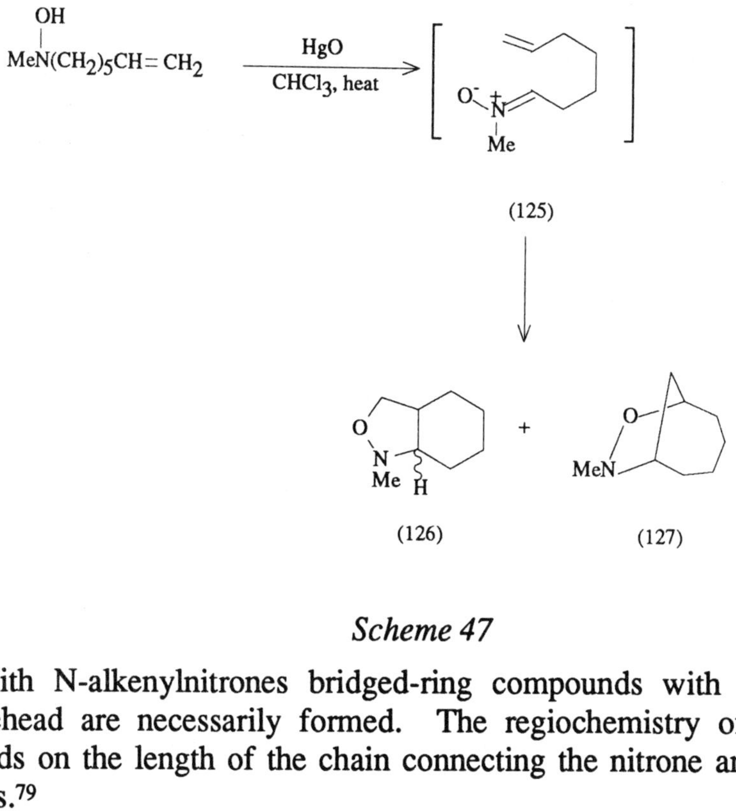

OH
MeN(CH$_2$)$_5$CH=CH$_2$ —HgO / CHCl$_3$, heat→ [ (125) ] → (126) + (127)

*Scheme 47*

With N-alkenylnitrones bridged-ring compounds with nitrogen at a bridgehead are necessarily formed. The regiochemistry of the addition depends on the length of the chain connecting the nitrone and the olefinic groups.[79]

Nitrones for cycloaddition reactions are generally produced *in situ* and not isolated, although some can be. Two commonly used methods are by oxidation of a disubstituted hydroxylamine with yellow mercuric oxide and reaction of an aldehyde or ketone with a monosubstituted hydroxylamine. A disadvantage of the first method is that with unsymmetrical hydroxylamines mixtures of nitrones are obtained. For example, 1-hydroxy-2-pentylpiperidine on oxidation with mercuric oxide affords a mixture of the aldo- and keto-nitrones (128) and (129).[80] A useful modification entailing direct oxidation of the piperidine with hydrogen peroxide in the presence of sodium tungstate or selenium dioxide affords the keto-nitrone only.[81]

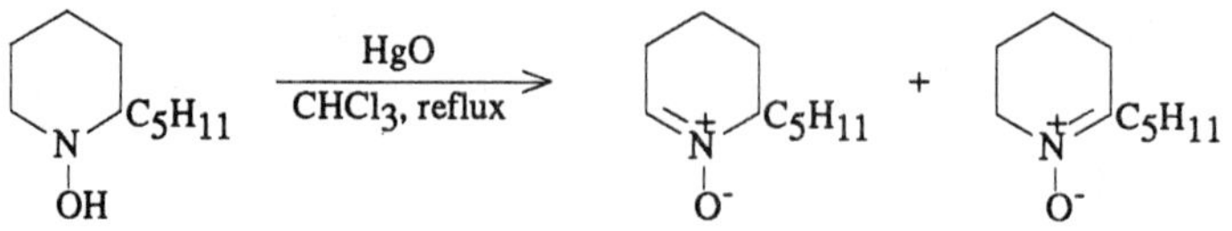

HgO
$CHCl_3$, reflux
$C_5H_{11}$
N
OH
+
(128)
(129) (1:3)

*Scheme 48*

The alternative route from an aldehyde or ketone and a monosubstituted hydroxylamine avoids this difficulty and leads regiospecifically to a single nitrone. Thus, the aldo-nitrone (131) was obtained regiospecifically from the aldehyde (130)[82] and in an example in which formation of the nitrone is followed by an intramolecular cycloaddition, the oxime (132) treated with acetaldehyde gave the isoxazolidine (134) by way of the transient nitrone (133). Further manipulation of (134) gave (±)-carpamic acid (135).[83]

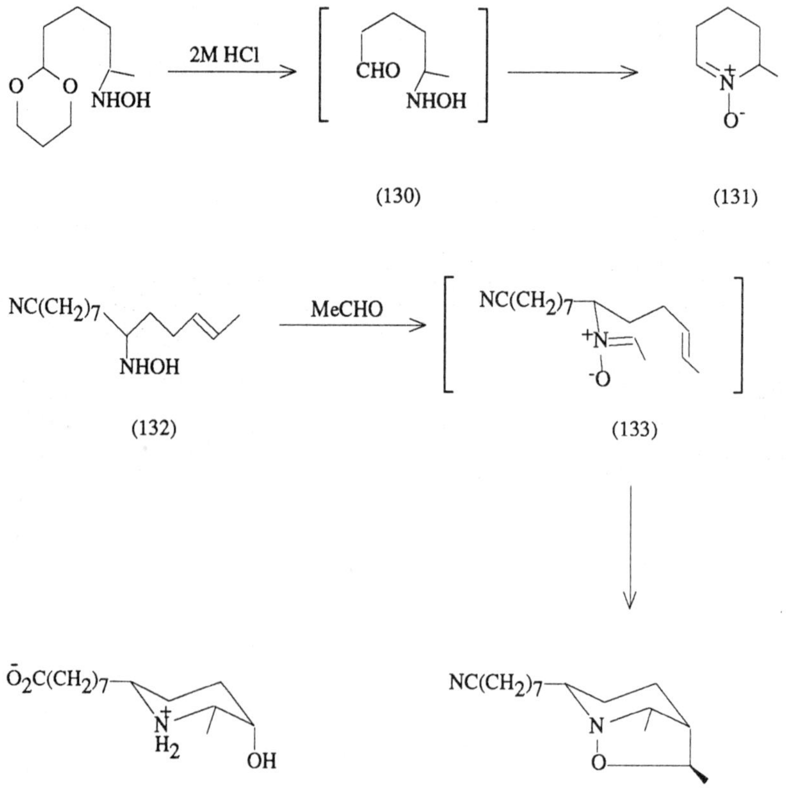

2M HCl
CHO
NHOH
(130)
(131)

$NC(CH_2)_7$
NHOH
MeCHO
(132)
(133)

$\bar{O}_2C(CH_2)_7$
$H_2$
OH
(135)
(134)

*Scheme 49*

In a useful alternative route, nitrones (eg.136) have been obtained *in situ* from oximes and electrophilic olefines in a tandem process involving initial Michael addition of the oxime to the olefin, followed by a 1,2-hydrogen shift to form the nitrone,[84] and cyclic nitrones (eg.137) have been prepared by silver ion-catalysed cyclisation of γ- and δ-oximino-allenes.[85]

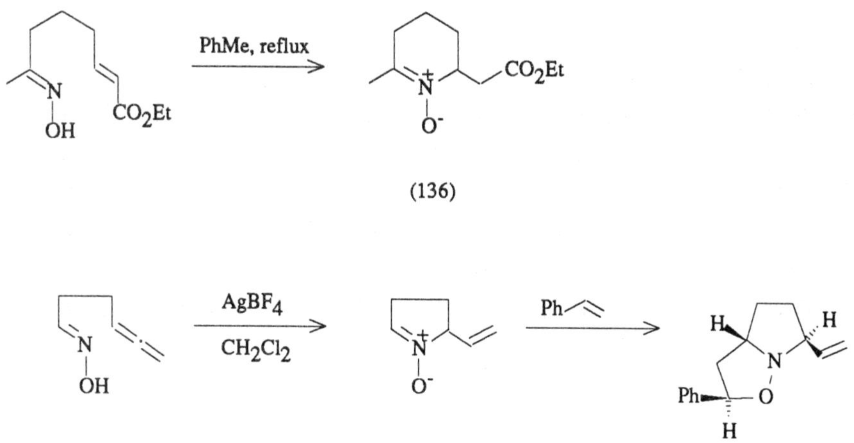

PhMe, reflux

$CO_2Et$

(136)

$AgBF_4$

$CH_2Cl_2$

Ph

(137)

*Scheme 50*

1,3-Dipolar cycloaddition reactions of nitrones are valuable in synthesis because the N-O bond of the adducts is easily cleaved and, since the initial cycloadditions are usually stereoselective, the sequence of cycloaddition followed by ring cleavage provides a route for the stereocontrolled synthesis of a variety of acyclic and substituted cyclic compounds. Thus, the adducts formed from nitrones and vinyltrimethylsilane are readily converted into αβ-unsaturated aldehydes in an excellent alternative to the Wittig reaction which avoids the strongly basic conditions associated with Wittig chemistry[86], and the adducts formed from allyltrimethylsilane have been used in the synthesis of allylamines.[87]

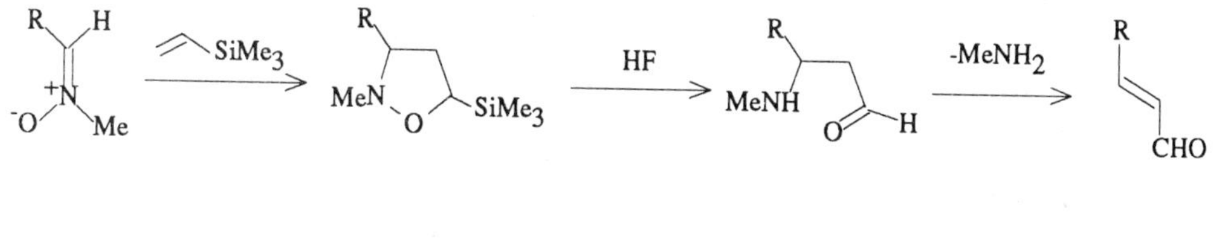

$SiMe_3$

HF

$-MeNH_2$

*Scheme 51*

Perhaps the most useful conversion of isoxazolidines is their reductive cleavage to 1,3-amino-alcohols, an alternative to the route from nitrile oxides described above; this reaction also has been widely employed in the synthesis of natural products, including alkaloids[88] and amino sugars. For example, the 2,3,6-trideoxy-3-aminohexose daunosamine (141, OH for OMe) was synthesised using a sequence in which the stereochemistry of the sugar backbone was established by a [3+2] cycloaddition of ethyl vinyl ether to the optically active nitrone (138).[89]

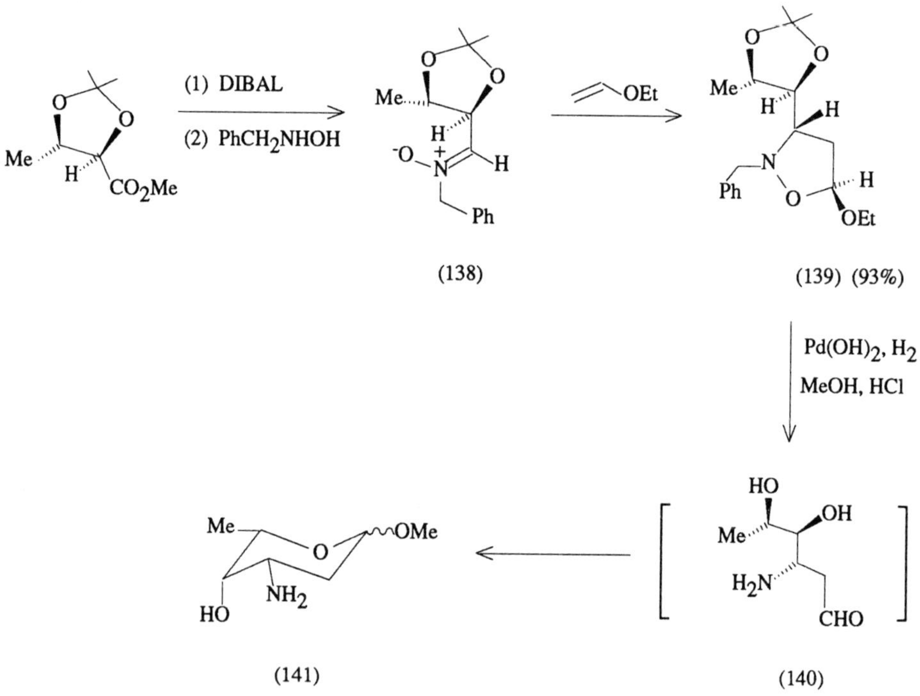

(1) DIBAL (2) $PhCH_2NHOH$

(138)

(139) (93%)

$Pd(OH)_2$, $H_2$ MeOH, HCl

(141)

(140)

*Scheme 52*

The cycloaddition to the chiral nitrone (138) took place with excellent facial- and *endo*-selectivity to give the adduct (139) exclusively in 93 per cent yield. Hydrogenolysis of the N–O bond and removal of the acetal protecting groups led directly to the methyl glycoside of daunosamine (141) by way of the aldehyde (140). In this synthesis the isoxazolidine ring serves as a masked β-amino aldehyde, and cycloaddition of nitrones to enol ethers followed by ring cleavage provides a useful alternative to the Mannich reaction for the preparation of compounds of this class.

1,3-Amino-alcohols are also obtained by reductive cleavage of the $\Delta^2$-isoxazolines formed in cycloaddition reactions of nitrile oxides. Whether the nitrile oxide or nitrone route is adopted will depend on circumstances, but the former may have advantages in certain cases because of the control it allows in generation of the chiral centre on the carbon which bears the nitrogen substitutent.

Attempts to achieve diastereoselective cycloadditions with optically active nitrones have given variable results. Not all nitrones bearing an optically active group on the nitrone carbon atom react as selectively as (138). The closely related nitrone (142) for example, prepared from (*R*)-glyceraldehyde acetonide, showed no facial selectivity in reaction with vinylene carbonate, giving an equimolecular mixture of (143) and (144) by *endo* addition to both faces of the (*Z*)-nitrone. These adducts were converted into the triacetates of the amino sugars 3-epigentosamine (145) and 2-epigentosamine (146).[90]

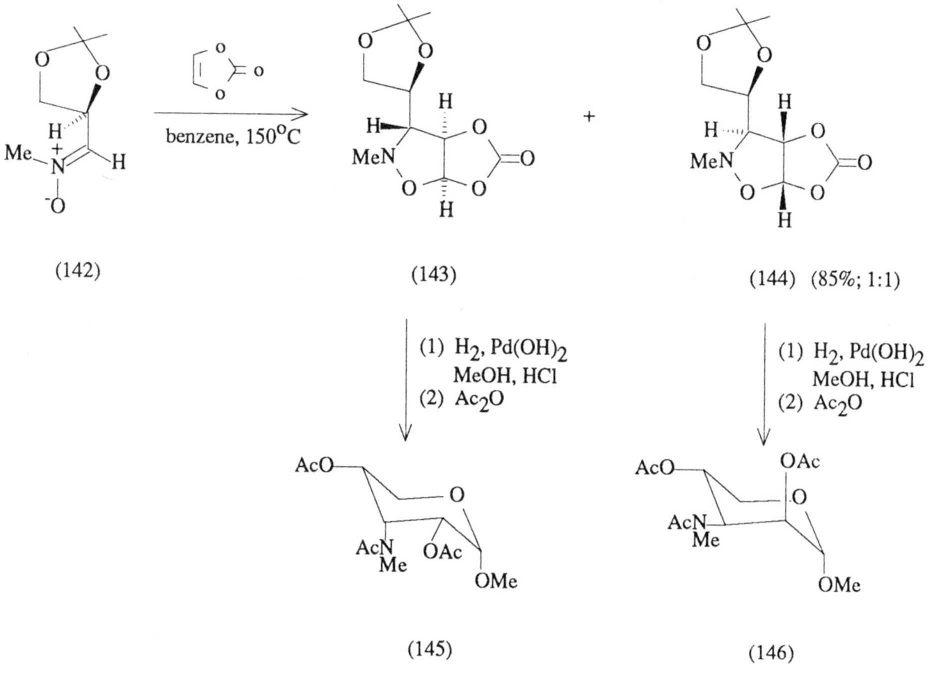

benzene, 150°C

(142) (143) (144) (85%; 1:1)

(1) $H_2$, $Pd(OH)_2$ MeOH, HCl (2) $Ac_2O$

(145) (146)

*Scheme 53*

Similarly, the nitrone (147) bearing the optically active auxiliary (–)-menthyl group gave an equimolecular mixture of the adducts (148) and (149) by *endo* and *exo* addition of *trans*-benzyl crotonate to the (*E*)-nitrone. The *endo* product was converted into the β-lactam (150).[91]

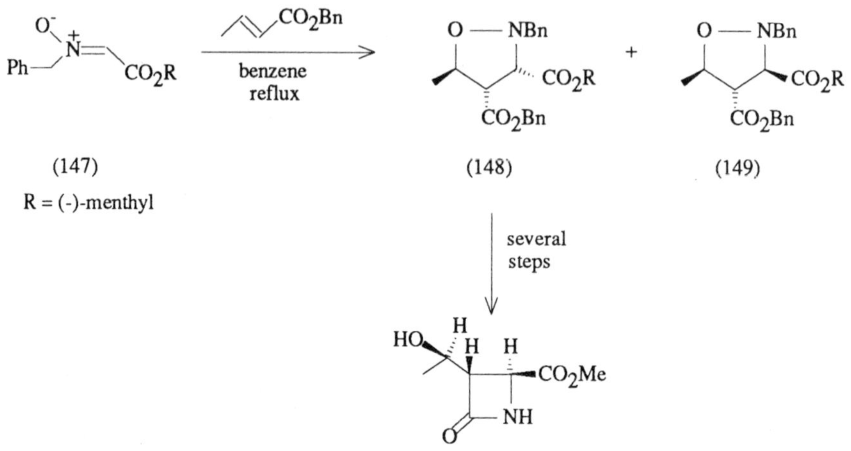

(147) R = (-)-menthyl → benzene, reflux → (148) + (149); (148) → several steps → (150)

*Scheme 54*

Greater selectivity was achieved in the intramolecular cyclisation of nitrone (151) in which the chiral centre is nearer the site of reaction than in (147). It cyclised in refluxing t-amyl alcohol to give exclusively the isoxazolidine (152), with four contiguous chiral centres, in 57 per cent yield. This was subsequently converted into the β-lactam (153).[92]

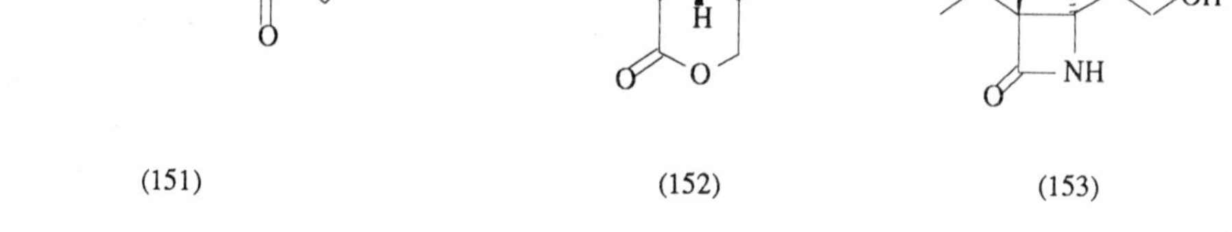

(151) → (152) (153)

*Scheme 55*

Similarly, cyclisation of the optically active nitrone (154) gave the single diastereomer (155) in 80 per cent yield.[93] In these reactions the constraints imposed on the transition states by the connecting chain contribute to the high diastereofacial selectivities observed.

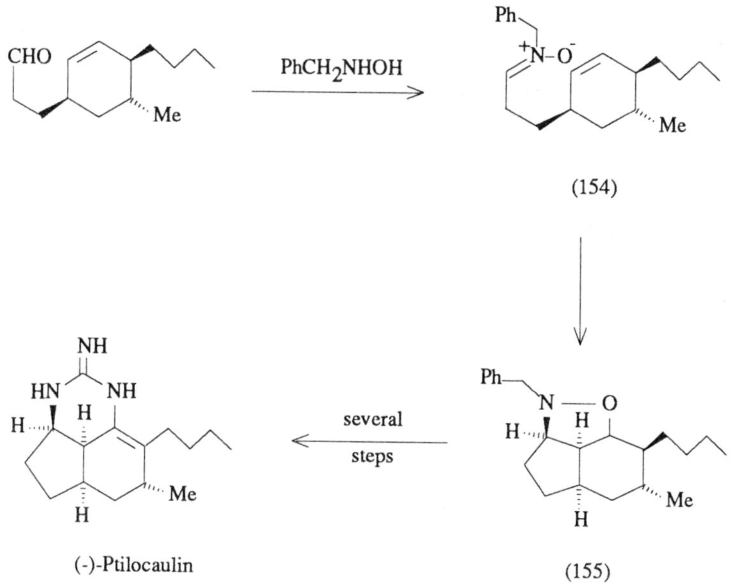

CHO
Me
$PhCH_2NHOH$
Ph
$^+N-O^-$
Me
(154)

NH
HN NH
H H
H
Me
several
steps
Ph
N — O
H H
Me
H
(-)-Ptilocaulin
(155)

*Scheme 56*

Nitrones bearing optically active auxiliary groups on the nitrogen atom have also been employed with some success. Thus, reaction of the optically active nitrone (156) with vinyl acetate gave a 7:3 mixture of the adducts (157) and (158) in 68 per cent yield. These were separated and converted individually into optically pure (*R*)-(–)-β-lysine and (*S*)-(+)-β-lysine (Scheme 57).[94]

So far, the cycloaddition of nitrones to enantiomerically pure olefins has received much less attention. High optical yields have been obtained in the reactions of nitrones with (*R*)-(+)-*para*-tolyl vinyl sulphoxide,[95] and a series of intramolecular cycloadditions to optically pure allylic acetals again proceeded with high facial selectivity.[96]

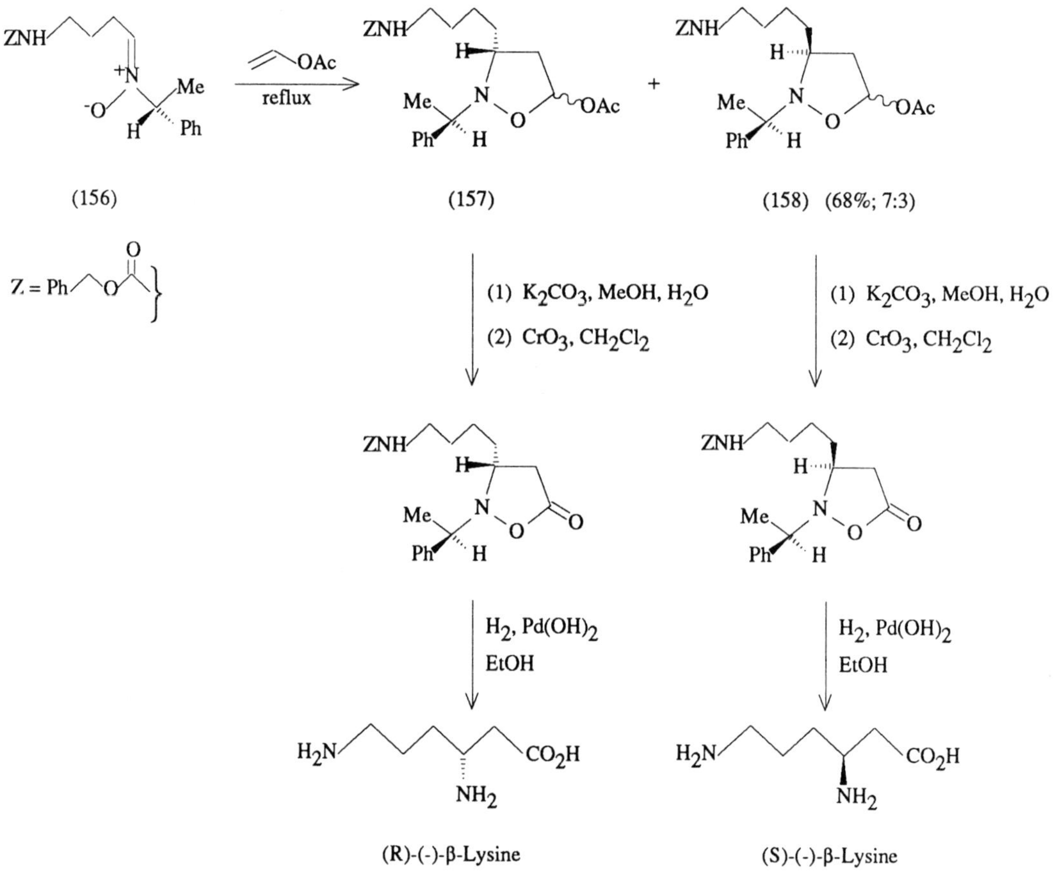

(156) (157) (158) (68%; 7:3)

Z = PhCH$_2$OC(O)–

reflux

(1) $K_2CO_3$, MeOH, $H_2O$
(2) $CrO_3$, $CH_2Cl_2$

$H_2$, $Pd(OH)_2$
EtOH

(R)-(-)-β-Lysine (S)-(-)-β-Lysine

*Scheme 57*

An interesting development has been the operation of double asymmetric induction in the reactions of a series of optically active nitrones with derivatives of L-vinylglycine.[97] The best results were obtained in the reaction of the nitrone (159), derived from 2,3-O-isopropylidene-5-O-trityl-D-ribose oxime, with the L-glycine derivative (160). Removal of the chiral auxiliary group from the isoxazolidines first formed and oxidation with N-chlorosuccinimide gave the isoxazolines (161) and (162) in the ratio 19:1. This is a striking result, an intermolecular reaction unusually showing high facial selectivity. The nitrone (159) and the L-glycine derivative (160) form a matched pair for double asymmetric induction.[98] The main product (161) was converted in several more steps into the amino acid acivicin (163).

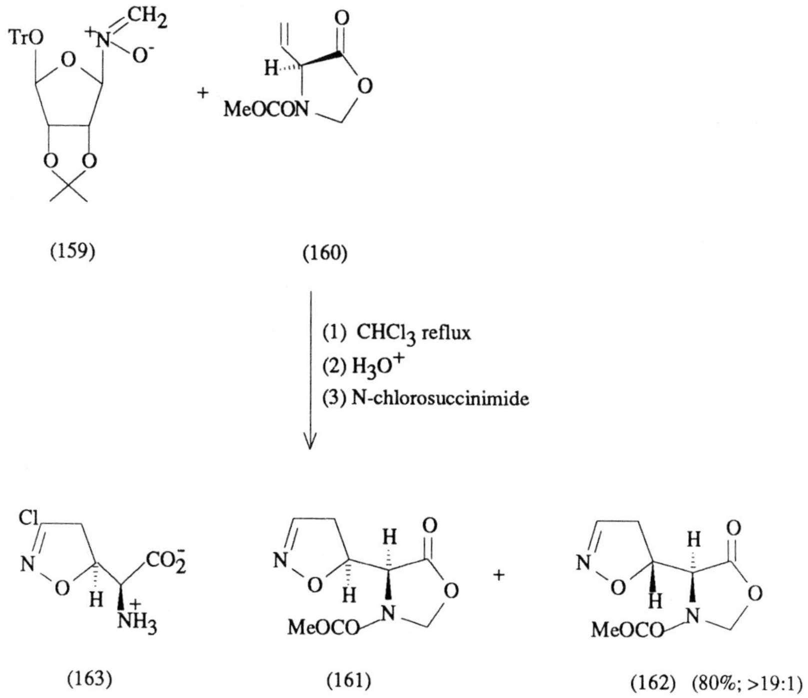

TrO O $^+N$ $CH_2$ $O^-$ O O + H O O MeOCON O

(159) (160)

(1) $CHCl_3$ reflux
(2) $H_3O^+$
(3) N-chlorosuccinimide

Cl N O H $CO_2^-$ $\overset{+}{N}H_3$ N O H H O O MeOCO N + N O H H O O MeOCO N

(163) (161) (162) (80%; >19:1)

*Scheme 58*

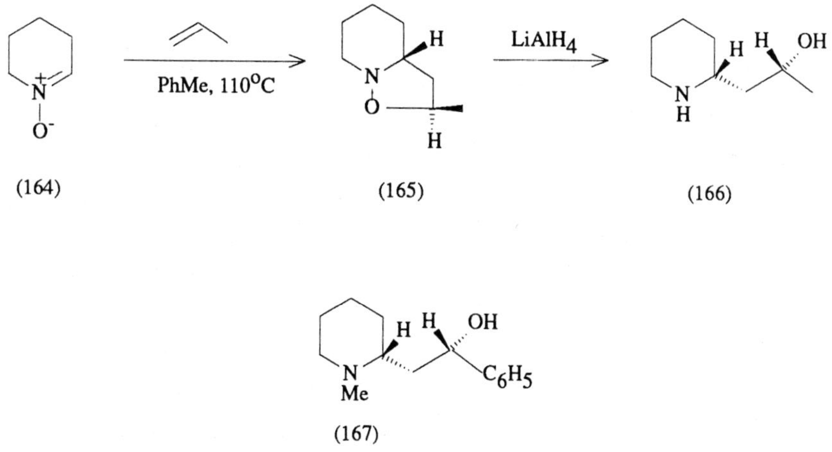

$\overset{+}{N}$ $O^-$ PhMe, 110°C N O H H $LiAlH_4$ N H H H OH

(164) (165) (166)

N Me H H OH $C_6H_5$

(167)

*Scheme 59*

The cyclic nitrones (164) and (168) have been used in the synthesis of a variety of piperidine, pyrrolidine, pyrrolizidine and quinolizidine alkaloids and of some tropane alkaloids, therein demonstrating the high regio- and stereo-selectivities shown in cycloaddition reactions of these nitrones.[99] Thus, reaction of tetrahydropyridine oxide (164) and propene in toluene at 110°C led to the single cycloadduct (165) in 53 per cent yield by *exo* addition. Reductive cleavage of the isoxazolidine ring with lithium aluminium hydride then gave the piperidine alkaloid dl-sedridine (166).[100] A similar sequence starting from the adduct from (164) and styrene gave dl-allosedamine (167).[101]

With 1-phenylbutadiene and the five-membered ring nitrone (168) reaction took place at the unsubstituted double bond of the diene to give a mixture of (169) and (170); the proportions varied with the reaction conditions, but the *trans* isomer (169) was always the main product. N-Methylation and reductive cleavage with zinc and acetic acid then gave darlinine (171) and epidarlinine (172).[102]

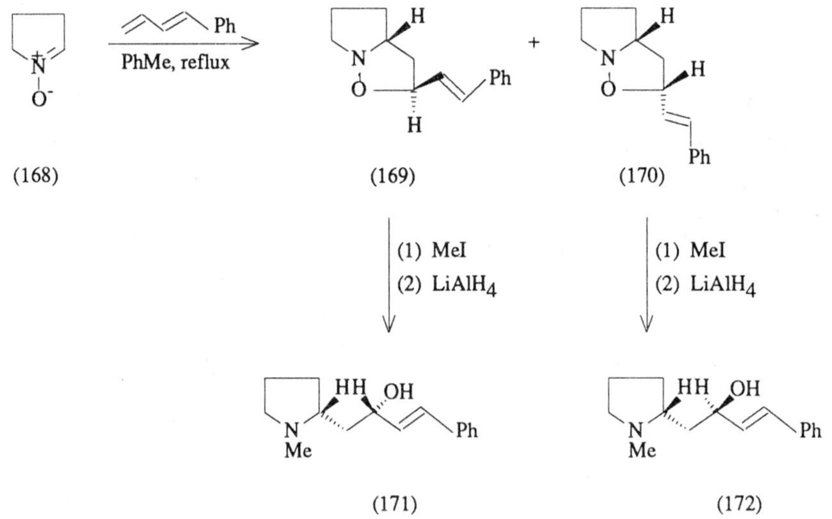

PhMe, reflux

(168) (169) (170)

(1) MeI
(2) $LiAlH_4$

(171) (172)

*Scheme 60*

With the monosubstituted cyclic nitrone (173) addition of dipolarophiles is highly stereoselective giving largely the *trans* substituted piperidines by orthogonal approach of the dipolarophile to the nitrone in a conformation in which the substituent is *quasi* equatorial. This provides a route to *trans*-2,6-dialkylpiperidines, which have not been easy to obtain. For example, reaction of (173) with 1-undecene gave isoxazolidine (174) almost entirely. Reductive cleavage of the N–O bond and dehydroxylation of the side chain then led to the ant-venom constituent solenopsin (175) free from the corresponding *cis* isomer.[82]

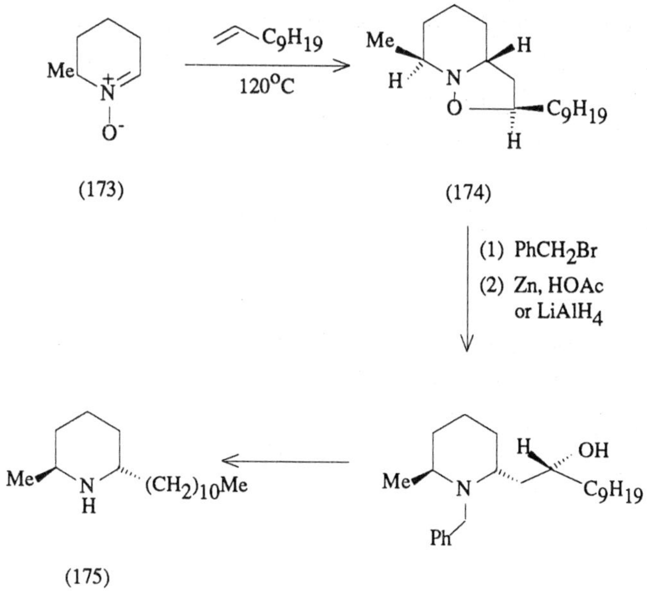
Me
C9H19
120°C
Me
H
N
O
H
C9H19
H
(173)
(174)
(1) PhCH2Br
(2) Zn, HOAc
or LiAlH4
Me
N
H
(CH2)10Me
Me
N
H
OH
C9H19
Ph
(175)

*Scheme 61*

In the pyrrolidine series, isoxazolidines such as (176), on oxidation with peroxy acids, give rise selectively to the less substituted nitrones such as (177).[103] Addition of a second dipolarophile and cleavage of the isoxazolidine ring leads to the corresponding *trans*-2,5-dialkylpyrrolidine.[104]

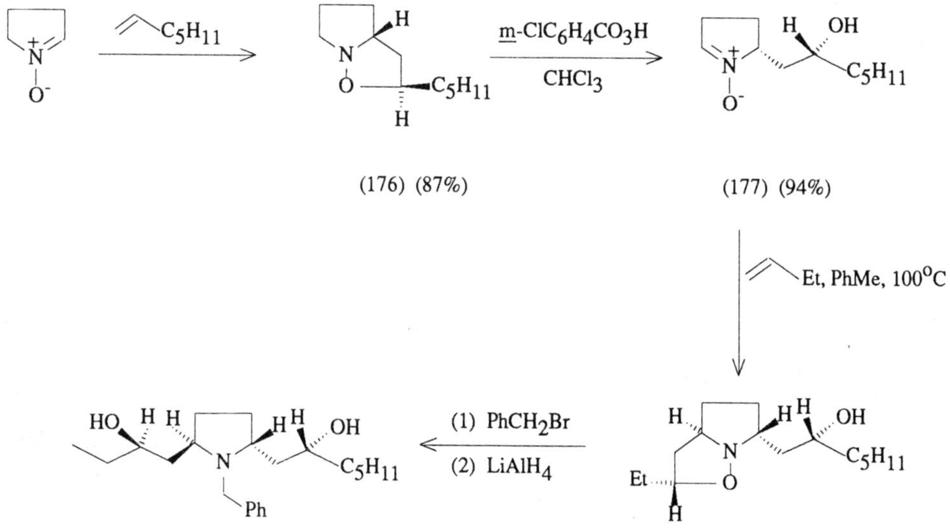
C5H11
H
N
O
C5H11
H
m-ClC6H4CO3H
CHCl3
H
OH
C5H11
(176) (87%)
(177) (94%)
Et, PhMe, 100°C
HO
H H
H H
OH
N
C5H11
Ph
(1) PhCH2Br
(2) LiAlH4
H
H H
OH
N
C5H11
Et
O
H

*Scheme 62*

Pyrrolizidine and quinolizidine alkaloids have been synthesised by similar methods with appropriate choices of starting materials. For example dl-supinidine (181) was readily obtained from the adduct (178) prepared from (168) and methyl 3-hydroxycrotonate. Reductive cleavage of the N–O bond of the derived mesylate gave (180) and thence, by dehydration and reduction of the ester, supinidine.[105]

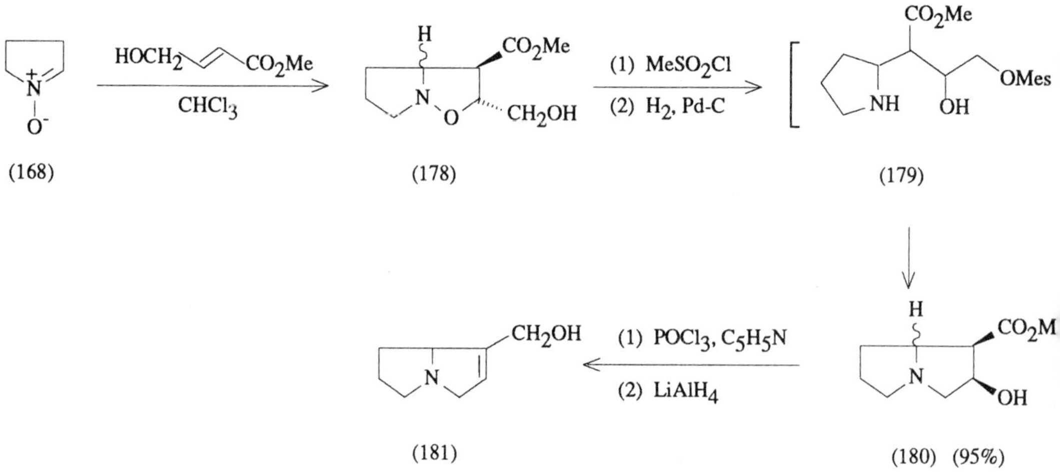

(168) → $HOCH_2CH{=}CHCO_2Me$, $CHCl_3$ → (178) → (1) $MeSO_2Cl$ (2) $H_2$, Pd-C → [(179)] → (180) (95%) → (1) $POCl_3$, $C_5H_5N$ (2) $LiAlH_4$ → (181)

*Scheme 63*

The regio- and stereo-selectivity of nitrone cycloadditions can be exploited to control stereochemistry in the synthesis of non-nitrogenous products, after elimination of the nitrogen. A good example is found in the construction of the side chain in compound (186),[106] where the stereochemistry and location of the double bond and the configuration of the C-24 methyl substituent are controlled in this way. Reaction of the (*Z*)-nitrone (182) with methyl 3,3-dimethylacrylate gave, regioselectively, a separable 1:1 mixture of isoxazolidine esters (183) and (184), formed by virtually exclusive *endo* addition to the (*Z*)-nitrone. The required isomer (183) was converted in several further steps into the amino-diol (185) which by Hofmann degradation gave, regio- and stereo-specifically, the $\Delta^{22}$-*trans* olefin (186). The course of this elimination is controlled by the configuration of the dimethylamino substituent and thus ultimately by the stereochemical course of the cycloaddition. With the (*S*)-configuration of the dimethylamino group Hofmann elimination *via* the most favoured transition state leads to (186).

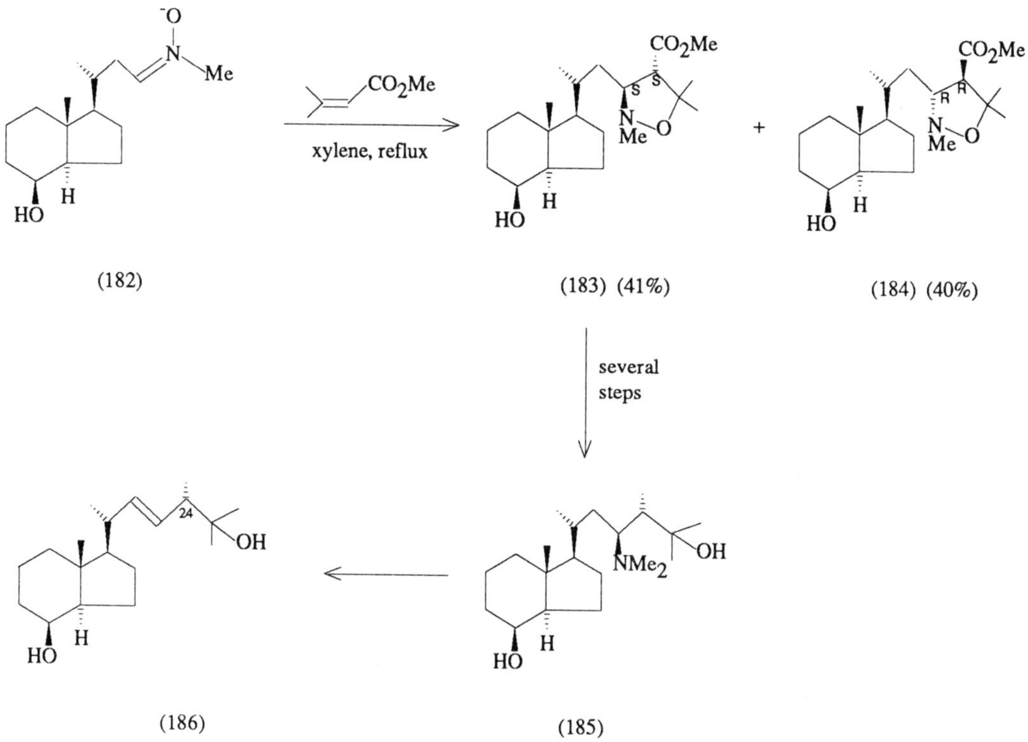

(182) (183) (41%) (184) (40%)

(186) (185)

*Scheme 64*

Nitrones also add readily to alkynes conjugated with electron-attracting substituents, forming $\Delta^4$-isoxazolines. The reactions are less regioselective than those of the corresponding olefins and frequently a large proportion of the C-4 substituted isoxazoline is formed.[70] Reactions involving alkynes as dipolarophile appear not to have been widely employed in synthesis. In a useful transformation, the adducts are readily converted into αβ-unsaturated ketones by oxidation with peroxyacids (Scheme 65).[107]

Many 1,3-dipolar cycloadditions are reversible and the reactions of nitrones are no exception.[108] Many isoxazolidines dissociate with more or less ease into the component nitrone and dipolarophile on thermolysis, so that the initial kinetic product of a cycloaddition can sometimes be converted into the thermodynamic one on further heating.[109]

Formation of cycloadducts with methyl acrylate has been employed to mask the nitrone functional group, which is regenerated by thermolysis at a later stage of the synthesis.[110]

## Azides, Azomethine Imines and Diazoalkanes

The remaining common 1,3-dipoles have not yet been widely employed in synthesis, apart from their straightforward application in the preparation of five-membered heterocyclic compounds. Azides were among the first 1,3-dipoles to be employed; they react with olefinic dipolarophiles to form $\Delta^2$-triazolines (4,5-dihydrotriazoles) and with acetylenic dipolarophiles they give triazoles.[111] As with nitrile oxides and nitrones, the reactions may be either LUMO (dipole) or HOMO (dipole)-controlled depending on the nature of the substituents on each component, and this affects the substitution pattern in the cycloadducts obtained. With mono-substituted olefins as dipolarophiles, either 1,4- or 1,5-disubstituted triazolines may be formed. Terminal olefins and olefins bearing electron-donating substituents such as enamines and vinyl ethers, generally give mainly the 1,5-disubstituted triazoline, while dipolarophiles with electron-withdrawing substituents form varied amounts of the 1,4-disubstituted isomers as well. Thus, phenyl azide and methyl acrylate gave largely 1-phenyl-1,2,3-triazoline-4-carboxylic ester (187)[112] while 1-hexene and ethyl vinyl ether formed the 1,5-disubstituted compounds (188). Reactions with acetylenic dipolarophiles are less regioselective.

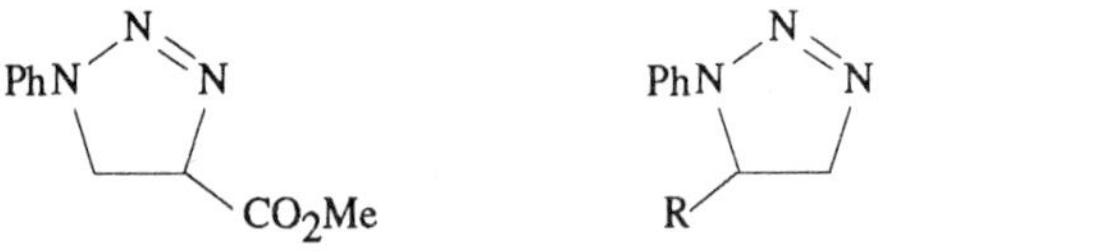

PhN N N CO$_2$Me

(187)

PhN N N R

(188: R = $C_4H_9$, OEt)

*Scheme 65*

Triazolines bearing an electron-attracting substituent at C-4 readily isomerise to β-aminodiazo compounds on base catalysis (Scheme 66).

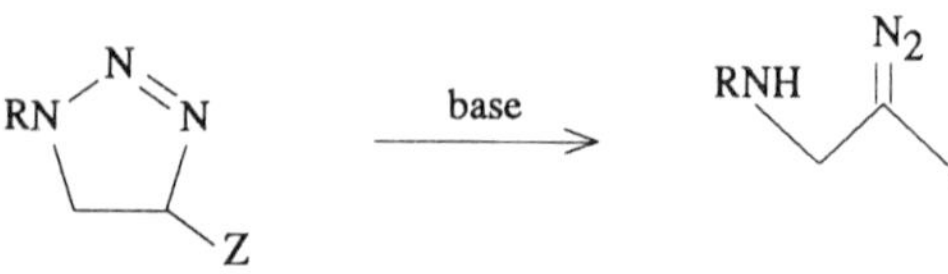

RN N N Z —base→ RNH $N_2$ Z

Z = electron-withdrawing group

*Scheme 66*

This has been exploited in a neat enantioselective synthesis of chiral hydroxypyrrolidines.[113] Thus, the ester (189) derived from D-erythrose, was converted into the 6-triflate and thence, *via* the 6-azide, directly into the dihydrotriazole (191) in an intramolecular cycloaddition. Ring-opening with sodium ethoxide gave the pyrrolidine diazo-ester (192) which on hydrogenolysis formed the optically pure pyrrolidinyl acetate (193). The corresponding (*Z*)-ester was similarly converted into the isomeric pyrrolidine (194).

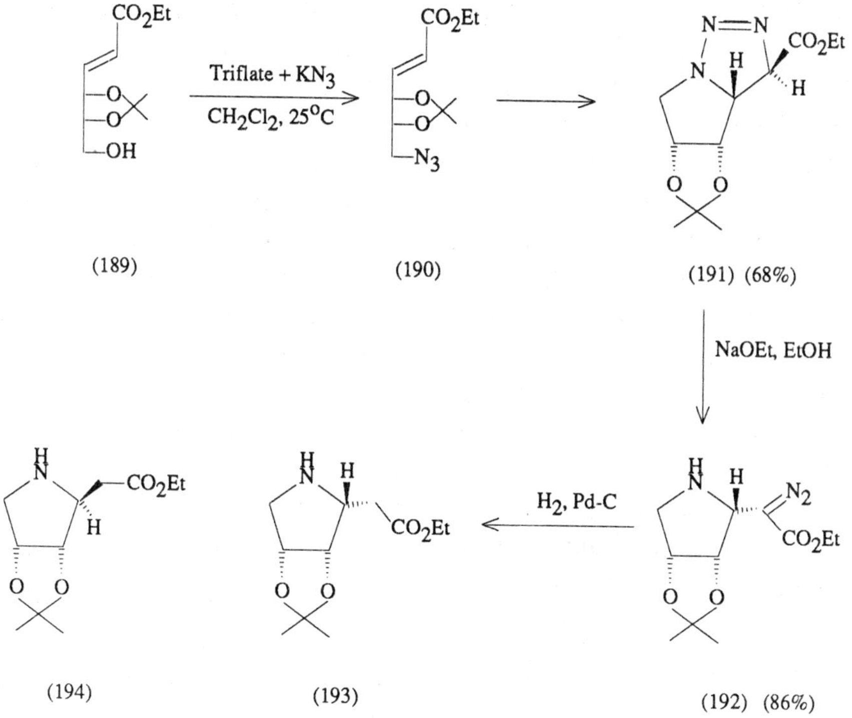

Triflate + $KN_3$
$CH_2Cl_2$, 25°C

(189) (190) (191) (68%)

NaOEt, EtOH

$H_2$, Pd-C

(194) (193) (192) (86%)

*Scheme 67*

The thermal decomposition of olefinic azides (195) gives cyclic imines (as 196) and 1-azabicyclo[3,1,0]hexanes (as 197) by way of the initially-formed but labile triazolines (198).[114]

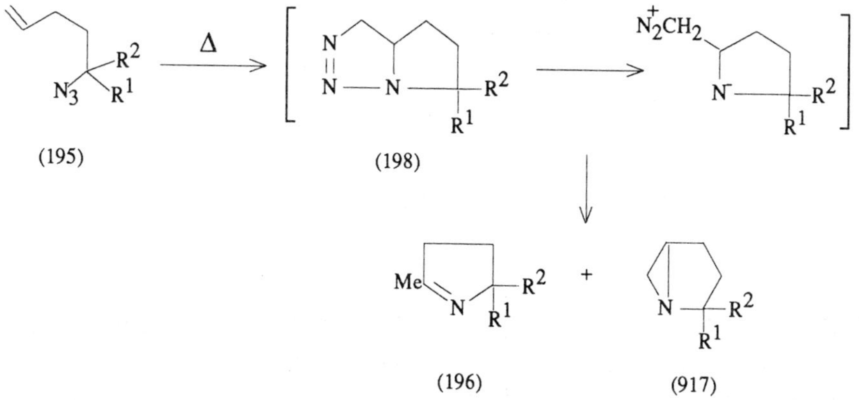

Δ

(195) (198)

(196) (917)

A reaction of this kind was employed in a synthesis of the indole derivative (201) in model experiments aimed at the ergot alkaloid clavicipitic acid.[115] The azide (199) heated in *ortho*-dichlorobenzene gave the single imine (200) in 62 per cent yield in a completely regiospecific reaction. The intermediate triazoline was not detected.

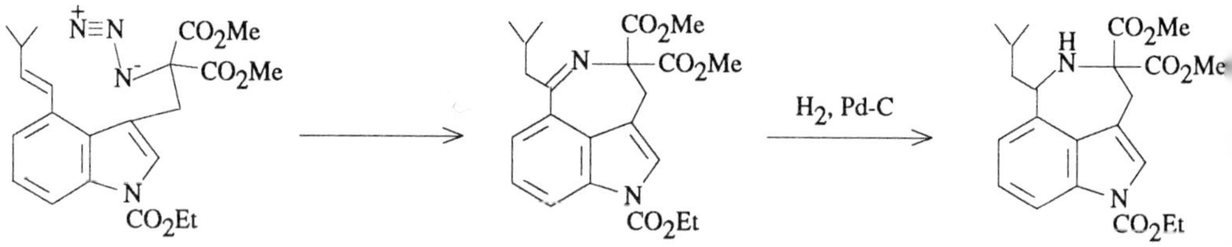

(199) (200) $H_2$, Pd-C (201) (92%)

*Scheme 69*

Other intramolecular cycloaddition reactions of azides have been reviewed.[116]

Azomethine imines are normally not isolable, but they can be prepared conveniently *in situ* by reaction of N-acyl-N′-alkylhydrazines with an aldehyde. Generated in the presence of an alkene they form pyrazolidines. With monosubstituted alkenes 3-substituted pyrazolidines (202) are formed preferentially, but with increasing electronegativity of the substituent the proportion of the 4-substituted isomer (as 203) increases.[117] Alkynes also react readily to give pyrazolines.

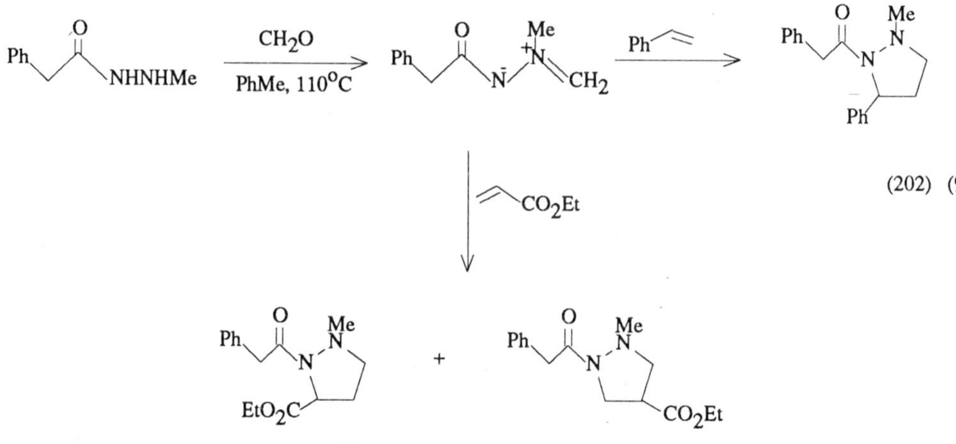

$CH_2O$, PhMe, 110°C

(202) (98%)

65% (203) 7%

*Scheme 70*

Intramolecular reactions take place readily.[118] Thus reaction of the N-alkenoyl-N′-alkylhydrazine (204) with benzaldehyde gave the bicyclic pyrazolidine (205) in 70 per cent yield. With N-acyl-N′-alkenylhydrazines bridged ring compounds are formed.[119] A reaction of the former type was the key step in a synthesis of saxitonin, the paralytic agent of the California mussel.[120]

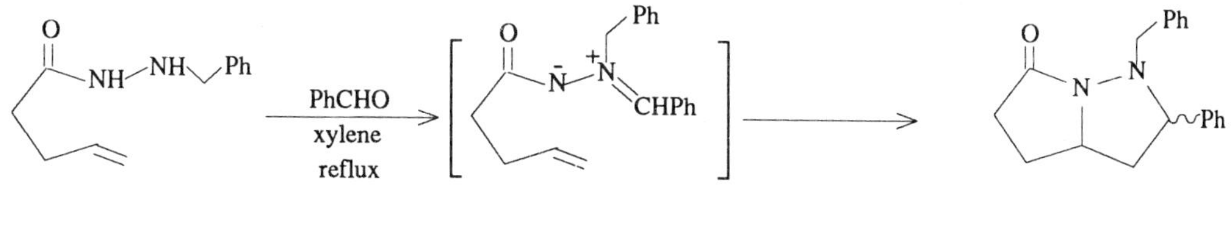

PhCHO xylene reflux

(204) (205) (70%)

*Scheme 71*

The N-imides of pyridine, quinoline and isoquinoline form another long-known class of azomethine imines.[121] They can be obtained *in situ* by deprotonation of N-amino salts; in the presence of acetylenic dipolarophiles they form adducts which spontaneously undergo dehydrogenation to the aromatic product (eg. 206).

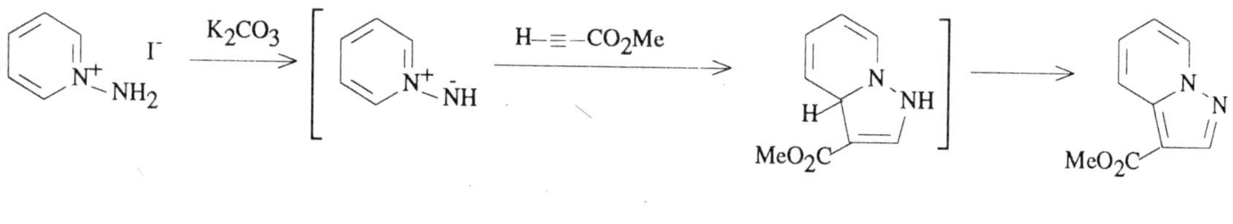

$K_2CO_3$ H–≡–$CO_2Me$

(206)

*Scheme 72*

Cycloaddition of diazoalkanes to olefins leads first to $\Delta^1$-pyrazolines which may isomerise to the $\Delta^2$-isomers. Acetylenic dipolarophiles give pyrazoles. With monosubstituted olefins reaction leads in nearly all cases to C-3 substituted pyrazolines. With simple diazoalkanes the reactions are HOMO (dipole)-controlled and the rate of reaction is increased by electron-withdrawing substituents on the dipolarophiles. Methyl acrylate, for example, reacts with diazomethane $10^6$ times faster than 1-hexene, to give the adduct (207). This general pattern is changed in the reactions of diazoalkanes bearing electron-withdrawing substituents, such as diazoacetic esters, with electron-rich olefins such as enamines. Reactions are now LUMO (dipole)-controlled. Thus, reaction of methyl diazoacetate with N-(1-propenyl)pyrrolidine gave the 3,5-disubstituted pyrazole (209), formed by elimination of pyrrolidine from the 4-aminopyrazoline (208). The addition of diazoalkanes to acetylenic dipolarophiles follows the same pattern; diazomethane and propyne furnish mainly 3-methylpyrazole, but methyl diazoacetate and 1-dimethylaminopropyne give the 4-dimethylaminopyrazole exclusively.[122]

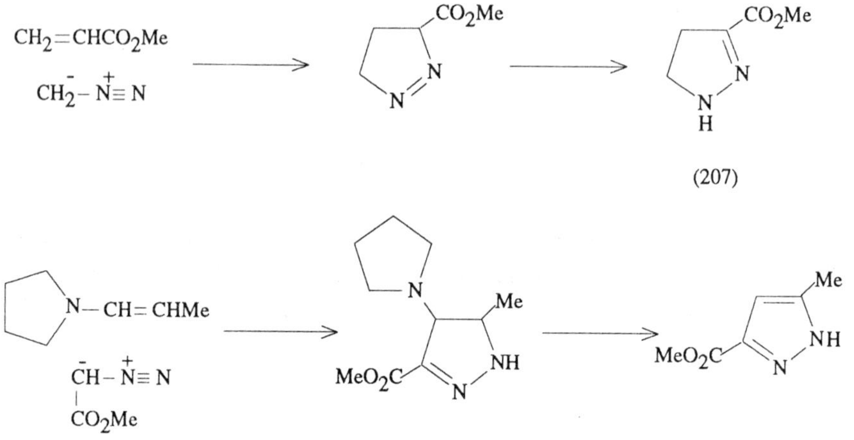

$CH_2=CHCO_2Me$ + $\bar{C}H_2-\overset{+}{N}\equiv N$ → (pyrazoline) → (207)

(pyrrolidinyl)$N-CH=CHMe$ + $\bar{C}H(CO_2Me)-\overset{+}{N}\equiv N$ → (208) → (209)

*Scheme 73*

Pyrazolines formed in cycloaddition reactions of diazoalkanes with olefins eliminate nitrogen on thermolysis or photolysis to give cyclopropanes. This useful method has been employed in the synthesis of a number of cyclopropane natural products. Thus, reaction of the vinyldiazoalkane (210) with (*E,Z*)-1,3,5-octatriene gave the pyrazoline (211) by attack on the terminal double bond. On photolysis this formed the naturally-occurring cyclopropane dictyopterene B (212) along with the *cis* isomer (213). Cope rearrangement of the latter provided easy access to another natural product ectocarpene (214).[123]

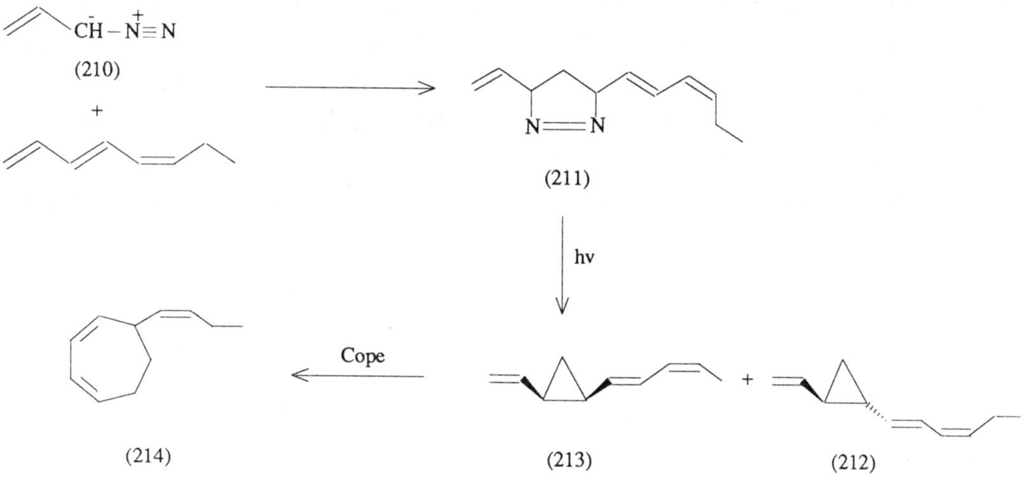

(210)

(211)

hv

Cope

(214) (213) (212)

*Scheme 74*

In an intramolecular example, the sesquiterpene (–)-cyclocopacamphene (217) was obtained by photolysis of the pyrazoline (216), itself formed by intramolecular 1,3-cycloaddition of the diazo compound (215).[124]

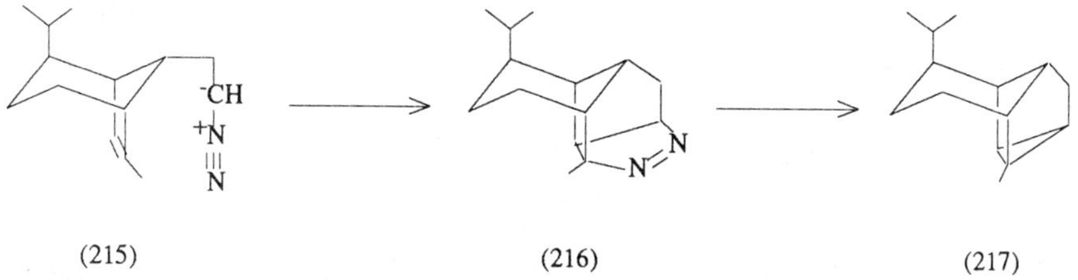

(215) (216) (217)

*Scheme 75*

Some intramolecular reactions of diazoalkanes, where the double bond of the dipolarophile and the 4π system of the dipole cannot attain the parallel plane arrangement required for normal cycloaddition, take a different course. The terminal nitrogen of the diazo compound behaves more like a nitrene and a 1,2-diazabicyclo[3,1,0]hex-2-ene is formed.[125] (Scheme 78)

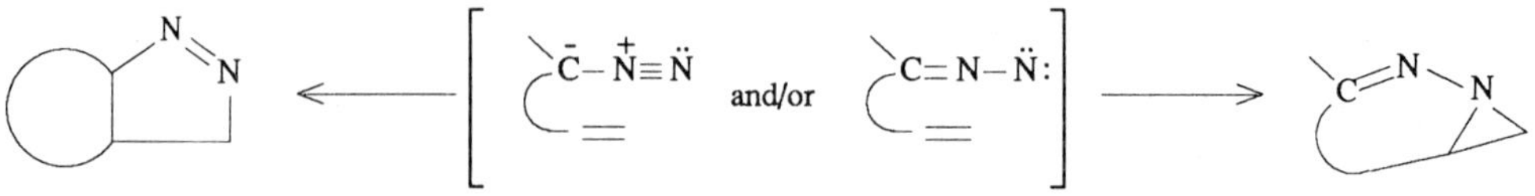

N N and/or C—N≡N C=N—N: C=N N

*Scheme 76*

Thus, while tosylhydrazone (218) on treatment with sodium hydride (to form the diazo compound) gave the expected pyrazoline (219) a similar 1,3-dipolar cycloaddition is not geometrically possible with the diazo compound from (220) and the alternative nitrene addition compound (221) is obtained.

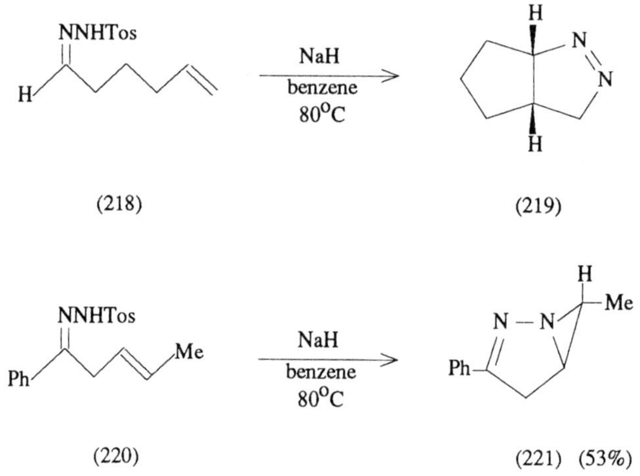

NNHTos H NaH benzene 80°C H N N H

(218) (219)

NNHTos Ph Me NaH benzene 80°C H Me N — N Ph

(220) (221) (53%)

*Scheme 77*

## References

1 R. Huisgen, *Angew. Chem. internat. edn.*, 1963, **2**, 565; *J. Org. Chem.*, 1968, **33**, 2291.

2 R. Huisgen in *1,3-Dipolar Cycloaddition Chemistry*, ed. A. Padwa, Wiley-Interscience, 1984, Vol.1, p.1.

3 R. Huisgen, *J. Org. Chem.*, 1976, **41**, 403.

4 See for example, R. Huisgen and R. Weinberger, *Tetrahedron Lett.*, 1985, **26**, 5119; K.N. Houk, R.A. Firestone, L.L. Munchausen, P.H. Mueller, B.H. Arison and L.A. Garcia, *J. Am. Chem. Soc.*, 1985, **107**, 7227; J.J.W. McDouall, M.A. Robb, U. Niazi, F. Bernardi and H.B. Schlegel, *J. Am. Chem. Soc.*, 1987, **109**, 4642.

5 See, for example, K.N. Houk, J. Sims, R.E. Duke, R.W. Strozier and J.K. George, *J. Am. Chem. Soc.*, 1973, **95**, 7287; K.N. Houk, J. Sims, C.R. Watts and L.J. Luskus, *J. Am. Chem. Soc.*, 1973, **95**, 7301.

6 R. Huisgen, G. Szeimies and L. Möbius, *Chem. Ber.*, 1966, **99**, 475; P. Scheiner, *Tetrahedron*, 1968, **24**, 349.

7 For example, R. Huisgen and R. Weinberger, *Tetrahedron Lett.*, 1985, **26**, 5119; R. Huisgen, M. Seidl, G. Wallbillich and H. Knupfer, *Tetrahedron*, 1962, **17**, 3.

8 R. Huisgen, W. Scheer and H. Hüber, *J. Am. Chem. Soc.*, 1967, **89**, 1753; R. Huisgen, W. Scheer, H. Mäder and E. Brunn, *Angew. Chem. internat. edn.*, 1969, **8**, 604.

9 A. Padwa, L. Fisera, K.F. Koehler, A. Rodriguez and G.S.K. Wong, *J. Org. Chem.*, 1984, **49**, 276.

10 Compare G. Roussi and J. Zhang, *Tetrahedron Lett.*, 1988, **29**, 3481.

11 K.N. Houk, J. Sims, R.E. Duke, R.W. Strozier and J.K. George, *J. Am. Chem. Soc.*, 1973, **95**, 7287; K.N. Houk, J. Sims, C.R. Watts and L.J. Luskus, *loc.cit.*, 7301.

12 R. Huisgen, W. Scheer and H. Huber, *J. Am. Chem. Soc.*, 1967, **89**, 1753; R. Huisgen, *Angew. Chem. internat. edn.*, 1963, **2**, 633; D. Wenkert, S.B. Ferguson, B. Porter, A. Qvarnstrom and A.T. McPhail, *J. Org. Chem.*, 1985, **50**, 4114.

13 P. DeShong, D.A. Kell and D.R. Sidler, *J. Org. Chem.*, 1985, **50**, 2309; P. DeShong and D.A. Kell, *Tetrahedron lett.*, 1986, **27**, 3979.

14 S. Takano, Y. Iwabuchi and K. Ogasawara, *J. Am. Chem. Soc.*, 1987, **109**, 5523.

15 quoted in E. Vedejs and J.W. Grisson, *J. Org. Chem.*, 1988, **53**, 1882.

16 E. Vedejs and J.W. Grisson, *J. Am. Chem. Soc.*, 1986, **108**, 6433; 1988, **110**, 3238; see also E. Vedejs and J.W. Grisson, *J. Org. Chem.*, 1988, **53**, 1876.

17 R. Grigg, H.Q.N. Gunaratne and J. Kemp, *J. Chem. Soc. Perkin 1*, 1984, 41; R. Grigg and H.Q.N. Gunaratne, *Tetrahedron Lett.*, 1983, **24**, 4457.

18 R. Grigg, *Chem. Soc. Rev.*, 1987, **16**, 89.

19 R. Grigg and J. Kemp, *J. Chem. Soc. Chem. Commun.*, 1977, 125.

20 P. Armstrong, R. Grigg, M.W. Jordan and J.F. Malone, *Tetrahedron*, 1985, **41**, 3547.

21 P.N. Confalone and E.M. Huie, *J. Am. Chem. Soc.*, 1984, **106**, 7175; P.N. Confalone and R.A. Earl, *Tetrahedron Lett.*, 1986, **27**, 2695.

22 C-L.J. Wang, W.C. Ripka and P.N. Confalone, *Tetrahedron Lett.*, 1984, **25**, 4613.

23 Y. Terao, M. Aono and K. Achiwa, *Heterocycles*, 1988, **27**, 981.

24 E. Vedejs and G.R. Martinez, *J. Am. Chem. Soc.*, 1979, **101**, 6452; cf. also A. Padwa, G. Haffmanns and M. Tomas, *J. Org. Chem.*, 1984, **49**, 3314.

25 E. Vedejs and G.A. Martinez, *J. Am. Chem. Soc.*, 1980, **102**, 7993; E.

Vedejs, S. Larsen and F.G. West, *J. Org. Chem.*, 1985, **50**, 2170.

26 A. Padwa and Y-Y. Chen, *Tetrahedron Lett.*, 1983, **24**, 3447.

27 R. Smith and D. Livinghouse, *J. Org. Chem.*, 1983, **48**, 1554; *Tetrahedron*, 1985, **41**, 3559.

28 M. Westling, R. Smith and T. Livinghouse, *J. Org. Chem.*, 1986, **51**, 1160.

29 R. Grigg and S. Thianpatanagul, *J. Chem. Soc. Chem. Commun.*, 1984, 180; R. Grigg, S. Surendrakumar, S. Thianpatanagul and D. Vipond, *J. Chem. Soc. Chem. Commun.*, 1987, 47; see also F. Orsini, F. Pelizzoni, M. Forte, R. Destro and P. Gariboldi, *Tetrahedron*, 1988, **44**, 519; R. Grigg, *Chem. Soc. Rev.*, 1987, **16**, 89.

30 R. Grigg, J. Idle, P. McMeekin and D. Vipond, *J. Chem. Soc. Chem. Commun.*, 1987, 49; H. Ardill, R. Grigg, V. Sridharan and S. Surendrakumar, *Tetrahedron*, 1988, **44**, 4953.

31 R. Grigg, M.F. Aly, V. Sridharan and S. Thiapatanagul, *J. Chem. Soc. Chem. Commun.*, 1984, 182.

32 D.A. Barr, R. Grigg, H.Q.N. Gunaratne, J. Kemp, P. McMeekin and V. Sridharan, *Tetrahedron*, 1988, **44**, 557.

33 C. Grundmann and P. Grünanger, *The Nitrile Oxides*, Springer, Berlin, 1971; M. Christle and R. Huisgen, *Chem. Ber.*, 1973, **106**, 3345; R. Huisgen, *J. Org. Chem.*, 1976, **41**, 403; S.A. Lang, Jr., and Y. Lin in *Comprehensive Heterocyclic Chemistry*, Eds. A.R. Katritzky and C.W. Rees, Vol.6, Pergamon, 1984, p.1; A. Padwa in *1,3-Dipolar Cycloaddition Chemistry*, Ed. A. Padwa, Wiley, 1984, Vol.2, p.368.

34 cf. V. Jäger, I. Muller, R. Schohe, M. Frey, R. Ehrler, B. Häfele and D. Schröter, *Lect. Heterocycl. Chem.*, 1985, **8**, 79 and references cited there.

35 cf. P.G. Baraldi, A. Barco, S. Benetti, G.P. Pollini and D. Simoni, *Synthesis*, 1987, 857.

36 K.N. Houk, J. Sims, C-R. Watts and L.J. Luskus, *J. Am. Chem. Soc.*, 1973, **95**, 7301; K.N. Houk and K. Yamaguchi in *1,3-Dipolar Cycload-*

*dition Chemistry*, ed. A. Padwa, Wiley Interscience, New York, 1984, Vol.2, p.407.

37 eg. C. Grundmann and R. Richter, *J. Org. Chem.*, 1968, **33**, 476; A.P. Kozikowski and X-M. Cheng, *Tetrahedron Lett.*, 1987, **28**, 3189.

38 R. Huisgen, *Angew. Chem. internat. edn.*, 1963, **2**, 565.

39 T. Mukaiyama and T. Hoshino, *J. Am. Chem. Soc.*, 1960, **82**, 5339; see also S.C. Sharma and K.B.G. Torssell, *Acta Chem. Scand.*, 1979, **B33**, 379.

40 G.A. Lee, *Synthesis*, 1982, 508; K.E. Larson and K.B.G. Torssell, *Tetrahedron*, 1984, **40**, 2985; K.B.G. Torssell, A.C. Hazell and R.G. Hazell, *Tetrahedron*, 1985, **41**, 5569.

41 Compare V. Jäger, H. Grund, V. Buss, W. Schwab, I. Müller, R. Schohe, R. Franz and R. Ehrler, *Bull. Soc. Chim. Belg.*, 1983, **92**, 1039.

42 A.P. Kozikowski and M. Adamczyk, *J. Org. Chem.*, 1983, **48**, 366.

43 Compare also I. Müller and V. Jäger, *Tetrahedron*. 1985, **41**, 4777.

44 K.B.G. Torssell, A.C. Hazell and R.G. Hazell, *Tetrahedron.*, 1985, **41**, 5569.

45 A.P. Kozikowski and M. Adamczyk, *Tetrahedron Lett.*, 1982, **23**, 3123; D.P. Curran, *J. Am. Chem. Soc.*, 1982, **104**, 4024.

46 D.P. Curran, *Tetrahedron Lett.*, 1983, **24**, 3443.

47 M. Asaoka, M. Abe and H. Takei, *Bull. Chem. Soc. Japan*, 1985, **58**, 2145; for other examples of the application of nitrile-oxide cycloaddition in the synthesis of natural products see A.P. Kozikowski and X-M. Cheng, *Tetrahedron Lett.*, 1987, **28**, 3189; A.P. Kozikowski and B.B. Mugrage, *J. Chem. Soc. Chem. Commun.*, 1988, 198; A.P. Kozikowski and P. Park, *J. Am. Chem. Soc.*, 1985, **107**, 1763; D.L. Comins and A.H. Abdullah, *Tetrahedron Lett.*, 1985, **26**, 43; A.P. Kozikowski and P.D. Stein, *J. Am. Chem. Soc.*, 1982, **104**, 4023; A.P. Kozikowski, B.B. Mugrage, B.C. Wang and Z. Xu, *Tetrahedron Lett.*, 1983, **24**, 3705; A.P. Kozikowski, S.H. Jung and J.P. Springer, *J. Chem. Soc. Chem.*

*Commun.*, 1988, 167; D.P. Curran, P.B. Jacobs, R.L. Elliot and B.H. Kim, *J. Am. Chem. Soc.*, 1987, **109**, 5280; A.P. Kozikowski and J.G. Scripko, *J. Am. Chem. Soc.*, 1984, **106**, 353.

48 A.P. Kozikowski and A.K. Ghosh, *J. Am. Chem. Soc.*, 1982, **104**, 5788; *J. Org. Chem.*, 1984, **49**, 2762.

49 see, e.g. A.P. Kozikowski and C-S. Li, *J. Org. Chem.*, 1985, **50**, 778.

50 K.N. Houk, S.R. Moses, Y-D. Wu, N.G. Rondan, V. Jäger, R. Schohe and F.R. Fronczek, *J. Am. Chem. Soc.*, 1984, **106**, 3880.

51 cf. K.N. Houk, H-Y. Duh, Y-D. Wu and S.R. Moses, *J. Am. Chem. Soc.*, 1986, **108**, 2754; A.P. Kozikowski and Y.Y. Chen, *Tetrahedron Lett.*, 1982, **23**, 2081.

52 K.N. Houk, S.R. Moses, Y-D. Wu, N.G. Rondan, V. Jäger, R. Schohe and F.R. Fronczck, *J. Am. Chem. Soc.*, 1984, **106**, 3880; K.N.Houk, H-Y. Duh, Y-D. Wu and S.R. Moses, *J. Am. Chem. Soc.*, 1986, **108**, 2754; A.P. Kozikowski and A.K. Ghosh, *J. Org. Chem.*, 1984, **49**, 2762; V. Jäger, R. Schohe and E.F. Paulus, *Tetrahedron Lett.*, 1983, **24**, 5501.

53 K.N. Houk, H-Y. Duh, Y-D. Wu and S.R. Moses, *J. Am. Chem. Soc.*, 1986, **108**, 2754.

54 see, for example A.P. Kozikowski and Y.Y. Chen, *Tetrahedron Lett.*, 1982, **23**, 2081; R. Annunziata, M. Cinquini, F. Cozzi and L. Raimondi, *J. Chem. Soc. Chem. Commun.*, 1987, 529; R. Annunziata, M. Cinquini, F. Cozzi, C. Gennari and L. Raimondi, *J. Org. Chem.*, 1987, **52**, 4674.

55 V. Jäger, V. Buss and W. Schwab, *Tetrahedron Lett.*, 1978, **34**, 3133.

56 A.P. Kozikowski and A.K. Ghosh, *J. Org. Chem.*, 1984, **49**, 2762.

57 see, for example, R.H. Jones, G.C. Robinson and E.J. Thomas, *Tetrahedron.*, 1984, **40**, 177.

58 A.P. Kozikowsky, Y. Kitagawa and J.P. Springer, *J. Chem. Soc. Chem. Commun.*, 1983, 1460; for another example see A.P. Kozikowski and X-M. Cheng, *Tetrahedron Lett.*, 1987, **28**, 3189.

59 D.P. Curran, *J. Am. Chem. Soc.*, 1982, **104**, 4024.

60 D.P. Curran, *J. Am. Chem. Soc.*, 1983, **105**, 5826.

61 D.P. Curran, *J. Am. Chem. Soc.*, 1982, **104**, 4024; 1983, **105**, 5826; see also V. Jäger and W. Schwab, *Tetrahedron Lett.*, 1978, 3129.

62 V. Jäger, V. Buss and W. Schwab, *Tetrahedron Lett.*, 1978, **34**, 3133.

63 V. Jäger, H. Grund, V. Buss, W. Schwab, I. Müller, R. Schohe, R. Franz and R.Ehrler, *Bull. Soc. Chim. Belg.*, 1983, **92**, 1039.

64 W. Schwab and V. Jäger, *Angew. Chem. internat. edn.*, 1981, **20**, 601,603.

65 cf. V. Jäger, I. Müller, R. Schohe, M. Frey, R. Ehrler, B. Häfell and D. Schröter, *Lect. Heterocycl. Chem.*, 1985, **8**, 79.

66 cf. A.P. Kozikowski, *Acc. Chem. Res.*, 1984, **17**, 410.

67 V. Jäger and R. Schohe, *Tetrahedron*, 1984, **40**, 2199.

68 I. Müller and V. Jäger, *Tetrahedron Lett.*, 1982, 4777.

69 A.P. Kozikowski, Y-Y. Chen, B.C. Wang and Z-B. Xu, *Tetrahedron*, 1984, **40**, 2345.

70 P.N. Confalone and E.M. Huie, *Organic Reactions*, 1988, **36**, 1; D. St. C. Black, R.F. Crozier and V.C. Davis, *Synthesis*, 1975, 205; J.J. Tufariello in *1,3-Dipolar Cycloaddition Chemistry*, Ed. A. Padwa, Wiley, 1984, Vol.2, p.83.

71 J. Sims and K.N. Houk, *J. Am. Chem. Soc.*, 1973, **95**, 5798; K.N. Houk, A. Bimanand, D. Mukherjee, J. Sims, Y-M. Chang, D.C. Kaufman and L.N. Domelsmith, *Heterocycles*, 1977, **7**, 293; D. Mukherjee, L.N. Domelsmith and K.N. Houk, *J. Am. Chem. Soc.*, 1978, **100**, 1954; A.Z. Bimananda and K.N. Houk, *Tetrahedron Lett.*, 1983, **100**, 1954; K.N. Houk, *Acc. Chem. Res.*, 1975, **8**, 361.

72 R. Huisgen, R. Grashey, H. Hauck and H. Seidl, *Chem. Ber.*, 1968,

**101**, 2043, 2548; R. Huisgen, R. Grashey, H. Seidl and H. Hauck, *Chem. Ber.*, 1968, **101**, 2559; A. Padwa, L. Fisera, K.F. Koehler, A. Rodriguez and G.S.K. Wong, *J. Org. Chem.*, 1984, **49**, 276.

73 R. Sustmann, *Pure Appl. Chem.*, 1974, **40**, 569; K.N. Houk, *Acc. Chem. Res.*, 1975, **8**, 361; K.N. Houk and K. Yamaguchi in *1,3-Dipolar Cycloaddition Chemistry*, Ed. A. Padwa, 1984, Wiley, New York, Vol.2, p.407.

74 cf. J.J. Tufariello and J.M. Puglis, *Tetrahedron Lett.*, 1986, **27**, 1265 and references cited there.

75 A. Padwa, L. Fisera, K.F. Koehler, A. Rodriguez and G.S.K. Wong, *J. Org. Chem.*, 1984, **49**, 276.

76 J.J. Tufariello and J.M. Puglis, *Tetrahedron Lett.*, 1986, **27**, 1263.

77 A. Toy and W.J. Thompson, *Tetrahedron Lett.*, 1984, **25**, 3533.

78 N.A. LeBel and J.J. Whang, *J. Am. Chem. Soc.*, 1959, **81**, 6334; A.C. Cope and N.A. LeBel, *J. Am. Chem. Soc.*, 1960, **82**, 4656; N.A. LeBel, *Trans. N.Y. Acad. Sci.*, 1965, **27**, 858; M. Raban, F.B. Jones, E.H. Carlton, E. Banucci and N.A. LeBel, *J. Org. Chem.*, 1970, **35**, 1496; N.A. LeBel, M.E. Post and J.J. Whang, *J. Am. Chem. Soc.*, 1964, **86**, 3759; N.A. LeBel and T.A. Lajiness, *Tetrahedron Lett.*, 1966, 2173; S.W. Baldwin, J.D. Wilson and J. Aubé, *J. Org. Chem.*, 1985, **50**, 4432.

79 cf. W. Oppolzer, S. Siles, R.L. Snowden, B.H. Bakker and M. Petrzilka, *Tetrahedron*, 1985, **41**, 3497.

80 E. Gossinger and B. Witkop, *Monat. für Chemie.*, 1980, **111**, 803; see also, E. Gossinger, *Tetrahedron Lett.*, 1980, **21**, 2229.

81 S.I. Murahashi, T. Oda, T. Sugahara and Y. Masui, *J. Chem. Soc. Chem. Commun.*, 1987, 1471; S.I. Murahashi and T. Shiota, *Tetrahedron Lett.*, 1987,**28**, 2383.

82 D.R. Adams, W. Carruthers, P.J. Crowley and M.J. Williams *J. Chem. Soc. Perkin 1*, 1989, 1507; see also R. Zschiesche and H-U. Reissig, *Tetrahedron Lett.*, 1988, **29**, 1685.

83 A.B. Holmes, C. Swithenbank and S.F. Williams, *J. Chem. Soc. Chem. Commun.*, 1986, 265.

84 R. Grigg, *Chem. Soc. Rev.*, 1987, **16**, 89; P. Armstrong, R. Grigg, S. Surendrakumar and W.J. Warnock, *J. Chem. Soc. Chem. Commun.*, 1987, 1327.

85 D. Lathbury and T. Gallagher, *Tetrahedron Lett.*, 1985, **26**, 6249; *J. Chem. Soc. Chem. Commun.*, 1986, 1017.

86 P. DeShong and J.M. Leginus, *J. Org. Chem.*, 1984, **49**, 3421.

87 P. DeShong, J.M. Leginus and S.W. Lander, *J. Org. Chem.*, 1986, **51**, 574.

88 J.J. Tufariello, *Acc. Chem. Res.*, 1979, 396.

89 P. DeShong, C.M. Dicken, J.M. Leginus and R.R. Whittle, *J. Am. Chem. Soc.*, 1984, **106**, 5598.

90 P. DeShong, C.M. Dicken, J.M. Leginus and R.R. Whittle, *J. Am. Chem. Soc.*, 1984, **106**, 5598.

91 T. Kametani, S-D. Chu and T. Honda, *J. Chem. Soc. Perkin 1*, 1988, 1593.

92 M. Ihara, M.Takahashi, K. Fukumoto and T. Kametani, *J. Chem. Soc. Chem. Commun.*, 1988, 9.

93 W.R. Roush and A.E. Watts, *J. Am. Chem. Soc.*, 1984, **106**, 721.

94 D. Keirs, D. Moffat and K. Overton, *J. Chem. Soc. Chem. Commun*, 1988, 654; for other examples see H. Iida, K. Kasahara and C. Kibayashi, *J. Am. Chem. Soc.*, 1986, **108**, 4647; T. Kametani, T. Nagahara and T. Honda, *J. Org. Chem.*, 1985, **50**, 2327; C. Belzecki and I. Panfil, *J. Org. Chem.*, 1979, **44**, 1212; A. Vasella and R. Voeffray, *Helv. Chim. Acta*, 1982, **65**, 1983; C.M. Tice and B. Ganem, *J. Org. Chem.*, 1983, **48**, 5048.

95 T. Koizumi, H. Hirai and E. Yoshii, *J. Org. Chem.*, 1982, **47**, 4005.

96 R. Annunziata, M. Cinquini, F. Cozzi and L. Raimondi, *Tetrahedron*, 1987, **43**, 4051.

97 S. Mzengeza, C.M. Yang and R.A. Whitney, *J. Am. Chem. Soc.*, 1987, **109**, 276.

98 S. Masamune, W. Choy, J.S. Petersen and L.R. Sita, *Angew. Chem. internat. edn.*, 1985, **24**, 1.

99 See, e.g. J.J. Tufariello, *Acc. Chem. Res.*, 1979, **12**, 396; S.A. Ali, J.H. Khan, M.I.M. Wazeer, *Tetrahedron*, 1988, **44**, 5911.

100 J.J. Tufariello and S.A. Ali, *Tetrahedron Lett.*, 1978, 4647.

101 cf. also C. Hootelé, W. Ibebeke-Bomangwa, F. Driessens and S. Sabil, *Bull. Soc. Chim. Belg.*, 1987, **96**, 57; W. Ibebeke-Bomangwa and C. Hootelé, *Tetrahedron*, 1987, **43**, 935.

102 J.J. Tufariello and J.M. Puglis, *Tetrahedron Lett.*, 1986, **27**, 1265; for other reactions involving addition to substituted butadienes see J.J. Tufariello and R.C. Gatrone, *Tetrahedron Lett.*, 1978, 2753; J.J. Tufariello and A.D. Dyszlewski, *J. Chem. Soc. Chem. Commun.*, 1987, 1138.

103 cf. N.A. LeBel and N.A. Spurlock, *J. Org. Chem.*, 1964, **29**, 1337; N.A. LeBel, *Trans. N.Y. Acad. Sci.*, 1965, **27**, 858; N.A. LeBel, M.E. Post and D. Huang, *J. Org. Chem.*, 1979, **44**, 1819; J.J. Tufariello and G.B. Mullen, *J. Am. Chem. Soc.*, 1978, **100**, 3638.

104 see, for example, J.J. Tufariello and J.M. Puglis, *Tetrahedron Lett.*, 1986, **27**, 1489.

105 J.J. Tufariello and J.P. Tette, *J. Org. Chem.*, 1975, **40**, 3866; cf. also J.J. Tufariello and K. Winzenberg, *Tetrahedron Lett.*, 1986, **27**, 1645; T. Iwashita, T. Kusumi and H. Kakisawa, *J. Org. Chem.*, 1982, **47**, 230 for other examples.

106 E.G. Baggiolini, J.A. Jacobelli, B.M. Hennessy, A.D. Batcho, J.F. Sereno and M.R. Uskokovic, *J. Org. Chem.*, 1986, **51**, 3098; see also P.M. Wovkulich, F. Barcelos, A.D. Batcho, J.F. Sereno, E.G. Baggiolini, B.M. Hennessy and M.R. Uskokovic, *Tetrahedron*, 1984, **40**, 2283.

107 A. Padwa, U. Chiacchio, D.N. Kline and J. Perumattan, *J. Org. Chem.*, 1988, **53**, 2338.

108 cf. G. Bianchi, C. de Micheli and R. Gandolfi, *Angew. Chem. internat. edn.*, 1979, **18**, 721.

109 see e.g. N.A. LeBel and T.A. Lajiness, *Tetrahedron Lett.*, 1966, 2173; N.A. LeBel and E.G. Banucci, *J. Org. Chem.*, 1971, **36**, 2440.

110 J.J. Tufariello and G.B. Mullen, *J. Am. Chem. Soc.*, 1978, **100**, 3638; J.J. Tufariello, G.B. Mullen, J.J. Tegeler, E.J. Trybulski, S.C. Wong and S.A. Ali, *J. Am. Chem. Soc.*, 1979, **101**, 2435.

111 W. Lwowski in *1,3-Dipolar Cycloaddition Chemistry*, Ed. A. Padwa, Wiley Interscience, New York, 1984, Vol.1, p.559; P.K. Kadaba, B. Stanovnik and M. Tisler in *Advances in Heterocyclic Chemistry*, Ed. A.R. Katritzky, Academic Press, Vol.37, 1984, p.217.

112 R. Huisgen, G. Szeimies and L. Möbius, *Chem. Ber.*, 1966, **99**, 475.

113 J.G. Buchanan, A.R. Edgar and B.D. Hewitt, *J. Chem. Soc. Perkin 1*, 1987, 2371.

114 A.L. Logothetis, *J. Am. Chem. Soc.*, 1965, **87**, 749.

115 A.P. Kozikowski and M.N. Greco, *Tetrahedron Lett.*, 1982, **23**, 2005.

116 A. Padwa, *Angew. Chem. internat. edn.*, 1976, **15**, 123.

117 W. Oppolzer, *Tetrahedron lett.*, 1970, 2199.

118 cf. A.Padwa in *1,3-Dipolar Cycloaddition Chemistry, ed. A. Padwa, Wiley Interscience, New York, 1984, Vol.2, p.277; W. Oppolzer, Tetrahedron Lett.*, 1970, 3091.

119 W. Oppolzer, *Tetrahedron Lett.*, 1972, 1707.

120 P.A. Jacobi, A. Browstein, M. Martinelli and K. Grozinger, *J. Am. Chem. Soc.*, 1981, **103**, 239.

121 cf. R. Grashey in *1,3-Dipolar Cycloaddition Chemistry*, Ed. A. Padwa,

Wiley-Interscience, New York, 1984, Vol.1, p.733.

122 cf. R. Huisgen in *1,3-Dipolar Cycloaddition Chemistry*, ed. A. Padwa, Wiley-Interscience, 1984, Vol.1, p.1; M. Regitz and H. Heydt, *loc.cit.*, p.393.

123 M.P. Schneider and M. Goldbach, *J. Am. Chem. Soc.*, 1980, **102**, 6114.

124 E. Piers, R.W. Britton, R.J. Keziere and R.D. Smillie, *Canad. J. Chem.*, 1971, **49**, 2623.

125 A. Padwa, A. Rodriguez, M. Tohida and T. Fukumaga, *J. Am. Chem. Soc.*, 1983, **105**, 933.

# 7 [2+2] CYCLOADDITION REACTIONS

[2+2] Cycloadditions give rise to four-membered rings, and may be either concerted or non-concerted.[1] Thermal concerted [2+2] cycloadditions have to be antarafacial on one component,[2] and the geometrical and orbital constraints thus imposed ensure that they are encountered only in special circumstances. Most thermal [2+2] cycloadditions of alkenes take place by a stepwise pathway involving diradical or zwitterionic intermediates.[3] Photochemical [2+2] cycloadditions, suprafacial on both components, and thus geometrically feasible, can be concerted, but in practice they do not always follow a concerted pathway.

## Ketenes

Concerted thermal [2+2] cycloadditions are encountered in the dimerisation of strained alkenes, where a strained, twisted double bond allows easier antarafacial addition of one component, and also, more importantly from the synthetic point of view, in the reactions of cumulenes, especially ketenes, with olefinic and other multiple bonds. Ketenes have been widely employed in synthesis in this way to prepare cyclobutanones and β-lactams by reaction with alkenes and imines.[4]

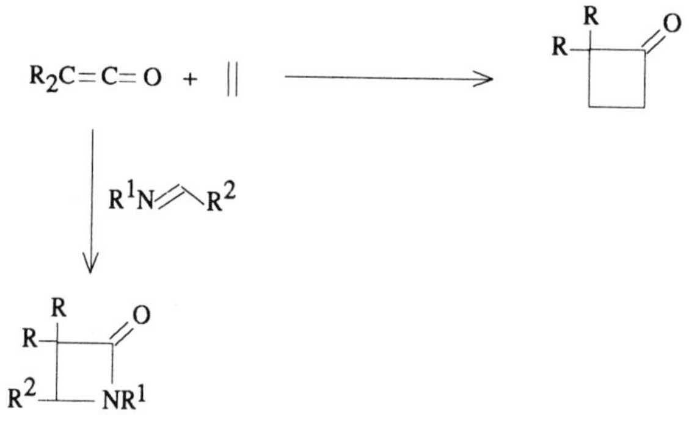

$R_2C{=}C{=}O$ + ||

$R^1N$ $R^2$

R R O NR1 R2

*Scheme 1*

These reactions have been regarded as concerted cycloadditions involving antarafacial addition to the ketene[2], but it appears that there may be a range of mechanisms of various degrees of concertedness.[5] In favour of the concerted pathway is the fact that the reactions are frequently highly stereoseletive, but the regiochemistry of many adducts is in line with an asymmetric transition state (as 1) in which the carbonyl carbon of the ketene is more closely bound to the more nucleophilic carbon of the alkene. (Scheme 2).

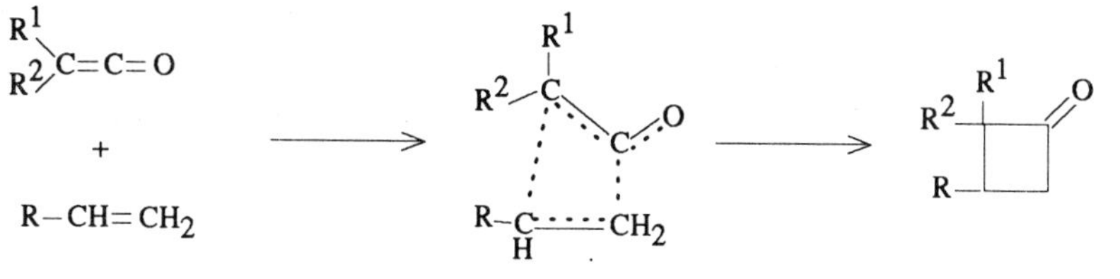

$R^1R^2C{=}C{=}O$ + $R{-}CH{=}CH_2$ → (1) → cyclobutanone

(1)

*Scheme 2*

There have been numerous applications of the cycloaddition of ketenes to alkenes to form cyclobutanone derivatives. Thus, reaction between dimethylketene, liberated from the acetal (2) and 2-*p*-tolylpropene gave the cyclobutanone (3) which was transformed into (±)-α- and (±)-β-cuparenones.[6]

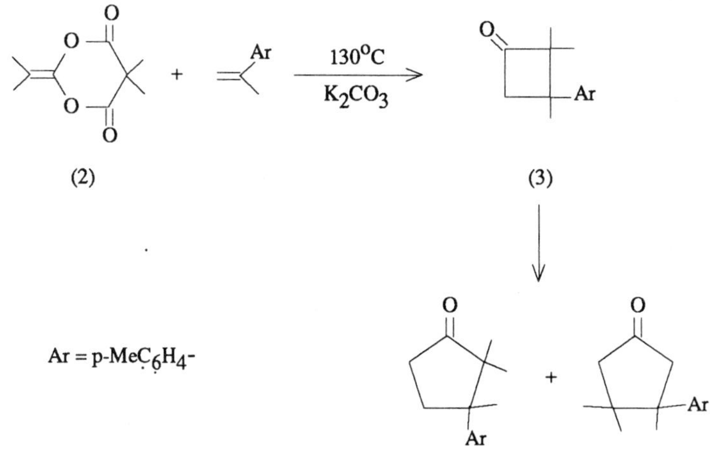

130°C, $K_2CO_3$

(2) (3)

Ar = p-$MeC_6H_4$-

*Scheme 3*

As this synthesis illustrates, the value of the [2+2] cycloaddition reactions of ketene in synthesis goes beyond the straightforward formation of four-membered rings, for in many cases subsequent ring-expansion or ring-contraction or rearrangement provides access to otherwise difficultly accessible compounds.

In intermolecular reactions it is frequently found advantageous to use chloro- or dichloro-ketenes rather than the less reactive alkyl- or aryl-ketenes which often give only poor yields of the desired adduct.[7] Halogenated ketenes are readily available, for example by dehydrohalogenation of α-halo acid chlorides, and generated *in situ* in the presence of an alkene they afford good yields of cycloadducts. The halogen in the α-halocyclobutanone is easily replaced by hydrogen if desired (Scheme 4).

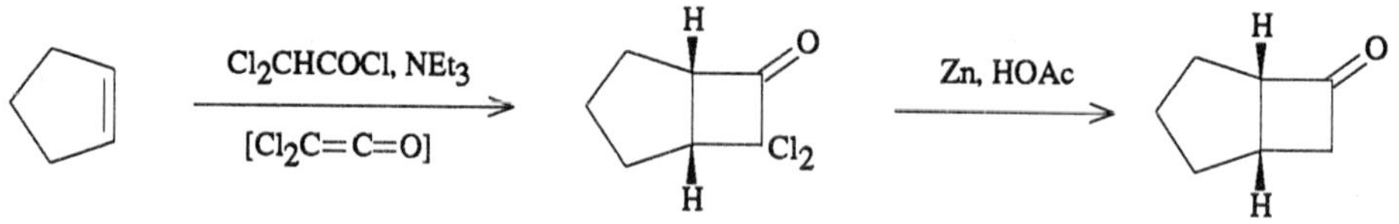

$Cl_2CHCOCl$, $NEt_3$ [$Cl_2C{=}C{=}O$] → (67%) → Zn, HOAc

*Scheme 4*

Alternatively, the halogen may be used to trigger a ring-contraction to a cyclopropane derivative. Thus, in the key step in a synthesis of the sesquiterpene (±)-sirenin (7), the α-chlorocyclobutanone (5), obtained by cycloaddition of chloromethylketene to the less hindered double bond of the diene (4), rearranged stereospecifically to the cyclopropanecarboxylic ester (6) on treatment with silver nitrate in methanol.[8]

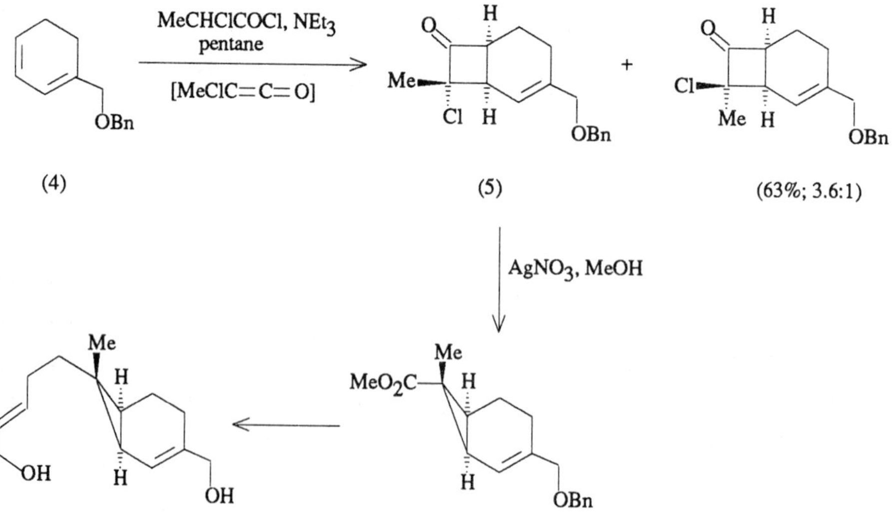

(4) — $MeCHClCOCl$, $NEt_3$, pentane [$MeClC{=}C{=}O$] → (5) + isomer (63%; 3.6:1)

(5) — $AgNO_3$, MeOH → (6) (53%) → (7)

*Scheme 5*

A number of difficultly accessible fused-ring cyclopropane derivatives has been made by this route.[9] The reaction is stereospecific; *endo* chloroketones yield the corresponding *endo* esters and *exo* chloroketones the *exo* esters.

In another useful transformation, modified Beckmann rearrangement of the N-methylnitrones derived from cyclobutanones followed by reduction of the resultant lactams, for example (10), gave *cis*-fused pyrrolidines (as 11) and this was used in the synthesis of the *Sceletium* alkaloid (±)-mesembrine.[10] Straightforward Beckmann rearrangement of the oxime of (9) gave the isomeric lactam (12).

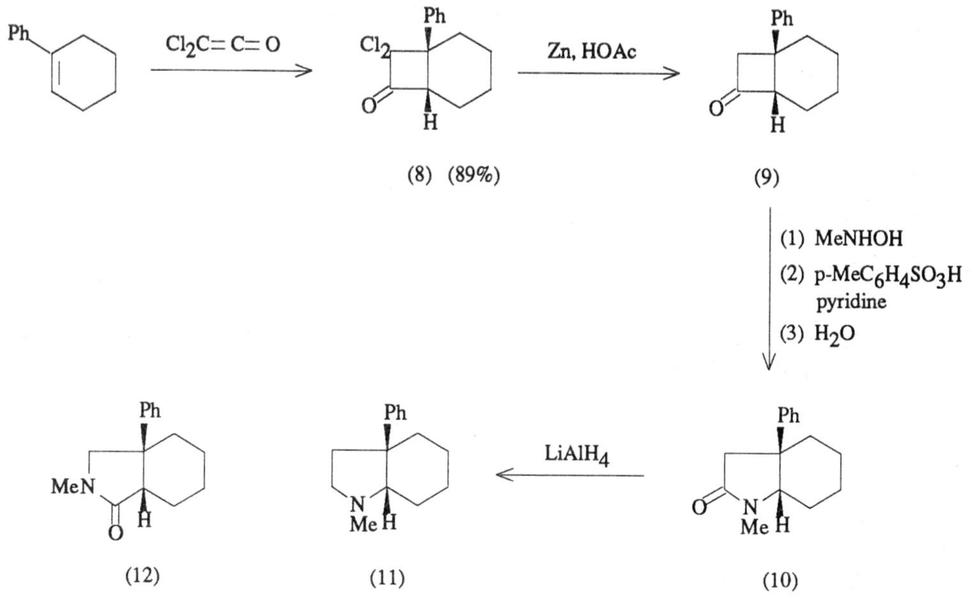

*Scheme 6*

Halogenated ketenes have also been employed in the synthesis of tropolones,[9] and eight-membered rings have been obtained by Cope rearrangement of 1,2-divinylcyclobutanones, themselves formed by [2+2] cycloaddition of vinylketenes to conjugated dienes.[11] The vinylketenes are generated by electrocyclic ring-opening of cyclobutenones or by 1,4-dehydrochlorination of αβ-unsaturated acid chlorides (Scheme 7).

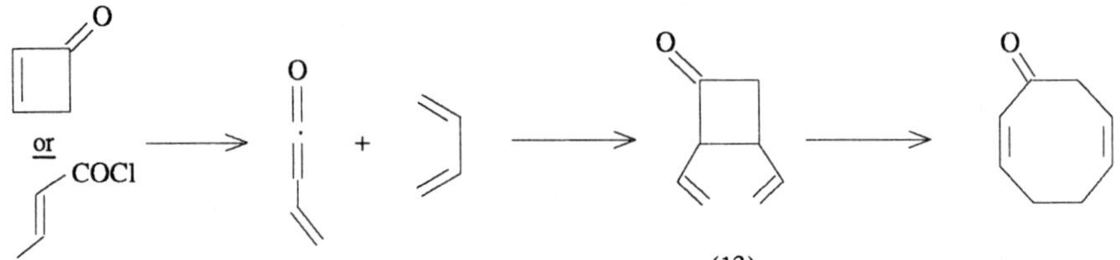

O

or

COCl

O

+

O

O

(13)

*Scheme 7*

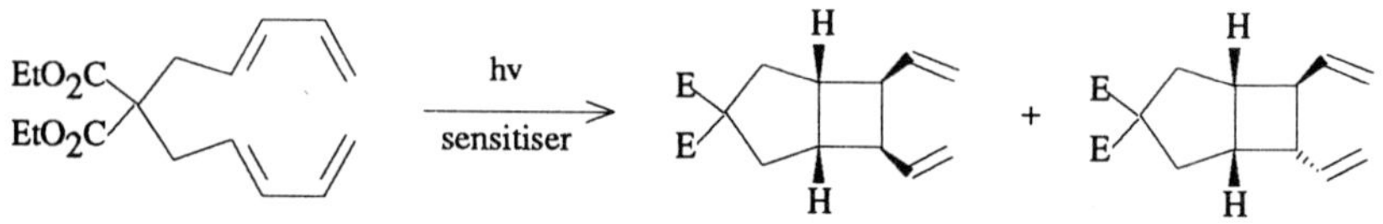

$EtO_2C$

$EtO_2C$

hv

sensitiser

H

E

E

H

+

H

E

E

H

(14)

(15)

(16)

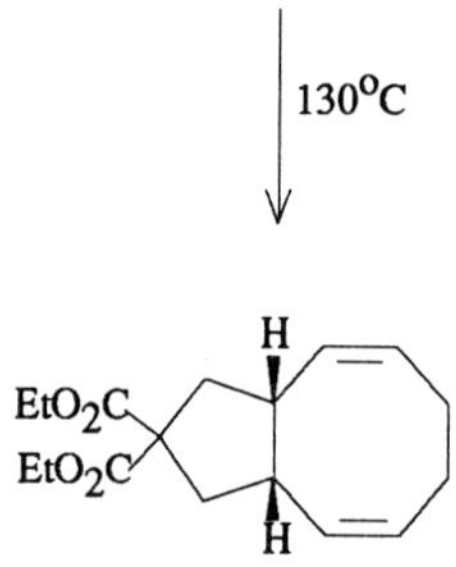

$130^oC$

H

$EtO_2C$

$EtO_2C$

H

(17)

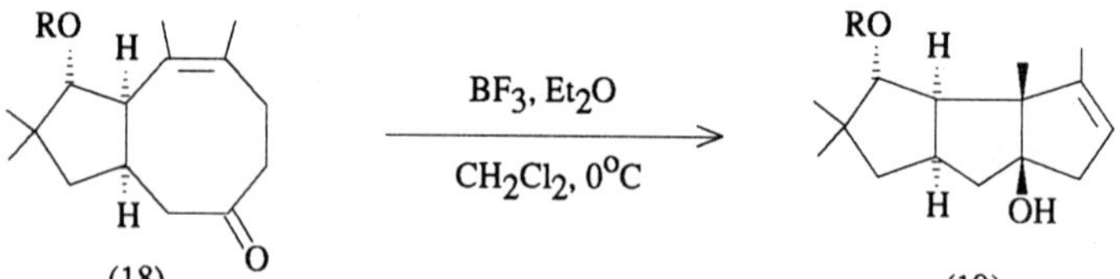

RO

H

H

O

(18)

$BF_3$, $Et_2O$

$CH_2Cl_2$, $0^oC$

RO

H

H

OH

(19)

HO

H

O

O

H

OH

O

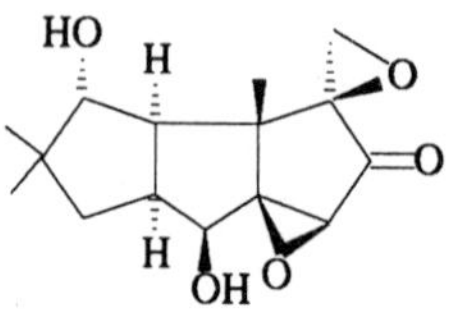

(20)

*Scheme 8*

In the cyclobutenone version of the reaction the enone is heated with the 1,3-diene in an inert solvent. Reaction proceeds through a succession of pericyclic reactions. Electrocyclic opening of the cyclobutenone generates a vinylketene derivative which combines with the conjugated diene in a regiospecific [2+2] cycloaddition. At the elevated reaction temperature the resulting 2,3-divinylcyclobutanone undergoes a [3+3] sigmatropic rearrangement to give the cyclo-octadienone. In some cases the intermediate divinylcyclobutanones can be isolated. Alkylated 2-vinylcyclobutanones have been obtained by reaction of methyl- and ethyl-vinylketene with alkenes; they can be rearranged to cyclopentenones with acid.[12]

In a related sequence, photolysis of the bis-diene (14) gave the 1,2-divinylcyclobutanes (15) and (16) which afforded the fused-ring cyclo-octadiene (17) on thermolysis.[13] The cyclo-octenone (18) derived in this way, on treatment with boron trifluoride etherate gave the tricyclic (19) which was converted in several more steps into (±)-coriolin (20) (Scheme 8).

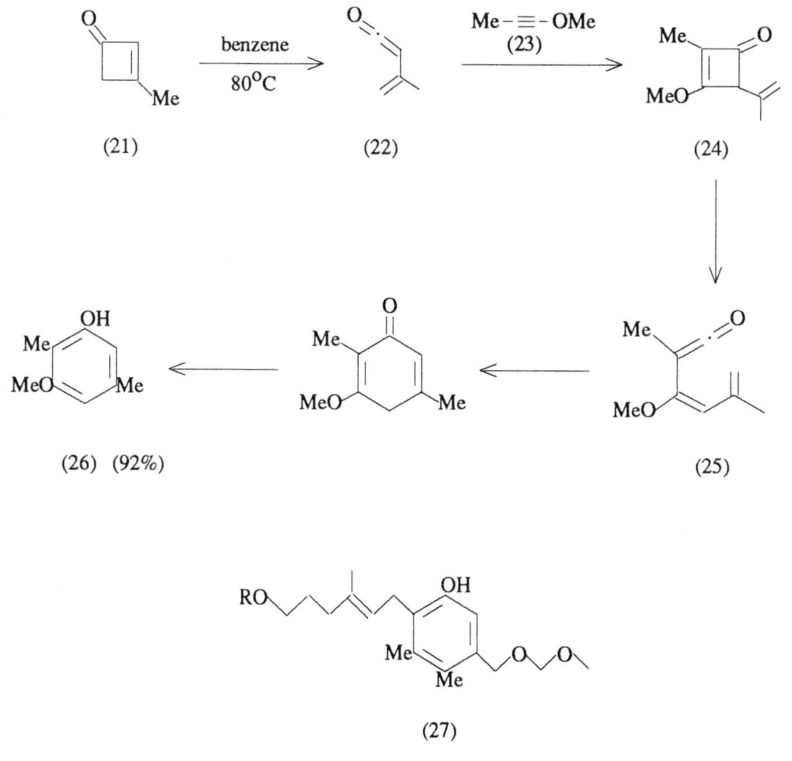

O Me benzene 80°C Me–≡–OMe (23) Me O MeO

(21) (22) (24)

OH Me MeO Me O Me MeO Me Me O MeO

(26) (92%) (25)

RO OH Me Me O O

(27)

*Scheme 9*

Ketenes also react with acetylenes to form cyclobutenones, and such a reaction forms a stage in an intriguing regiocontrolled ring synthesis of highly substituted benzene derivatives by a one-step thermal combination of a heterosubstituted alkyne with a cyclobutenone, followed by three successive pericyclic reactions. Thus, the vinylketene (22), obtained by thermal cleavage of the cyclobutenone (21), combines with the ketenophilic acetylene (23) to give the cyclobutenone (24). 4-Electron cycloreversion to (25) followed by 6-electron cyclisation gave the cyclohexadienone and thence the substituted phenol (26).[14] A similar sequence was used to make the highly substituted benzene derivative (27) required as an intermediate in the synthesis of mycophenolic acid (Scheme 9).[15]

Diastereoselective [2+2] cycloadditions of ketenes have been effected using both ketenes and alkenes bearing optically active auxiliary groups. Thus, the enol ethers (28), carrying the chiral auxiliary groups (31) and (32) reacted with dichloroketene to give the cyclobutanones (29) and (30) with up to 80 per cent diastereomeric excess. These were transformed into the corresponding cyclopentanones with diazomethane and thence into α-chlorocyclopentenones.[16]

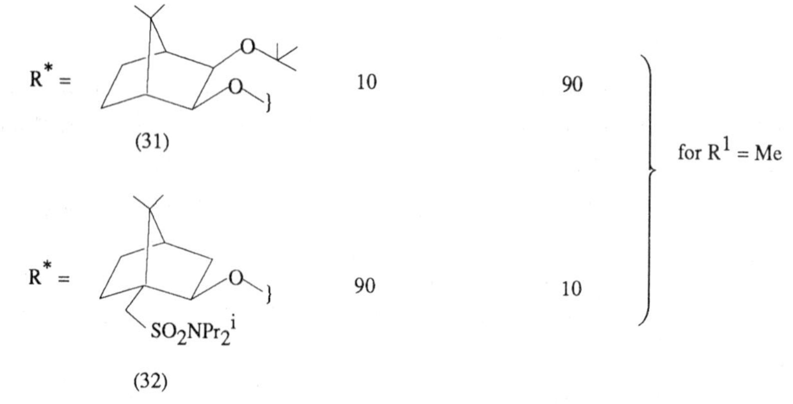

*Scheme 10*

Even better results were obtained with the enol ether (33) derived from (1*S*,2*R*)-2-phenylcyclohexanol. With dichloroketene it gave the adduct (34) with greater than 95 per cent diastereomeric excess by selective attack of the ketene on the back (*re*) face of the double bond. A single crystallisation gave the pure diastereomer. It is suggested that the enol ether reacts in the conformation (33) in which the front (*si*) face of the double bond is shielded by the phenyl substituent. Ring expansion of (34) with diazomethane and further transformations gave (–)-α-cuparenone (35) and (+)-β-cuparenone (36).[17]

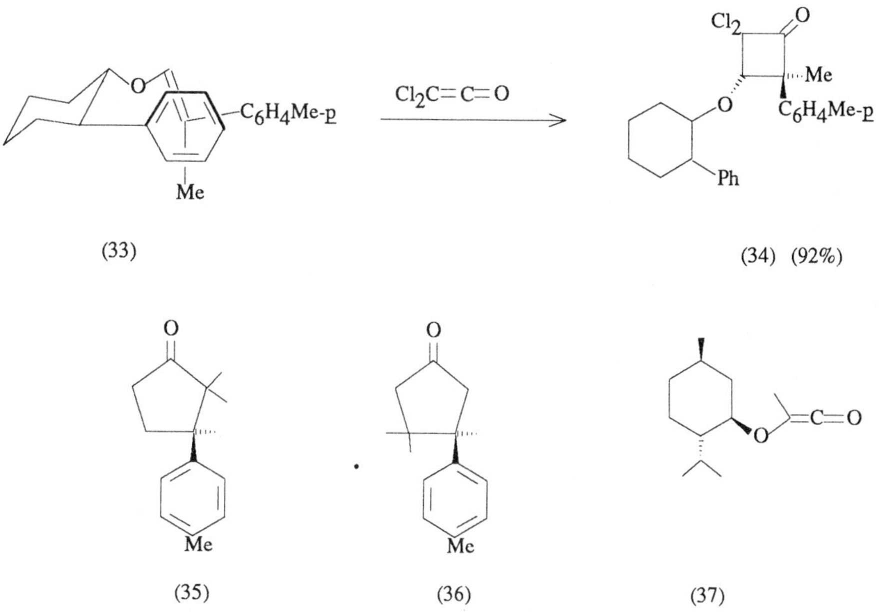

(33) $\xrightarrow{Cl_2C{=}C{=}O}$ (34) (92%)

(35) (36) (37)

*Scheme 11*

The optically active ketene (37), with the optically active auxiliary menthyloxy group at C-2 of the ketene, gave cyclobutanones with diastereomeric excess of 50-70 per cent in reactions with *cis* enol ethers.[18]

Ketenes add to other multiple bonds besides carbon-carbon double bonds.[4] Imines, for example, react readily to give β-lactams, although there is some doubt that these are truly concerted reaction.[19] With (*E*)-imines the reaction generally affords the *cis*-disubstituted azetidinones with high enantioselectivity, a discovery made by Hubschwerlen[19] and utilised in a highly enantioselective synthesis of the monobactam antibiotics. This work has been extended by others[20] and utilises optically-active auxiliary groups. Thus, reaction of oxazolidone (38), derived from (*S*)-phenylglycine, with a range of benzylimines (39) in presence of triethylamine, proceeded with very high

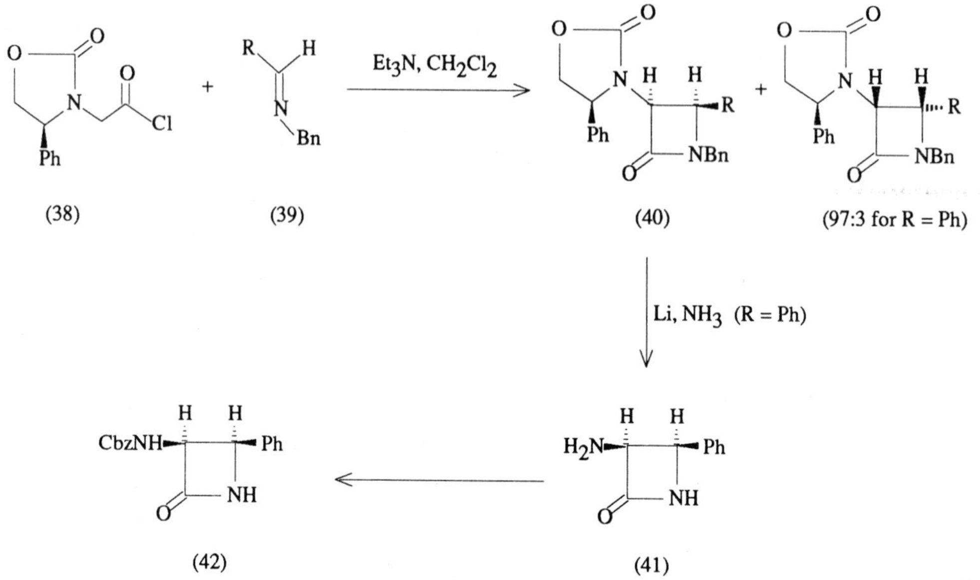

O O O N Cl Ph + R H N Bn

Et$_3$N, CH$_2$Cl$_2$

O O N H H R Ph NBn O + O O N H H R Ph NBn O

(38) (39) (40) (97:3 for R = Ph)

Li, NH$_3$ (R = Ph)

H H CbzNH Ph NH O

H H H$_2$N Ph NH O

(42) (41)

*Scheme 12*

H O O H NAr

MeOCH$_2$COCl
NEt$_3$, CH$_2$Cl$_2$
[MeOCH=C=O]
(44)

H H H MeO O O NAr O

(43) (45) (54%; only product)

Ar = p-MeOC$_6$H$_4$-

CF$_3$CO$_2$H

HOCH$_2$ O NHAr H O H H OMe

H H MeO CHOHCH$_2$OH NAr O

(47) (46)

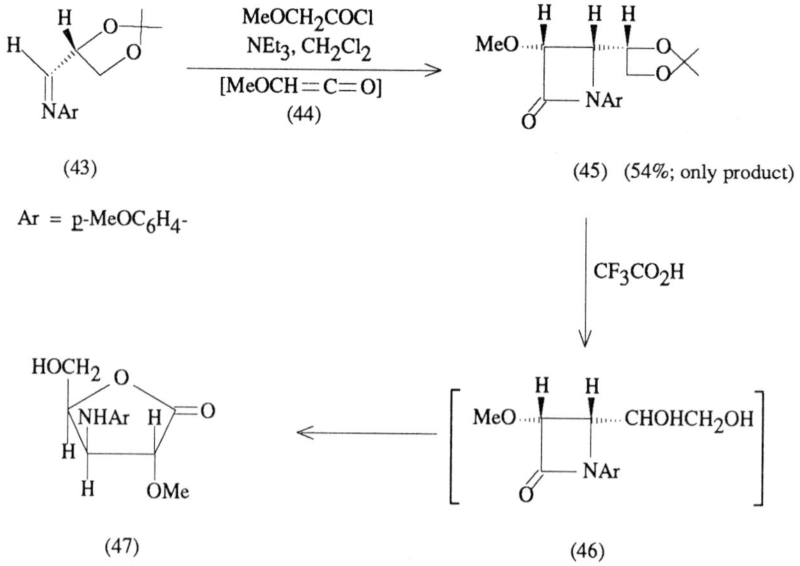

*Scheme 13*

levels of asymmetric induction to form cycloadducts (40).[20] Thus, for (39,R=Ph) the reaction was almost completely diastereoselective. The main product (40,R=Ph) was obtained pure after one crystallisation. Subsequent cleavage of the chiral auxiliary group and protection of the free amine with benzyl chloroformate gave the optically pure β-lactam (42) (Scheme 12).

Again, reaction of the optically active imine (43) and variously substituted ketenes, including (44), gave β-lactam adducts with very high diastereoselectivity. The absolute configuration of (45) was established by conversion into the lactone (47) (Scheme 13).[21] β-Lactams are also obtained by cycloaddition of chlorosulphonyl isocyanate to alkenes.[22]

As an alternative to free ketenes, ketene-iminium salts have given excellent results in cycloaddition reactions with alkenes.[23] They are easily prepared by reaction of tertiary amides with trifluoromethanesulphonic (triflic) anhydride and collidine, or from α-chloro enamines, and they have the advantage that they are more electrophilic than ketenes and do not dimerise. They react readily with monosubstituted and 1,2-substituted olefins, giving [2+2] cycloadducts which form cyclobutanones on hydrolysis. Yields are generally good, and in this respect keteneiminium salts are superior to the corresponding ketenes which react only sluggishly with unactivated olefins or acetylenes. Thus the dimethylamide of isobutyric acid, treated with triflic anhydride and collidine in the presence of styrene gave the adduct (48) which afforded the cyclobutanone (49) on hydrolysis in 70 per cent overall yield. Acetylenes similarly give cyclobutenones.

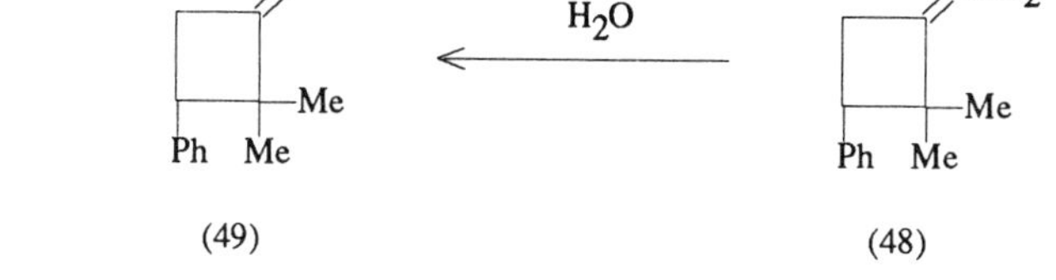

$Me_2CHCONMe_2$ → ($(CF_3SO_2)_2O$, collidine, $CHCl_3$, reflux) → $Me_2C{=}C{=}\overset{+}{N}Me_2$

→ (Ph⁄⁄) → (48) [cyclobutane iminium: $\overset{+}{N}Me_2$, Me, Ph, Me]

(48) → ($H_2O$) → (49) [cyclobutanone: O, Me, Ph, Me]

*Scheme 14*

With 1,1-disubstituted alkenes only poor yields of cycloadducts are obtained in intermolecular cases, because of competitive Friedel-Crafts reactions.

Cycloadditions of keteneiminium salts are stepwise reactions and not concerted, and occasionally result in loss of stereochemistry of the olefin component.[24]

Using amides derived from the optically active pyrrolidine (50) cyclobutanones have been obtained with high optical purity.[25] The best results so far have been obtained with the iminium salt (53) derived from (–)-(50) and isobutyric acid. Reaction with cyclopentene followed by hydrolysis gave the fused ring cyclobutanone (54) almost optically pure; with styrene, the cyclobutanone (55) was obtained in 80 per cent enantiomeric excess.

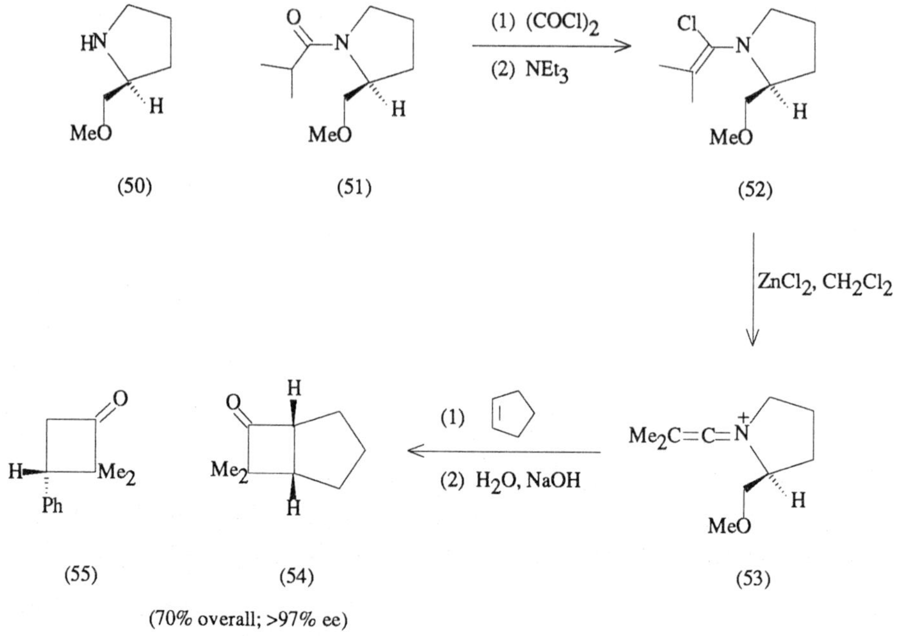

(70% overall; >97% ee)

*Scheme 15*

Intramolecular [2+2] cycloaddition reactions of ketenes are being increasingly employed for the synthesis of polycyclic cyclobutanones and products which can be derived from them.[26] The reactions proceed best with ketenes in which the ketene and olefin are connected by a chain of three atoms; this appears to offer the best compromise between strain energy in the product and the entropy of activation. Thus, whereas 6-methylheptenoyl chloride (56) gave the bicyclo[3,2,0]heptanone (57) in 80 per cent yield on treatment with triethylamine, the corresponding bicyclo[4,2,0]octanone (59) was obtained from 7-methyloctenoyl chloride (58) in only 3 per cent yield.[27]

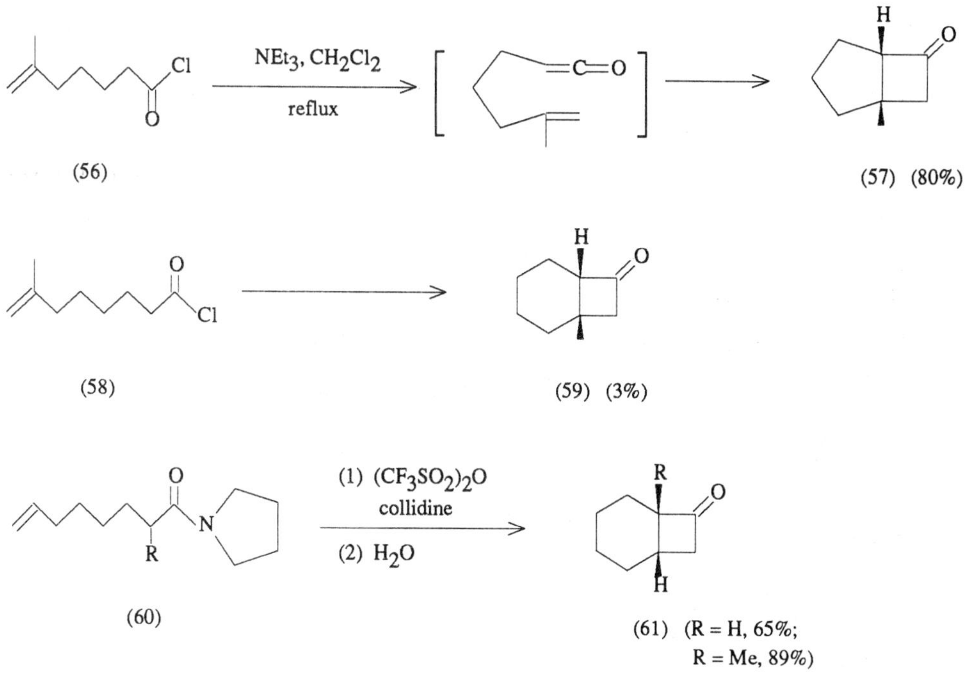

(56) $NEt_3$, $CH_2Cl_2$ reflux (57) (80%)

(58) (59) (3%)

(60) (1) $(CF_3SO_2)_2O$ collidine (2) $H_2O$ (61) (R = H, 65%; R = Me, 89%)

*Scheme 16*

Better results are sometimes obtained using keteneiminium salts.[23,27,28] These frequently give good yields of bicyclo[4,2,0]octanones, for example (61), where the ketenes themselves are ineffective.

In principle intramolecular cycloaddition reactions of ketenes could give rise to bicyclo[n,2,0] fused ring compounds or to bicyclo[n,1,1] bridged-ring compounds. In practice the course of the reactions appears to depend on the substitution pattern of the olefin. Terminal olefins afford bicyclo[n,2,0]-alkanones (as 61), but substrates in which the terminal olefinic carbon atom is disubstituted give bridged bicyclo[n,1,1]alkanones; 1,2-disubstituted olefins give mixtures. This is in line with the general mechanistic picture of ketene cycloadditions. Although thermal [2+2] cycloaddition reactions of ketenes are generally regarded as concerted, in practice the products are frequently more readily accounted for by an asynchronous process in which the bond from the carbonyl carbon of the ketene to the more nucleophilic carbon atom of the olefin is more fully developed than the other in the transition state.[29] In line with this picture yields are frequently higher with more nucleophilic olefins.

Cycloaddition reactions of ketenes have been exploited in the synthesis of a number of natural products. For example, an intramolecular reaction leading to a fused-ring cyclobutanone formed the key step in a synthesis of (±)-clovene.[30] The ketene (63) generated directly from the carboxylic acid (62), gave the cyclobutanone (64) in 47 per cent yield. Ring expansion of the cyclobutanone and *gem*-dimethylation gave (±)-clovene (66) by way of the ketone (65).

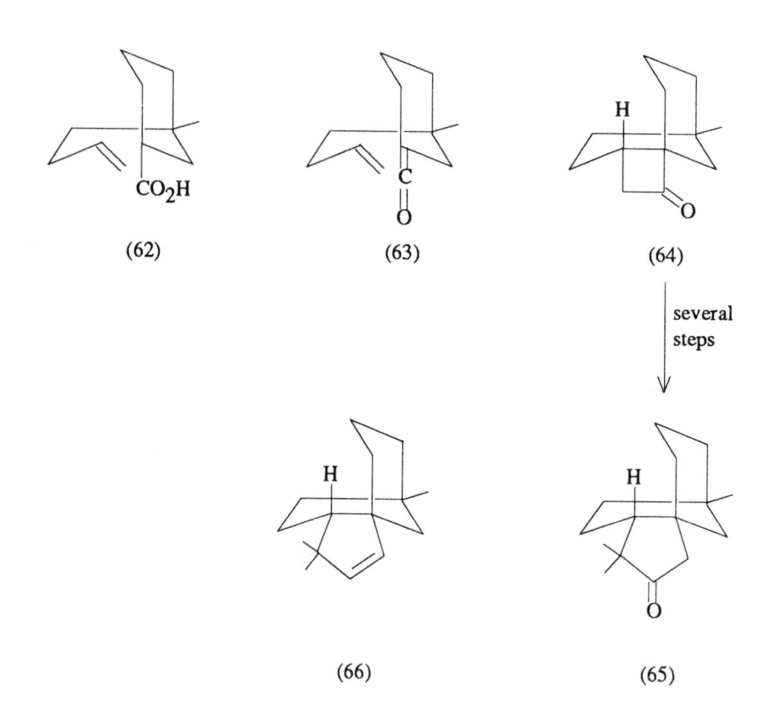


*Scheme 17*

A synthesis of both enantiomers of grandisol was achieved from the ketene (68) derived, ultimately, from (*R*)-3-hydroxybutyric acid.[31] Cycloaddition was effected in 70 per cent yield, but in this case with only moderate diastereoselectivity. The separated diastereomers were converted in several more steps into (+)- and (–)-grandisol.

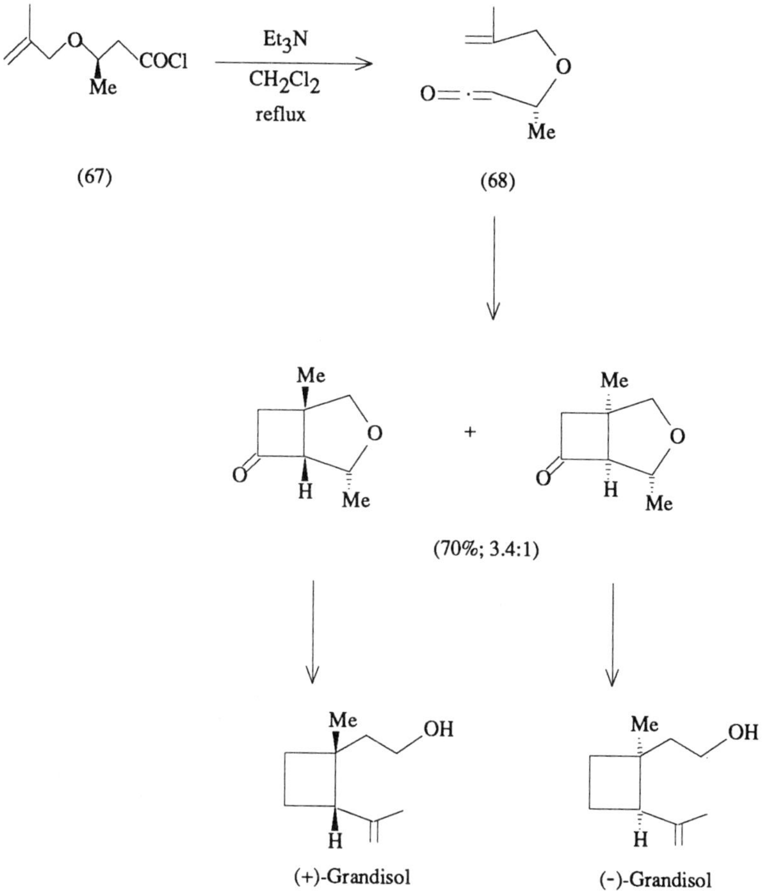

$Et_3N$
$CH_2Cl_2$
reflux

(67) (68)

(70%; 3.4:1)

(+)-Grandisol (-)-Grandisol

*Scheme 18*

Some intramolecular cycloadditions with alkylketenes proceed in only poor yield, but it has been found that certain substituents, notably chlorine and alkoxy, increase the rate of cycloaddition leading to improved yields of cycloadducts.[28,32] Excellent results have also been obtained with aryl-ketenes[33] and with vinylketenes.[34] The vinylketenes are formed *in situ* from the appropriate αβ-unsaturated acid chloride. Thus, the acid chloride (69) gave the vinylbicyclo[3,2,0]heptanone (71) in 50 per cent yield by way of the ketene (70). But the homologue (72) did not give (73). The longer connecting chain appears to decrease the rate of cycloaddition, allowing oligomerisation to intervene.

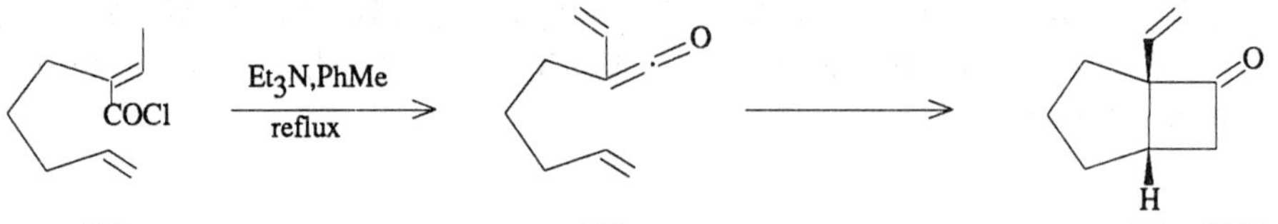

(69) $Et_3N$,PhMe reflux (70) (71) (50%)

(72) (73)

*Scheme 19*

With the ketene (74) the electrophilic carbonyl carbon atom of the ketene attacks the internal less substituted carbon of the olefin to give the bridged bicyclo[3,1,1]heptanone (75). Catalytic migration of the double bond gave chrysanthenone (77) and Wolff-Kishner reduction afforded β-pinene (76).[35] β-*cis*-Bergamotene and β-*trans*-bergamotene were obtained in the same way as β-pinene, starting respectively from nerylacetone and geranylacetone.[29]

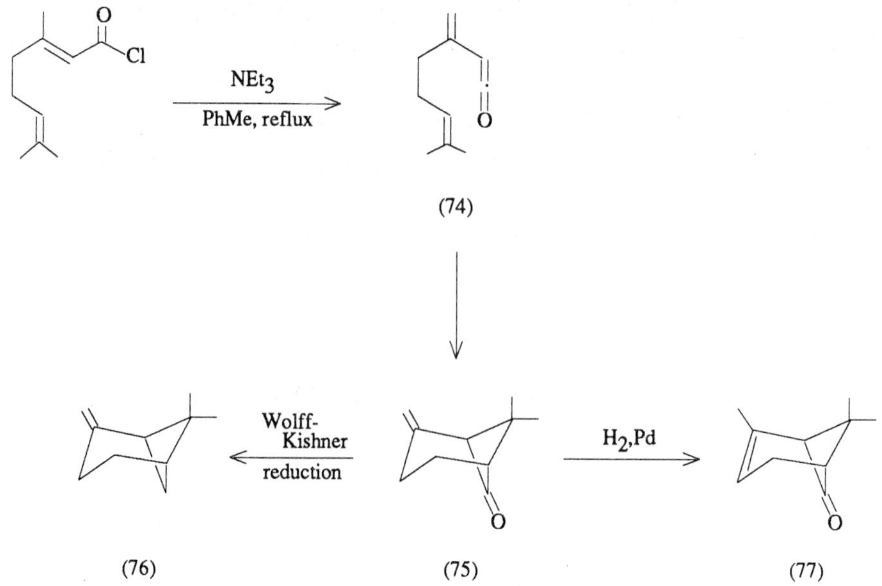

$NEt_3$ PhMe, reflux (74)

Wolff-Kishner reduction $H_2$,Pd

(76) (75) (77)

*Scheme 20*

## [2+2] Photocycloaddition to Enones

Photocycloaddition of alkenes and alkynes to multiple bonds provides another good route to four-membered rings. By far the most important reaction of this kind synthetically is the addition of alkenes to enones to form acylcyclobutanes.

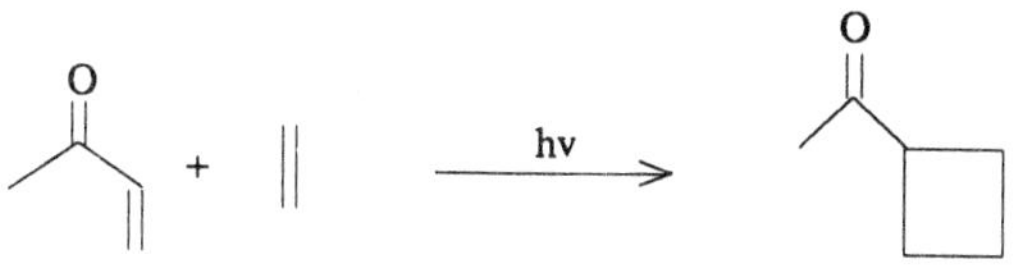

Scheme 21

Concerted, suprafacial [2+2] cycloadditions of this kind are "allowed" reactions,[2] but in fact photocycloaddition reactions of enones proceed by a stepwise pathway *via* triplet biradical intermediates.[36] Consequently the stereochemistry of the olefin is generally lost in the cyclobutane products because of rotational equilibration in the biradical intermediate. In the cycloaddition of alkenes to cyclopentenones to form bicyclo[3,2,0]heptanones it is found that the ring fusion is always *cis*, but with cyclohexenones the results are more varied. Monosubstituted acyclic alkenes generally give mixtures of *endo* and *exo* products.

Scheme 22

It is uncertain whether initial bond formation takes place at the α- or β-position of the enone.

In intermolecular reactions the regiochemistry of addition to unsymmetrical olefins depends on their structure. In general, electron-rich alkenes, such as ethyl vinyl ether, form mainly the head-to-tail isomer, while electron-deficient olefins give mainly the head-to-head product (Scheme 23).

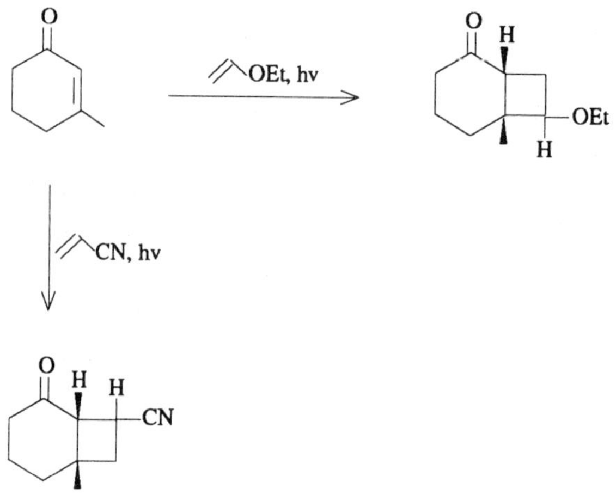

*Scheme 23*

It is generally considered that addition takes place to the n->π* photo-excited state of the enone in which the dipole is reversed compared with that in the ground state. The regiochemistry is then accounted for by initial formation of a π-complex with the olefin, the orientation of which depends largely on the electronic nature of the olefin but also, to some extent, on the experimental conditions, particularly the solvent.

In intermolecular reactions the double bond of the enone must be part of a ring of six or fewer members. Open-chain enones, on excitation, undergo energy-wasting *cis-trans* isomerisation at the expense of cycloaddition, and the same applies to cyclic enones of seven or more atoms, unless *cis-trans* isomerisation of the double bond is constrained in some way. Intramolecular addition to open chain enones, on the other hand, takes place readily. For example, the diene (78), on irradiation, afforded the aza[2,1,1]-bicyclohexane derivative (79), subsequently converted into the amino acid (80).[37]

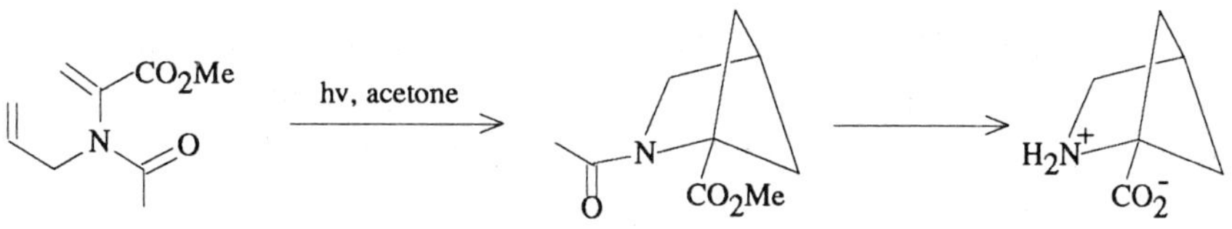

$CO_2Me$ hv, acetone N O N $CO_2Me$ O $H_2N^+$ $CO_2^-$

(78) (79) (55%) (80)

*Scheme 24*

The general rules given above for predicting the regiochemitry of intermolecular photochemical cycloaddition reactions of enones do not necessarily apply to intramolecular cases, where strain and steric interactions may outweigh the orientational influence of π-complex formation. In intramolecular additions the general rule appears to be that the major regioisomeric product is that formed by initial cyclisation of the biradical intermediate to give a five-membered ring where that is possible. For example, the dienone (81) gave (82) not (83).

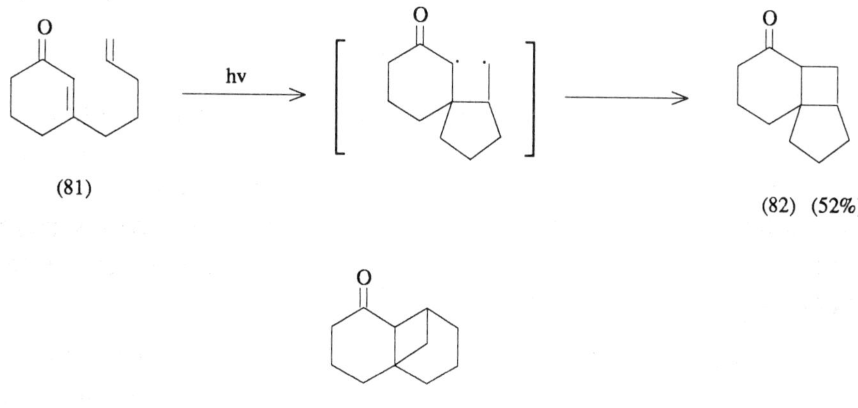

O hv O O

(81) (82) (52%)

O

(83)

*Scheme 25*

Within those limitations [2+2] photocycloaddition to enones has featured as an important step in the synthesis of a number of natural products, and it is particularly useful in these cases where rearrangement or fragmentation of the four-membered ring provides access to otherwsie difficultly accessible polycyclic systems.[38] A good early example is seen in the synthesis of α-caryophyllene alcohol (87).[39] The photo-addition product (84) on reaction with methyl-lithium gave specifically the alcohol (85). Wagner-Meerwein rearrangement with sulphuric acid led directly to α-caryophyllene alcohol by hydration of the intermediate carbonium ion (86) from the less hindered side. In this synthesis a tertiary carbonium ion rearranges, unusually, to a secondary one, encouraged, presumably, by the relief of strain in the four-membered ring of (85).

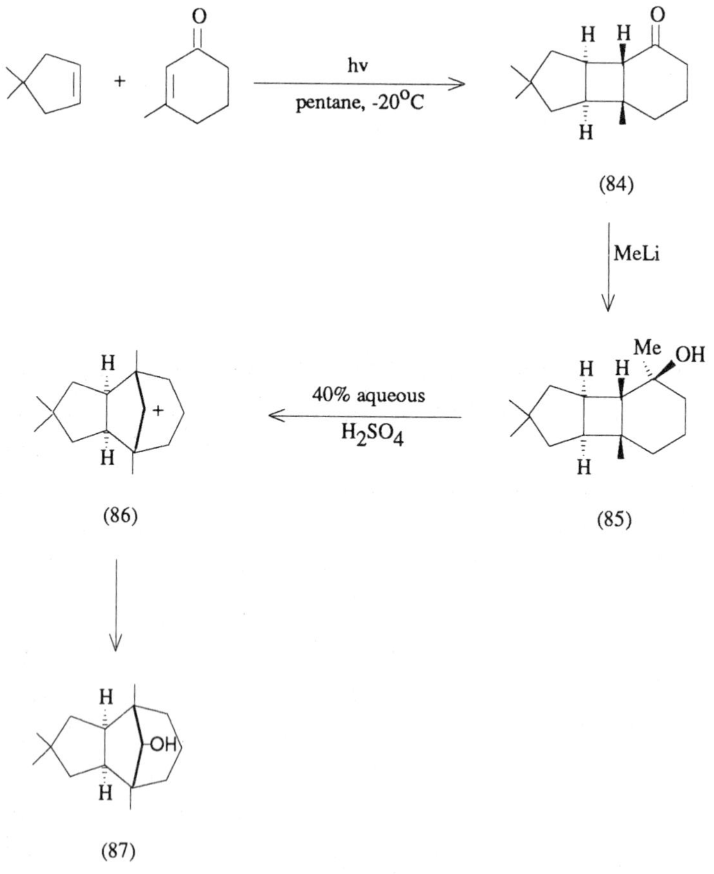

hv

pentane, -20°C

(84)

MeLi

40% aqueous

$H_2SO_4$

(86)

(85)

(87)

*Scheme 26*

Again, a key step in a synthesis of the propellane sesquiterpene modhephene (91) entailed photocycloaddition of 1,2-dichloroethylene, employed as an acetylene equivalent, to the enone (88), followed by Cargill rearrangement[40] of the strained βγ-unsaturated ketone (89) to (90).[41]

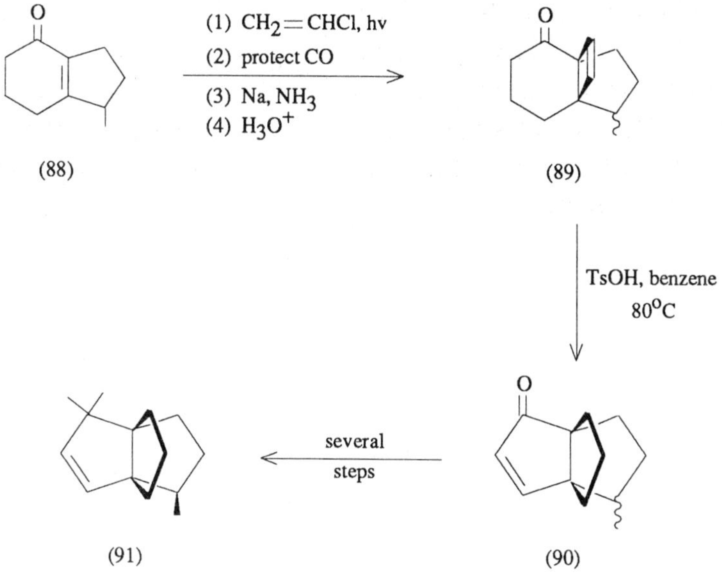

(1) $CH_2{=}CHCl$, hv
(2) protect CO
(3) Na, $NH_3$
(4) $H_3O^+$

(88) (89)

TsOH, benzene
80°C

several steps

(91) (90)

*Scheme 27*

Fragmentation of the initial photo-adducts may be exploited in various ways.[42] In one example, irradiation of piperitone and the cyclobutenes (92, R=Me or $CO_2Me$) gave the *cis,anti,cis* adducts (93, R=Me,$CO_2Me$) in highly regioselective reactions.[38] Thermal rearrangement then led to the cyclodecadienones (94, R=Me,$CO_2Me$) which, under the reaction conditions, cyclised to the cadinane precursors (95, R=Me,$CO_2Me$) by an ene-type ring closure (Scheme 28).[43]

Intramolecular reactions also have been exploited in the synthesis of sesquiterpenes. The regiochemistry of the cycloadditions generally follows the 'rule of five' (p.349). Thus, in a synthesis of (±)-isocomene (99), irradiation of the enone (96) gave the single adduct (97) in 77 per cent yield. In this single reaction a compound with three contiguous quaternary chiral centres is produced stereospecifically and effectively all the stereochemistry of the natural product is established. Wittig reaction gave the alkene (98) which was converted into (±)-isocomene by toluenesulphonic acid in benzene.[44] Photocycloaddition of an alkene to an enone has also formed an important step in syntheses of hirsutane[45] and of acorane[43,46] sesquiterpenes.

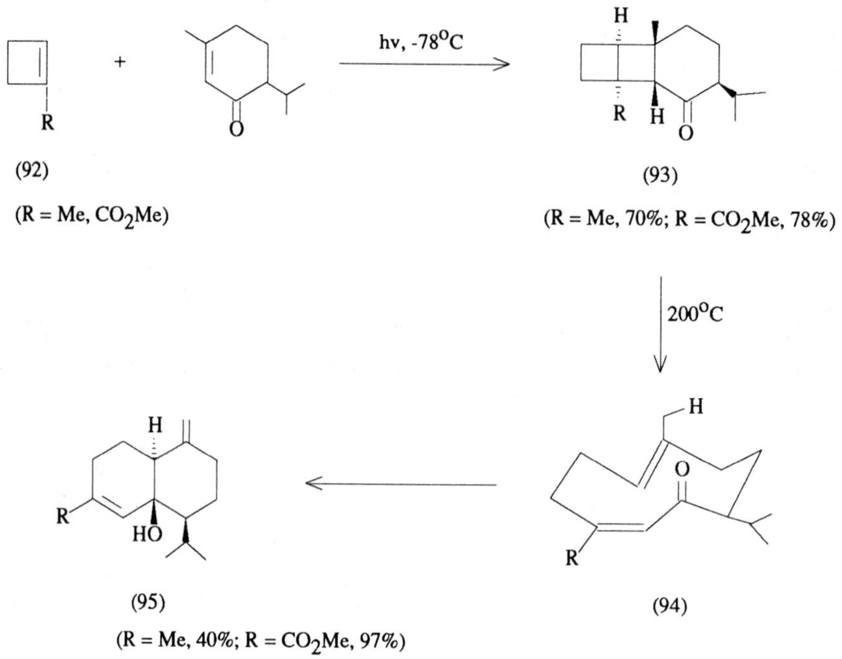

hν, -78°C

H

R H O

(92)

(R = Me, $CO_2$Me)

(93)

(R = Me, 70%; R = $CO_2$Me, 78%)

200°C

H

O

R

(94)

H

R

HO

(95)

(R = Me, 40%; R = $CO_2$Me, 97%)

*Scheme 28*

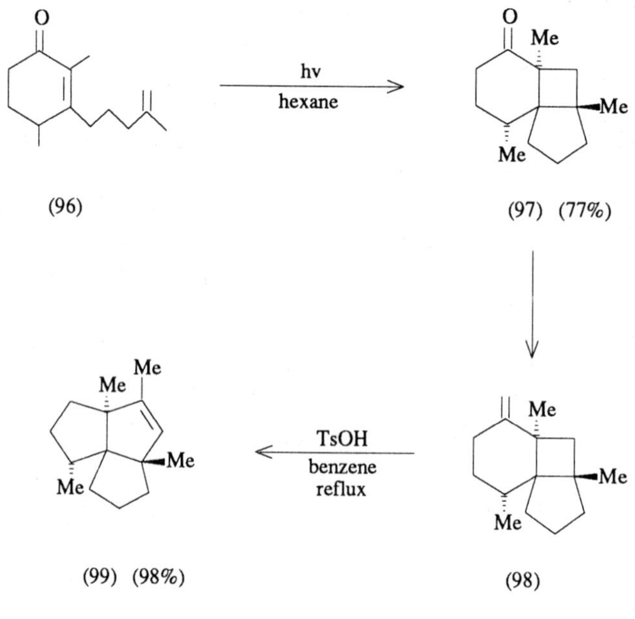

O

hν

hexane

(96)

O Me

Me

Me

(97) (77%)

Me

Me

Me

TsOH

benzene

reflux

Me Me

Me

Me

(99) (98%)

(98)

*Scheme 29*

## The De Mayo Reaction

One of the most useful reactions in this class is the photocycloaddition of olefins to the enolic form of 1,3-diketones. The initial addition leads to a β-acylcyclobutanol, but under the conditions of the reaction this undergoes retro-aldolisation with ring cleavage to give a 1,5-diketone.[47] Both open chain and cyclic 1,3-diketones can be employed in the reaction. Thus, irradiation of a solution of acetylacetone in cyclohexene affords the 1,5-diketone (101) by spontaneous retro-aldolisation of the intermediate β-hydroxyketone (100). The reaction proceeds by excitation of the enolic form of the diketone, and adduct formation to acyclic enones competes effectively with *cis-trans* isomerisation because of intramolecular hydrogen bonding in the enol.

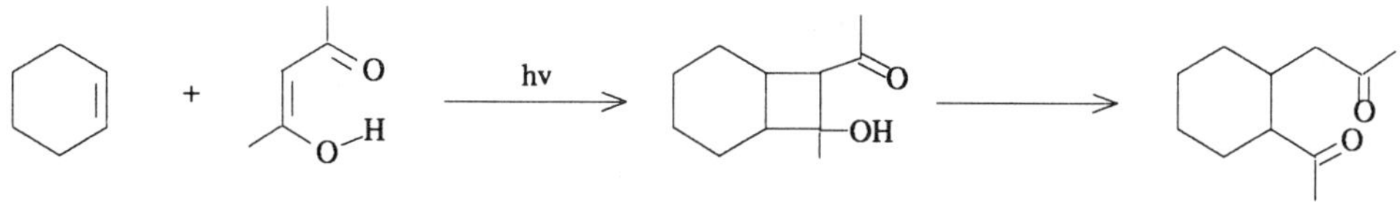

hν

(100) (101)

*Scheme 30*

With cyclic 1,3-diketones the sequence results in ring expansion by two carbon atoms, and reactions of this kind have been exploited in the synthesis of a number of polycyclic natural products.[36,42] With cyclic 1,3-diketones it is often found advantageous for solubility reasons, to employ the corresponding enol acetates or enol silyl ethers rather than the free enols. Thus, photolysis of the silyl enol ether (102) gave the single photo-adduct (103) in 68 per cent yield. Modification to (104) and fragmentation triggered by fluoride ion gave the azulenone intermediate (105) (Scheme 31).[48]

A valuable application of this reaction is in the synthesis of iridoids through addition of alkenes to methyl diformylacetate, which leads directly to the tetrahydrocoumalate ring system (106).[49]

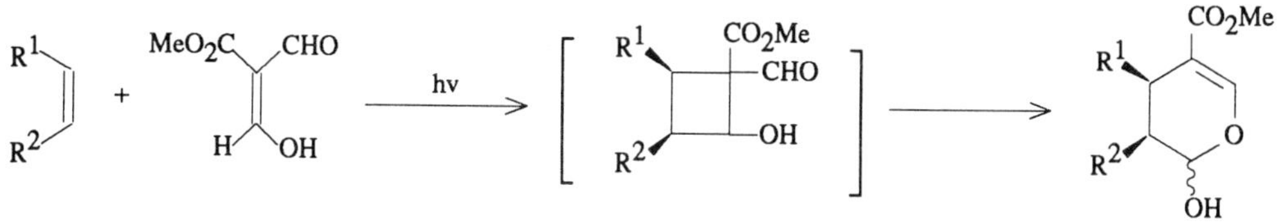

$R^1$ $R^2$ + $MeO_2C$ CHO H OH hν [ $R^1$ $R^2$ $CO_2Me$ CHO OH ] → $R^1$ $R^2$ $CO_2Me$ O OH

(106)

*Scheme 32*

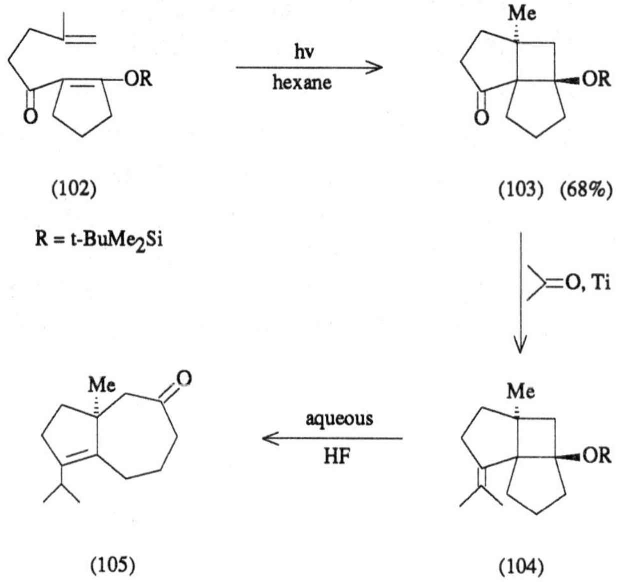

(102)

R = t-BuMe$_2$Si

hv / hexane

(103) (68%)

=O, Ti

aqueous HF

(105)

(104)

*Scheme 31*

Thus, irradiation of a mixture of optically active acetate (107) and methyl diformylacetate gave the loganin aglucone acetate (109) regio- and stereoselectively by attack of the enol on the less-hindered face of the olefin. The intermediate cyclobutanol (108) was not detected.[50]

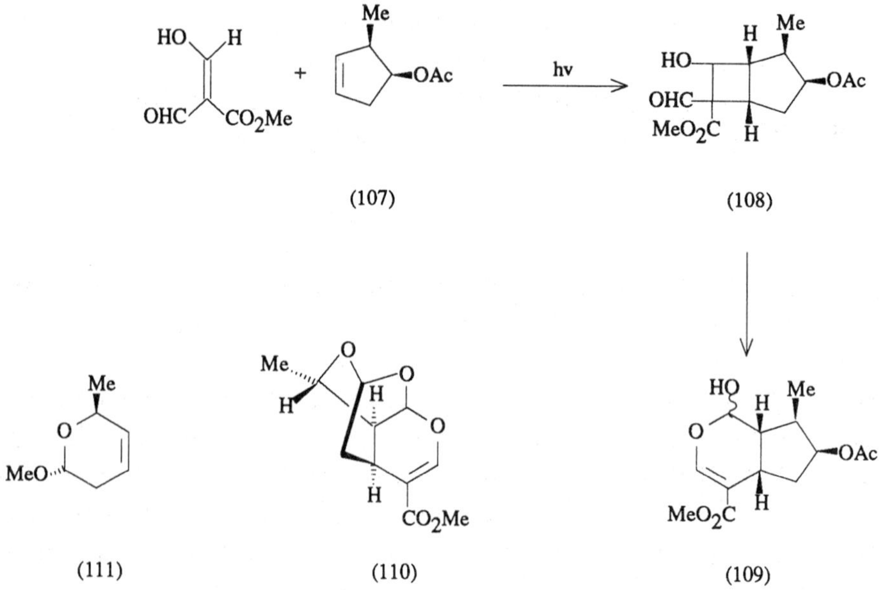

(107)

(108)

(111)

(110)

(109)

*Scheme 33*

The seco-iridoid (–)-sarracenin (110) was similarly obtained from methyl diformylacetate and the optically active acetal (111).[51]

Attempts to effect asymmetric syntheses of cyclobutanes by photocycloaddition to enones bearing optically active auxiliary ester[52] or acetal[53] groups had only limited success. Better results were obtained with the optically active enone (112), prepared from (*S*)-(+)-valinol. Photo-addition of ethylene gave largely the *exo* addition product (113). Cleavage of the chiral auxiliary group by acid hydrolysis gave the optically pure cyclobutane derivative (114) which was converted readily into (–)-grandisol.[54]

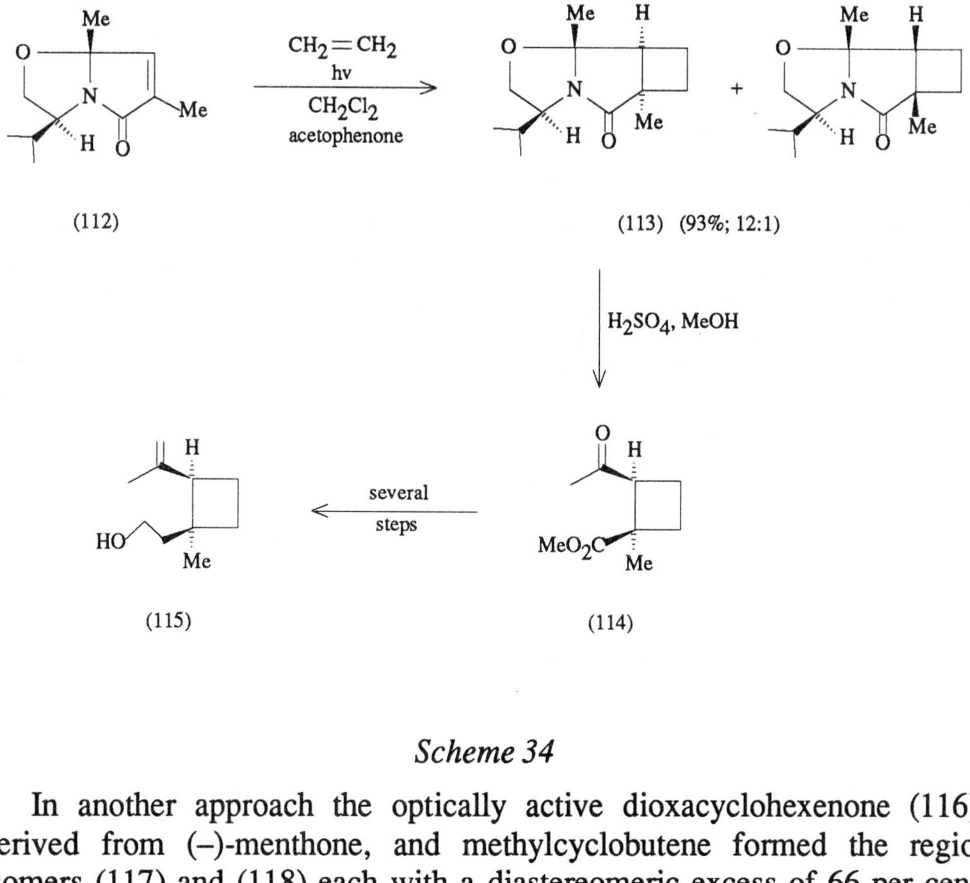

Me O N Me H O

$CH_2{=}CH_2$ hν $CH_2Cl_2$ acetophenone

Me H O N H O Me + Me H O N H O Me

(112) (113) (93%; 12:1)

$H_2SO_4$, MeOH

H HO Me — several steps — O H $MeO_2C$ Me

(115) (114)

*Scheme 34*

In another approach the optically active dioxacyclohexenone (116), derived from (–)-menthone, and methylcyclobutene formed the regio-isomers (117) and (118) each with a diastereomeric excess of 66 per cent, corresponding to the facial selectivity of the addition. Hydrolysis of purified (117) furnished the optically pure cyclobutane (119) which was converted into (+)-grandisol (120). (–)-Grandisol was similarly synthesised from (121).[55]

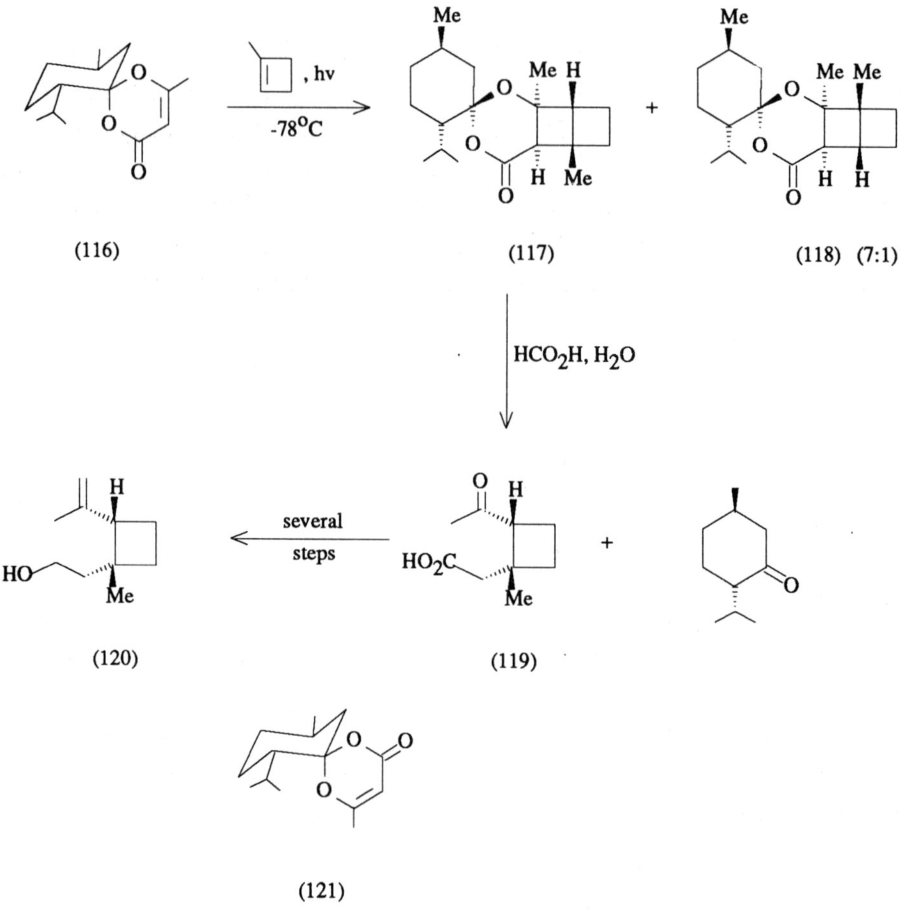

, hv

-78°C

Me

Me H

H Me

Me

Me Me

H H

(116)

(117)

(118) (7:1)

$HCO_2H$, $H_2O$

H

Me

HO

several

steps

H

$HO_2C$

Me

(120)

(119)

(121)

*Scheme 35*

## Copper-catalysed Photocycloaddition of Hepta-1,6-dienols

It has been found that alkenes add readily to the double bond of allylic alcohols on irradiation in the presence of copper(I) trifluoromethane-sulphonate (triflate). Intramolecular reactions are particularly effective, hepta-1,6-dienols (as 122) giving bicyclo[3,2,0]heptan-2-ols in good yield.[56] The derived ketones fragment cleanly at 500°C providing a novel route to cyclopent-2-enones (Scheme 44). In contrast, the corresponding 1,7-octa-dienols, which would give bicyclo[4,2,0]octanols, do not cyclise.

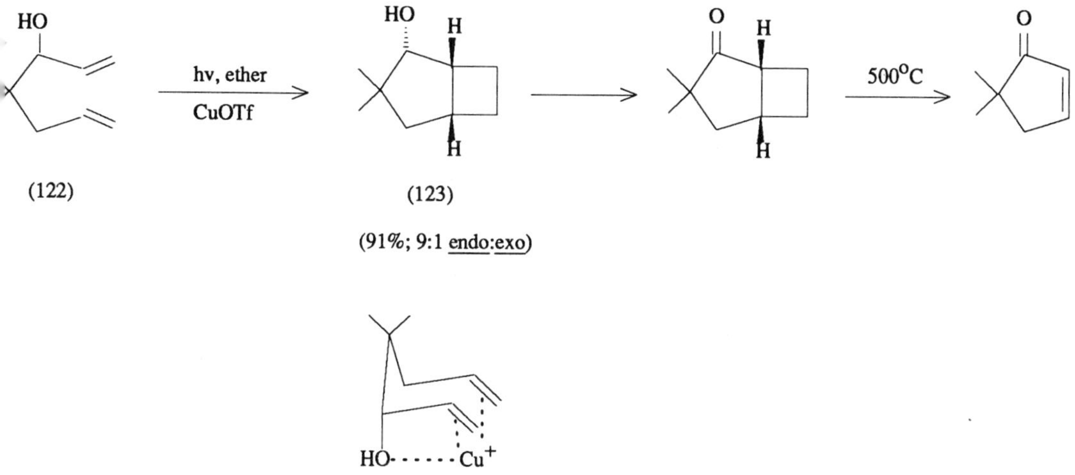

HO

hv, ether

CuOTf

HO H

H

O H

H

500°C

O

(122)

(123)

(91%; 9:1 endo:exo)

HO······Cu$^+$

(124)

*Scheme 36*

Little is known about the mechanism of the reaction. Since the *endo* hydroxy isomers (as 123) nearly always predominate, it has been suggested that co-ordination of the two olefinic bonds and the hydroxyl group with copper in a complex such as (124) is important.[56,58] The procedure has advantages in some cases for the preparation of bicyclo[3,2,0]heptan-2-ones where intermolecular addition of an alkene to a cyclopentenone either does not take place or gives mixtures of regio-isomers. Thus, photo-addition of propene to cyclopentenone affords a mixture of regio-isomers (125) and (126), but catalysed intramolecular cycloaddition of the diene (127) and oxidation of the product gives only the regio-isomer (125).[57]

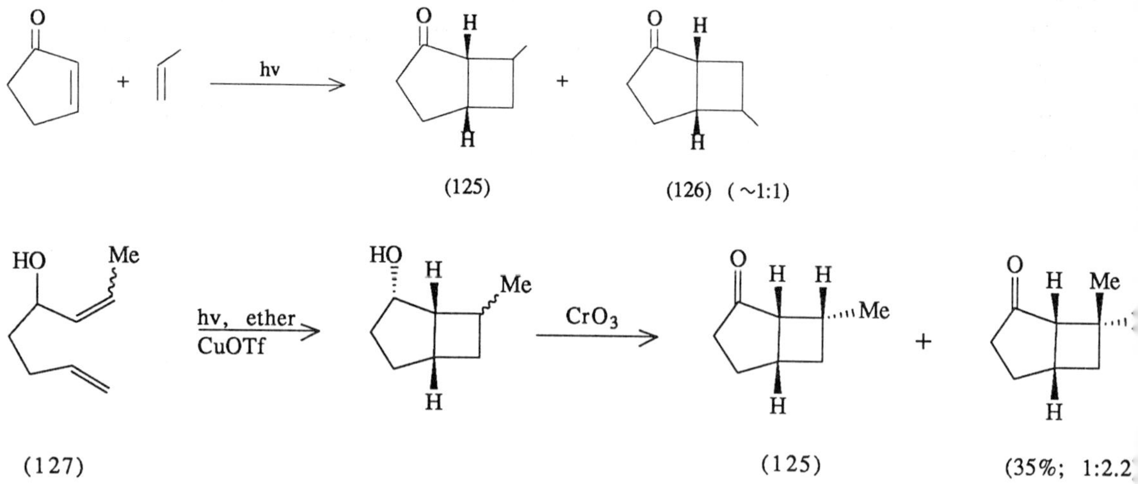

(125) (126) (~1:1)

(127) (125) (35%; 1:2.2)

*Scheme 37*

The reaction has been employed in a direct synthesis of the sesquiterpenes α- and β-panasinsenes (130) and (131) by way of the ketone (129) obtained as a single product on irradiation of the diene alcohol (128) and oxidation of the cyclobutanol formed.[58]

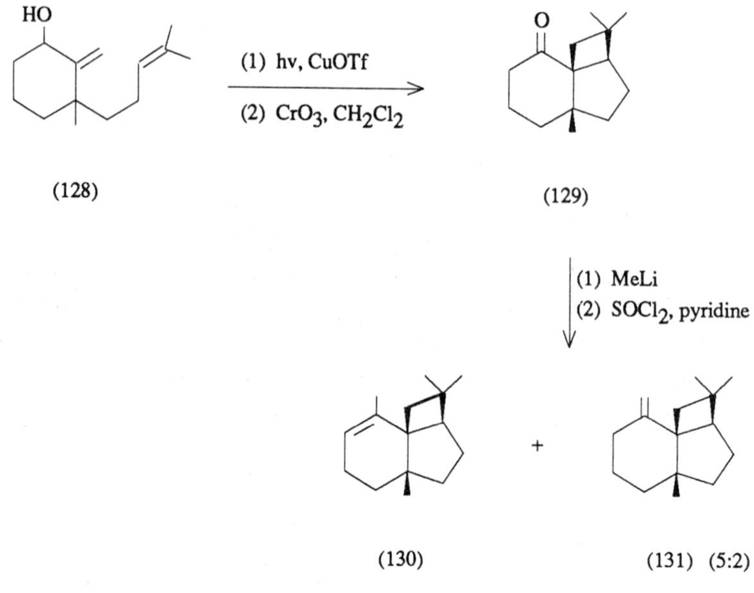

(1) hv, CuOTf
(2) $CrO_3$, $CH_2Cl_2$

(128) (129)

(1) MeLi
(2) $SOCl_2$, pyridine

(130) (131) (5:2)

*Scheme 38*

The hydroxyl substituents in the immediate products of cycloaddition can facilitate useful transformations of the photoproducts through cyclobutylcarbinol — cyclopentyl rearrangement.[59] Thus, solvolytic ring expansion of the tosylate of the *exo*-tricyclodecane (133), obtained as the main product from cyclisation of (132), followed by catalytic hydrogenation of the olefin produced, gave the hydrocarbon (134), found in the skeleton of the sesquiterpenes isocomene and pentalenic acid.

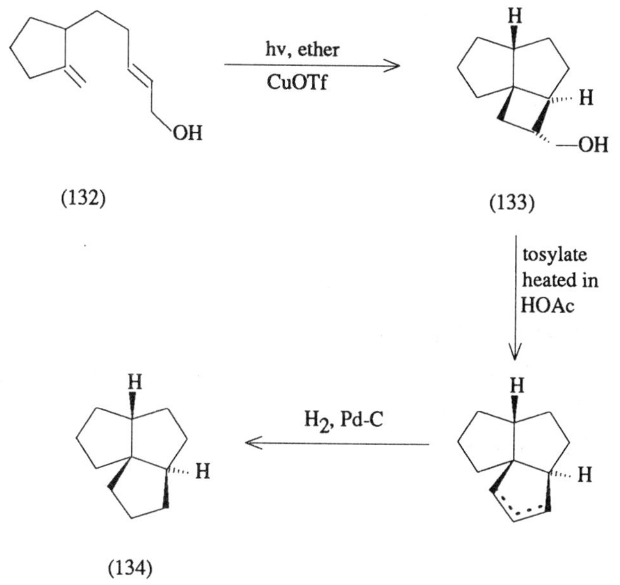

*Scheme 39*

## The Paternò-Büchi Reaction

Photochemical cycloaddition of olefins to the carbon-oxygen double bond of aldehydes and ketones also takes place readily in the well-known Paternò-Büchi synthesis of oxetanes.[60] A stepwise radical mechanism is suggested for these reactions; formation of the more stable diradical intermediate accounts for the regiochemistry of the products in most cases (Scheme 40).

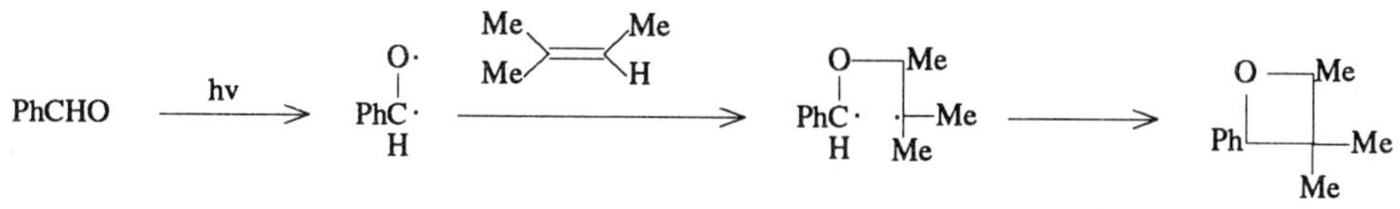

*Scheme 40*

Reactions of this kind have not been widely employed in complex synthesis, but recently photo-adducts obtained from furan derivatives and aldehydes have been shown to be valuable precursors for the synthesis of aldols and of highly oxygenated natural products containing five-membered oxygen rings. Photocycloaddition of aldehydes to furans leads to the head-to-head *exo* products (135) with high regio- and stereo-selecctivity.[61] Hydrolysis of the acetals formed affords *threo* aldols of 1,4-dicarbonyl compounds, and the *cis*-fused dioxabicyclo[3,2,0]heptene skeleton of the adducts can be functionalised in various ways, with high stereoselectivity, to give, after hydrolysis, acyclic chains containing several chiral centres.[62] Thus, the adduct (136) from 2-methylfuran and acetaldehyde, gave the aldol (137) on hydrolysis, and the adduct (138) from 2,5-dimethylfuran and benzaldehyde, by hydroboration and oxidation, was converted into the tetrahydrofuran derivative (139) with total stereocontrol of five contiguous chiral centres, in 85 per cent yield.

R, $R^1$, O + $R^2$CHO —hv→ H, H, R, $R^2$, O, O, $R^1$

(135)

H, H, Me, Me, O, O, H —0.1M HCl→ Me, O, OH, Me, CHO

(136) (137)

H, H, Me, Ph, O, O, Me —(1) $BH_3$, THF; (2) $H_2O_2$, NaOH→ OH, H, OH, Me, Ph, O, Me

(138) (139)

*Scheme 41*

Similarly, perhydroxylation of the olefinic bond in (140) took place mainly from the α-face to give, after reorganisation of the hemiacetal system, 3-deoxy-DL-streptose (141), isolated as the 1,2-O-isopropylidene derivative in 33 per cent yield.[62a] Spontaneous epimerisation of the carbon atom bearing the formyl group took place during the reaction.

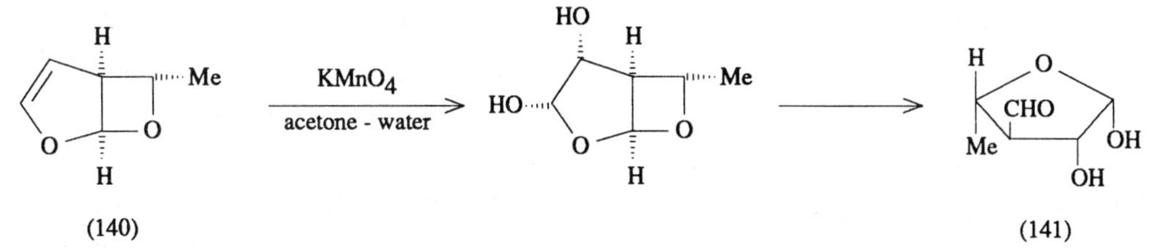

H Me O O H (140) $KMnO_4$ acetone - water HO H Me HO O O H H O CHO Me OH OH (141)

*Scheme 42*

The way in which these adducts can be used as key elements in more complex syntheses is illustrated in the synthesis of the *bis*-lactone avenaciolide (147).[63]

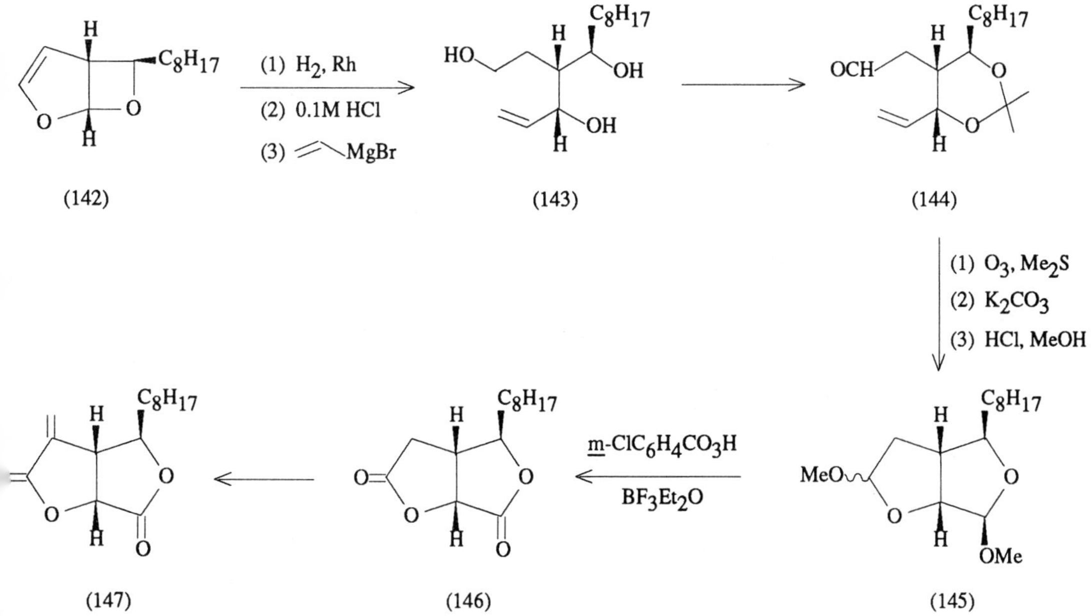

H $C_8H_{17}$ O O H (142) (1) $H_2$, Rh (2) 0.1M HCl (3) MgBr HO H $C_8H_{17}$ OH OH H (143) OCH H $C_8H_{17}$ O O H (144) (1) $O_3$, $Me_2S$ (2) $K_2CO_3$ (3) HCl, MeOH MeO H $C_8H_{17}$ O O H OMe (145) m-$ClC_6H_4CO_3H$ $BF_3Et_2O$ O H $C_8H_{17}$ O O H O (146) H $C_8H_{17}$ O O H O (147)

*Scheme 43*

Here the photoadduct (142), obtained as a single isomer in nearly quantitative yield from furan and nonanal, served as the readily available springboard in which two of the three stereocentres of the target are already present. Hydrogenation, hydrolysis of the acetal and reaction with vinylmagnesium bromide gave (143) and thence (144). Ozonolysis of (144), base-catalysed epimerisation and acidification led directly to (145) as a mixture of methoxy anomers. This was converted into bis-lactone (146) and thence into avenaciolide (147).

In related reactions the photoadduct (148), from 3,4-dimethylfuran and 3-benzyloxypropanal, was used as starting material for the synthesis of asteltoxin (150),[64] and the ginkgolide analogue (151) was prepared from the photoadduct from 2-tributylstannylfuran and butyl glyoxylate.[65]

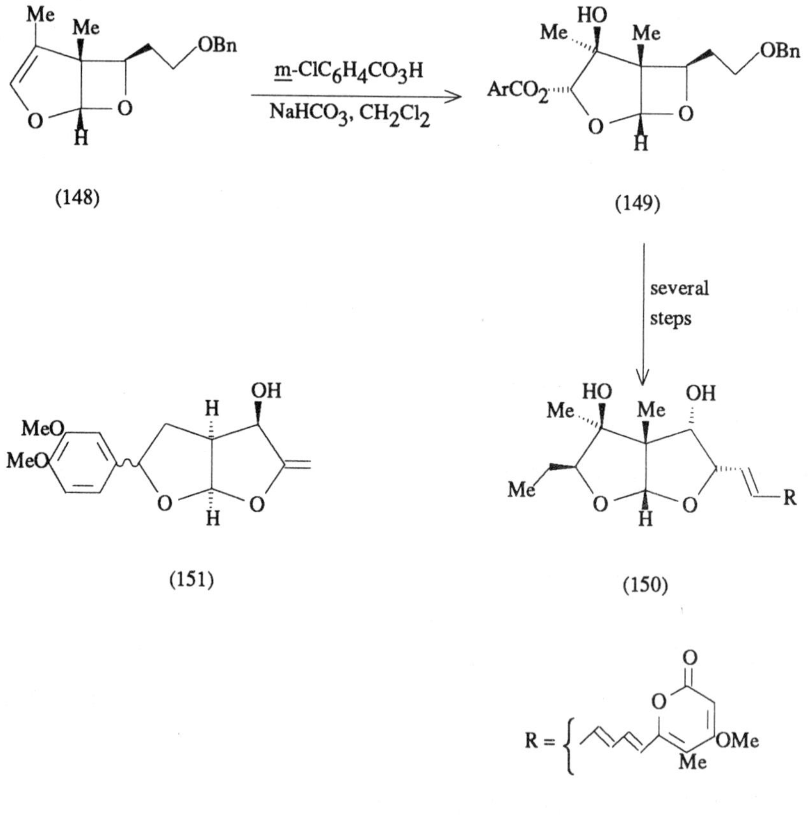

(148) $\xrightarrow[\text{NaHCO}_3,\ \text{CH}_2\text{Cl}_2]{m\text{-ClC}_6\text{H}_4\text{CO}_3\text{H}}$ (149) $\xrightarrow{\text{several steps}}$ (150)

(151)

*Scheme 44*

Attempts to bring about asymmetric photocycloaddition of furans to optically active aldehydes have met with only modest success.[66]

## References

1 cf. P.D. Bartlett, *Science*, 1968, **159**, 833.

2 R.B. Woodward and R. Hoffmann, *The Conservation of Oribital Symmetry*, Academic Press, 1970.

3 cf. T. Gilchrist and R.C. Storr, *Organic Reactions and Orbital Symmetry*, Cambridge University Press, 2nd edn., 1979.

4 see, for example, W.S. Patai, *Chemistry of Ketenes, Allenes and Related Compounds*, Interscience, New York, 1980; H. Ulrich, *Cycloaddition Reactions of Heterocumulenes*, Academic Press, London, 1967; W.T. Brady, *Tetrahedron*, 1981, **37**, 2949.

5 cf. A.H. Al-Husaini and H.W. Moore, *J. Org. Chem.*, 1985, **50**, 2595; R.W. Holder, N.A. Graf, E. Duesler and J.C. Moss, *J. Am. Chem. Soc.*, 1983, **105**, 2929; E.J. Corey and M.C. Desai, *Tetrahedron Lett.*, 1985, **26**, 3535.

6 P. Leriverend, *Bull. Soc. Chim. France*, 1973, 3498.

7 L. Ghosez, R. Montaigne and P. Mollet, *Tetrahedron Lett.*, 1966, 135.

8 K.E. Harding, J.B. Strickland and J. Pommerville, *J. Org. Chem.*, 1988, **53**, 4877.

9 W.T. Brady, *Tetrahedron*, 1981, **37**, 2949.

10 P.W. Jeffs, G. Molina, N.A. Cortese, P.R. Hauck and J. Wolfram, *J. Org. Chem.*, 1982, **47**, 3876; P.W. Jeffs, N.A. Cortese and J. Wolfram, *J. Org. Chem.*, 1982, **47**, 3881.

11 R.L. Danheiser, S.K. Gee and H. Sard, *J. Am. Chem. Soc.*, 1982, **104**, 7670.

12 D.A. Jackson, M. Rey and A.S. Dreiding, *Tetrahedron Lett.*, 1983, **24**, 4817.

13 P.A. Wender and C.R.D. Correia, *J. Am. Chem. Soc.*, 1987, **109**, 2523.

14 R.L. Danheiser and S.K. Gee, *J. Org. Chem.*, 1984, **49**, 1672; R.L. Danheiser, A. Nishida, S. Savariar and M.P. Trova, *Tetrahedron Lett.*, 1988, **29**, 4917.

15 R.L. Danheiser, S.K. Gee and J.J. Perez, *J. Am. Chem. Soc.*, 1986, **108**, 806.

16 A.E. Greene and F. Charbonnier, *Tetrahedron Lett.*, 1985, **26**, 5525.

17 A.E. Greene, F. Charbonnier, M-J. Luche and A. Moyano, *J. Am. Chem. Soc.*, 1987, **109**, 4752.

18 G. Fräter, U. Müller and W. Günther, *Helv. Chim. Acta*, 1986, **69**, 1858.

19 cf. C. Hubschwerlen and G. Schmid, *Helv. Chim. Acta*, 1983, **66**, 2206 and references cited there.

20 D.A. Evans and E.B. Sjogren, *Tetrahedron Lett.*, 1985, **26**, 3783, 3787; see also I. Ojima and H-J.C. Chen, *J. Chem. Soc. Chem. Commun.*, 1987, 625.

21 A.K. Bose, V.R. Hegde, D.R. Wagle, S.S. Bari and M.S. Manhus, *J. Chem. Soc. Chem. Commun.*, 1986, 161; see also Y. Ito, T. Kawabata and S. Terashima, *Tetrahedron Lett.*, 1986, **27**, 5751; C. Hubschwerlen and G. Schmid, *Helv. Chim. Acta*, 1983, **66**, 2206.

22 cf. E.J. Moriconi and W.C. Meyer, *J. Org. Chem.*, 1971, **36**, 2841.

23 J-B. Falmagne, J. Escudero, S. Taleb-Sahraoui and L. Ghosez, *Angew. Chem. internat. edn.*, 1981, **20**, 879.

24 H. Saimoto, C. Houge, A-M. Hesbain-Frisque, A-M. Mockel and L. Ghosez, *Tetrahedron Lett.*, 1983, **24**, 2251.

25 C. Houge, A-M. Hesbain-Frisque, A. Mockel, L. Ghosez, J.P. Declerq, G. Germain and M. Van Meersche, *J. Am. Chem. Soc.*, 1982, **104**, 2920.

26 B.B. Snider, *Chem. Rev.*, 1988, **88**, 793.

27 I. Marko, B. Ronsmans, A-M. Hesbain-Frisque, S. Dumas and L. Ghosez, *J. Am. Chem. Soc.*, 1985, **107**, 2192.

28 B. Snider and R.A.H.F. Hui, *J. Org. Chem.*, 1985, **50**, 5167.

29 cf. E.J. Corey and M.C. Desai, *Tetrahedron Lett.*, 1985, **26**, 3535.

30 K.L. Funk, P.M. Novak and M.M. Abelman, *Tetrahedron Lett.*, 1988, **29**, 1493.

31 K. Mori and M. Mujake, *Tetrahedron*, 1987, **43**, 2229.

32 B.B. Snider and Y.S. Kulkarni, *J. Org. Chem.*, 1988, **52**, 307; see also W.T. Brady and Y.F. Giang, *J. Org. Chem.*, 1985, **50**, 5177.

33 B.B. Snider and M. Niwa, *Tetrahedron Lett.*, 1988, **29**, 3175.

34 S.Y. Lee, Y.S. Kulkarni, B.W. Burbaum, M.I. Johnston and B.B. Snider, *J. Org. Chem.*, 1988, **53**, 1848; B.B. Snider, E. Ron and B.W. Burbaum, *J. Org. Chem.*, 1987, **52**, 5413; S.Y. Lee, M. Niwa and B.B. Snider, *J. Org. Chem.*, 1988, **53**, 2356.

35 Y.S. Kulkarni and B.B. Snider, *J. Org. Chem.*, 1985, **50**, 2809.

36 cf. S.W. Baldwin in *Organic Photochemistry*, Ed. A. Padwa, Marcel Dekker, New York, 1981, Vol.5, p.123; A.C. Weedon in *Synthetic Organic Photochemistry*, Ed. W.M. Horspool, Plenum Press, New York, 1984.

37 M.C. Pirrung, *Tetrahedron Lett.*, 1980, 4577; P. Hughes, M. Martin and J. Clardy, *Tetrahedron Lett.*, 1980, 4579.

38 cf. M. Vanderwalle and P. De Clercq, *Tetrahedron*, 1985, **41**, 1767; M.T. Crimmins, *Chem. Rev.*, 1988, **88**, 1453.

39 E.J. Corey and S. Nozoe, *J. Am. Chem. Soc.*, 1965, **87**, 5733.

40 cf. R.L. Cargill, T.E. Jackson, N.P. Peet and D.M. Pond, *Acc. Chem. Res.*, 1974, **7**, 106.

41 A.B. Smith and P.J. Jerris, *J. Org. Chem.*, 1982, **47**, 1845.

42 cf. W. Oppolzer, *Acc. Chem. Res.*, 1982, **15**, 135.

43 J.R. Williams and J.F. Callaghan, *J. Chem. Soc. Chem. Commun.*, 1979, 404, 405; P.A. Wender and J.C. Hubbs, *J. Org. Chem.*, 1980, **45**, 365; P.A. Wender and L.J. Letendre, *J. Org. Chem.*, 1980, **45**, 367; P.A. Wender and S.L. Eck, *Tetrahedron Lett.*, 1982, **23**, 1871.

44 M.C. Pirrung, *J. Am. Chem. Soc.*, 1981, **103**, 82.

45 K. Tatsuta, K. Akimoto and M. Kinoshita, *J. Am. Chem. Soc.*, 1979, **101**, 6117; K. Tatsuta, K. Akimoto and M. Kinoshita, *Tetrahedron*, 1981, **37**, 4365; J.S.H. Kueh, M. Mellor and G. Pattenden, *J. Chem. Soc. Chem. Commun.*, 1978, 5.

46 W. Oppolzer, L. Gorrichon and T.G. Bird, *Helv. Chim. Acta*, 1981, **64**, 186.

47 P. de Mayo, *Acc. Chem. Res.*, 1971, **4**, 41.

48 G. Pattenden and G.M. Robertson, *Tetrahedron Lett.*, 1986, **27**, 399.

49 G. Büchi, J.A. Carlson, J.E. Powell and L-F. Tietze, *J. Am. Chem. Soc.*, 1970, **92**, 2165; 1973, **95**, 540; see also R.T. Chaudhuri, T. Ikeda and C.R. Hutchinson, *J. Am. Chem. Soc.*, 1984, **106**, 6004.

50 J.J. Partridge, N.K. Chadha and M.R. Uskokovic, *J. Am. Chem. Soc.* 1973, **95**, 532.

51 S.W. Baldwin and M.T. Crimmins, *J. Am. Chem. Soc.*, 1982, **104**, 1132; see also L-F. Tietze, K-H. Glusenkamp, M. Nakane and C.R. Hutchinson, *Angew. Chem. internat. edn.*, 1982, **21**, 70.

52 G.L. Lange, C. Decicco, S.L. Tan and G. Chamberlain, *Tetrahedron Lett.*, 1985, **26**, 4707; G.L. Lange and M. Lee, *Tetrahedron Lett.*, 1985, **26**, 6163.

53 G.L. Lange and C.P. Decicco, *Tetrahedron Lett.*, 1988, **29**, 2613.

54 A.I. Meyers and S.A. Fleming, *J. Am. Chem. Soc.*, 1986, **108**, 306.

55 M. Demuth, A. Palomer, H-D. Sluma, A.K. Dey, C. Krüger and Yi-Hung Tsay, *Angew. Chem. internat. edn.*, 1986, **25**, 1117.

56 R.G. Salomon, *Tetrahedron.*, 1983, **39**, 485; R.G. Salomon and S. Ghosh, *Organic Syntheses*, 1984, **62**, 125; R.G. Salomon, D.J. Coughlin, S. Ghosh and M.G. Zagorski, *J. Am. Chem. Soc.*, 1982, **104**, 998.

57 C. Shih, E.L. Fritzen and J.S. Swenton, *J. Org. Chem.*, 1980, **45**, 4462.

58 J.E. McMurry and W. Choy, *Tetrahedron Lett.*, 1980, **21**, 2477.

59 R.G. Salomon, S. Ghosh, M.G. Zagorski and M. Reitz, *J. Org. Chem.*, 1982, **47**, 829; K. Avasti and R.G. Salomon, *J. Org. Chem.*, 1986, **51**, 2556.

60 G. Jones in *Organic Photochemistry*, Ed. A. Padwa, Dekker, New York, 1981, Vol.5, p.123.

61 S. Toki, K. Shima and H. Sakurai, *Bull. Chem. Soc. Japan*, 1965, **38**, 760; K. Shima and H. Sakurai, *ibid.*, 1966, **39**, 1806; E.G. Whipple and G.R. Evenega, *Tetrahedron*, 1968, **24**, 1299.

62 (a) T. Kozluk and A. Zamojski, *Tetrahedron*, 1983, **39**, 805; (b) S.L. Schreiber, A.H. Hoveyda and H-J. Wu, *J. Am. Chem. Soc.*, 1983, **105**, 660.

63 S.L. Schreiber and A.H. Hoveyda, *J. Am. Chem. Soc.*, 1984, **106**, 7200.

64 S.L. Schreiber and K. Satake, *J. Am. Chem. Soc.*, 1983, **105**, 6723; 1984, **106**, 4186.

65 S.L. Schreiber, D. Desmaele and J.A. Porco, *Tetrahedron Lett.*, 1988, **29**, 6689.

66 S. Jarosz and A. Zamojski, *Tetrahedron*, 1982, **38**, 1447; S.L. Schreiber and K. Satake, *Tetrahedron Lett.*, 1986, **27**, 2575.

# INDEX